Ethnobiology

Series editors:

Ethnobiology is the study of the dynamic relationship between plants, animals, people, and the environment. Academic and applied interests include ethnobotany, ethnozoology, linguistics, paleoethnobotany, zooarchaeology, ethnoecology, and many others. The field lies at a dynamic intersection between the social and biological sciences. The major contribution from the biological sciences has come from economic botany, which has a rich historical and scientific tradition. Indeed, the objectives of the colonial enterprise were as much about the quest for "green gold" –herbal medicines, spices, novel cultivars, and others—as it was for precious metals and sources of labor. The view that ethnobiology concerns mostly the discovery of new and useful biota extended into the 20th century. The social sciences have contributed to the field in both descriptive studies but also within quantitative approaches in cognitive anthropology that have led to general principles within ethnobiological classification. Ethnobiological research in recent years has focused increasingly on problem solving and hypothesis testing by means of qualitative and especially quantitative methods. It seeks to understand how culturally relevant biotas are cognitively categorized, ranked, named, and assigned meaning. It investigates the complex strategies employed by traditional societies to manage plant and animal taxa, communities, and landscapes. It explores the degree to which local ecological knowledge promotes or undermines resource conservation, and contributes to the solution of global challenges, such as community health, nutrition, and cultural heritage. It investigates the economic value and environmental sustainability to local communities of non-timber forest products, as well as the strategies through which individual ecological knowledge and practices encourage resilience to change—modernization, climate change, and many others. Most importantly, contemporary ethnobiological research is grounded in respect for all cultures, embracing the principles of prior informed consent, benefit sharing, and general mindfulness.

More information about this series at http://www.springer.com/series/11551

Bernardo Urbani • Manuel Lizarralde
Editors

Neotropical Ethnoprimatology

Indigenous Peoples' Perceptions of and Interactions with Nonhuman Primates

Springer

Editors
Bernardo Urbani
Centro de Antropología
Instituto Venezolano de Investigaciones Científicas
Caracas, Venezuela

Manuel Lizarralde
Department of Botany and Environmental Studies Program
Connecticut College
New London, CT, USA

ISSN 2365-7553 ISSN 2365-7561 (electronic)
Ethnobiology
ISBN 978-3-030-27503-7 ISBN 978-3-030-27504-4 (eBook)
https://doi.org/10.1007/978-3-030-27504-4

This Springer imprint is published by the registered company Springer Nature Switzerland AG
The registered company address is: Gewerbestrasse 11, 6330 Cham, Switzerland

To Ana María and Anne Marie, our daughters and son, family, and students…

To our fathers, Franco Urbani and Roberto Lizarralde†, lunch companions, and long-lasting friends who shared their impressions of far-away travels in Venezuelan forests throughout the hallways of their alma mater, *the Universidad Central de Venezuela…*

To the indigenous peoples of the Neotropics and the nonhuman primates that shared their ancestral lands…

Foreword

Since at least 1863, when "Darwin's Bulldog" Thomas Henry Huxley published his classic book *Evidence as to Man's Place in Nature*, primates have been increasingly studied by scientists because they are our closest living relatives in the animal kingdom; our prehistoric ancestors were primates; and we are primates (Kelley and Sussman 2007; Sussman 2007). Ethnoprimatology reveals yet a fourth kind of relevance of primates; humans relate to nonhuman primates in a diversity of fascinating and important ways, especially where there is ecological sympatry which increasingly occurs within the geographical range of most nonhuman primate species.

Linda Wolfe and Agustin Fuentes (2007: 701) define ethnoprimatology as "the study of the multifarious interaction of human and nonhuman primates, including, but not exclusive to, the image of primates in folklore, legends, and myths; influence of cultural beliefs on the hunting of primates; conservation ecology and the management of primate populations in their natural habitats; and so forth." Some would extend ethnoprimatology to even include settings where nonhuman primates are managed by humans, such as zoos (Palmer and Malone 2018). In addition, another special relevance of ethnoprimatology with very practical significance is the bidirectional transmission of parasites and diseases between humans and other primates (e.g., Cormier 2012).

Furthermore, from the perspective of primate conservation, as Phyllis Dolhinow and Agustín Fuentes (1999: 146–147) assert, "biodiversity and conservation-related themes have become a critical part of primate studies. It is no longer possible to study a group or population of free-ranging nonhuman primates without coming into contact with human disturbance, manipulation, or destruction of habitat. It has become readily apparent that no form of conservation action is possible without taking into account the human role in local utilization of protected areas." Human disturbance increasingly includes the impact of global climate change. As the Primate Specialist Group of the International Union for the Conservation of Nature (IUCN) periodically reports, many primate populations are endangered due to increasing human population and economic and ecological pressures (http://www.primate-sg.org/). Indeed, an impending extinction crisis is recognized, with 60% of primate species threatened (Estrada et al. 2017). Thus, ethnoprimatology is increasingly becoming important (see also Fuentes and Wolfe 2002; Waller 2016).

Ethnoprimatology is also relevant as it challenges several long-established tendencies which can be counterproductive. For decades the ideal of nonhuman primate field studies has been to make unintrusive observations on the undisturbed naturalistic behavior and ecology of free-ranging nonhuman primates in their natural habitat independent of humans, although habituation and sometimes provisioning may be involved. This approach remains essential but impractical because various kinds and degrees of human influences appear to be becoming nearly ubiquitous as part of the Anthropocene. Disregarding the human factor could result in biased and distorted results; accordingly, ethnoprimatology focuses on interrelationships between humans and other primates. It challenges the counterintuitive tendency to view *Homo sapiens* as part of nature in an evolutionary sense, but not in an ecological sense. Ethnoprimatology considers the human population as part of the animal community in an ecosystem and the associated ecological processes. It problematizes persistent ontological dualisms, such as human/animal and culture/nature, replacing them with a relational and processual perspective. It can evaluate the meanings of primate and human (Riley 2018). Finally, ethnoprimatology challenges, transcends, and even integrates the often arbitrary, and sometimes antagonistic, separation of biological and cultural anthropology, as well as anthropology and biology, wherever their interests converge. A holistic approach, mixed-methods biosocial tool kit, and collaborative multidisciplinary team research are often desirable (see Dore et al. 2018; Eben and Helmreich 2010; Mullin 1999; Parathian et al. 2018; Riley 2006, 2013, Robinson and Remis 2018).

Since some ideas about ethnoprimatology were first developed (Sponsel 1997), there has been a rapidly growing impressive accumulation of research on the subject with well over a thousand publications (e.g., McKinney and Dore 2018). In the Neotropics, Loretta A. Cormier (2003) published a pioneering account of the variety of interrelationships between the Guajá foragers and monkeys in the Brazilian Amazon. Literature reviews on ethnoprimatology were made (see Carter and Carter 1999; Urbani 2002, 2005; Cormier 2006; Wolfe and Fuentes 2007; McKinney and Dore 2018; Riley 2018), including comparisons on how different cultures relate to the same kind of monkey (Cormier and Urbani 2008; Urbani and Cormier 2015). Agustín Fuentes and Linda Wolfe (2002) edited a substantial anthology on ethnoprimatology as did James D. Patterson and Janette Wallis (2005). A research guide is even available (Dore et al. 2017). Indeed, by now the amount of literature dealing with ethnoprimatology, in content if not explicitly identified as such, is sufficient to facilitate a whole university course entirely focused on this subject with journal articles, book-length case studies, and anthologies, albeit so far there is no single textbook and no focused journal.

Previously there have been relatively few studies in Neotropical ethnoprimatology, in large part following the anthropocentric assumption that African primates are more relevant because they are closer to human evolution. With this interesting book, the editors, Bernardo Urbani and Manuel Lizarralde, and authors offer a treasure trove of 18 cases from nine Latin American countries encompassing multiple species of genera, including capuchins, spider monkeys, howlers, night monkeys, sakis, marmosets, squirrel monkeys, tamarins, titi monkeys, uakaris, and wooly monkeys. Importantly, at last many of the authors are Latin Americans. This book is unprecedented as a regional compilation. The wealth of information, analyses, perspectives, and insights in this edited volume should be of particular interest to any primatologist and many biologists, biological and cultural anthropologists, and specialists in human/animal studies and multispecies ethnography. It provides a historical benchmark for all subsequent research in ethnoprimatology in the Neotropics and beyond.

Department of Anthropology
University of Hawai'i at Mānoa,
Honolulu, HI, USA

Leslie E. Sponsel

References

Carter A, Carter C (1999) Cultural representations of nonhuman primates. In: Dolhinow P, Fuentes A (eds) The nonhuman primates. Mayfield Publishing Company, Mountain View

Cormier LA (2003) Kinship with monkeys: the Guajá Foragers of Eastern Amazonia. Columbia University Press, New York

Cormier LA (2006) A preliminary review of neotropical primates in the subsistence and symbolism of indigenous lowland South American peoples. Ecol Environm Anthropol 2:14–32

Cormier LA (2012) The ten-thousand year fever: rethinking human and wild-primate malarias. Left Coast Press, Walnut Creek

Cormier LA, Urbani B (2008) The ethnoprimatology of the spider monkeys (*Ateles* spp.): from past to present. In: Campbell CJ (ed) Spider monkeys: behavior, ecology and evolution of the genus *Ateles*. Cambridge University Press, Cambridge

Dolhinow P, Fuentes A (eds) (1999) The nonhuman primates. Mayfield Publishing Company, Mountain View

Dore KM, Riley EP, Fuentes A (eds) (2017) Ethnoprimatology: a practical guide to research at the human-nonhuman primate interface. Cambridge University Press, Cambridge

Dore KM, Radford L, Alexander S, Waters S (2018) Ethnographic approaches in primatology, Folia Primatol 89:5–12

Eben KS, Helmreich S (2010) The emergence of multispecies ethnography. Cultural Anthropol 25:545–576

Estrada A, Garber PA, Rylands AB, Roos C, Fernandez-Duque E, Di Fiore A, Nekaris AI, Nijman V, Heymann EW, Lambert JE, Rovero F, Barelli C, Setchell JM, Gillespie TR, Mittermeier RA, Arregoitia LV, Boyle S, Fuentes A, MacKinnnon KC, Amato KR, Meyer ALS, Wich S, Sussman RW, Pan R, Kone I, Baoguo L (2017) Sci Adv 3:e1600946

Fuentes A, Wolfe LD (eds) (2002) Primates face to face: conservation implications of human and nonhuman primate interconnections. Cambridge University Press, Cambridge

Fuentes A (2012) Ethnoprimatology and the anthropology of human-primate interface. Ann Rev. Anthropol 41:101–117

Huxley TH (1863) Evidence as to man's place in nature. Appleton and Company, London

Kelley EA, Sussman RW (2007) An academic genealogy on the history of American field primatologists. Am J Physical Anthropol 132:406–425

McKinney T, Dore KM (2018) The state of ethnoprimatology: its use and potential in today's primate research. Int J Primatol 39:730–748

Mullin, MH (1999) Mirrors and windows: sociocultural studies of human-animal relationships. Ann Rev. Anthropol 28:201–224

Palmer A, Malone N (2018) Extending ethnoprimatology: Human-alloprimate relationships in managed settings. Int J Primatol 39:831–851

Parathian HE, McLennan MR, Hill CM, Frazão-Moreira A, Hockings KJ (2018) Breaking through disciplinary barriers: Human-wildlife interactions and multispecies ethnography. Int J Primatol 39:749–775

Patterson JD, Wallis J (eds) (2005) Commensalism and conflict: the primate-human interface. American Society of Primatologists, Norman

Riley EP (2006) Ethnoprimatology: toward reconciliation of biological and cultural anthropology. Ecol Environm Anthropol 2:75–86

Riley EP (2013) Contemporary primatology in anthropology: beyond the epistemological abyss. Am Anthropol 115: 411–422

Riley EP (2018) The maturation of ethnoprimatology: theoretical and methodological pluralism. Int J Primatol 39:705–729

Robinson CA, Remis MJ (2018) Engaging holism: exploring multispecies approaches in primatology. Int J Primatol 39:776–796

Sponsel LE (1997) The human niche in Amazonia: explorations in ethnoprimatology. In: Kinzey WG (ed) New World Primates. Ecology, evolution, and behavior. Aldine de Gruyter, New York

Sussman, RW (2007) A brief history of primate field studies. In: Campbell CJ, Fuentes A, MacKinnon K, Panger M, Bearder SK (eds) Primates in perspective. Oxford University, Oxford

Urbani B (2002) Neotropical ethnoprimatology: an annotated bibliography. Neotrop Primates 10: 24–26

Urbani B (2005) The targeted monkey: a re-evaluation of predation on New World primates. J Anthropol Scienc 83:89–109

Urbani B, Cormier L (2015) The ethnoprimatology of the howler monkeys (*Alouatta* spp.): from past to present. In: Kowalewski, MM, Garber PA, Cortés-Ortiz L, Urbani B, Youlatos D (eds) Howler monkeys: behavior, ecology, and conservation. Springer Press, New York

Waller MT (ed) (2016) Ethnoprimatology: primate conservation in the 21st century. Springer Press, New York

Wolfe LD, Fuentes A (2007) Ethnoprimatology: contextualizing human and nonhuman primate interactions. In: Campbell CJ, Fuentes A, MacKinnon K, Panger M, Bearder SK (eds) Primates in perspective. Oxford University, Oxford

Neotropical Ethnoprimatology: An Introduction

Studies on the interconnection between human and nonhuman primates in the Neotropics have been expanding progressively and exponentially since the term *ethnoprimatology* was coined by Leslie E. Sponsel in 1997 after completing anthropological field research among Venezuelan indigenous peoples (e.g., Sponsel 1981) and being a visiting scholar at the Center for Anthropology of the Venezuelan Institute for Scientific Research in Caracas. After 5 years of such disciplinary inauguration, to our knowledge, only two studies mentioned this term explicitly (Cormier 2000; Urbani and Gil 2001). The rise of this primatological branch for the Neotropics started when three chapters published in the book *Primates Face to Face. The Conservation Implications of Human-Nonhuman Primate Interconnections* (2002) edited by Agustín Fuentes and Linda D. Wolfe, emerged (Cormier 2002, Lizarralde 2002, Shepard 2002), and later the hallmark monograph by Cormier (2003) appeared. Prior to this moment, publications devoted fully to Neotropical ethnoprimatology, under Sponsel's definition, were counted around a couple of dozen (see review: Urbani 2002), including the pieces of Castro et al. (1975), Queiroz and Kipnis (1991), Townsend (1995), and Fleck et al. (1999). This novel line of ethnographical and primatological research received limited support, and it was not until a decade passed since 1997 when intellectual paradigms changed and a number of novel ethnoprimatological publications from the tropical Americas were launched (e.g., Voss and Fleck 2011; Stafford 2015; Roncal et al. 2018). Now, by presenting this book, our hope, as editors, is to stimulate further research on ethnoprimatology in this region, engaging new theoretical and methodological questions.

Nevertheless, before, several regional ethnographical works clearly showed that indigenous societies are intrinsically linked to the primates living in their lands in many elements of their culture, such as their cosmology, diet, pets, and specific body adornments (e.g., head- or arm-bands and necklaces with monkey teeth), or the use of skins as shaman pouches and leather for drums. Their bones are used as tools to pierce earlobes in adulthood ceremonies or in looms to make clothing and stems for their tobacco pipes. In some societies, such as the Toba-Qom (Medrano and Suárez, this volume), Lokono, Kari'na, and Warao (Rybka this volume), monkeys are even believed to predict weather changes or to anticipate rain. Today, this belief of indigenous root is so widespread that current peasants from Central and South America indicate that howler monkeys are howling when rain is coming. Additionally, monkey meat was also considered very important in the diet of indigenous societies in the past according to British explorer Henry Walter Bates or anthropologist Michael J. Harner (1973), who also presented a rich body of images of monkeys and the Jivaro. Another iconic ethnography is the work of the ethnographer Charles Wagley who began his Amazonian research in the late 1930s. He recorded that the Tapirapé of Brazil use sharp monkey bones to pierce the lower lip of males at birth (Wagley 1977). He also highlighted the reference of monkey in socialization in this statement: "When a boy became *churangí* (young adolescent), his behavior was likened to that of a monkey" (Wagley 1977: 1949); thus a common name for boys who are mischievous is *kai*, meaning "monkey" in Tapirapé (Wagley 1977). These are a few of the many examples illustrating how diluted the information about monkeys in the literature on indigenous peoples has become. One might argue that given the own nonindigenous background of past authors had a preference to particularly show the exotic animals of the forests of indigenous societies, such as primates, but actually the literature serves as an empirical evidence of these abundant and rich interactions between indigenous peoples and monkeys in their lands (e.g., Urbani 2005; Cormier 2006). However, most of these publications provided a glimpse of these interactions until the first ethnoprimatological works from the Neotropics emerged as indicated in the first paragraph of this introductory chapter. This volume is the first compilation that provides examples of the richness of these interactions in 25 different indigenous cultures in 10 countries of the tropical regions of the American continent.

This edited volume has a total of 18 chapters (Table 1). They are written by authors of different cultural backgrounds and with multiple perspectives. As can be observed in the table of contents, majority of the chapters are led by Latin American scholars or permanent non-Latin American residents in the region (13/18; 72%), and almost two-thirds of the first authors are women (11/18; 61%). The ethnoprimatological studies presented here referred to indigenous peoples inhabiting their ancestral territories from southern Mexico to northern Argentina (Fig. 1). Again, as can be seen in the table of contents, most of the chapters are based on research conducted in Mexico and Venezuela (three entries each), followed by studies from Guyana, Ecuador, and Peru with two pieces per country and single chapters from Argentina, Bolivia, Brazil, Colombia, and Guatemala. There is also a research on Yanomami ethnoprimatology in their binational territory between Brazil and Venezuela. Indigenous lands with monkey communities of two species in Argentina up to 14 in the Peruvian Amazon are ethnoprimatologically explored, but overall indigenous landscapes in the tropical Americas with multiple cultural and natural challenges are examined in the context of complex human/nonhuman primate interfaces. These chapters provide an exceptional sample of the nearly 500 indigenous societies in the Neotropics interacting with many of the 171 monkey species of the New World (Oviedo et al. 2000; Estrada et al. 2017).

Table 1 Synopsis of the ethnoprimatological studies present in this edited volume

# in Fig. 1	Indigenous society	Linguistic family[a]	Country	Chapter author(s)
1	Popoluca	Mixe-Zoquean	Mexico	M. Pinto-Marroquin and J. C. Serio-Silva
2	Maya, Tzeltal/Chol, Zoque, Totonac, and Creole	Mayan, Mixe-Zoquean, Totonacan, and composite	Mexico	E. Urquiza-Haas, R. I. Ojeda Martínez, and K. Kotrschal
3	Lacandon Maya	Mayan	Mexico	Y. García del Valle, F. Ruan-Soto, F. Guerrero-Martínez, and F. Reyes-Escutia
4	Maya-Q'eqchi'	Mayan	Guatemala	M. Rosales-Meda and M. S. Hermes
5	Tikuna	Tukuna-Juri (Independent)	Colombia	A. Maldonado and S. Waters
6	Barí	Chibchan	Venezuela	M. Lizarralde
7	Mapoyo	Kariban	Venezuela	B. Urbani
8	Jotï	Independent	Venezuela	S. Zent and E. Zent
9	Yanomami	Yanomaman (Independent)	Brazil and Venezuela	J. P. Boubli, B. Urbani, H. Caballero-Arias, G. H. Shepard Jr., and M. Lizarralde
10	Waimiri Atroari	Kariban	Brazil	R. R. de Souza-Mazurek and A. C. Bruno
11	Kari'na, Lokono, and Warao	Kariban/Arawakan (Independent)	Guyana	K. Rybka
12	Wapishana	Arawakan	Guyana	T. Henfrey
13	Secoya	Tukanoan	Ecuador	S. de la Torre, P Yépez, and A. Payaguaje
14	Waorani	Sabala (Independent)	Ecuador	M. Franzen-Levin
15	Shawi	Kawapanan	Peru	L. González-Saavedra
16	Wampis (Huambisa)	Jivaroan	Peru	K. Swierk
17	Tacana	Tacanan	Bolivia	W. R. Townsend, R. B. Wallace, K. Lara-Delgado, and G. Miranda-Chumacero
18	Qom (Toba)	Guaycuruan	Argentina	C. Medrano and V. Suárez

[a]Lizarralde (1989) and Oviedo et al. (2000)

The first chapter included in this edited volume studies the relationship between the Popoluca and two primate species (*Alouatta palliata* and *Ateles geoffroyi*) found in Los Tuxtlas, Mexico. Marianna Pinto-Marroquin and Juan Carlos Serio-Silva found that this indigenous society has strong cosmological beliefs and that the Popoluca also use primates as pets and in medical alignments. In general, they suggest that empathy with local monkeys by the Popoluca might promote primate conservation in a highly endangered ecosystem of the Neotropics. Additionally, in

Fig. 1 Location of the ethnoprimatological studies present in this edited volume (map by M. Lizarralde, after an open-access base map from Wiki Commons)

Mexico, Esmeralda Urquiza-Haas and collaborators provide a provocative approach in ethnoprimatology: the study of mental state attributions. By interviewing participants of multiple ethnical backgrounds (Maya, Tzeltal/Chol, Zoque, Totonac, and Creole) at the Yucatan village of Conhuas, the authors explore the way the villagers attribute moral rights to primates and other animals. In doing so, concepts of emotion and intelligence are studied as they are perceived in howler and spider monkeys. A chapter about the Lacandon people in Chiapas (Mexico), written by

Yasminda García del Valle and collaborators, focuses on the rich interaction of this Mayan culture with two different monkey species (*Alouatta pigra* and *Ateles geoffroyi*), providing a deep perspective from pre-Hispanic times to the present. They show the importance of monkeys through mural painting in caves and pottery art work. In the *Popol Vuh*, a very important sacred Mayan text, the origin of monkeys and other folktales for their cosmological role are well described, where spider monkeys are shown as transcribers and important administrators. The authors also interview the Lacandon about the importance of animals, revealing that monkeys rank 9th and 12th in their cultural significance out of 35 taxa. At the end of their chapter, they explain that monkeys are no longer threatened by hunting since the Lacandon see them as important attractions to tourists as economic incentives to their protection. To conclude the section devoted to ethnoprimatological studies carried out in Mesoamerica, Marleny Rosales-Meda and María Susana Hermes share a stimulating study on the interconnection between the Maya-Q'eqchi' and primates of their ancestral lands in Guatemala. The authors show how tied and long cultural links among the indigenous people, howlers, and spider monkeys are positively affecting primate populations and forest conservation. Rosales-Meda and Hermes encourage decision-makers to inform themselves about the cosmological visions of indigenous societies prior to proposing resolutions regarding biodiversity preservation.

Landing in South America, Angela M. Maldonado and Siân Waters provide current views on the Tikuna ethnoprimatology of Colombia. The authors show that Tikuna insertion into current market economy has changed their relations with the monkeys of their territory. Food taboos on primates diminished, and overexploitation consequently increased. The authors suggest alternatives to generate incomes, such as community-based primate watching. Near the border of Colombia, Manuel Lizarralde provides a novel Barí ethnoprimatological study from the Venezuelan side of the Sierra de Perijá. The author presents extensive ethnoecological information on the plants used by the four primate species that sympatrically live in the Bari's lands. Based on this data, he suggests that Neotropical forests might be labeled as *primatogenic* forests created by human and nonhuman primates.

Also in Venezuela, Bernardo Urbani writes on the Mapoyo, an almost extinct Carib language but living culture. He addresses on perceptions and changes on their relationship with their monkeys. The Mapoyo no longer use the four species of monkeys as their ancestors did, but they remember how they were used in the past. Urbani gathered a rich ethnographic data from colonial time to recent anthropological research provides a robust body of information about monkeys' uses, distribution, ecology, and role in their cosmology. However, this chapter provides evidence on the relationship between the indigenous peoples and their monkeys in the future because of Mapoyo process of cultural changes. Stanford Zent and Egleé Zent, in their chapter, provide a rich ecological "multi-species" ethnography on the Jotï people residing in the Venezuelan Guayana, addressing six different monkey species. They provide a complex text on the cosmological role, as well as food, contributing about one third of all hunted animals. Also, Zent and Zent share a deep knowledge of the Jotï's ecology and taxonomy of their monkeys. For the Jotï, the most important primate is the spider monkey, from mythological to subsistence reasons. In Venezuela-Brazil border, Jean P. Boubli and collaborators present a comprehensive literature review on the Yanomami perception and the use of ten species of Amazonian primates, as well as field information from villages of both sides of the border. The interconnections between this indigenous society and monkeys are extensive and involve material culture, hunting, food taboos, cosmology, and the use of monkeys as pets. Hunting practices are increasingly revised within the context of current possible unsustainability of nonhuman primate populations.

Rosélis R. de Souza-Mazurek and Ana Carla Bruno examine the role of primates among the Waimiri Atroari of Brazil. Primates are fundamental subjects in the cosmology of these indigenous people, and three cebid species are preferred in hunting games. Taboos on the consumption of monkeys exist among the Waimiri Atroari. In Guyana, Konrad Rybka makes a linguistic and environmental comparison of three societies, the Lokono, Kari'na, and Warao, in the Moruka River. This research questions how cultures borrowed ethnoecological information in their environmental adaptation in a multiethnic region. Rybka compares the same indigenous societies and their ethnographic literature on regions where they are the dominant and the sole society. According to him, "languages are highly sensitive to environmental pressures" since sympatric monkeys might share names borrowed from other indigenous societies or drop or retain terms for primates "independently of the cultural import of their referents." In another study, also in Guyana, on the Wapishana, Thomas Henfrey makes a comparison between the local ecological knowledge and scientific knowledge. The Wapishana have empirical detailed information of six different diurnal monkeys.

In western Amazonia, Stella de la Torre and coauthors navigate into the realms of Secoya ethnoprimatology of Ecuador. The authors suggest that this indigenous society's knowledge on primates is at risk and that some primate species in the territory of the Secoya are locally extinct as unsustainable activities increased. As suggested by Maldonado and Waters, the authors also advocate for ecologically sustainable sources for economic viability. Also in Ecuador, Margaret Franzen Levin explores the relationship between the Waorani and primates in their communities. Monkeys are frequently used as game species. As hunting continues, large-bodied monkeys are at risk, and spider monkeys are highly vulnerable. From Peru, Luisa González-Saavedra presents information on the Shawi cosmovision about monkeys. She indicates that the Shawi have a close interconnection with a large primate community in their lands, including other four arboreal mammals also classified as "monkeys." González-Saavedra found that by understanding the cosmological origin of monkey species, it is also possible to find the cultural origin of the Shawi themselves. In another chapter also in Peru, Kacper Świerk provides an ethnoprimatological case of the Wampis coexisting with a rich primate community that includes 14 different species. He provides a deep description of the ecology and subsistence of these indigenous people as well as a complex body of Wampis mythologies associated with monkeys. Also, it includes a detailed biogeographical distribution of monkeys that the author was able to learn from the Wampis in a rapid biological assessment of their population in their territory, especially focusing on the Kampankis mountains where the Peruvian government is planning to establish a national park. However, Świerk points out that the Wampis would like to keep using this ancestral land for their own resources.

Looking at the southern part of the continent, Wendy R. Townsend and colleagues study the hunting practices of the Tacana of Bolivia. Various primates are game species, although spider monkeys are preferred as they are considered to be particularly tasty. The Tacana prefer to travel longer in order to hunt *Ateles chamek* even if other primate taxa are nearby their villages. As this and other monkeys are increasingly important in hunting, as they culturally are, the authors provide quantitative data for potential use in decision-making policies regarding the sustainability of this arboreal mammal group in northern Bolivia. The last chapter by Celeste Medrano and Valentín Suárez (a member of the indigenous society) on the Qom (Toba) of northeastern Argentina is on the cultural and cosmological interpretation of one primate species, the black-and-gold howler monkey. They examine the cultural perception of the Qom of their monkeys from an ontological and interpretative perspective. Also, Medrano and Suárez provide a comprehensive collection of myths in relation to monkeys from the literature.

Interestingly enough is that when M. Lizarralde was starting to write this introduction chapter in Manzano Alto, Merida, Venezuela, after 10 days of silence, he started hearing howler monkeys just half a kilometer from his mother's home. Similarly and at the same time, it occurred to B. Urbani when began to write this piece at a field site in a remaining northern Venezuelan rainforest. Hearing the howler monkeys was a relief to know that they still exist in those mountainous forests. The question we ask as editors is: what could be the eventual fate of monkeys in Neotropical forests? There has already been bad news regarding their disappearance. The Neotropical region holds 171 of the world's primate species, 33.9% of the 504 species known globally (Estrada et al. 2017). However, 36% of primate species are threatened with extinction, and 63% of them have declined due to deforestation, mostly to agriculture and cattle ranching as well as logging, mining, and fossil fuel extraction (Estrada et al. 2017). This volume and many other publications have highlighted that hunting primates has not been sustainable due to increasing human population, low fertility rate for larger monkeys, introduction of western technology, and decrease of indigenous people's territory due to colonization. However, there is also good news. Some indigenous societies, from the Lacandon Maya and Popoluca in Mexico to the Tikuna of Colombia and the Secoya of Ecuador (this volume), have been trying to conserve and protect monkey populations by not hunting them because they know that these monkeys not only play an important role in their ecosystem but could also be an ecotourism attraction in their communities. Hopefully, this is the beginning of a new path toward primate conservation and protection in these forests in the twenty-first century.

We, as humans, are not exempt of a cultural baggage that modulates our perceptions of nature. Looking at our closest relatives empathically and with a culturally driven view enables us to think holistically about the future of human and nonhuman primates alike. Primatologists, historically, have tended to view their discipline as nonhuman primate-centered, and on the contrary, ethnographers, sometimes, seem to focus mainly, or exclusively, anthropocentrically. Given these realms, ethnoprimatology is actually designated as an opportunity to balance both ways of approaching and socially appropriating our Order. Ethnoprimatology provides an ample understanding of nonhuman primate populations and human societies that are at risk for survival, after crossing into the twenty-first century, in which not only the knowledge about nonhuman primates and other organisms will be lost but also the societies. Figure 2 epitomizes the previous statement. This evocative image represents fragile colored feathers covering a cranium of an endangered woolly monkey (*Brachyteles arachnoides*) made by a member of a possibly extinct Tupi society living in the threatened Atlantic Forest of Brazil. Still, there is time to change the unbalance, and ethnoprimatologists have a fundamental role in this endeavor.

To conclude, we envision this volume as a novel forum for thinking ethnoprimatologically. Therefore, the content of this edited volume provides a wide range of approaches and perspectives that form an excellent collection of cases which mainly

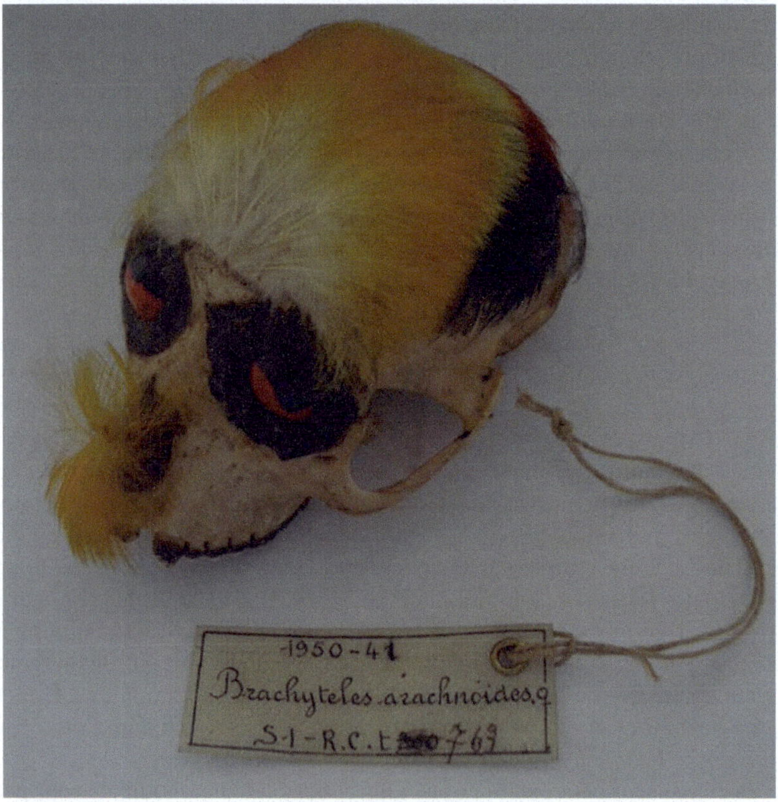

Fig. 2 Fabaceae seeds fixed with bee wax and feathers on a cranium of *Brachyteles arachnoides* from the Brazilian Atlantic Forest (Piece #1950–41 of the *Collections Mammifères et Oiseaux – Anatomie comparée* at the *Muséum national d'Histoire naturelle de Paris*. Photograph by B. Urbani)

focus on how indigenous societies relate to Neotropical primates and vice versa. We hope that the chapters in this book will serve as a framework for future ethnoprimatological research that, as stated previously, will necessarily need to ask novel theoretical and methodological inquiries as well as to follow multi-faceted approaches.

Centro de Antropología, Instituto Venezolano de Investigaciones Científicas
Caracas, Venezuela
Bernardo Urbani

Department of Botany and Environmental Studies Program, Connecticut College
New London, CT, USA
Manuel Lizarralde

Acknowledgments Thanks to Erika Wagner for her comments on the text and to Jacques Cuisin for his hospitality at the Museum of Natural History of Paris (BU) and for allowing the publication of Fig. 2.

References

Bates, HW (1863) The naturalist on the river Amazon. John Murray, London

Castro N, Revilla J, Neville M (1975) Carne de monte como una fuente de proteínas en Iquitos, con referencia especial a monos. Rev. Forestal Perú 19–32

Cormier LA (2000) The ethnoprimatology of the Guajá indians of Maranhão, Brazil. PhD Dissertation. Tulane University, New Orleans

Cormier LA (2002) Monkey as food, monkey as child: Guajá symbolic cannibalism. In: Fuentes A, Wolfe LD (eds) Primates face to face: conservation implications of human and nonhuman primate interconnections. Cambridge University Press, Cambridge

Cormier LA (2003) Kinship with monkeys: the Guajá Foragers of Eastern Amazonia Columbia University Press, New York

Estrada A, Garber PA, Rylands AB, Roos C, Fernandez-Duque E, Di Fiore A, Nekaris AI, Nijman V, Heymann EW, Lambert JE, Rovero F, Barelli C, Setchell JM, Gillespie TR, Mittermeier RA, Arregoitia LV, Boyle S, Fuentes A, MacKinnnon KC, Amato KR, Meyer ALS, Wich S, Sussman RW, Pan R, Kone I, Baoguo L (2017) Sci Adv 3:e1600946

Fleck DW, Voss RS, Patton JL (1999) Biological basis of saki (*Pithecia*) folk species recognized by the Matses Indians of Amazonian Peru. Int J. Primatol 20:1005–1028

Fuentes A, Wolfe LD (eds) (2002) Primates face to face: conservation implications of human and nonhuman primate interconnections. Cambridge University Press, Cambridge

Harner MJ (1973) The Jívaro: people of the sacred waterfalls. Doubleday Anchor, Garden City

Lizarralde M (1989) Indice y Mapa de Gupos Etnolingüisticos Autoctonos de America del Sur. Antropológica (Suplemento No 5). Fundación La Salle de Ciencias Naturales, Caracas

Lizarralde M (2002) Ethnoecology of monkeys among the Barí of Venezuela: perception, use and conservation. In: Fuentes A, Wolfe LD (eds) Primates face to face: conservation implications of human and nonhuman primate interconnections. Cambridge University Press, Cambridge

Oviedo G, Maffi L, Lizarralde M (2000) Indigenous and traditional peoples in the global 200 ecoregions. World Wildlife Fund, Washington

Queiroz, Helder L de, Kipnis R (1991) Os indios Guajá e os primatas da Amazonia Maranhense: un caso de sustentabilidade de caça?. In: Ferrari SP, Schneider H (eds) A primatologia no Brasil 5. Sociedade Brasileira de Primatologia and Universidade Federal do Pará, Belém

Roncal CM, Bowler M, Gilmore MP (2018) The ethnoprimatology of the Maijuna of the Peruvian Amazon and implications for primate conservation. J Ethnobiol Ethnomed 14:19

Shepard G (2002) Primates in Matsigenka subsistence and worldview. In: Fuentes A, Wolfe LD (eds) Primates face to face: conservation implications of human and nonhuman primate interconnections. Cambridge University Press, Cambridge.

Sponsel LE (1981) The Hunter and the hunted in the Amazon: an integrated biological and cultural approach to the behavioral ecology of human predation. PhD Dissertation. Cornell University, Ithaca

Sponsel LE (1997) The human niche in Amazonia: explorations in ethnoprimatology. In: Kinzey WG (ed) New world primates. Ecology, evolution, and behavior. Aldine de Gruyter, New York

Stafford CA, Alarcon-Valenzuela J, Patiño J, Preziosi RF, Sellers WI (2016) Know your monkey: identifying primate conservation challenges in an indigenous Kichwa community using an ethnoprimatological approach. Folia Primatol 87:31–47

Townsend W (1995) Cultural teachings as an ecological data base: Murui (Witoto) knowledge about primates. Latinamericanist 31:1–7

Urbani B (2002) Neotropical ethnoprimatology: an annotated bibliography. Neotrop Primates 10:24–26

Urbani B, Gil E (2001) Consideraciones sobre restos de primates de un yacimiento arqueológico del Oriente de Venezuela (América del Sur): Cueva del Guácharo, estado Monagas. Munibe (Antropol-Arkeolo) 53:135–142

Voss RS, Fleck DW (2011) Mammalian diversity and Matses ethnomammalogy in Amazonian Peru. Part 1, Primates. Bull Am Mus Nat Hist 351:1–81

Wagley C (1977) Welcome of tears: the Tapirapé Indians of Central Brazil. Oxford University, New York

Acknowledgments

We would like to express our gratitude to a *pléyade* of top-notch academics, from pioneer to younger scholars, who kindly served as external reviewers in this edited volume. They are, in alphabetical order, William Balée (Tulane University), Alexandre Beaudoin-Duquette (Universidad Nacional Autónoma de México), Stephen Beckerman (Pennsylvania State University), Brent Berlin (University of Georgia), Gay Biery-Hamilton (Rollins College), Hortensia Caballero-Arias (Instituto Venezolano de Investigaciones Científicas), Alessandra Caputo-Jaffé (Universidad Adolfo Ibañez), Luiz Costa (Universidade Federal de Rio de Janeiro), Zoe Díaz-Martín (Tulane University), Louis C. Forline (University of Nevada), Jeremy Kosler (University of Cincinnati), Francia Medina (Universidad Central de Venezuela), Juan Luis Rodríguez (City University of New York), Krisna Ruette (Instituto Venezolano de Investigaciones Científicas), Isabel Scarborough (Parkland College), Glenn H. Shepard Jr. (Museu Paraense Emilio Goeldi), Harry Walker (London School of Economics and Political Science), and Julie L. Williams (Universidad San Francisco de Quito). Thanks to Leslie E. Sponsel for writing the foreword of this book and for being an inspirational figure in ethnoprimatology since its inception back in 1997. We would also like to thank the editors of the Springer Ethnobiology Series Robert Voeks and John Richard Stepp and the Springer staff members Abitha Pradeep Coumar, Rahul Sharma, Eric Hardy, Rivka Kantor, Anthony L. Dunlap, Nicholas DiBenedetto, and especially Eric Stannard for their support at all stages of this editorial task and also our friend and colleague Lori A. Cormier for being present at the beginning of this project. Last but not least, overall, we greatly appreciate the participation and patience of an academically challenging roster of authors who made this edited volume possible. Thus, once again, to everyone above, thanks a lot, *muito obrigado*, *muchas gracias*.

Contents

Contributors

Jean P. Boubli School of Environment and Life Sciences, University of Salford, Salford, UK

Ana Carla Bruno NPA-Instituto Nacional de Pesquisas da Amazônia, Manaus, Brazil

Coordenação de Sociedade, Ambiente e Saúde – COSAS, Manaus, Brazil

Hortensia Caballero-Arias Centro de Antropología, Instituto Venezolano de Investigaciones Científicas, Caracas, Venezuela

Rosélis R. de Souza-Mazurek Coordenação de Biodiversidade, Instituto Nacional de Pesquisas da Amazônia, Manaus, Brazil

Local Heritage, Environment and Globalization Laboratory – IRD MNHN, Paris, France

Takahi Socioambiental, São Paulo, SP, Brazil

Stella de la Torre Universidad San Francisco de Quito, Colegio de Ciencias Biológicas y Ambientales. Av. Interoceánica y Jardines del Este, Cumbayá, Quito, Ecuador

Yasminda García del Valle Instituto de Biología, Universidad Nacional Autónoma de México, México D.F., Mexico

Margaret Frazen-Levin Graduate Group in Ecology, University of California, Davis, CA, USA

Luisa González-Saavedra Miembro del Grupo Etnografía de América, Universidad Complutense de Madrid, Madrid, Spain

Fernando Guerrero-Martínez Centro de Investigaciones Multidisciplinarias sobre Chiapas y la Frontera Sur. Universidad Nacional Autónoma de México, San Cristóbal de Las Casas, Mexico

Thomas Henfrey Centre for Ecology, Evolution and Environmental Change (cE3c), Lisbon University, Schumacher Institute for Sustainable Systems, Bristol, UK

María Susana Hermes Organización para la Conservación de la Naturaleza y Desarrollo Comunitario (ORCONDECO), Alta Verapaz, Guatemala

Kurt Kotrschal Department of Behavioural Biology, University of Vienna, Vienna, Austria

Kantuta Lara-Delgado Wildlife Conservation Society, Bolivia Program, La Paz, Bolivia

Manuel Lizarralde Department of Botany and Environmental Studies Program, Connecticut College, New London, CT, USA

Angela M. Maldonado Fundacion Entropika, Leticia, Colombia

Celeste Medrano Instituto de Ciencias Antropológicas CONICET, Buenos Aires, Argentina

Guido Miranda-Chumacero Wildlife Conservation Society, Bolivia Program, La Paz, Bolivia

Rosa Icela Ojeda Martínez Escuela de Antropología e Historia del Norte de México, Instituto Nacional de Antropología e Historia, Chihuahua, Mexico

Alfredo Payaguaje Nacionalidad Secoya del Ecuador. Comunidad Secoya San Pablo de Catesiaya, San Pablo, Ecuador

Marianna Pinto-Marroquin Posgrado, Instituto de Ecología AC, Xalapa, Veracruz, Mexico

Red de Biología y Conservación de Vertebrados, Instituto de Ecología AC, Xalapa, Veracruz, Mexico

Felipe Reyes-Escutia Laboratorio de Educación Ambiental y Sustentabilidad, Instituto de Ciencias Biológicas, Universidad de Ciencias y Artes de Chiapas, Tuxtla Gutiérrez, Mexico

Marleny Rosales-Meda Organización para la Conservación de la Naturaleza y Desarrollo Comunitario (ORCONDECO), Alta Verapaz, Guatemala

Felipe Ruan-Soto UNAM, Becario del Programa de Becas Posdoctorales en la UNAM, Centro de Investigaciones Multidisciplinarias sobre Chiapas y la Frontera Sur. Universidad Nacional Autónoma de México, San Cristóbal de Las Casas, Mexico

Konrad Rybka Berkeley Linguistics Department, University of California at Berkeley, Berkeley, CA, USA

Juan Carlos Serio-Silva Red de Biología y Conservación de Vertebrados, Instituto de Ecología AC, Xalapa, Veracruz, Mexico

Glenn H. Shepard Jr Coordenação de Ciências Humanas, Museu Paraense Emílio Goeldi, Belém do Pará, Brazil

Leslie Sponsel Department of Anthropology, University of Hawai'i. Mānoa, Honolulu, HI, USA

Valentín Suárez Comunidad Riacho de Oro, Formosa, Argentina

Kacper Świerk Institute of Ethnology and Cultural Anthropology, University of Szczecin, Szczecin, Poland

Wendy R. Townsend (deceased) Noel Kempff Mercado Natural History Museum, Santa Cruz de la Sierra, Bolivia

Bernardo Urbani Centro de Antropología, Instituto Venezolano de Investigaciones Científicas, Caracas, Venezuela

Esmeralda Gabriela Urquiza-Haas Department of Behavioural Biology and Department of Cognitive Biology, University of Vienna, Vienna, Austria

Robert B. Wallace Wildlife Conservation Society, Bolivia Program, La Paz, Bolivia

Siân Waters Department of Anthropology, Durham University, Durham, UK

Pablo Yépez Fundación Raíz Ecuador. Vía Intervalles, Pasaje Benjamín Carrión, Quito, Ecuador

Egleé Zent Centro de Antropología, Instituto Venezolano de Investigaciones Cientificas, Caracas, Venezuela

Stanford Zent Centro de Antropología, Instituto Venezolano de Investigaciones Cientificas, Caracas, Venezuela

About the Editors

Bernardo Urbani is an Associate Researcher at the Center for Anthropology of the Venezuelan Institute for Scientific Research. His research interests are Neotropical primate ecology and behavior, history of primatology, ethnoprimatology/archeoprimatology, and primate conservation. He has received the Early Career Achievement Award of the American Society of Primatologists and the Martha J. Galante Award of the International Society of Primatology. He has also been elected member of the Global Young Academy.

Manuel Lizarralde is an Associate Professor at the Department of Botany and Environmental Studies Program at Connecticut College. His principal research focus is the botanical and ecological knowledge of indigenous peoples of the tropical rainforest. His research focus is the Barí people of Venezuela with 34 months of fieldwork over the last 29 years. He has also done ethnobotanical research with the Matsigenka of Peru and is the author of an index and map of South American indigenous languages.

Part I
Mesoamerica

Chapter 1
Perception and Uses of Primates Among Popoluca Indigenous People in Los Tuxtlas, Mexico

Marianna Pinto-Marroquin and Juan Carlos Serio-Silva

1.1 Introduction

Traditions, symbolic attributes, and religion influence people's perception of primates (Loudon et al. 2006b; Hill and Webber 2010). These perceptions have a direct effect on the attitude of human communities toward primates and impact the survival of these animals (Baker 2013; Hill and Webber 2010). Perception is defined as the personal notion that is held about an entity or phenomenon and is articulated to a collective worldview, which is developed in social, cultural, and historical contexts (Allot 2001; Ceballos-Mago and Chivers 2010). Studying perceptions helps to understand how humans view their environment and how they appreciate the natural environment and the animals; therefore, it helps to understand the attitudes and decisions that are made toward its use and management (Lefebvre 1991; Arizpe et al. 1993; Sotelo et al. 2003; Fernández 2008). It also provides critical information to define viable strategies for the management of ecosystems and to facilitate social participation in conservation (Castillo et al. 2009).

M. Pinto-Marroquin
Posgrado, Instituto de Ecología AC, Xalapa, Veracruz, Mexico

Red de Biología y Conservación de Vertebrados, Instituto de Ecología AC, Xalapa, Veracruz, Mexico

J. C. Serio-Silva (✉)
Red de Biología y Conservación de Vertebrados, Instituto de Ecología AC, Xalapa, Veracruz, Mexico
e-mail: juan.serio@inecol.mx

B. Urbani, M. Lizarralde (eds.), *Neotropical Ethnoprimatology*, Ethnobiology,
https://doi.org/10.1007/978-3-030-27504-4_1

The region of Los Tuxtlas, Mexico (18°37′–18°35′ N, 95°08′–95°05′ W), is the northernmost limit of tropical rainforest in the Americas where an area of 1551 km^2 has been protected as a Biosphere Reserve since 1998 (Dirzo and Miranda 1991; Laborde 2004; Cristóbal-Azkarate and Dunn 2013). In this region, at least 88% of the original rainforest have been lost and fragmented mainly due to cattle ranging (Guevara-Sada et al. 2004; Cristóbal-Azkarate and Dunn 2013; Castillo-Campos and Laborde 2004). Here different ethnic groups coexist with mestizos and *criollos* of different origin, and there is a cultural mosaic of traditions, religious practices, and forms of production (CONANP 2006).

Los Tuxtlas also correspond to the northernmost distribution of two of the three Mexican primate species: the mantled howler monkeys (*Alouatta palliata mexicana*) and the Mexican spider monkeys (*Ateles geoffroyi vellerosus*) (Cristóbal-Azkarate and Dunn 2013). According to the International Union for the Conservation of Nature, this species of howler monkey is listed in the low concern category of threatened species, while this species of spider monkey is listed as endangered (IUCN 2017). In CITES (2018), both species are in the Appendix I, and in the Mexican context, both are considered endangered by the NOM-059-SEMARNAT-2010 (SEMARNAT 2010). Habitat loss and fragmentation has strongly affected the populations of these primates in Los Tuxtlas (Estrada and Coates-Estrada 1996).

Historically, these primates shared their territory with the Popoluca indigenous people, the native ethnic group of Los Tuxtlas region, who are also known as Zoque-Popoluca and Mixe-Popoluca (Baéz-Jorge 1992; Rodríguez-Luna et al. 2011; CDI 2015). They are direct descendants of the Olmecs, the most ancient culture of Mesoamerica, having received influence of Teotihuacan, Totonaca, and Maya cultures (Baéz-Jorge 1973; CONANP 2006). The Popoluca mainly inhabit the Sierra de Santa Marta. They consider themselves descendants of *Homshuk*, "the god of maize" who "was born from an egg and was the original seed of the maize that has fed the humans" (Baéz-Jorge 1992). Although they are strongly influenced by the catholic religion, they retain some ceremonies of pre-Hispanic origin, and many still speak their native language (CDI 2015). Popolucan use and manage their natural resources according to their ancient traditions; however, the demographic and economic transformations that are occurring in the region put these traditions at risk (CONANP 2006). They are organized in *ejidos*, a group of houses forming a village with its surrounding parcels of land where members carry out their subsistence activities (Covarrubias 1980). The *ejido* is a juridical figure, coming from the Mexican revolution, through which a community of people owns a large territorial extension to distribute it among its members (Flores 2008).

Taking into account that Los Tuxtlas is an important region for conservation of spider and howler monkeys in Mexico (Estrada and Mandujano 2003; Oropeza and Rendón 2012) and that Popoluca is the ethnic group who has populated the region since pre-Hispanic times (Rodríguez-Luna et al. 2011), we studied the perception and use of primates by two Popolucan communities. We expect that this information will be useful in constructing primate conservation strategies in Los Tuxtlas region.

The objectives of this study are to (a) recognize the role of primates in Popolucan culture and traditions, (b) understand the perceptions of primates in popoluca communities, (c) identify the interest in primate conservation among the Popolucans, and (d) provide recommendations for conservation of primates in Los Tuxtlas region.

1.2 Methods

This study was conducted in two Popoluca indigenous communities located within the boundaries of the Reserva de la Biósfera Los Tuxtlas, Mexico (Fig. 1.1). The first community, Los Mangos (18°13′ N, 95°08′ W), is located in the influence zone of the reserve and is part of the municipality of Hueyapan de Ocampo. It has a population of 2722 habitants (IMSS 2015), of which 32.6% are bilingual, speaking Popoluca and Spanish (PROGEDER 2013). The second community, Piedra Labrada (18°23′ N, 94°46′ W), is located in the buffer zone of the reserve and is part of the municipality of Tatahuicapan de Juárez. It has a population of 510 habitants, of which 94% live in Popoluca indigenous homes (IMSS 2015). Both communities have howler monkeys (*A. palliata*) and spider monkeys (*A. geoffroyi*) in their neighboring forests.

The study was performed between 2016 and 2017 and was divided into three phases, (1) the recognition visit, (2) the ethnographic period, and (3) the collection of qualitative data. During the recognition visit, we contacted local leaders and authorities and formally requested a prior, free, and informed consent (*consentimiento previo, libre e informado*) to carry out the research. During the ethnographic period, we lived for 2 weeks in each community to understand the cultural context and to encourage people to exchange personal and collective stories. The contact with the people was made with the help of a local leader who introduced the researcher to the villagers. During this period, an ethnobiological tour was made with members of the community in the area where monkeys were present, and finally, a pilot interview was conducted to verify if the questions of the semi-structured interviews that were to be used were appropriate. Observations and reflections were recorded on a field diary.

During collection of qualitative data phase, 46 semi-structured interviews were formally conducted, 23 in each community (Los Mangos, 8 women, 15 men; Piedra Labrada, 11 women, 12 men). The sample size was defined based on redundancy, i.e., sampling ends when no new relevant information is received from new interviewees (Patton 2002; Peña-Mondragon 2011). To identify key informants, the snowball method was applied. This method consists in asking each interviewee to suggest another person from the community who can provide valuable information regarding the research topic (Patton 2002; Tójar 2006; Cantoni 2009). The selection of interviewees was done according to four conditions (after Cantoni 2009): (1) people frequently visiting areas with primates, (2) people who used natural resources

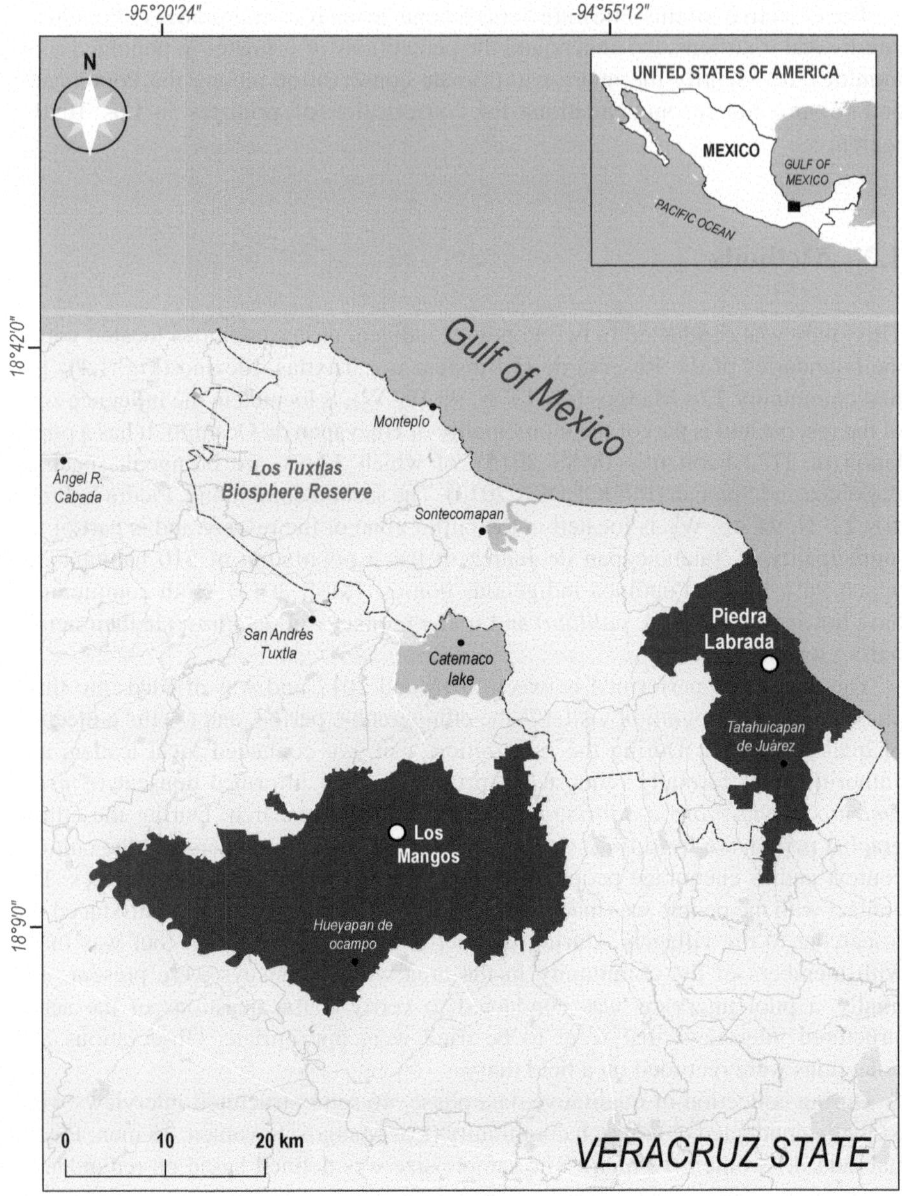

Fig. 1.1 Location of the study sites at Los Tuxtlas, Mexico

in the forest where monkeys where found, (3) leaders who knew the socio-ecological processes of their territories, and (4) people who had detailed knowledge of the Popolucan culture and history. With these conditions, all of the people interviewed were older than 40 years. The semi-structured interviews consisted in open questions and were carried out through conversations (Galindo-Cáceres 1998). Interviews were recorded on digital audio files with prior verbal consent. Each interview lasted approximately 50 minutes. Interviews were transcribed, and information was organized in categories and analyzed using Atlas-ti 7.5.4 (Scientific Software Development GmbH 2013) following the guidelines of qualitative analysis (Tárres 2004; Taylor and Bogdan 1987).

1.3 Results

In both communities, the local names used for spider monkey (*A. geoffroyi*) are *chango* or *chango araña* in local Spanish and *uutzu* in Popoluca language, while the howler monkey (*A. palliata*) is known as *mono*, *monos zambo*, and *monos saraguato* in local Spanish and *buhurum uutzu* in Popoluca language. According to people interviewed, *uutzu* means monkey and *buhurum* does not have any particular meaning as a single noun or adjective. When they only refer to *mono*, they are referring to howler monkey, while they almost never refer to spider monkeys in this way. Spider monkeys are generally called *chango*.

1.3.1 Cultural and Traditional Aspects About Primates

Popoluca origin myths are associated to *Homshuk*, the god of maize. Primates do not appear in those myths and neither are related to religious beliefs because they do not have a relationship with maize crops. Nevertheless, they are related to *dueños* and *chaneques*, spirits of the forests that protect those forests, watching the animals and taking care of them.

The most significant cultural aspect about primates we found is a widespread and popular belief that, when howler monkeys' vocalization is more loud and frequent than normal, they announce changes in the weather. They are announcing the arrival of the cold air and heavy rain season known as *nortes* and also hot and dry seasons that are known as *suradas*. Popoluca people can identify the coming season according to the kind of calls howler monkeys are emitting, as well as the time of the day when they do it. If it is in the morning, a *surada* is coming, but if it is in the afternoon, then a *norte* is coming. The change of weather occurs 1–3 days after the howler monkeys announce it. The interviewees claim that this "weather service alert" provided by howler monkeys is very accurate and useful to plan daily activities. They also say that the Popoluca ascendants in ancient times knew how to interpret the relationship between howler monkeys' vocalizations about weather better than people do today.

It is also widely believed that, sometimes, when these animals see a woman, they pursue her with the aim of wooing her and kidnapping her in the forest. There are different variations of a myth about a howler monkey that kidnapped a woman and sexually abused her. The women got pregnant and had a half monkey and half human as a baby. She was rescued by people who heard her ask for help. These stories are told to the young women by their parents, who recommend they never go alone to the forest and rivers. Among the interviewees, these stories were mentioned mainly by women and are better known by the older women. In the community of Piedra Labrada, it was reported that a howler monkey did approached a group of women and was killed by a man. He did it to prevent this animal from kidnaping or hurting these women.

For spider monkeys, beliefs and myths are less common than for howler monkeys. The spider monkey is mainly associated with joy and fun. There is a traditional dance in the Popolucan culture that is performed in honor to *Homshuk* to ask for a successful cultivation and abundant harvest without problems. In this dance, people dress in animal skin or costumes of animals that can harm the maize crops. Although the spider monkey does not harm crops, it must participate because it represents the joy that should not be missed in dances. This dance is still performed in Popolucan communities located in San Pedro Soteapan, which is the place of origin of the Popoluca and where their culture is most traditional.

Another myth that is less known is about a forest's spirit that appears to them near rivers when they have worries about their own bad behavior. These supernatural creatures can kidnap people and take them to the bottom of rivers to drown them. They appear in the form of a "water spider monkey," which looks like a small spider monkey made of water with intense red lips, very long arms, and no tail. In dreams, spider monkeys represent policemen and mean that there will be a problem or someone is going to jail. Another belief is that spider monkeys used to be humans, which is a similar story to the "two twin heroes" of *Popol Vuh* (a body of mythological-historical narrative of the Quiché Mayan of Guatemala [Recinos 1978]), which was also told by the interviewees. A few people also stated that spider monkeys' screams announce earthquakes. Likewise, a story was shared about a man who, after eating a spider monkey, acquired its appearance and behavior. Some said that grabbing a spider monkey by the tail will bring bad luck with money issues. Finally, it was also said that both howlers and spider monkeys like small children, especially the naughty ones, and for this reason, they try to kidnap them. This, apparently, is a story normally told by parents to prevent bad behavior in children.

With respect to artistic representations, one of the interviewees found a ceramic container of indigenous origin that had a monkey at its base in Los Mangos. Unfortunately, this object was lost. In Piedra Labrada, there are numerous representations of rock art of Olmec culture origin. In fact, the name of the *Piedra Labrada* community literally means "carved stone" in Spanish. At this site, there are petroglyphs with different representations among which there is a monkey (Fig. 1.2) as well as monkey-like stone sculptures. Archaeological studies have been carried out in the area, and work has been done to preserve and recover this cultural heritage (Córdova 2004; Cárdenas 2005; Budar 2008; Budar et al. 2008; Ortiz 2008).

Fig. 1.2 Petroglyph, probably of Olmec origin, representing a monkey in the locality of Piedra Labrada

1.3.2 Emotive Perceptions About Primates

In both communities, most respondents stated that spider and howler monkeys produce emotions related to sympathy and curiosity. Both species were associated mainly with the emotions of joy because it is widely considered that primates "bring happiness to the forest." Popoluca consider a "privilege" to be able to observe these animals, especially because they realize that their populations are decreasing. Many interviewees consider that monkeys, especially spider monkeys, are joyful because they look and behave very similar to "happy humans." Most informants indicated that they are habituated to the presence of howler monkeys because they are used to hearing their vocalizations every day early in the morning and late afternoon.

People also associate howler monkey vocalizations with joy and company in the forest, and they enjoy emulating their howler monkey long call, because the males sometimes respond to the imitated call by howling back. Some of the interviewees commented that howler monkeys were useful to people because their calls announce when people should wake up in the morning and when there are intruders of their land. Many people agreed that, because of the power of the howls, they appear to be large animals that can cause fear to those who are not familiar with their vocalizations. However, despite the fact that a good perception about primates is more frequent, many women interviewed felt afraid of howler monkeys due to the local belief that howler monkeys kidnap and abuse women as indicated above.

1.3.3 Medicinal Use of Primates

In the past, spider monkeys were frequently used for medicinal purposes, but this is currently uncommon due to legal restrictions for hunting and the reduction of primate communities. The Popoluca use the spider monkey mainly to prevent and cure skin diseases, like infections, welts, rashes, hives, pimples, and rheumatism. Other diseases that have been treated with this species are cancer, epilepsy, physical paralysis, heart disease, and flu. Some people mentioned they have eaten them roasted. Also, when they were babies, their grandparents gave them broth of spider monkey to prevent skin diseases. It was also said that older people used spider monkeys' fat to treat rheumatism. Howler monkeys were not used for medicinal purposes by the Popoluca.

1.3.4 Use of Primates as Pets

In the past, it was common for Popoluca community members to have pet spider monkeys. Currently, no one has primate pets because it is illegal since the 1990s and because their populations have drastically diminished. The Popolucans consider taming juvenile spider monkeys to be easy; however, to capture juvenile, they normally kill the mother. Pet spider monkeys were tied to posts in the houses and fed with vegetables such as papaya and bananas or items like tortillas. Once they were habituated to live with people, they were released, but they would return to their captors' homes. Popolucans prefer to have these monkeys as pets without depriving them of their freedom. Also, they consider that spider monkeys are difficult to keep as they easily acquire diseases, and consequently, they die. For this reason, they need a clean place to live. Popolucans do not have howler monkeys as pets because they consider them to be difficult to tame and raise and because they bite and attack people.

1.3.5 Use of Primates as Food

The consumption of primates as food is not a Popolucan tradition, mainly because of their resemblance with humans. However, members of both communities learned from outsider poachers how to hunt and prepare spider monkeys. Most of the Popolucans interviewed had tasted spider monkeys under the influence of non-Popolucan friends. The people who have tried spider monkey meat agree that it has a very good taste similar to pork, beef, or deer. In contrast, howler monkey is not consumed because it is considered insipid, strange, and having an unusual taste.

1.3.6 Economic Uses and Perceptions About Primates

According to the informants of both communities, hunting of spider monkeys is mainly conducted by non-Popolucans. For economic purposes, this activity has had a significant negative impact on their local populations. Approximately 30 years ago, people from the neighboring municipality of Catemaco used to arrive at Los Mangos community to capture spider monkeys and sell them as bush meat to restaurants. The consumption of spider monkeys in Catemaco is traditional, and their meat is very appreciated. Even some restaurants still offer *chango* (spider monkey) meat in their menus, although currently this is pork. Some Popolucans from Los Mangos sold spider monkeys to people from Catemaco as food and, in some isolated cases, to outsiders as pets. In Piedra Labrada, the meat of spider monkey was intensively used as bait for shrimp fishing by the inhabitants of a neighboring community. These people are not Popolucans, but immigrants from another area of Veracruz that settled there because of governmental land policies. In the neighbor community, the shrimp market was quite important for the local economy, but the shrimp populations was overharvested, as well as the spider monkey population. The Popolucans from Piedra Labrada did not agree with the intensive use and extraction of natural resources carried by their neighbors.

In both communities, people perceive that primates have a high potential as a tourist attraction and consequently represents an economic opportunity. Close to both communities, there are touristic projects or places inhabited by primates which are attractive to tourists. Therefore, for these communities, they also become a tourist attraction, and the inhabitants believe that it is necessary to protect the current forests where primates are present and expand them to promote their population increase. It was also acknowledged that it is necessary to promote the cultivation of tree species important in their diet to have suitable areas for monkeys' survival. Popolucans believe that it is also necessary to reconnect those forest remnants islands with other forest areas where primates are present, so more and different groups would come close to people.

1.3.7 Perceived Ecological Importance of Primates

People recognize that primates play a relevant role in the persistence of the forest, but they do not know exactly how. Few people mentioned that primates are dispersers of seeds, contributing to the maintenance of native vegetation and consequently to the conservation of other animals. In the same way, primates are indicators of the quality of the ecosystem because forests where they are present are always in good conditions.

1.3.8 Perceived Abundance and Distribution of Primates

According to the interviewees in both localities, the original population of spider monkeys used to be equal or larger than those of howler monkeys. But because of deforestation and hunting, populations of both species have been decreasing gradually since ~5 to 15 years ago. At that time, howler monkeys' populations slowly started to increase, while those of spider monkeys have not. In Piedra Labrada, populations of spider monkeys are still low, while in Los Mangos they have almost completely disappeared.

In both communities, primates are mainly found near rivers where the vegetation has been protected by a common agreement to conserve water resources, as well as in the adjacent land where there are fruiting trees that monkeys depend on. In Los Mangos, howler monkeys are concentrated in the southwestern zone on the slopes of the Hueyapan River. This area corresponds to a community forest reserve that covers an area of 322 ha. In Piedra Labrada, howler monkeys live in the protected vegetation near the Tecuanapa River. With respect to the spider monkeys, individuals that managed to survive to deforestation and hunting migrated to areas more isolated from humans. In Los Mangos, there are rumors that spider monkeys have been recently seen in the Hueyapío Stream Canyon. In Piedra Labrada, spider monkeys can be found in parcels belonging to local people bordering the core area of Los Tuxtlas Biosphere Reserve and have also been seen occasionally near the Tecuanapa River. Some interviewees consider that spider monkeys are slowly returning to Piedra Labrada because they are hunted in other localities close to the core area of Los Tuxtlas Biosphere Reserve. People have seen spider monkeys walking on the ground due to the lack of arboreal vegetation, exhibiting an atypical terrestrial behavior.

1.3.9 Perceived Threats to the Conservation of Primates

According to the Popolucans, cattle ranching is the greatest threat to primates, as the native vegetation has been removed to graze cattle. Commercial agriculture and the extraction of timber are perceived to threaten the primates, but to a lesser extent, because ranching is more economically profitable than agriculture as it requires less human labor while larger areas of forests are converted into pastures rather than crops. Cattle ranching is gradually displacing agriculture in the region.

The main current problem expressed in Los Mangos is that, for approximately 22 years, outsiders just interested in economic benefits have been buying lands and have been cutting down considerable areas of native vegetation for cattle ranching. This has been happening due to the new land title laws that were implemented in the late 1990s, which allowed *ejidatarios* to sell their land to outsiders. On the other hand, in Piedra Labrada, another threat is the lack of environmental commitment of the surrounding *ejidos* since they do not have regulations for the use and management of their natural resources. Additionally, species of trees on which primates feed have diminished. Several timber species that are also part of primate's diet, such as *zapote mamey* (*Manilkara zapota*), have declined because of their hard and high-quality wood that is widely used for building houses and furniture.

1.3.10 Interest in Primate Conservation

In both communities, all interviewees believe that it is important that primates survive and their populations increase. They were very emphatic about their wish to have future generations know these animals. Also, they consider primates have the right to live in good conditions, just like people. For this to happen, they believe that it is necessary to protect existing forest and reforest new areas with trees that primates feed on, mainly spider monkeys that require extensive forest areas and are more sensitive to habitat transformation. Some *ejidatarios* maintain a small portion of native vegetation in their parcels, and some of them are visited by primates. They allow these animals to be there and protect them from people who try to harm them.

In Los Mangos, governmental institutions have promoted the idea of protecting the wild animals and reforesting. Thus, areas of vegetation belonging to native Popolucans are currently more stable than in the past. On the other hand, in Piedra Labrada, people have seen howler monkey populations recovering gradually, thanks to regulations imposed by the *Ejido* Community Council which prohibited hunting wild animals and encouraging the preservation of native vegetation for 8 years. Respondents consider that Piedra Labrada is currently a safer place for animals. Finally, in both communities, some people consider that tourism, apart from being an economic opportunity, could also be a strategy to promote primate conservation.

1.4 Discussion

In the context of the Popoluca cosmogony, the primates do not have a relevant totemic position, possibly because they are not present in the myths of origin as maize is the basic element of their cultural identity (Baéz-Jorge 1992; López 1992; CDI 2015) and howler and spider monkeys have no interactions with maize crops. Nevertheless, primates are present in widespread popular beliefs among the Popolucans. Beliefs about howler monkeys are more popular than those about spider monkeys, possibly because people are more familiar with their presence. People often hear them howling, making this monkeys more notorious, even if people do not see them. In the same way, people perceive that howler monkeys are more abundant than spider monkeys. This perception coincides with the results of an ecological study conducted by Estrada and Coates-Estrada (1996) in Los Tuxtlas region, which found that howler monkeys have larger populations than spider monkeys, as howler monkeys can survive in disturbed habitats and small fragments of forest (Bicca-Marques 2003; Arroyo-Rodríguez and Dias 2010; Aristizábal-Borja et al. 2011), while spider monkeys require continuous larger sections of forests in better conditions (Fahrig 1998; Boyle and Smith 2010; Ordóñez-Gómez et al. 2014). In both study sites, the presence of howler monkeys was reported in areas next to the rivers. Also, Estrada and Coates-Estrada (1996) reported that it is common to find this species in the region of Los Tuxtlas inhabiting residual vegetation of tropical forest, located on the edges of rivers and streams, because it facilitates the movement of these species between fragments.

The most popular belief is that howler monkeys announce changes in weather. The fact that some people use this "monkey weather service" to plan their activities shows that it has a pragmatic utility in their life. A similar situation was reported by Ruoso (2012) in Santa Cruz do Sul, Brazil, where howler monkeys are considered predictors of heavy rains as they howl drastically and insistently because they possibly are sensitive to atmospheric pressure. Voss and Fleck (2011) also reported that Matse people in Peru believe that vocalizations of red howler monkeys in the morning are predictors of a clear day. In Tanzania, the presence of primates close to farmhouses, fields, and villages is used as a predictor of rainy seasons (Mahoo et al. 2015; Chengula and Nyambo 2016). There is also evidence that, with changes in barometric and hydrostatic pressure, humidity, and air temperature, animals' nervous system becomes more active than normal, and they change their behavior (Acharya 2011). Not only the Popolucans but also many cultures around the world have systems to predict weather conditions based on bioindicators which include unusual animal behavior (Ravi et al. 2008; Acharya 2011; Okonya and Kroschel 2013; Rautela and Karki 2015; Alves and Barboza 2018). On the other hand, bearing in mind that Nahuatl indigenous people also live in Los Tuxtlas region, this belief could have a relationship with the god of wind, *Ehecátl-Quetzalcóatl*, who, according to the Nahuatl cosmology, blows the winds that precede the rains and is represented by a monkey (CONABIO 2011). Further research about the relationship between the weather changes and howler monkeys' behavior would likely bring interesting results.

The myth about a howler monkey kidnapping a woman has become a widespread belief and influences the perception women have about this species. Many of them are afraid when they see a howler monkey. A similar myth exists among the Mundurucu indigenous people in Brazil, in which a man marries a female howler monkey which produces children with him (Murphy 1958). The report of a Popolucan man who killed a spider monkey to protect some women is an evidence that local beliefs influence people's actions toward primates (Simons and Meyers 2001; Loudon et al. 2006a, b; Saj et al. 2006; Hill and Webber 2010; Baker 2013). It also shows that the perceived similarity between primates and humans creates expectations and has implications on how people view and respond to certain kinds of primate behaviors (Hill and Webber 2010).

The fact that spider monkeys are perceived to be closely related to the *Homo sapiens* because of their morphological resemblance and that they are also highly associated to joy and fun echoes the work of Bruner and Cucina (2005), who found that, historically in Mesoamerica, spider monkeys have attracted people's attention because of their morphology and movements similar with humans. The authors also emphasized that the Mayas associated spider monkeys with histrionic and artistic characteristics (Bruner and Cucina 2005). Also, for the Aztecs, primates were a symbol of happiness (CONABIO 2011). This similarity in cultural perception about primates could also be because the Popolucans have Maya culture influence and share their territory with indigenous people speaking Aztec language, the Nahuatl (CONABIO 2011).

With respect to the uses, we found that spider monkeys have been used traditionally by the Popolucans as medicine and pets and nontraditionally as food or for economical purposes (bush meat and pets). As a consequence, hunting of primates became illegal in 1992, when the Mexican Government signed the CITES. In contrast, for howler monkeys, there were no reported uses by Popolucans. Most of the interviewees are reluctant to consume both primates for food because of their resemblance to humans. Something similar has been documented in two regions of Brazil, where people abstain from consuming spider monkeys due to their physical similarity to humans (Kracke 1978; Basso 1973). Likewise, Cormier and Urbani (2008) report that, in the border area between Colombia and Venezuela, spider monkeys (*Ateles hybridus*) are considered to be very similar to children and therefore are not consumed or hunted. Ellwanger et al. (2015) observed that perceptions toward monkeys are positive because of their similarity of behavior to humans, and therefore, people abstain from hunting and consuming them. However, even though most Popolucans abstained from using spider monkeys as food, some have done it due to the influence of outsiders and perceive this meat as tasting good. It shows an induced change in the cultural perception that, without any restrictions imposed by law, could have affected the populations of these species in the communities. In Mexico, the consumption of spider monkeys has been reported for the Lacandon indigenous people in Chiapas (Baer and Merrifield 1972; García del Valle et al. 2015). Is has been referenced that the species of the genus *Ateles* spp. are one of the groups of primates most valued at a nutritional level in different Neotropical traditional societies (Urbani 2005; Cormier and Urbani 2008; Voss and Fleck 2011). For example, in Peru, howler monkeys are the least preferred primates at a nutritional level for the Matsigenka indigenous people because of its taste (Shepard Jr 2002). The Cashinahua amerindians in Peru consider that howler monkeys are not edible (Kensinger et al. 1975), and the Matses believe that just older people can eat them because, if some young person do it, they become lazy (Voss and Fleck 2011). Likewise, Cormier (2003) found that, due to their cultural position, howler monkeys are one of the most prohibited species for consumption by Amazonian indigenous people. Nevertheless, this contrasts with what was reported by Urbani (2005), who found that howler monkeys are one of the most consumed species in the Neotropics. This reflects that howler monkeys are subject to a wide range of restrictions and preferences that vary considerably between cultures (Urbani and Cormier 2015).

About the traditional use of spider monkeys as medicine by the Popolucans, it is not just curative but also preventive, mainly to treat skin diseases and rheumatic problems. Although Popolucans assure that it is effective, it has not been medically proved. The medicinal use of spider monkeys was also reported in Colombia by farmers to treat malaria (Cormier and Urbani 2008). On the other side, the absence of use of howler monkeys as medicine and food could be related to its bad taste, as was reported by some respondents. This is also one of the reasons why many Amerindian cultures abstain from consuming this species (Urbani and Cormier 2015).

Duarte-Quiroga and Estrada (2003) reported 179 cases of primates used as pets in Mexico City of which 70% corresponded to spider monkeys. For the Popolucans, it corresponds to a process of traditional taming in which the monkeys are free around the houses. In this regard, Cormier and Urbani (2008) acknowledge that the way in which indigenous communities keep primates as pets is very different from the way western societies do it.

The Popolucans believe that primates are important for the ecosystem, but their knowledge about the ecological role of these species is not well understood. With respect to that, it is necessary to develop educational strategies for people in Los Tuxtlas, not just to Popolucans, to understand the role of primates as seed dispersers and in the regeneration of the forest and how they indirectly benefit human communities and at the same time to promote preservation of traditional or local ecological knowledge and culture. In the same way, it is relevant to make people become more conscious of the importance of conserving trees that are part of the diet of primates and are appreciated as timber, as well as to promote the creation of community agreements for the use and management of natural resources in other localities. Finally, it is important to prevent the sale of their land to outsiders since it is promoting the deforestation for cattle ranching that is reducing the habitat for monkeys.

Regarding the perceived threats that primates face in Los Tuxtlas, the Popolucans have an accurate knowledge, as they reported as the main threat the loss of habitat, due to expanding cattle ranching and hunting. These threats have been reported also as the principal ones in ecological studies, along with selective logging (Estrada and Coates-Estrada 1996). Cattle ranching is gradually displacing agriculture, mainly maize crops, which could affect Popolucan cultural values regarding the importance of the maize in their cosmology.

For the conservation of primates in Los Tuxtlas, it is recommended to implement ecotourism programs that can be an economical alternative to cattle ranching. These programs should be ecologically and culturally responsible, use primates as flagship species, and take advantage of the archaeological heritage of Piedra Labrada. These programs could also be designed with the participation of the communities and take into account the particular cultural and socio-ecologic contexts of each site. Finally, the Popolucans can be allies for the conservation of primates as they are interested and committed to this objective for their children and future generations. In this direction, they have created agreements for responsibly using and managing natural resources that, in general, have a positive impact to primates. To conserve primates in Los Tuxtlas, it is recommended that the issues explored in this study should be taken into account when approaching the local communities, while acknowledging these perspectives.

Acknowledgments We are deeply grateful to the local authorities of both localities and to the Popoluca people who have shared their time and valuable knowledge with us, as well as Carlos Castaño-Uribe and Luciana Porter-Bolland for the academic support during this research and Concepción Acosta, Lorenzo Arteaga, John Aristizabal, Flavio Albino, and Bertha and Beto Palma for their invaluable support in the fieldwork. We are also grateful to Adriana Sandoval Comté for the help with the mapping, to Rolando González-Trápaga for the technical support and, finally, to Colin A. Chapman and Julien Rouyrre for the language editing. The research was partially supported by CONACYT's MSc. Beca # 671519 and by the Instituto de Ecología A.C.

References

Acharya S (2011) Presage biology: lessons from nature in weather forecasting. Indian J Tradit Knowl 10:114–124

Allot R (2001) Language and evolution: language papers. Perception. http://www.percepp.demon.co.uk/perceptn.htm. Accessed 3 June 2017

Alves RRN, Barboza RRD (2018) Animals as ethnozooindicators of weather and climate. In: Alves RRN, Albuquerque UP (eds) Ethnozoology: animals in our lives. Elsevier Academic Press, Kidlington, pp 283–420

Aristizábal-Borja JF, Pozo-Montuy G, Pérez-Torres J, Serio-Silva JC (2011) Anotaciones de la ecología alimentaria de monos aulladores negros en un fragmento con condiciones de hacinamiento (Balancán, Tabasco, México). Univ Sci 16:140–146

Arizpe L, Paz F, Velázquez M (1993) Cultura y cambio global: percepciones sociales sobre la deforestación en la selva lacandona. Centro Regional de Investigaciones Multidisciplinarias, Porrúa

Arroyo-Rodríguez V, Dias PA (2010) Effects of habitat fragmentation and disturbance on howler monkeys: a review. Am J Primatol 72:1–16

Baer P, Merrifield WR (1972) Los Lacandones de México, Dos Estudios. Instituto Nacional Indigenista–Secretaria de Educación Publica, Mexico City

Baéz-Jorge F (1973) Los Zoque-Popoluca, Estructura Social. Instituto Nacional Indigenista y Secretaría de Educación, México City

Baéz-Jorge F (1992) Homshuk y el simbolismo de la ovogénesis en Mesoamérica. In: Esponda VM, Pincemin S, Rosas M (eds) Antropología mesoamericana, Homenaje a Alfonso Villa Rojas. Gobierno del Estado de Chiapas, Consejo Estatal de Fomento a la Investigación y Difusión de la Cultura, DIF-Chiapas, Instituto Chiapaneco de Cultura, Tuxtla Guliérrez, pp 303–332

Baker LR (2013) Links between local folklore and the conservation of Sclater's monkey (*Cercopithecus sclateri*) in Nigeria. Afr Primates 8:17–24

Basso EB (1973) The Kalapalo Indians of Central Brazil. Holt, Rinehart and Winston, New York

Bicca-Marques JC (2003) How do howler monkeys cope with habitat fragmentation? In: Marsh LK (ed) Primates in fragments: ecology and conservation. Kluwer Academic/Plenum Publishers, New York, pp 283–303

Boyle SA, Smith AT (2010) Can landscape and species characteristics predict primate presence in forest fragments in the Brazilian Amazon? Biol Conserv 143:1134–1143

Bruner E, Cucina A (2005) *Alouatta, Ateles*, and the ancient Mesoamerican cultures. J Anthropol Sci 83:111–117

Budar L (2008) Detrás de los cerros, en el último rincón de Los Tuxtlas: Piedra Labrada. In: Budar L, Ladrón de Guevara S (eds) Arqueología, paisaje y cosmovisión en Los Tuxtlas. Universidad Veracruzana, Xalapa, pp 105–116

Budar L, Becerra G, Melgarejo M (2008) Mensajes Silenciosos en piedras grabadas. Una descripción de las manifestaciones gráficas rupestres de Piedra Labrada. In: Budar L, Ladrón de Guevara S (eds) Arqueología, paisaje y cosmovisión en Los Tuxtlas. Universidad Veracruzana, Xalapa, pp 117–130

Cantoni NM (2009) Técnicas de muestreo y determinación del tamaño de la muestra en investigación cuantitativa. Revista Argentina de Humanidades y Ciencias Sociales 7(2). http://www.sai.com.ar/metodologia/rahycs/rahycs_v7_n2_06.htm. Accessed 10 Jan 2018

Cárdenas J (2005) Propuesta museográfica para la sala arqueológica del ecomuseo en Piedra Labrada, Veracruz. B. Dissertation. Universidad Veracruzana, Xalapa

Castillo A, Corral-Verdugo V, González-Gaudiano E, Paré L, Paz MF, Reyes J, Schteigart M (2009) Conservación y sociedad en México. In: Sarukhán J (ed) Capital Natural de México, Estado de conservación y tendencias al cambio, vol II. CONABIO, México, pp 761–801

Castillo-Campos G, Laborde J (2004) La Vegetación. In: Guevara S, Laborde J, Sanchex G (eds) Los Tuxtlas: el paisaje de la sierra. Instituto de Ecología, A. C., Unión Europea, pp 231–265

Ceballos-Mago N, Chivers DJ (2010) Local knowledge and perceptions of pet primates and wild Margarita capuchins on Isla de Margarita and Isla de Coche in Venezuela. Endanger Species Res 13:63–72

CDI (2015) Popoluca. Comisión Nacional para el Desarrollo de los Pueblos Indígenas de México. http://www.cdi.gob.mx/pueblos_mexico/Popoluca.pdf. Accessed 7 Jan 2018

Chengula F, Nyambo B (2016) The significance of indigenous weather forecast knowledge and practices under weather variability and climate change: a case study of small holder farmers on the slopes of Mount Kilimanjaro. Int J Agric Ext 2:31–43

CITES - Convention on International Trade in Endangered Species of Wild Fauna and Flora (2018). https://www.cites.org/. Accessed 7 Jan 2018

CONABIO (2011) Fichas de especies prioritarias: Mono Aullador Pardo (*Alouatta palliata*). Comisión Nacional para el Conocimiento y Uso de la Biodiversidad, México

CONANP (2006) Programa de Conservación y Manejo de la Reserva de la Biosfera Los Tuxtlas. Comisión Nacional de Áreas Naturales Protegidas, México

Córdova A (2004) Cabeza de piedra y el cuerpo que se revienta. B. Dissertation. Universidad Veracruzana, Xalapa

Cormier LA (2003) Kinship with monkeys: the Guajá foragers of eastern Amazonia. Columbia University Press, New York

Cormier LA, Urbani B (2008) The ethnoprimatology of the spider monkey: from past to present. In: Campbell C (ed) Spider monkeys: the biology and behavior of the genus *Ateles*, Cambridge studies in biological and evolutionary anthropology. Cambridge University Press, Cambridge, pp 377–403

Covarrubias M (1980) El sur de México. In: Rodríguez I, Valdez F (eds) Clásicos de la Antropología Mexicana. Libros de México S.A, México

Cristóbal-Azkarate J, Dunn JC (2013) Lessons from Los Tuxtlas: 30 years of research into primates in fragments. In: Marsh LK, Chapman CA (eds) Primates in fragments. Springer, New York, pp 505–525

Dirzo R, Miranda A (1991) El limite boreal de la selva tropical húmeda en el continente americano: Contracción de la vegetación y solución de una controversia. Interciencia 16:240–247

Duarte-Quiroga A, Estrada A (2003) Primates as pets in Mexico City: an assessment of the species involved, source of origin, and general aspects of treatment. Am J Primatol 61:53–60

Ellwanger AL, Riley EP, Niu K, Tan CL (2015) Local people's knowledge and attitudes matter for the future conservation of the endangered Guizhou Snub-Nosed Monkey (*Rhinopithecus brelichi*) in Fanjingshan National Nature Reserve, China. Int J Primatol 36:33–54

Estrada A, Coates-Estrada R (1996) Tropical rain forest fragmentation and wild populations of primates at Los Tuxtlas. Int J Primatol 5:59–783

Estrada A, Mandujano S (2003) Investigaciones con *Alouatta* y *Ateles* en México. Neotrop Primates 11(3):147–156

Fahrig L (1998) When does fragmentation of breeding habitat affect population survival? Ecol Model 105:273–292

Fernández Y (2008) ¿Por qué estudiar las percepciones ambientales? Una revisión de la literatura mexicana con énfasis en Áreas Naturales Protegidas. Espiral 15(43):179–202

Flores CE (2008) Suelo ejidal en México. Cuaderno de Investigación Urbanística Ciur 57

Galindo-Cáceres J (1998) Técnicas de investigación en sociedad, cultura y comunicación. Addison Wesley-Longman, Mexico

García del Valle Y, Naranjo JE, Caballero J, Martorell C, Ruan-Soto F, Enríquez PL (2015) Cultural significance of wild mammals in Mayan and mestizo communities of the Lacandon Rainforest, Chiapas, Mexico. J Ethnobiol Ethnomed 11:1–13. https://doi.org/10.1186/s13002-015-0021-7
Guevara-Sada S, Laborde J, Sánchez-Ríos G (2004) Los Tuxtlas: El paisaje de la sierra. Instituto de Ecología A. C. and European Union, Xalapa
Hill CM, Webber AD (2010) Perceptions of nonhuman primates in human–wildlife conflict scenarios. Am J Primatol 72:919–924
IMSS (2015) Censo oficial: Delegación regional Veracruz Sur, Región 02. Instituto Mexicano del Seguro Social, Coatzacoalcos
IUCN (2017) IUCN red list of threatened species. http://www.iucnredlist.org. Accessed 08 Jan 2018
Kensinger KM, Rabineau P, Tanner H, Ferguson SG, Dawson A (1975) The Cashinahua of eastern Peru. In: Dwyer JP (ed) Studies in anthropology and material culture. The Haffenreffer Museum of Anthropology, Brown University, Providence
Kracke WH (1978) Force and persuasion, leadership in an Amazonian society. University of Chicago Press, Chicago
Laborde J (2004) La Reserva de la Biosfera. In: Laborde J, Sacnhez G (eds) Los Tuxtlas: El paisaje de la sierra. Instituto de Ecología, A. C., Unión Europea, Xalapa, pp 271–279
Lefebvre H (1991) The production of space. Blackwell, Cambridge
López A (1992) Homshuk, Análisis temático del relato. Anales de Antropología 29:281–283
Loudon JE, Howells ME, Fuentes A (2006a) The importance of integrative anthropology: a preliminary investigation employing primatological and cultural anthropological data collection methods in assessing human-monkey co-existence in Bali, Indonesia. Ecol Environ Anthropol 2(1):2–13
Loudon JE, Sauther ML, Fish KD, Hunter-Ishikawa M, Ibrahim YJ (2006b) One reserve, three primates: applying a holistic approach to understand the interconnections among ring-tailed lemurs (*Lemur catta*), Verreaux's sifaka (*Propithecus verreauxi*), and humans (*Homo sapiens*) at Beza Mahafaly Special Reserve, Madagascar. Ecol Environ Anthropol 2(2):54–74
Mahoo H, Mbungu W, Yonah I, Recha J, Radeny M, Kimeli P, Kinyangi J (2015) Integrating indigenous knowledge with scientific seasonal forecasts for climate risk management in Lushoto District in Tanzania. CGIAR Research Program on Climate Change, Agriculture and Food Security - CCAFS, Copenhagen
Murphy RF (1958) Mundurucu religion, vol 49. University of California Publications in Archaeology and Ethnology, Berkeley
Okonya JS, Kroschel J (2013) Indigenous knowledge of seasonal weather forecasting: a case study in six regions of Uganda. Agric Sci 4:641–648
Ordóñez-Gómez JD, Arroyo-Rodríguez V, Nicasio-Arzeta S, Cristóbal-Azkarate J (2014) Which is the appropriate scale to assess the impact of landscape spatial configuration on the diet and behavior of spider monkeys? Am J Primatol 77(1):56–65
Oropeza P, Rendón E (2012) Programa de Acción para la Conservación de las Especies: Primates, Mono araña (*Ateles geoffroyi*) y Monos aulladores (*Alouatta palliata, Alouatta pigra*). Secretaría de Medio Ambiente y Recursos Naturales, Comisión Nacional de Áreas Naturales Protegidas, México
Ortiz D (2008) Tsa'atsech+ Ity+k: aprendizajes en la creación de un ecomuseo en el sur de Veracruz. In: Budar L, Ladrón de Guevara S (eds) Arqueología, paisaje y cosmovisión en Los Tuxtlas. Universidad Veracruzana, Xalapa, pp 117–130
Sotelo P y, Susana L, Luna AM (2003) Percepción y conocimiento ambiental en la costa de Quintana Roo: una caracterización a través de encuestas. Investigaciones Geográficas 52:99–116
Patton M (2002) Qualitative research & evaluation methods, 3rd edn. Sage, Thousand Oaks
Peña-Mondragon JL (2011) Daños económicos al ganado y percepciones sociales sobre el jaguar (*Panthera onca veraecrucis* Nelson and Goldman, 1993) en la Gran Sierra Plegada, Nuevo León, México. MSc. Dissertation. Universidad Nacional Autónoma de México, México
PROGEDER (2013) Evaluación rural participativa del ejido Los Mangos, Hueyapan de Ocampo, Ver. Promotora de Gestión y Enlace para el Desarrollo Rural, México

Rautela P, Karki B (2015) Impact of climate change on life and livelihood of indigenous people of higher Himalaya in Uttarakhand, India. J Geogr Environ Earth Sci Int 3(3):1–14

Ravi SK, Pochaiah M, Murthy VRK, Ramakrishna YS (2008) In: Central Research Institute for Dry land Agriculture – CRIDA (ed) Indigenous rain forecasting in Andhra Pradesh, Santoshnagar, Hyderabad

Recinos A (1978) El Popol Vuh (Las Antiguas Historias del Quiché), 9th edn. Editorial Universitaria Centroamericana, Costa Rica

Rodríguez-Luna E, Gómez-Pompa A, López J, Velázquez RN, Aguilar Y, Vázquez M (2011) Atlas de los espacios naturales protegidos de Veracruz, México. Gobierno del Estado de Veracruz, Universidad Veracruzana, Xalapa

Ruoso D (2012) The climatic perception of urban population of Santa Cruz do Sul/RS. Raega 25:64–91

Saj TL, Mather C, Sicotte P (2006) Traditional taboos in biological conservation: the case of *Colobus vellerosus* at the Boabeng-Fiema Monkey Sanctuary, Central Ghana. Soc Sci Inf 45(2):285–310

Scientific Software Development GmbH (2013) Atlas.ti Software Version 7. Berlin. http://www.atlasti.com/

SEMARNAT (2010) Norma Oficial Mexicana, NOM-059-SEMARNAT-2001: "Protección ambiental, especies nativas de México de flora y fauna silvestres. Categorías de riesgo y especificaciones para su inclusión, exclusión o cambio: lista de especies en riesgo". Secretaria del Medio Ambiente y Recursos Naturales, Diario Oficial, Mexico

Shepard G Jr (2002) Primates in Matsigenka subsistence and worldview. In: Fuentes A, Wolfe LD (eds) Primates face to face: the conservation implications of human-nonhuman primate interconnections. Cambridge University Press, Cambridge, pp 101–136

Simons EL, Meyers DM (2001) Folklore and beliefs about the aye aye (*Daubentonia madagascariensis*). Lemur News 6:11–16

Tárres ML (2004) Observar, escuchar y comprender. Sobre la tradición cualitativa en la investigación social. Flacso, Colegio de México, México

Taylor SJ, Bogdan R (1987) Introducción a los métodos cualitativos de investigación; la búsqueda de significados. Paidós, Barcelona

Tójar J (2006) Investigación cualitativa. Comprender y actuar. La Muralla, Madrid

Urbani B (2005) The targeted monkey: a re-evaluation of predation on New World primates. J Anthropol Sci 83:89–109

Urbani B, Cormier L (2015) The ethnoprimatology of the howler monkeys (*Alouatta* spp.): from past to present. In: Kowalewski MM, Garber PA, Cortes-Ortiz L, Urbani B, Youlatos D (eds) Howler monkeys: behavior, ecology, and conservation. Springer, New York, pp 259–280

Voss RS, Fleck DW (2011) Mammalian diversity and Matses ethnomammalogy in Amazonian Peru. Part 1: Primates. Bull Am Mus Nat Hist 351:1–81

Chapter 2
Mental State Attribution to Nonhuman Primates and Other Animals by Rural Inhabitants of the Community of Conhuas Near the Calakmul Biosphere Reserve in the Yucatan Peninsula, Mexico

Esmeralda Gabriela Urquiza-Haas, Rosa Icela Ojeda Martínez, and Kurt Kotrschal

2.1 Introduction

Throughout the human evolutionary history until today, people have been surrounded by other animal species. They compete with them or establish a range of relationships, from mutually cooperative to exploitative. Today, understanding the complexities of our relationships to other species is more important than ever, given the high rate of biodiversity loss. Positive attitudes about animal welfare and feelings of empathic concern toward them seem to reflect their inclusion in people's moral realm; such attitudes have been at the core of those human–animal relations, which yield positive outcomes for the animals (Ellingsen et al. 2010; Taylor and Signal 2005; Furnham et al. 2003). The "moral realm" or "scope of justice" refers to boundaries within which individuals are deserving a fair treatment (Opotow and Weiss 2000). Therefore, they represent a key concept in affecting success in conservation and animal welfare efforts.

Numerous studies suggest that human attitudes and emotional responses to nonhuman animals are shaped by the interplay of inherited dispositions (Jacobs 2009;

E. G. Urquiza-Haas (✉)
Department of Behavioural Biology and Department of Cognitive Biology, University of Vienna, Vienna, Austria
e-mail: esmeralda.urquiza@univie.ac.at

R. I. O. Martínez
Catedrática CONACyT, Escuela de Antropología e Historia del Norte de México, Instituto Nacional de Antropología e Historia, Chihuahua, Mexico

K. Kotrschal
Department of Behavioural Biology, University of Vienna, Vienna, Austria

B. Urbani, M. Lizarralde (eds.), *Neotropical Ethnoprimatology*, Ethnobiology,
https://doi.org/10.1007/978-3-030-27504-4_2

Kotrschal 2009; Barrett 2005), demographical and psychological characteristics (Swami et al. 2008; Signal and Taylor 2006; Knight et al. 2004; Furnham et al. 2003; Driscoll 1995; Broida et al. 1993; Rajecki et al. 1993; Kellert and Berry 1987), cultural beliefs (Dickman 2010; Serpell 2004), the animal's physical and behavioral characteristics (Batt 2009; Horowitz and Bekoff 2007; Simons and Meyers 2001; Kellert et al. 1996; Merckelbach et al. 1987; Lorenz 1950), and the kind of interaction and relationships between human societies and animals (Liu et al. 2011; Dickman 2010; Distefano 2005). Among all those factors, mind attribution seems a key variable for the readiness of people to grant rights to animals (Gray et al. 2012; Gray et al. 2007; Mameli and Bortolotti 2006; Knight et al. 2004; Hills 1995).

Mind attribution or mind perception reflects the degree to which people assign conspecifics and any other animals the capacity to experience a range of emotions and cognitive capacities (Waytz et al. 2010; Gray et al. 2007). While recognizing and attributing mental states to other humans is usually regarded as "empathy" or "theory of mind," the attribution of mental states to animals or any other living or nonhuman living entity is generally labeled as "anthropomorphism" which is defined as the attribution of "human characteristics or behavior to a god, animal, or object" (Soanes and Stevenson 2005). Two traits that have commonly defined humanness are the ability to think (reason, choose, deliberate, etc.) and the ability to feel (Waytz et al. 2010; Farah and Heberlein 2007; Gray et al. 2007). The attribution of mind and internal mental states to animals may be considered a human universal insofar as it represents a historical constant and has been therefore regarded an inevitable feature of human thinking about animals (Kennedy 1992). This inevitability might indeed be the result of a series of automatic and reflective cognitive processes occurring in the human brain, triggered by the perception of a living organism or animated nonliving agent (for review, see Urquiza-Haas and Kotrschal 2015).

Why do people attribute mental states to nonhumans? Studies have pointed at a diverse set of factors that play a key role in mind perception. Among these are motivational factors, personal traits, nonhuman agent characteristics, and nonhuman agent behavior. Waytz et al. (2010) proposed two motivational precursors of mind perception: the first one refers to the motivation of people to understand and predict behavior of others and the second to the motivation to establish social bonds. Personal traits are also relevant for the attribution of mind to nonhuman agents. Males are less likely than females to believe that animals experience emotional states like depression, anxiety, love, and grief (Walker et al. 2014). In line with the former, women are generally more emphatic toward animal suffering than males (Angantyr et al. 2011). Independently of gender, more empathic individuals exhibit higher subjective empathy ratings and corrugator EMG activity in response to film stimuli depicting different animal species in negative circumstances (Westbury and Newman 2008). Hills (1995) showed a positive relation between empathy toward animals and belief in the mental experiences of animals, while Paul (2000) found a positive relationship between empathy toward humans and toward other animals.

It has been shown that the nature of the human–animal association also affects mind attribution to animals. For example, animal ownership increases emotion attribution to animals (Wilkins et al. 2015), and people in habitual contact with animals tend to attribute the capacity to experience secondary emotions (Morris et al. 2008) to these animals. Primary emotions such as anger, disgust, fear, happiness, sadness, and surprise (Ekman 1999) are shared at least within the mammals, as are the basic emotional systems of seeking, rage, fear, lust, care, panic/grief, and play (Panksepp 2005). Secondary emotions, on the other hand, are defined as "products of social construction through the attachment of social definitions, labels, and meanings to differentiated conditions of interactions and social organization" (Kemper 1987:276). Morris et al. (2012), for example, found that participants who have not experienced regular contact with animals attribute far fewer emotions to them than participants that have habitual contact; also, keepers of a particular species always report more emotions for that species than non-keepers. In addition, owning a companion animal increased the likelihood of attributing some animals the capacity to experience grief, as compared with respondents who did not own a companion animal (Walker et al. 2014).

The characteristics of animals play a crucial role in the perception of animal mind as well. Perceived similarity and phylogenetic relatedness of animals have been consistently found to be positively associated with the attribution of higher mental processes and complex thinking (Herzog and Galvin 1997; Rasmussen et al. 1993), intelligence (Nakajima et al. 2002), cognitive abilities like self-recognition, intention recognition and ability to deceive (Eddy et al. 1993), and the attribution of empathic and communicative abilities to nonhuman animals (Harrison and Hall 2010). Moreover, human–animal similarity is involved not only in the deliberate attribution of mental states but also in more automatic processes, such as empathic responses to animal ill-being or suffering. Plous (1993) found that college students watching videos of apparent abuse of a monkey showed greater than average and greater maximum skin conductance scores (a measure of arousal), as well as greater self-reported difficulty in watching the video as compared to watching the abuse of a bull frog. Westbury and Newman (2008) also showed that phasic skin conductance responses to watching videos of different species in negative circumstances increased with their phylogenetic closeness to humans.

In addition to perceived similarity, animal behavior per se is obviously recognized as an important trigger of attributing emotions and mind to other animals or to animated stimuli. For example, Mitchell and Hamm (1997) showed that people rely more on details of behavior and context than on morphological similarity or phylogenetic closeness when assessing emotions (jealousy) and intentions (deception) in nonhuman animals. Morris et al. (2000) assessed the consistency of peoples' anthropomorphic explanations of dog behavior and their behavioral triggers, as they observed short videos featuring human–pet dog interactions. The authors found a remarkable consistency in people's anthropomorphic accounts of the dog's behavior in describing the observed interactions. A similar result was reached by Morris et al. (2008) who asked 40 dog owners to report instances of jealousy in their dogs; participants consistently included four elements, namely, a certain context, which almost always involved a social triad (the owner, the dog, and the "other"), a behavior that elicited the jealousy, and the type of behavior signaling jealousy. The "others" almost always consisted of other persons or other dog/animal, the eliciting behavior was paying close attention to the "other," and the behavioral expression of jealousy was the dog pushing itself between the owner and the other, the dog barking/growling/whining, as well as the dog showing aggressive behaviors. These studies indicate a common "syntax" and "semantics" in the folk interpretations of animal behavior. It has been suggested that in the case of animal behavior, people simply project their own experience/theories of mental states in certain situations to other animals: "In domestic settings people and their companion animals frequently face similar situations. To the extent that a companion animal's reaction to a situation has something in common with that of the human observer, the tendency would be to interpret the animal behavior in human subjective terms" (Rasmussen and Rajecki 1995:132). Such projections involve several underlying assumptions: For example, one can only assume that a particular mental state in oneself is the same mental state in others only if (a) it has the same or similar observable properties, i.e., your own expression of happiness is the same or similar to the expression of happiness of the other, and (b) that what causes one's own mental state is of a similar nature as what causes that of the other; this implies that (c) you and me do not differ in some fundamental way regarding the expected causal relation between the cause and the mental state experienced.

2.2 The Present Study

The aim of the present study was to identify the structure of folk psychological explanations mediating mind attribution to primates and other animals. More specifically, we wanted to identify if there is a specific set of behaviors and circumstances that prompt the attribution of emotions and cognitive abilities to other species. To achieve this, we interviewed a group of people living in the vicinity of the Calakmul Biosphere Reserve with respect to their beliefs about the ability of domestic and wild animals to think and to experience mental states like anger, fear, pain, and joy and the capacity to deceive. We refer to both emotions and cognitions as mental states because both are mental phenomena taking place in the brain (Oosterwijk et al. 2012; Panksepp 2005; LeDoux 2000).

The study group consisted of a convenience sample comprised of 9 women and 14 men (23 total), between 18 and 82 years of age, with different cultural, educational, and occupational backgrounds (Appendix 1). Most of the participants (17) were living in the community of Conhuas by the time of the interview, while the rest (6) came from other rural communities within the state of Campeche. The reason for selecting this community to do our interviews was twofold. First, we had to look for participants who were potentially acquainted with a diversity of wild and domestic animals to capture their subjective experiences and ideas about their cognitive and emotional abilities. Most importantly, we had to select a group of participants who were somehow exposed to spider monkeys and preferably howler monkeys as well. The second reason was that one of the authors had a long-standing working relation with some members of this community, thereby facilitating access to participants.

Conhuas is a small community of 503 inhabitants (238 women and 265 men) from different cultural backgrounds situated in the vicinity of the Calakmul Biosphere Reserve (Fig. 2.1; INEGI 2015). Participants had multiple linguistic backgrounds: Tzeltal and Chol (3 individuals), Maya (3), Totonac (3), Zoque (1), and Spanish (12). The Calakmul Biosphere Reserve is a natural reserve located (17° 09′ -19° 12' N and 89° 09′ -90° 08' O) within the state of Campeche, Mexico. It includes an area of 723,185 hectares covered by short-to-medium-stature forests with significant dry season leaf loss, tall humid forest, and seasonally inundated, short-stature forest (SEMARNAP 2000). The reserve hosts a large biodiversity including 18 species of fish, 16 of amphibians, 50 of reptiles, close to 300 of birds, and 94 of mammals (Carabias-Lillo et al. 2000). Included among these are two primate species: the spider monkey (*Ateles geoffroyi*) and the howler monkeys (*Alouatta pigra*). Even though the community of Conhuas lies outside the reserve polygon, a portion of the *ejidal* lands of the community, used for agricultural purposes, is situated within the reserve. The main economic activity of the community is semi-subsistence farming, some still practice subsistence hunting (although prohibited within the reserve), and in

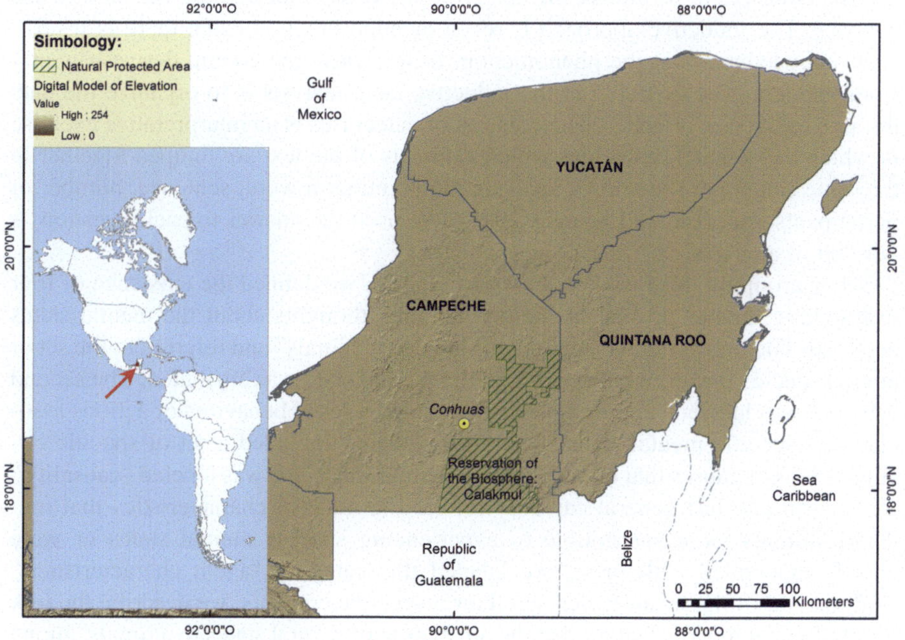

Fig. 2.1 Location of the study site

recent times, there is an increased economic dependency on the provision of a wide range of touristic services (INEGI 2015).

Data was obtained through semi-structured interviews. During the first part of the interview, we explored how people represent two closely related mental state concepts, thinking and intelligence, and how these are attributed to certain animal species including the spider and howler monkey, jaguar, peccary, dog, birds, snake, ant, and fly. The second part of the interview consisted of a series of questions aimed to understand the triggers of mental state attribution. Among the mental states included were four basic emotions (anger, pain, fear, joy) and one complex mental state (ability to deceive). For this section of the interview, a set of cards with the pictures of 21 animal species belonging to 12 orders (Appendix 2) were shown to participants who were asked to sort out the ones capable of experiencing the mental state in question. Participants were then encouraged to elaborate on how they could recognize these subjective experiences of animals.

All interviews were carried out in Spanish and were audio and video recorded with explicit permission of the participants. The recorded interviews were then transcribed and analyzed via a qualitative content analysis which is described by Mayring (2000) as a mixed method approach for the analysis of textual data. The central instrument for the analysis relies on the specification of categories, indexes, or codes. According to Elo and Kyngäs (2008), the definition of categories for the analysis of the verbal content of interviews can be established through an inductive or deductive approach. The deductive approximation involves the use of theoretical-derived concepts to formulate the categories or codes representing the base of the analysis. The inductive approach is recommended in cases when there is insufficient knowledge about the phenomenon. In this case, the coding derives directly from the data. The first step in the inductive data analysis is to organize the data through categories or codes. The creation of categories is an interpretative exercise in which the researcher decides which elements of the text are lumped together in the same category. The unit of analysis can be either a word, sentence, number of participants, etc. (Elo and Kyngäs 2008). We used the answer to each question as our unit of analysis.

After grouping the answers of all participants, we defined the existence of four overarching themes present in the participants' thoughts about the mental states assessed. The first general category was labeled "animals" and referred to the set of animal species toward which each specific mental state was attributed. The second category was labeled "expression" as it included a set of behaviors or actions associated to specific mental states. The third category included a set of specific circumstances or causes that elicited certain mental states and was labeled "causality." The fourth and last general category referred to a set of characteristics that rendered animals more susceptible to experiencing specific mental states or were closely associated with these; we labeled this category "agent characteristics." With the aid of these categories and their respective contents, we describe the folk psychological models that guided the attribution of mental states to animals among this group of participants.

2.3 Results

2.3.1 The Folk Psychology of Basic Emotions

2.3.1.1 Anger

The Oxford Dictionary (Soanes and Stevenson 2005) defines anger as an unpleasant emotion caused by the threat of danger, pain, or harm. From an emic perspective, anger was defined as an internal state that arises as a response to fear and to a motivation to defend oneself/a territory/a mate/a relative, which is expressed through aggressive behavior, specific vocalizations, and piloerection. According to participants, most animals can feel anger. As can be appreciated in Table 2.1, the most common ways in which animals express anger are through a series of aggressive behaviors that include biting, stinging, attacking, and chasing. Anger is also recognized by certain body signals that include erected hair or feathers, screaming, and growling (Table 2.1). Different species have their own behavioral repertoire to express anger. Spider monkeys are well known to throw objects at people, usually branches but also pieces of bark, excrement, or whatever they find. Dogs and snakes bite, scorpions and bees sting, chickens and wild turkeys peck and chase, while peccaries, pigs, and cows chase people when angry. Participants believed there are many causes that trigger anger in animals. For example, instances of intraspecies interaction that result in aggressive outbursts include males fighting with other males for a territory, a prey, or a mate. Animals in heat, in reproductive period, or with litter are recognized as animals prone to anger. Especially females (chickens and cows) with litter are prone to show aggressive behavior when approached. Aggressive behavior (signaling anger) is also recognized as a defense mechanism to being bothered, having their space invaded, and trying to grab or corner them. The mere presence of humans might trigger an aggressive response resulting from feeling threatened (Table 2.1). Anger seems to be an emotional response of which almost all animals are capable of, from primates to arthropods, as well as wild and domestic animals. Nevertheless, some species received a higher frequency of mentions which included the dog, snake, jaguar, peccaries, cow, scorpion, pig, spider monkey, and chicken (Fig. 2.2).

Table 2.1 Content of the categories for the capacity of animals to experience anger

Expression	Causality	Agent characteristics
Bite Attack Sting Throw objects Hurl at you Follow you Confront you Scream Growl Show erected hair or feathers	Fight for a mate, a territory, or a prey Fight with other males to show who is the best Female attacks to protect their litter As a reaction to fear As a defense mechanism when the animal is attacked or bothered, when its territory is invaded, when cornered, and when people try to grab or handle them As a reaction to the mere presence of humans	Females with litter Animals in heat or reproductive period Males

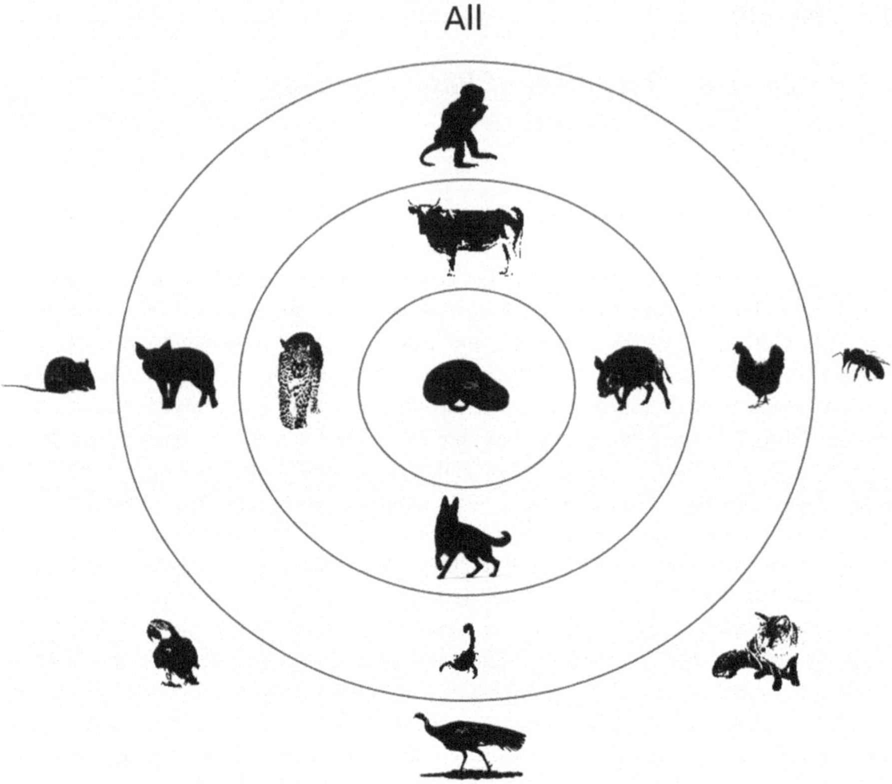

Fig. 2.2 Animals most frequently mentioned as capable of feeling anger. Note: Inner circle (animals mentioned by more than ten participants), first concentric circle (>7<9), second concentric circle (>4<6), outside (>1<3)

2.3.1.2 Fear

Fear is defined by the Oxford Dictionary (Soanes and Stevenson 2005) as an unpleasant emotion caused by the threat of danger, pain, or harm. From an emic perspective, fear can be described as an internal state that arises as a response to threat or danger infringed by either humans or predators and that is expressed through evasive, aggressive, or passive behavior. Almost all animals are entitled with the capacity to feel fear but especially forest-dwelling animals and among them prey animals even more so (Fig. 2.3). Monkeys were not perceived as fearful animals since they do not hide nor seem to avoid encounters with humans (unless they are hunted as M. Lizarralde has observed among the Barí people in Venezuela, pers. comm. in 14 March 2018). To the contrary, spider monkeys were perceived as bold by chasing people while throwing branches at them. Deer, jaguars, and

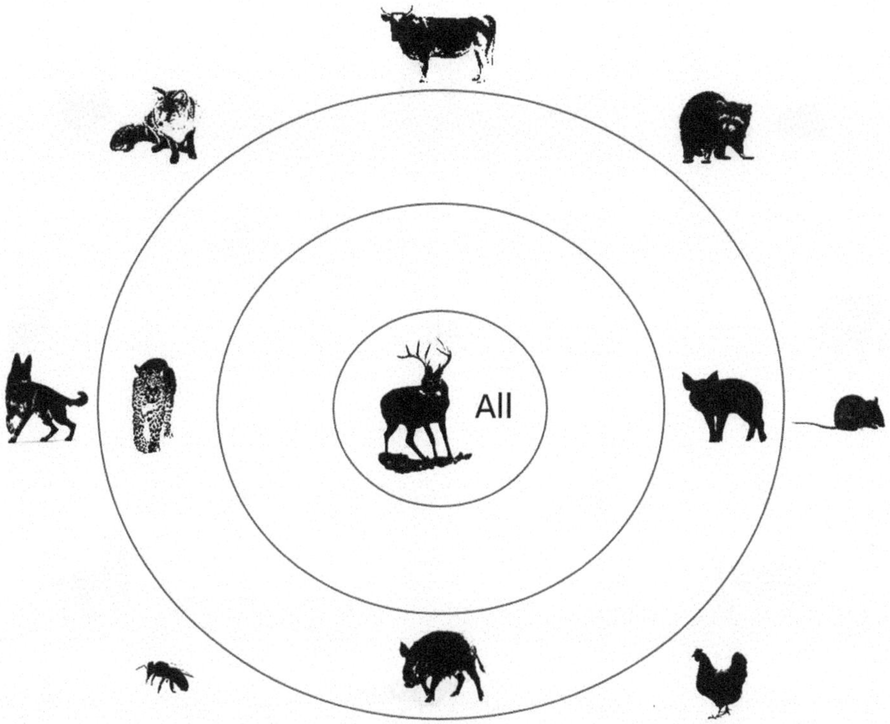

Fig. 2.3 Animals most frequently mentioned as capable of feeling fear. Note: Inner circle (animals mentioned by more than ten participants), first concentric circle (>7<9), second concentric circle (>4<6), outside (>1<3)

Table 2.2 Content of the categories for the capacity of animals to experience fear

Expression	Causality	Agent characteristics
Run	Sight of humans	Wild animals
Flee	Approached by humans	Prey animals
Hide	Approached by a predator	
Attack	If threatened	
Paralyze	Facing danger	
Tremble		

peccaries were the most commonly mentioned animals with the ability to feel fear. Encountering humans or predators triggers fear, which is easily recognized by animals running away, hiding, fleeing, being paralyzed, or trembling (Table 2.2). Wild animals more likely than domesticated animals respond with fear to the mere sight of a person; others react with fear to an evident threat, like cornering or trying to grab the animal. Attacking is recognized by some as a sign of fear, but others interpret it as the opposite, that is, a sign of lack of fear.

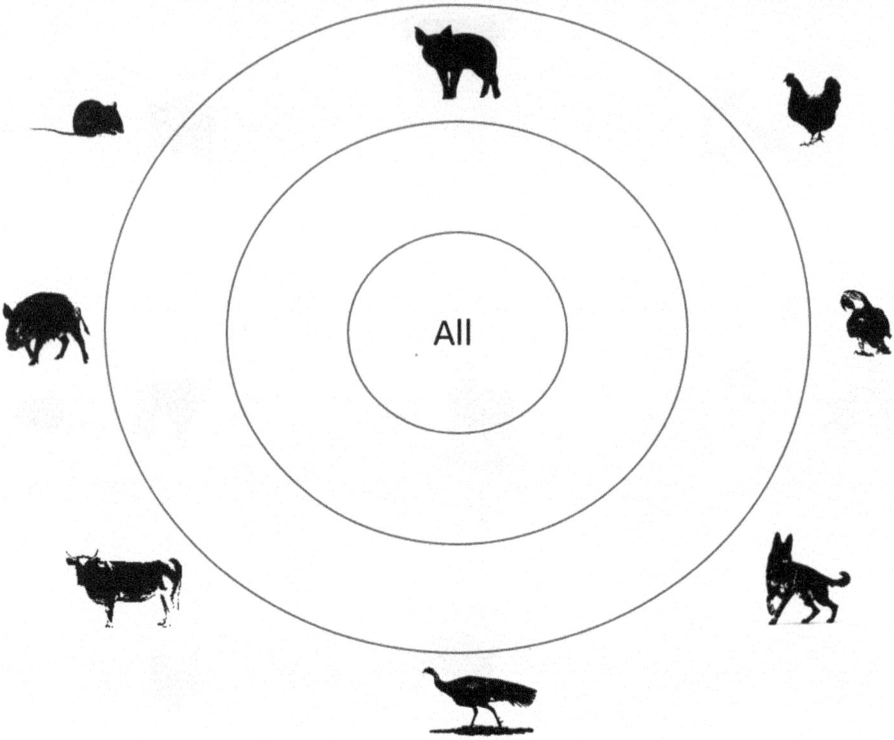

Fig. 2.4 Animals most frequently mentioned as capable of feeling pain. Note: Inner circle (animals mentioned by more than ten participants), first concentric circle (>7<9), second concentric circle (>4<6), outside (>1<3)

2.3.1.3 Pain

Oxford Dictionary (Soanes and Stevenson 2005) defines pain as a highly unpleasant physical sensation caused by illness or injury. From a local perspective, pain can be defined as an internal state that is caused by illness or injury and is expressed through passive or abnormal behavior and specific vocalizations. According to almost all participants, the ability to feel pain is a mental state that all living beings are capable of, as they are all made from the same materials as human beings, flesh and blood. Therefore, being alive is basically the only prerequisite to experience pain. Participants who listed a set of species included almost exclusively animals found within human communities like the dog; animals consumed as food like the cow, pig, and chicken; wild prey animals like the peccary and wild turkey; as well as animals like the mouse, possibly considered vermin (Fig. 2.4). Therefore, these animals are the most common targets of pain infliction by humans. Expressions of pain include a change in behavior (e.g., from lively and active to secluded and

Table 2.3 Content of the categories for the capacity of animals to experience pain

Expression	Causality	Agent characteristics
Lay down	If hurt	All beings that are alive
Limp	If sick	All beings that are made of flesh and blood
Tremble		Prey animals
Scream		Pet animals
Moan		Animals used as food
Abnormal behavior		Animals considered vermin

Table 2.4 Content of the categories for the capacity of animals to experience joy

Expressions	Causality	Agent characteristics
Play	Finding food	Birds
Shake their tail	Mating	Pet animals
Wallow	Being free and safe	
Run around	Dogs when seeing their owners	
Jump		
Sing		
Scream		

apathetic), a specific kind of vocalization (e.g., scream, moan), and certain bodily expressions (e.g., trembling, inability to walk, limping; Table 2.3). Causes of pain were also well recognized and include getting hurt or being sick.

2.3.1.4 Joy

The Oxford Dictionary defines joy as a feeling of great pleasure and happiness (Soanes and Stevenson 2005). On the other hand, the emic definition of joy can be expressed as an internal state that originates by having one's basic needs met (i.e., food, mating, safety, and freedom) and is conveyed through energetic movement, specific vocalizations, and playful behavior. In contrast to the experience of pain, not all animals are believed to feel joy. The experience of joy seemed to be attributed in two different ways. Some participants believed that most animals were able to feel joy as a result of finding food and mates and being free and safe, while others considered that some joy is manifested through behaviors like playing, running, and jumping around and screaming (Table 2.4). Monkeys were said to express their joy through vocalizations, jumping, scratching their belly, and playful behavior. Even howler monkeys were thought to express their joy through their characteristic "howl." Dogs express their joy through jumping, wallowing, and wagging their tails, which is generally associated with encountering the owner or other family member. Dogs and parrots were mentioned most frequently for their ability to feel joy, followed by the wild turkey and the monkey (Fig. 2.5). It is notable that of all animals these are closest to people, not only because of their continuous presence within human communities in the Yucatan Peninsula but also given their behavioral similarity (i.e., their capacity to learn) and close emotional bonds (i.e., as house pets).

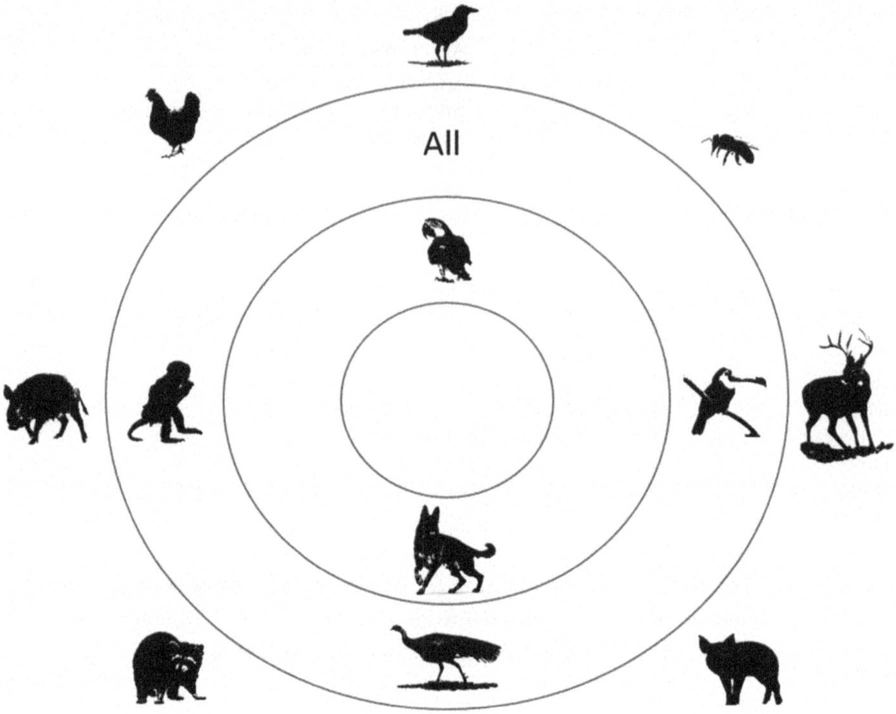

Fig. 2.5 Animals most frequently mentioned as capable of feeling joy. Note: Inner circle (animals mentioned by more than ten participants), first concentric circle (>7<9), second concentric circle (>4<6), outside (>1<3)

2.3.2 *The Folk Psychology of Complex Mental States*

2.3.2.1 Thinking, Intelligence, and Deceit

The concepts of thinking and intelligence were sometimes used as interchangeable terms. When participants were asked about the ability of animals to think, they frequently included the term intelligence in their responses. The ability to think was awarded to almost all animals except for insects. There were several different exceptions related to the term. Thinking was equated to a basic surviving skill that allowed animals to face everyday challenges of all living beings like acquiring food and avoiding becoming the food of others. It is frequently mentioned that prey animals must think how to avoid being killed and predators must think how to get their prey (Table 2.5.a). Thinking was also related to taking an appropriate or expected action when faced with certain circumstances, especially dangerous encounters with people. For example, participants reasoned that animals running away from humans were acting as a person would, that is, correctly assessing the dangerous situation and getting away from it (Table 2.5.b). Animals with the ability to learn and

Table 2.5 Participant quotes: intelligence and thinking

(a)	"Every living being has a degree of intelligence or thinking capacity...." Female, 35, housewife "I think all living beings have a degree of thinking." Male, 37, farmer, *Totonac* "I think that all living beings have to think how to spend their time." Female, 30, housewife and beekeeper, *Tzeltal*
(b)	"Many animals are smart, because they all have to defend themselves, some have, and others don't…. The wild Turkey has to take care of himself if not they are going to eat him, the same happens with the deer…they are all clever because they have to be able to get their food and take be aware of their enemies." male, 45, employee, *Maya* "I think they (jaguars) do (think) because if they wouldn't, how would they hunt?" male, 37, farmer, *Totonac* "I think they think like us trying to cover their needs." male, 23, museum worker
(c)	"Deer are intelligent, when they see people they run away, they know that people will try to kill them." female, 36, housewife "I think all animals are intelligent because when they see you they run… (are they as intelligent as us?) ... Yes, because they do the same as us, we would also run if confronted with a dangerous animal." male, 31, farmer "I don't think they do, because if they could, they wouldn't let themselves get killed so easy." male, 66, farmer, *Totonac* "The deer is intelligent. We used to see them at the *milpa*, we could not grab them, they would see us and ran away… the jaguar too…once we saw one by the river and fled…that means they are intelligent." male, 34, farmer "This animal (deer) is the first one to flee when it sees you, even if you are just sitting, if they see you they escape, and they are very agile, they are very clever." male, 31, farmer "I guess that deer do (think), because if they wouldn't think because when you get to see them or try to approach them they run away. So, I think that they either have a presentiment or they think, one of the two things, because if they did neither they would just stay there." male, 37, farmer, *Totonac*
(d)	"Well, the parrot is clever because they learn to speak…." male, 35, farmer "We had one here (a monkey) ...And we taught him, and (he learned) like a person, he did not talk but he understood…." male, 82, farmer, *Maya* "One could say that some animals are more intelligent than others...For example, parrots can learn to say some words with training." male, 45, gardener "I think dogs do think because there are very intelligent dogs who you can teach almost anything…." male, 37, farmer, *Totonac* "I believe they (dogs) think because when they are talked to them understand, and that is why I think they think…when you call hem they come and when you tell them not to do something they don't." female, 22, *Tzeltal and Chol*
(e)	"Howler monkeys are more like us, they are bigger and have a beard, when I see them I believe that they may think as a human being." male, 82, farmer, *Maya* "Monkeys are the most intelligent, they use tools." male, 28, *Totonac* "Monkeys are one the most intelligent species of all animals…in fact they have the same behavior as us." male, 23, museum worker "Monkeys are the most intelligent, if they see you and feel threatened they start throwing branches at you, and if you dare to (come close) they'll piss on you (laughs)." male, 34, farmer "…in one occasion I saw a spider monkey couple…the husband left and the female stayed behind…when he saw that she wasn't following him he returned and put his arm around her, and I saw like he whispered something after which she went with him…I always tell that story, everybody finds it very amusing, they are very intelligent, he must have preached her or said something…." male, 45, employee, *Maya*
(f)	"The bee is very intelligent, but in its world, in its own way, not like us." male, 33, farmer "I think all living beings have a degree of thinking." male, 35, farmer "All animals are intelligent maybe not to the degree of humans." female, 35, housewife

Table 2.6 Content of the categories for the capacity of animals to think

Expression	Agent characteristics
Learn	Prey animals
Respond appropriately	Predators
Trick others	Primates
Understand human communication	Pet animals
Infer the intentions of others	
Successful in acquiring food and avoiding danger	

understand human intentional communication were also considered to have the capacity to think. In line with the former, mostly pet animals such as dogs and parrots were accredited with the ability to learn (Table 2.5.c). Perceived human–animal similarity, either physical or behavioral, appeared also to increase people's perception about the ability of animals to think (Table 2.5.d). Monkeys were considered to look and behave like humans, and dogs think because of their capacity to understand and respond to human communicative actions and parrots for their ability to learn to talk. In fact, spider monkeys and dogs were the animals with the most frequent mentions for their capacity of thinking. Nevertheless, the quality of thinking and intelligence attributed to animals was not necessarily analogous to the quality of human thinking. Many participants stated that animals do possess their "own kind of thinking," which was either expressed as a different kind or a "thinking to a certain degree" (Table 2.5.e). Also, even though most participants of both genders agreed that animals were capable of thinking, a higher proportion of men (six of 14) denied animals this ability when compared to women (1 of 9). However, even though these participants didn't believe that animals were capable to think, they nevertheless awarded some species like the jaguar, snake, and *ocellated* wild turkey the ability to deceive. It must be noted that this mental state was also considered a marker of thinking insofar as it was conceived as a surviving skill that allowed animals to get a prey or avoid being predated (see below).

The behavioral expressions of thinking and intelligence were not as straightforward, immediate, or evident as the expressions of emotions. Instead the attribution of thinking and intelligence appeared to be exclusively triggered by inferential processes. Participants reasoned that to survive, learn, and trick others, animals must be able to think or be intelligent. Therefore, instead of a list of overt expressions directly reflecting inner subjective phenomena as in the case of emotions, thinking generated a list of events or outcomes which allowed to infer the ability to think (Table 2.6).

To dwell a little deeper into the folk attribution of intelligence, we asked people if they considered the spider or howler monkey to be more intelligent. Most participants considered the spider monkey more intelligent than the howler monkey; only a few is said to be unable to express an opinion. Diverse reasons were given for this choice, and only two types of arguments were repeatedly expressed: (1) because spider monkeys move faster than howlers and are more agile and restless and (2) because they throw objects at people. In addition, some participants reported that they had spider monkeys as pets or have seen spider monkeys dressed as people.

Table 2.7 Content of the categories for the capacity of animals to deceive

Expression	Causality	Agent characteristics
They hide and come out after you are gone Plays dead but is not Make sounds that are heard as if they came from different directions Distracts others They camouflage Hide to attack Hide to protect themselves Hide their food	To defend themselves from anyone trying to harm them or their litter To get a prey To avoid getting preyed	Predators Preys

2.3.2.2 Deceit

The Oxford Dictionary (Soanes and Stevenson 2005) defines deceit as the action of deliberately causing someone to believe something that is not true, especially for personal gain. From a local perspective, deceit can be defined as intentional behavior aimed at misrepresenting or concealing truth motivated by the need to preserve oneself as well as one's biological relatives. Deceit was mostly attributed when animals manage to "gain the upper hand" either by hiding, by surprising others, by doing something unexpected, or by distracting others to get what they want or to get out of a dangerous situation (Table 2.7). Deception is an ability that some animals use to escape predation or to enhance their predatory capacities. For this reason, the jaguar, the fox, and the wild turkey were the most commonly mentioned animals to possess the ability to deceive (Fig. 2.6). The jaguar's ability to deceive manifests in his sneaky way to approach his prey. In the case of the wild turkey, the most commonly mentioned behavior reflecting the ability of this animal to deceive is through the production of a special kind of vocalization referred to as *pujidos* which is translated as moans and which the animal produces to generate confusion regarding its location. Participants refer that ocellated wild turkey produces vocalizations which are sometimes heard in one location and then in another even opposite direction, thereby misleading hunters.

2.4 Discussion

The results of the interviews indicate that, from a folk psychological point of view, mental states are not exclusive of human beings, but that most animals, from insects to primates, possess them to some degree. Mental states are seemingly considered part of the makeup of an organism, as they aid to navigate the world and to master the basics to survive. For our participants, emotions are also shared with a wide range of animal species, which is also a common finding in previous studies

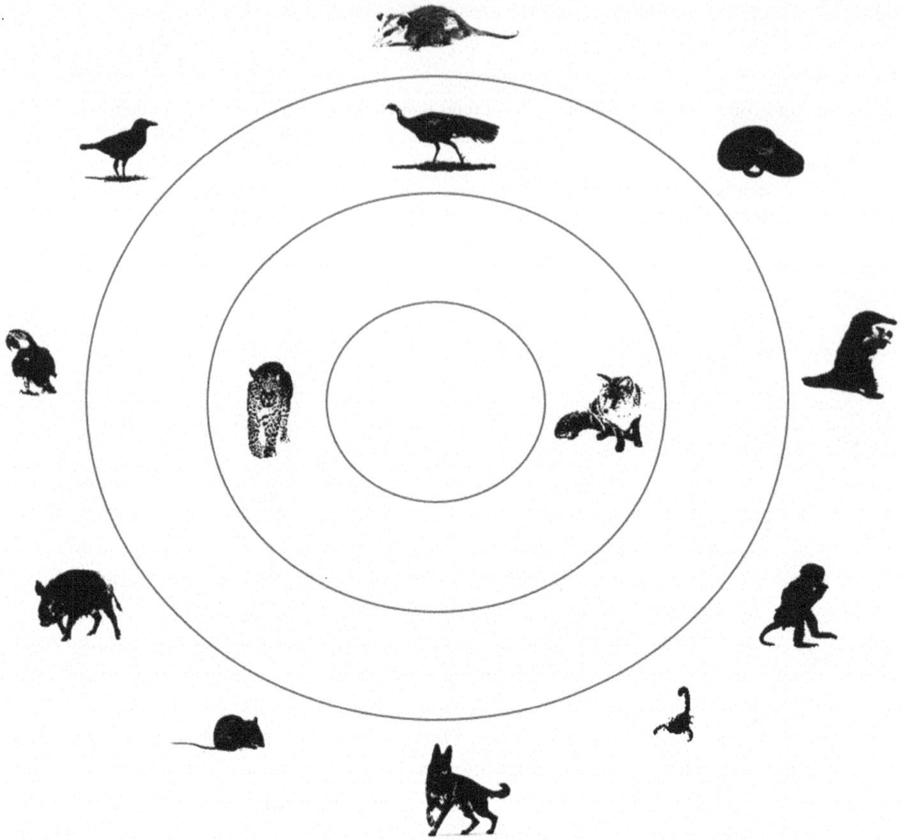

Fig. 2.6 Animals most frequently mentioned as capable of deceiving. Note: Inner circle (animals mentioned by more than ten participants), first concentric circle (>7<9), second concentric circle (>4<6), outside (>1<3)

(Martens et al. 2016; Konok et al. 2015; Morris et al. 2012; Rasmussen et al. 1993). Despite this, some animals received more mentions for certain emotions than others. That is, we found that animals that were mentioned for their ability to feel fear, for example, are not mentioned in the same frequency for their ability to feel joy. This may indicate that those animals mentioned with a higher frequency for some emotion may either be considered more prone to experience the mental state in question or alternatively, in which this mental state has its clearest or most frequent behavioral expression. Fear, for example, was more pervasive among wild animals, especially frequently preyed species like the dear, peccary, and wild turkey, or among animals that face or used to face an important hunting pressure, like the jaguar. Although pain is attributed to all animals, those used as food (pigs, cows, wild turkey, chicken), pets (dog and parrot), and vermin animals (mice and fox) were specifically mentioned. These are probably the animals more frequently subjected to harm by humans and are therefore those providing evidence through different behavioral expressions of their capacity to feel pain, which is probably mediated by the activation of reflexive mechanisms of empathy in humans (Westbury and Newman 2008). Joy is apparently more commonly attributed to pet animals and to birds whose singing is considered an expression of joy.

Mental states, specifically primary emotions, are associated with specific and largely behavioral expressions. Take, for example, the bodily and behavioral expressions of anger and fear described by Charles Darwin in *The Expression of the Emotions in Man and Animals* (1872): with respect to anger, "erection of the dermal appendages, hairs, feathers... (p.83) ... scream (p.98) ... rush at each other (p.99) ... inflate themselves (p.105) ... roaring... growling (p.85) ... prepared to spring (p.116)," and fear, "... efforts to hide or escape... (p. 9) ... trembling... (p. 67) ... sweating (p.73) ... helpless prostration... attempts to escape (p.81) ... bristling of the hair (p.96) ... remaining motionless (p.144)" (Darwin 1872). Adolphs (2013) includes as behavioral expressions of fear in humans the following: attacking, running, freezing, screaming, hiding, and risk assessment (i.e., vigilance). Rats, when confronted to stress, avoid places where a potential threat was detected (Vazdarjanova and McGaugh 1998) but show defensive aggression in face of an imminent and inevitable threat (Reynolds and Berridge 2008). With regard to pain, there are clear guidelines for its recognition among laboratory animals: abnormal behavior (Morton and Griffiths 1985), change in body weight, external physical appearance, changes in behavior (Morton and Griffiths 1985), and altered posture and gait (i.e., limp) (Tabo et al. 1999), among others (see review in Carstens and Moberg 2000). In sum, folk and scientific accounts of the expression of emotions show remarkable similarities which are unsurprising, given that humans share with animals the basic neurological and physiological structures involved in coping with environmental, ecological, and social challenges (Julius et al. 2012). The hypothalamus–pituitary–adrenal axis (HPA), involved in the response to stress through the fight–flight–freeze response, is an ancient physiological system present in vertebrates (Denver 2009), and the subcortical circuits supporting basic emotional systems (seeking, rage, fear, lust, care, panic/grief, and play) are shared at least within the mammals (Panksepp 2005).

Another finding of the present study is the apparent reliance of mental state attribution on different cognitive processes: (1) observation of a behavior associated with a mental state (e.g., attack–anger, flee–fear) potentially involving implicit cognitive processes like motor matching mechanisms and evolved mental representations (Franklin et al. 2013; Barrett 2005; Barrett et al. 2005; Buccino et al. 2004; Blythe et al. 1999) and (2) inference of mental states via causal reasoning (e.g., finding food or mates provokes joy, getting hurt causes pain) (Carey 1995), category- and similarity-based induction (Miser and Sloutsky 2013), and conditional reasoning (Evans 2002). The different cognitive processes involved in the attribution of emotions and other complex mental states have been discussed previously (Urquiza-Haas and Kotrschal 2015; Barrett 2005; Person et al. 2000). Even though causal explanations were expressed for all emotions, participants made explicit allusions to an ontological similarity between animals and humans in the case of joy and pain. Being alive and having a body made of "flesh and blood" seemed to be the most important ontological similarity or condition to experience pain, while having the same physiological needs or drives seemed to be the condition for the ability to experience joy. The attribution of so-called "higher" or "complex" mental states or abilities like thinking, intelligence, and the ability to deceive seems to rely almost exclusively on inferential processes. Mental abilities like thinking and intelligence seem to be derived from two different processes, similarity-based induction and conditional reasoning. In similarity-based induction, the presence of an unobserved property (i.e., intelligence) is inferred in a novel object or subject based on the similarity that these have with the familiar objects/subjects for which the property is known (i.e., humans) (Sloutsky et al. 2007; Sloutsky and Fisher 2004; Welder and Graham 2001). Monkeys were the species most commonly mentioned for their capacity to think. In contrast with those arguments employed by participants to account for thinking processes in other species, those used for monkeys were mostly based on their physical and behavioral similarity to humans, which might indicate that similarity-based induction mediates the attribution of the referred capacity to this animal. Take, for example, the following: "(Do monkeys think?) …Monkeys are one of the most intelligent of all animals…In fact, monkeys behave like us" (Male, 23, museum worker), or "Monkeys act like a human being, but they don't talk, they just act, they throw sticks and follow you" (Male, 66, farmer, Totonac); "I believe that monkeys are one of the ones that think a bit more…because they dwell on top of the threes and are the ones that can spot where they can move" (Male, 30, gardener). Only one participant referred to the use of tools by monkeys as evidence of their thinking capacity. In contrast, other animals were credited with the capacity to think by using a series of "intelligence" or "thinking" markers. Dogs and parrots are credited with the ability to think for their capacity to understand and learn: "Dogs do think, there are very intelligent dogs to which you can teach anything, people teach them to jump, I saw that on TV…" (Male, 37, farmer, Totonac); "An animal that is so smart as humans? I wouldn't know…the only one that I have seen like that is the parrot…he is the only one that talks, sings, whistles and asks for food…he has some degree of intelligence" (Male, 34, farmer). On the other hand, wild animals were accredited with the capacity to think given their ability to survive by being clever enough to get food and avoid getting preyed. The former examples point to the use of a conditional reasoning as a base for these attributions (if p then q; Evans 2002).

Spider monkeys were considered more intelligent than howlers. Among the most common arguments given by participants to justify this judgment were that spider monkeys were more energetic and lively and they throw objects at people: "Spider monkeys are more intelligent because they are more playful. When they are raised by people they even play with the children and make their pranks … (the monkey) grabs you with both arms and walks on his two small feet" (Male, 37, farmer, Totonac); "The spider monkey is more agile, he is quicker" (Male, 31, farmer);

"The spider monkey moves faster" (Male, 28, unemployed, Totonac); "Well, the spider monkey is more intelligent because of his behavior...when he sees you he starts throwing branches at you, Ramon fruits, Zapote fruits, he is more restless and nimble. *Zaraguatos* (howler monkeys) are more peaceful, they see you and do nothing, they have another way of life, so to say" (Male, 45, farmer, Maya). Spider monkeys, more than howlers, seemingly possess a larger number of humanlike traits: they are curious, agile, and not fearful of humans, they get angry and show it through humanlike behavior (i.e., by throwing objects), and they are playful and care for their own like humans would do. The Barí of Venezuela show a similar contrasting view about the howlers and spider monkeys, in which the first are perceived as slow and less intelligent (Lizarralde 2002, cited in Urbani and Cormier 2015). Morewedge et al. (2007) found a timescale bias in the attribution of mind, that is, individuals tend to assume that agents (animals, robots, and animations) had intentions, consciousness, thought, and intelligence when they moved at a similar pace as humans when compared to agents that moved at a quicker or slower pace. In addition to the speed of movement, the characteristic behavior of throwing branches or other objects of spider monkeys might also increase their perceived similarity through the potential recruitment of motor matching mechanisms (di Pellegrino et al. 1992). Buccino et al. (2004) showed that motor matching processes, involving of their own motor areas in the brain, when people observed motor actions performed by humans (talking, reading, and biting), monkeys (lip-smacking and biting), or dogs (barking and biting), depended not on the species but on the actions shown. Hence, the motor cortex in peoples' brains is involved only when the actions performed by another agent are familiar to them. Spider monkeys are frequently taken from the wild to be used as pets. Duarte-Quiroga and Estrada (2003) found in a survey among 179 primate pet keepers in Mexico City that 67% and 15% of the 12 primate species reported were spider and howler monkeys, respectively. Among the participants in our study, no one owned howler monkeys as pets, in contrast to spider monkeys. Urbani and Cormier (2015) did an extensive review of the available ethnographic records about the ecological and cultural relation of howler monkeys with indigenous societies in Central and South America and found but a few references about their keeping as pets. In contrast, spider monkeys are found to be commonly held pets among the Guaymi of Costa Rica, the Barí of Venezuela, and Matsigenka of Peru, among others (see review in Cormier and Urbani 2008). Both Duarte-Quiroga and Estrada (2003) and Urbani and Cormier (2015) attribute this bias to the difficulty of keeping howler monkeys as pets given their highly specific diet.

In sum, we found that the attribution of mental states to animals relies on a series of observable and unobservable (inferred) properties that manifest in the anatomy and behavior of animals. As mentioned before, these attributions are not necessarily merely "anthropomorphic," especially in the case of primary emotions. Even though the accuracy of mental state attribution remains uncertain, there is sufficient evidence that mental state attribution to a wide range of species may be regarded a human universal. What remains to be explored is how mind attribution relates to the personification of animals and how humans negotiate the use of animals in the face of animal personhood. This is particularly intriguing in the case of primates. Urbani (2005) reviewed the available information related to predation of primates by human and nonhuman predators and found that humans are nowadays their main predator in the Neotropics. Large-bodied primates from the genera *Alouatta* (howler monkeys), *Ateles* (spider monkeys), and *Lagothrix* (woolly monkeys) tend to be the preferred prey (Urbani 2005). This certainly contradicts the hypothesis that animals perceived more humanlike are considered more minded and, therefore, would be worthy of moral concern. As part of our interviews, we asked participants if they consider monkeys (both howler and spider monkeys) as an acceptable food source. Just a couple of participants stated that they would be willing to try it. Most of them argued that they would not taste monkey meat. They gave three types of arguments to justify why people in their community do not eat primates: (1) cultural reasons as people in this community were not accustomed to or not taught to eat them; (2) aversion caused by their physical appearance, smell, or taste; and (3) aversion because they are humanlike. Some participants stated that monkeys are humanlike because of their close phylogenetical relationship or because they look and behave like people. The Parintintin of Brazil avoid eating them due to the same reason, while the Kalapalo and the Guajá ethnic groups also from Brazil eat them exactly because of this (Cormier 2003; Kracke 1978; Basso 1973, cited in Cormier 2006).

An alternative hypothesis is that people engage in a series of strategies to reduce the potential cognitive discomfort (i.e., cognitive dissonance) caused by holding two conflicting beliefs, values, or behaviors (Festinger 1957). Bastian et al. (2011) suggest that people deny mind to animals that are consumed as food; others argue that the mere categorization of animals as food minimizes the moral rights attributed to animals (Bratanova et al. 2011). In the present study, the capacity to think or being intelligent seemed unrelated to the categorization of animals as acceptable food sources. In fact, most of the species considered "clever" or "intelligent" by our participants were pets and predators, as well as prey animals commonly used as food. It is known that hunting societies from the Arctic, subarctic, and Amazonia attribute a certain personhood status to prey animals or animals that are considered dangerous, powerful, or similar to humans (Helander-Renvall 2010; Fausto 2007; Willerslev 2007; Viveiros de Castro 1998; Brightman 1993; Hallowell 1960); in some respect, these are considered ontologically similar to humans (Hill 2011). Fausto (2007:498) expresses this potential ontological conflict of killing "others" that are perceived as similar in simple terms: "If the predation of animals is equivalent to killing people, would hunting not immediately merge into warfare? And if both these phenomena are inscribed within a field of social relations between subjects imbued with intentionality, would not food consumption necessarily slip into cannibalism?" It has been proposed that in Artic and subarctic societies, the act of hunting is not violent or aggressive, but rather as a pre-established social exchange of favors between the hunter and the prey in which both parties benefit in some way (Nadasdy 2007; Ingold 1994; Bird-David 1990; but see Knight 2012, for a critique). The Sami of Norway believe the reindeer give themselves to humans in exchange of shelter (Helander-Renvall 2010); hunters from the Rock Cree in northern Manitoba believe that animals give themselves to hunters (Brightman 1993), as do Kluane people of the Southwest Yukon (Nadasdy 2007). In exchange of the self-sacrifice of the prey, hunters must comply with a series of obligations they have toward them: proper treatment of the animal's remains, enabling the parting and journey of the spirit of the animals, and observing a series of taboos (Brightman 1993; Sabo and Sabo 1985; Nelson 1983). The *Guajá* from Brazil and the *Barí* from Venezuela, who consume spider and howler monkeys, respectively, believe that their divine creator instructed them to eat them (Cormier 2003; Lizarralde 2002, cited in Cormier 2006). Perceiving animals as minded but "others/aliens/outgroups" (Willerslev 2007) as enemies (Fausto 1999) and objects (Epstein 2004) or holding the belief that an almighty god created animals for the use of humans could also be counted among these strategies to reduce the inherent guilt associated with the inevitable antagonistic engagement resulting from the use of animals and the sometimes automatic or unreflecting self-identification of humans with them.

In conclusion, attribution of emotions and other mental states to animals seem to be a common phenomenon triggered by behaviors expressed by them in response to certain circumstances. Some mental states were ascribed in a straightforward fashion by our participants based on an observed behavior; attacking, for example, is an incontrovertible sign of anger and sometimes fear. Other mental abilities, like intelligence, are not "observed directly" and are rather inferred based on a given context and outcome. If the possession of mental states does indeed define the inclusion of living or nonliving entities in the social sphere of humans, then monkeys, along with a plethora of domestic and wild animals, can be indeed understood as "other-than-human persons," not only from a symbolic but also from a cognitive point of view.

2.5 Acknowledgments

This work was supported by a grant for graduate studies provided by the National Council of Science and Technology (CONACYT) (No. 310752) and by the doctoral college program "Cognition & Communication" of the Austrian Science Fund FWF—http://www.fwf.ac.at/en/ (No. W1234-G17). We would like to express our gratitude to our kind participants for sharing with us their time and knowledge. We specially thank Isy Aguilera Jiménez for assisting us with the interviews during fieldwork, Ana Karen Caballero for her support with the construction of the database, and Daniel Ortiz Santa Maria for producing the map depicted in Fig. 2.1.

Appendix 1: Sociodemographic Information of Participants

Folio	Sex	Age	Occupation	State of origin	Actual place of residence[a]	Did one of your parents spoke an indigenous language? (indigenous language)	Do you speak an indigenous language?	Last study grade
1	Male	18	Farmer	Campeche	Conhuas	No	No	Secondary education
2	Female	26	Museum worker	Campeche	Conhuas	Yes (Chol and Tzeltal)	Yes	Secondary education
3	Male	35	Farmer	Chiapas	Concepcion	No	No	Secondary education
4	Male	28	Unemployed	Campeche	Conhuas	Yes (Totonac)	No	Secondary education
5	Female	21	Student	Campeche	Xpujil	No	No	None
6	Male	30	Gardener	Chiapas	Becam	No	No	Primary education
7	Female	54	Housewife	Tabasco	Conhuas	Yes	No	Primary education
8	Female	36	Housewife	Chiapas	Conhuas	No	No	Secondary education
9	Male	33	Farmer	Campeche	Conhuas	No	No	Secondary education
10	Male	31	Farmer	Veracruz	Conhuas	No	No	High school
11	Female	58	Housewife	Veracruz	Conhuas	No	No	Primary education
12	Male	37	Farmer	Veracruz	Conhuas	Yes (Totonac)	No	Secondary education
13	Male	82	Farmer	Campeche	Conhuas	Yes (Maya)	Yes	None
14	Female	22	Unemployed	Campeche	Conhuas	Yes (Tzeltal and Chol)	Yes	Secondary education
15	Male	40	Farmer	Chiapas	Conhuas	Yes (Zoque)	Yes	Primary education
16	Male	34	Farmer	Veracruz	Conhuas	No	No	Primary education
17	Male	23	Museum worker	Chiapas	Huehuejuez	No	No	None
18	Male	66	Farmer	Veracruz	Conhuas	Yes (Totonac)	No	Primary education
19	Female	36	Housewife	Campeche	Conhuas	Yes (Maya)	No	Secondary education

(continued)

Appendix 1 (continued)

Folio	Sex	Age	Occupation	State of origin	Actual place of residence[a]	Did one of your parents spoke an indigenous language? (indigenous language)	Do you speak an indigenous language?	Last study grade
20	Female	35	Housewife	Veracruz	Conhuas	No	No	Primary education
21	Male	45	Gardener	Campeche	Santa Lucía	No	No	Secondary education
22	Female	30	Housewife, beekeeper	Campeche	Conhuas	Yes (Tzeltal)	Yes	Secondary education
23	Male	45	Employee	Campeche	Timun	Yes (Maya)	Yes	High school

[a]All participants had their actual residence in the state of Campeche, Mexico

Appendix 2: List of Animal Cards Shown to Participants

Order	Common name	Taxonomic name
Primate	Howler monkey	*Alouatta pigra*
	Spider monkey	*Ateles geoffroyi yucatanensis*
Carnivora	Dog	*Canis lupus familiaris*
	Gray fox	*Urocyon cinereoargenteus*
	Jaguar	*Panthera onca*
	Raccoon	*Procyon lotor*
Artiodactyla	Cow	*Bos taurus*
	Deer	*Odocoileus virginianus*
	Pig	*Sus scrofa ssp. domesticus*
	White collared peccary	*Pecari tajacu*
Pilosa	Anteater	*Tamandua mexicana*
Galliformes	Chicken	*Gallus gallus domesticus*
	Ocellated Turkey	*Meleagris ocellata*
Piciformes	Toucan	*Ramphastos sulfuratus*
Psittaciformes	Yellow-lored amazon	*Amazona xantholora*
Passeriformes	Yucatan magpie	*Cyanocorax yucatanicus*
Rodentia	Yucatec mouse	*Peromyscus yucatanicus*
Squamata	Snake	*Bothrops asper*
Hymenoptera	Ant	*Atta cephalotes*
	Bee	*Apis mellifera*
Scorpiones	Scorpion	*Centruroides gracilis*

References

Adolphs R (2013) The biology of fear. Curr Biol 23(2):R79–R93
Angantyr M, Elkund J, Hansen EM (2011) A comparison of empathy for humans and empathy for animals. Anthrozoös 24(4):369–377
Barrett HC (2005) Cognitive development and the understanding of animal behavior. In: Ellis BJ, Bjorklund DF (eds) Origins of the social mind: evolutionary psychology and child. The Guilford Press, New York, pp 438–467
Barrett HC, Todd PM, Miller GF, Blythe PW (2005) Accurate judgments of intention from motion cues alone: a cross-cultural study. Evol Hum Behav 26(4):313–331
Basso EB (1973) The Kalapalo Indians of Central Brazil. Holt, Rinehart and Winston, New York
Bastian B, Loughnan S, Haslam N, Radke HRM (2011) Don't mind meat? The denial of mind to animals used for human consumption. Personal Soc Psychol Bull 38(2):247–256
Batt S (2009) Human attitudes towards animals in relation to species similarity to humans: a multivariate approach. Biosci Horiz 2(2):180–190
Bird-David N (1990) The giving environment: another perspective on the economic system of gatherer-hunters. Curr Anthropol 31(2):189–196
Blythe PW, Todd PM, Miller GF (1999) How motion reveals intention: categorizing social interactions. In: Gigerenzer G, Todd PM (eds) Simple heuristics that make us smart: evolution and cognition. Oxford University Press, New York, pp 257–285
Bratanova B, Loughnan S, Bastian B (2011) The effect of categorization as food on the perceived moral standing of animals. Appetite 57:193–196
Brightman R (1993) Grateful prey: Rock Cree human-animal relationships. University of California Press, Berkeley-Los Angeles-Oxford
Broida J, Tingley L, Kimball R, Miele J (1993) Personality differences between pro-and antivivisectionists. Soc Anim 1(2):129–144
Buccino G, Liu F, Canessa N, Patteri I, Lagravinese G, Benuzzi F et al (2004) Neural circuits involved in the recognition of actions performed by nonconspecifics: an FMRI study. J Cogn Neurosci 16(1):114–126
Carabias-Lillo J, Provencio E, de la Maza EJ, de la Gala Méndez JB (2000) Programa de manejo de la Reserva de la Biosfera Calakmul. Instituto Nacional de Ecología, D.F. México
Carey S (1995) On the origin of causal understanding. In: Sperber D, Premack D, Premack AJ (eds) Causal cognition: a multidisciplinary debate. Clarendon Press/Oxford University Press, New York, pp 268–308
Carstens E, Moberg GP (2000) Recognizing pain and distress in laboratory animals. ILAR J 41(2):62–71
Cormier LA (2003) Kinship with monkeys: the Guajá foragers of eastern Amazonia. Columbia University Press, New York
Cormier LA (2006) A preliminary review of Neotropical primates in subsistence and symbolism of indigenous lowland South American peoples. Ecol Environ Anthropol 2:14–32
Cormier LA, Urbani B (2008) The ethnoprimatology of the spider monkey (Ateles spp.): from past to present. In: Campbell C (ed) Spider monkeys: the biology, behavior and ecology of the genus Ateles. Cambridge University Press, Cambridge, p 377
Darwin C (1872) The expression of the emotions in man and animals. Oxford University Press, New York
Denver RJ (2009) Structural and functional evolution of vertebrate neuroendocrine stress systems. Ann N Y Acad Sci 1163:1–16
di Pellegrino G, Fadiga L, Fogassi L, Gallese V, Rizzolatti G (1992) Understanding motor events: a neurophysiological study. Exp Brain Res 91(1):176–180
Dickman A (2010) Complexities of conflict: the importance of considering social factors for effectively resolving human-wildlife conflict. Anim Conserv 13(5):458–466
Distefano E (2005) Human-Wildlife Conflict worldwide: collection of case studies, analysis of management strategies and good practices. FAO, Rome

Driscoll JW (1995) Attitudes towards animals: species ratings. Soc Anim 3(2):139–150
Duarte-Quiroga A, Estrada A (2003) Primates as pets in Mexico City: an assessment of the species involved, source of origin, and general aspects of treatment. Am J Primatol 61(2):53–60
Eddy TJ, Gallup GG, Povinelli DJ (1993) Attribution of cognitive states to animals: anthropomorphism in comparative perspective. J Soc Issues 49(1):87–101
Ekman P (1999) Basic emotions. In: Dalgleish T, Power M (eds) Handbook of cognition and emotion. Wiley, Chichester/New York, pp 45–60
Ellingsen K, Zanella AJ, Bjerkås E, Indrebø A (2010) The relationship between empathy, perception of pain and attitudes towards pet among Norwegian dog owners. Anthrozoös 23(3):231–243
Elo S, Kyngäs H (2008) The qualitative content analysis process. J Adv Nurs 62(1):107–115
Epstein RA (2004) Animals as objects, or subjects of right. In: Sunstein CR, Nussbaum MC (eds) Animal rights: current debates and new directions. Oxford University Press, New York, pp 143–161
Evans JS (2002) Logic and human reasoning: an assessment of the deduction paradigm. Psychol Bull 128:978–996
Farah MJ, Heberlein AS (2007) Personhood and neuroscience: naturalizing or nihilating? Am J Bioeth 7:37–48
Fausto C (1999) Of enemies and pets: warfare and shamanism in Amazonia. Am Ethnol 26(2):933–956
Fausto C (2007) Feasting on people: eating animals and humans in Amazonia. Curr Anthropol 48(4):497–530
Festinger L (1957) A theory of cognitive dissonance. Stanford University Press, Stanford
Franklin AJ, Nelson M, Baker JE, Beeney TK, Vescio A, Lenz-Watson A et al (2013) Neural responses to perceiving suffering in humans and animals. Soc Neurosci 8(3):217–227
Furnham A, McManus C, Scott D (2003) Personality, empathy and attitudes to animal welfare. Anthrozoös 16(2):135–146
Gray H, Gray K, Wegner D (2007) Dimensions of mind perception. Science 315(5812):619
Gray K, Young L, Waytz A (2012) Mind perception is the essence of morality. Psychol Inq 23(2):101–124
Hallowell AI (1960) Ojibwa ontology: behaviour and world view. In: Diamond S (ed) Culture and history. Essays in Honor of Paul Radin. Columbia University Press, New York, pp 19–52
Harrison MA, Hall AE (2010) Anthropomorpohism, empathy, and perceived communicative ability vary with phylogenetic relatedness to humans. JSEC 4:34–48
Helander-Renvall E (2010) Animism, personhood and the nature of reality: Sami perspectives. Polar Record 46(1):44–56
Herzog HA, Galvin S (1997) Anthropomorphism, common sense, and animal awareness. In: Mitchell RW, Thompson NS (eds) Anthropomorphism, anecdotes, and animals. SUNY Press, Albany, NY, pp 237–253
Hill E (2011) Animals as agents: hunting ritual and relational ontologies in prehistoric Alaska and Chukotka. Camb Archeol J 21(3):407–426
Hills AM (1995) Empathy and belief in the mental experience of animals. Anthrozoös 8(3):132–142
Horowitz A, Bekoff M (2007) Naturalizing anthropomorphism: behavioral prompts to our humanizing of animals. Anthrozoös 20(1):23–35
INEGI (2015) Catálogo de claves de entidades federativas, municipios y localidades. http://geoweb.inegi.org.mx/mgn2k/catalogo.jsp. Accessed Jan 2017
Ingold T (1994) What is an animal? Cambridge University Press, Cambridge, UK
Jacobs M (2009) Why do we like or dislike animals? Hum Dimens Wildl 14(1):1–11
Julius H, Beetz A, Kotrschal K, Turner D, Uvnäs-Moberg K (2012) Attachment to pets. An integrative view of human-animal relationships with implications for therapeutic practice. Hogrefe Publishing, Gottingen
Kellert SR, Berry JK (1987) Attitudes, knowledge, and behaviors toward wildlife as affected by gender. Wildl Soc Bull 15(3):363–371

Kellert SR, Black M, Rush CR, Bath AJ (1996) Human culture and large carnivore conservation in North America. Conserv Biol 10(4):977–990
Kemper TD (1987) How many emotions are there? Wedding the social and automatic components. Am J Sociol 93:263–289
Kennedy JS (1992) The new anthropomorphism. Cambridge University Press, Cambridge, UK
Knight J (2012) The anonymity of the hunt: a critique of hunting as sharing. Curr Anthropol 53(3):334–355
Knight S, Vrij A, Cherryman J, Nunkoosing K (2004) Attitudes towards animal use and belief in animal mind. Anthrozoös 17(1):43–62
Konok V, Nagy K, Miklósi A (2015) How do humans represent the emotions of dogs? The resemblance between the human representation of the canine and the human affective space. Appl Anim Behav Sci 162:37–46
Kotrschal K (2009) Die evolutionäre Theorie der Mensch-Tier-Beziehung. In: Otterstedt C, Rosenberger M (eds) Gefährten - Konkurrenten - Verwandte. Die Mensch-Tier-Beziehung im wissenschaftlichen Diskurs. Vandenhoeck & Ruprecht, Göttingen, pp 55–77
Kracke WH (1978) Force and persuasion, leadership in an Amazonian society. University of Chicago Press, Chicago
LeDoux JE (2000) Emotion circuits in the brain. Annu Rev Neurosci 23(1):155–184
Liu F, McShea WJ, Garshelis DL, Zhu X, Wang D, Shao L (2011) Human-wildlife conflicts influence attitudes but not necessarily behaviors: factors driving the poaching of bears in China. Biol Conserv 144(1):538–547
Lizarralde M (2002) Ethnoecology of monkeys among the Barí of Venezuela: perception, use and conservation. In: Fuentes A, Wolfe LD (eds) Primates face to face: the conservation implications of human-nonhuman primate interconnections. Cambridge University Press, Cambridge, MA, pp 85–100
Lorenz K (1950) Ganzheit und Teil in der tierischen und menschlichen Gemeinschaft. Reprinted in: Martin R (ed) Studies in Animal and Human Behaviour, 1971, vol. 2, Harvard University Press, Cambridge, MA, pp 115–195
Mameli M, Bortolotti L (2006) Animal rights, animal minds, and human mindreading. J Med Ethics 32:84–89
Martens P, Enders-Slegers MJ, Walker JK (2016) The emotional lives of companion animals: attachment and subjective claims by owners of cats and dogs. Anthrozoös 29(1):73–80
Mayring P (2000) Qualitative content analysis. Forum: qualitative social research, 1(2). Retrieved 10 Jan 2017, from http://www.qualitative-research.net/fqs-texte/2-00/02-00mayring-e.htm
Merckelbach H, Marcel A, Van Den H, Van der Molen GM (1987) Fear of animals: correlations between fear ratings and perceived characteristics. Psychol Rep 60(3, Pt 2):1203–1209
Miser T, Sloutsky VM (2013) Distinguishing between category-based and similarity-based induction. In: Knauff M, Pauen M, Sebanz N, Wachsmuth I (eds) Proceedings of the XXXV annual conference of the cognitive science society, pp 3056–3061
Mitchell R, Hamm M (1997) The interpretation of animal psychology: anthropomorphism or behavior reading? Behaviour 134(3):173–204
Morewedge CK, Preston J, Wegner DM (2007) Timescale bias in the attribution of mind. J Pers Soc Psychol 93(1):1–11
Morris P, Fidler M, Costall A (2000) Beyond anecdotes: an empirical study of "anthropomorphism.". Soc Anim 8(2):151–165
Morris P, Doe C, Godsell E (2008) Secondary emotions in non-primate species? Behavioural reports and subjective claims by animal owners. Cogn Emot 22(1):3–20
Morris P, Knight S, Lesley S (2012) Belief in animal mind: does familiarity with animals influence belief about emotions? Soc Anim 20(3):211–224
Morton DB, Griffiths PHM (1985) Guidelines on the recognition of pain and discomfort in experimental animals and an hypothesis for assessment. Vet Rec 116:431–436
Nadasdy P (2007) The gift in the animal: the ontology of hunting and human-animal sociality. Am Ethnol 34(1):25–43

Nakajima S, Arimitsu K, Lattal M (2002) Estimation of animal intelligence by university students in Japan and the United States. Anthrozoös 15(3):194–205
Nelson RK (1983) Make prayers to the raven: a Koyukon view of the northern forest. University of Chicago Press, Chicago
Oosterwijk S, Lindquist KA, Anderson E, Dautoff R, Moriguchi Y, Barrett LF (2012) States of mind: emotions, body feelings, and thoughts share distributed neural networks. NeuroImage 62(3):2110–2128
Opotow S, Weiss L (2000) Denial and the process of moral exclusion in environmental conflict. J Soc Issues 53(3):475–490
Panksepp J (2005) Affective consciousness: core emotional feelings in animals and humans. Conscious Cogn 14(1):30–80
Paul E (2000) Empathy with animals and with humans: are they linked? Anthrozoös 13(4):124–202
Person P, Laaksolahti J, Lonnqvist P (2000) Anthropomorphism – a multi-layered phenomenon. American Association for Artificial Intelligence, Kista
Plous S (1993) Psychological mechanisms in the human use of animals. J Soc Issues 49(1):11–52
Rajecki D, Rasmussen H, Craft H (1993) Labels and the treatment of animals: archival and experimental cases. Soc Anim 1(1):45–60
Rasmussen JL, Rajecki DW (1995) Differences and similarities in humans' perceptions of the thinking and feeling of a dog and a boy. Soc Anim 3(2):117–137
Rasmussen JL, Rajecki DW, Craft HD (1993) Humans' perceptions of animal mentality: ascriptions of thinking. J Comp Psychol 107(3):283–290
Reynolds SM, Berridge KC (2008) Glutamate motivational ensembles in nucleus accumbens: rostrocaudal shell gradients of fear and feeding. Eur J Neurosci 17(10):2187–2200
Sabo G, Sabo DR (1985) Belief systems and the ecology of sea mammal hunting among the Baffinland Eskimo. Arct Anthropol 22(2):77–86
SEMARNAP (2000) Programa de manejo del área natural protegida Reserva de la Biosfera de Calakmul. Diario Oficial de la Federación, Viernes 7 de abril de 2000
Serpell JA (2004) Factors influencing human attitudes to animals. Anim Welf 13(S1):145–151
Signal T, Taylor N (2006) Attitudes to animals in the animal protection community compared to a normative community sample. Soc Anim 14(3):265–274
Simons EL, Meyers DM (2001) Folklore and beliefs about the aye aye (*Daubentonia madagascariensis*). Lemur News, July, 6: 11–16
Sloutsky VM, Fisher AV (2004) Induction and categorization in young children: a similarity-based model. J Exp Psychol 133(2):166–188
Sloutsky VM, Kloos H, Fisher AV (2007) When looks are everything: appearance similarity versus kind information in early induction. Psychol Sci 18(2):179–185
Soanes C, Stevenson A (2005) Oxford dictionary of English, Revised edn. Oxford University Press, Oxford, UK
Swami V, Furnham A, Christopher A (2008) Free the animals? Investigating attitudes toward animal testing in Britain and the United States. Scand J Psychol 49(3):269–276
Tabo E, Jinks SL, Eisele JH Jr, Carstens E (1999) Behavioral manifestations of neuropathic pain and mechanical allodynia, and changes in spinal dorsal horn neurons, following L4-L6 dorsal root constriction. Pain 80:503–520
Taylor N, Signal TD (2005) Empathy and attitudes towards animals. Anthrozoös 18(1):18–27
Urbani B (2005) The targeted monkey: a re-evaluation of predation on New World primates. J Anthropol Sci 83:89–109
Urbani B, Cormier LA (2015) The ethnoprimatology of the howler monkeys (Alouatta spp.): from past to present. In: Kowalewski MM et al (eds) Howler monkeys, developments in primatology: progress and prospects. Springer, New York, pp 259–280
Urquiza-Haas EG, Kotrschal K (2015) The mind behind anthropomorphic thinking: attribution of mental states to other species. Anim Behav 109:167–176
Vazdarjanova A, McGaugh JL (1998) Basolateral amygdala is not critical for cognitive memory of contextual fear conditioning. Proc Natl Acad Sci U S A 95:15003–15007

Viveiros de Castro E (1998) Cosmological deixis and Amerindian perspectivism. J R Anthropol Inst 4(3):469–488

Walker JK, McGrath N, Handel IG, Waran NK, Phillips CJC (2014) Does owning a companion animal influence the belief that animals experience emotions such as grief? Anim Welf 23(1):71–79

Waytz A, Gray K, Epley N, Wegner D (2010) Causes and consequences of mind perception. Trends Cogn Sci 14(8):383–388

Welder AN, Graham SA (2001) The influences of shape similarity and shared labels on infants' inductive inferences about nonobvious object properties. Child Dev 72(6):1653–1673

Westbury H, Newman D (2008) Empathy-related responses to moving film stimuli depicting human and nonhuman animal targets in negative circumstances. Biol Psychol 78(1):66–74

Wilkins A, McCrae LS, McBride EA (2015) Factors affecting the human attribution of emotions toward animals. Anthrozoös 28(3):357–369

Willerslev R (2007) Soul hunters: hunting, animism, and personhood among the Siberian Yukaghirs. University of California Press, Berkeley-Los Angeles-Oxford

Chapter 3
Local Knowledge and Cultural Significance of Primates (*Ateles geoffroyi* and *Alouatta pigra*) Among Lacandon Maya from Chiapas, Mexico

Yasminda García del Valle, Felipe Ruan-Soto, Fernando Guerrero-Martínez, and Felipe Reyes-Escutia

3.1 Introduction

Mexico is part of some of the most bioculturally rich regions worldwide. Not only does it contain a highly diverse flora and fauna, but it also houses over 50 indigenous peoples (Toledo et al. 2003). The Mexican south-east, and the region known as the Lacandon rainforest in the state of Chiapas in particular, has been considered one of the most developed, preserved, and floristically and faunistically rich zones in the country (Castillo-Campos and Narave 1992). Land mammals are an important component of this biological diversity, accounting for 12% of the mammal species on Earth. Mexico scores first place in land mammal diversity in the American continent and second place worldwide (Ceballos and Ehrlich 2002; Dirzo 1992). Furthermore, this region is home to many ethnolinguistic groups of people, mainly Chol, Tseltal, Tsotsil, and Lacandon (Ruan-Soto et al. 2007).

Due to the interaction of peoples with their rainforests, they have generated knowledge about these complex systems and the biodiversity they contain (Toledo et al. 2003; Berkes 1999). Mammals in particular can be considered one of the most

Y. García del Valle (✉) · F. Reyes-Escutia
Laboratorio de Educación Ambiental y Sustentabilidad, Instituto de Ciencias Biológicas, Universidad de Ciencias y Artes de Chiapas, Tuxtla Gutiérrez, Mexico
e-mail: yasmindadelvalle@hotmail.com

F. Ruan-Soto
UNAM, Becario del Programa de Becas Posdoctorales en la UNAM, Centro de Investigaciones Multidisciplinarias sobre Chiapas y la Frontera Sur. Universidad Nacional Autónoma de México, San Cristóbal de Las Casas, Mexico

F. Guerrero-Martínez
Centro de Investigaciones Multidisciplinarias sobre Chiapas y la Frontera Sur. Universidad Nacional Autónoma de México, San Cristóbal de Las Casas, Mexico

B. Urbani, M. Lizarralde (eds.), *Neotropical Ethnoprimatology*, Ethnobiology,
https://doi.org/10.1007/978-3-030-27504-4_3

salient groups for societies because they have been put to diverse uses (fur, commerce, crafts, medicinal, ornamental, pets, food) (Ojasti 1993) and they also satisfy other needs, such as cultural, religious, symbolic, and intellectual (Pérez-Gil et al. 1995). This is to say, they are given both tangible and intangible properties ranging from usage to involvement in myths and rituals for their traditionally imposed symbolism, which gives them a remarkable place in cosmovision (Santos-Fita et al. 2009). In general, we could state that the relationship between human groups and different groups of organisms is multidimensional and a product of history (Ruan-Soto et al. 2013).

Among the species of mammals that stand out for their size and behavior are the howler monkeys (*Alouatta palliata* and *A. pigra)* and the spider monkeys (*Ateles geoffroyi*). Perhaps because of our proximity to their phylogenetics, biology, and behavior, the relationship between humans and monkeys has a special relevance (Alves et al. 2010). Ethnoprimatology is a branch of ethnobiology that studies the cultural significance of non-human primates in different human societies and cultures, which may have direct or indirect relationships with these organisms. Furthermore, it delves on the way in which this relationship has been built across history and the consequences of such a history in the conservation of primate species. While ethnoprimatology research is quite novel, most of the work that has been done focuses on monkey species under some degree of risk of extinction in order to plan actions for their conservation based on environmental education projects (Estrela 2009).

Contrasting with the situation in other regions of the world, where relationships between other primates and humans are deemed conflictive due to the damage these animals cause on crops to a degree that they are considered pests (Rocha and Fortes 2015; Lee and Priston 2005), in Mesoamerica, this relationship does not seem to have been conceived that way. In this region, non-human primates have had an important role in the symbolic spheres of societies, and they have been represented in diverse art objects (Bruner and Cucina 2005). While some interpretations of the meanings the three species of non-human primates present in the Mexican south-east could have been proposed (*Ateles geoffroyi*, *Alouatta pigra*, and *Alouatta palliata*), studies giving evidence of the relationship between these species and rural human groups with which they share spaces or attempting to comprehend the multidimensional aspects of this relationship are scarce.

This chapter is an overview of the relationship between Lacandon Mayan people from the communities of Naha and Metzabok and the two species of non-human primates in their territory (*Ateles geoffroyi* and *Alouatta pigra*). We reviewed and interpreted evidences from the presence of these species in pre-Hispanic art in the Mayan area to ethnographic data referring to people's ethno-ecological knowledge and an analysis of the local cultural significance of these species.

3.2 The Lacandon Maya from Naha and Metzabok and Their Environment

3.2.1 The Lacandon

Undoubtedly, Lacandon people are among the Mayan groups that have most fascinated researchers since the second half of the last century. Their origins and a great deal of their history during the Colonial period are uncertain and have been diversely interpreted. Regarding this, the term "Lacandon" was used to name various rebellious Mayan groups that fled into the Lacandon rainforest during the Colonial regime (Boremanse 1984). Some of these came originally from different zones of the states of Yucatan, Campeche, and Tabasco in Mexico and from the Guatemalan Peten. They spoke different linguistic dialects of Yucatec Maya, and they are the direct forefathers of the group currently called Lacandon (De Vos 1980). Regardless, the self-denomination of the Lacandon is *Hach Winik*, which has been translated as "true people." Furthermore, there exists a geographical division between northern Lacandonians (inhabitants of Naha and Metzabok) and southern Lacandonians (living in Lacanjá, Betel, and San Javier) (Cano et al. 2009; Eroza 2006; Baer and Merrifield 1972). According to Boremanse (1986), current northern Lacandonians are originally from the Yucatan Peninsula, while southern Lacandonians came from Peten, Guatemala.

Currently, the official organization is represented by authorities from the "Lacandon zone" community, an indigenous alliance including three ethnical groups (Lacandon Maya, Chol, and Tseltal). Their main authority is the communal goods commissary and the vigilance council composed exclusively of Lacandon Maya (CONANP 2006). The subsistence of the Lacandon has been based on knowledge of their environment and the development of a complex traditional agriculture system supplemented by the collection of fruits, seeds, and plants from the rainforest as well as hunting and lake fishing. From late 20th century on, the Lacandons have had strong influence from the outside. Nowadays, they have switched from traditional polyculture to monoculture, and they even pay the Tseltal to cultivate their land. Furthermore, the National Commission for Protected Natural Areas (CONANP), the federal institution in charge of the management of natural reserves, has had strong influence in decision-making by implementing preservation projects and promoting economic support for rainforest inhabitants to engage in conservation.

3.2.2 The Lacandon Environment

The Lacandon rainforest is located in east/northeast Chiapas, Mexico (Fig. 3.1). It is a hydrological basin of relevance for water capture. Among its permanent lakes are the Ocotal and Metzabok. The presence of limestone material, along with the effect

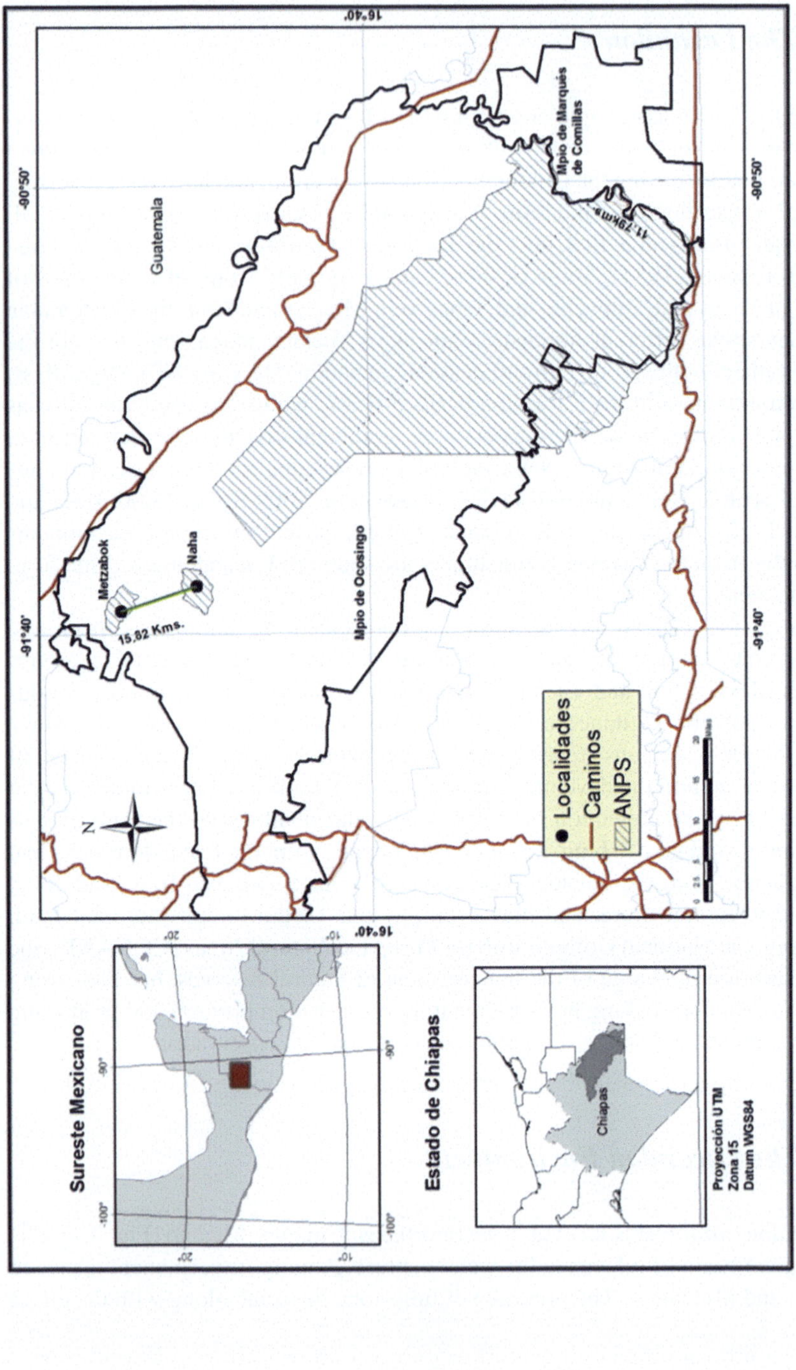

Fig. 3.1 Location of the Lacandon communities Naha and Metzabok, Chiapas, Mexico

of water filtration, accelerates the process of rock dilution, which originates crooks, filtrations, and subterranean circulation (García-Gil and Lugo 1992). The dominant climate is tropical humid with abundant rains in the summer and part of the fall (May through November) and a short dry season from March to May. Mean annual temperature is 25 °C with a medium annual rainfall between 2300 mm and 2600 mm (Quintana-Ascencio et al. 1990). The Lacandon territory has an altitude that ranges from 900 meters to barely above sea level. The vegetation is classified as high perennial rainforest (Pennington and Sarukhan 1998) with additional presence of medium sub-perennial rainforest (Castillo-Campos and Narave 1992). The first vegetation type is a very dense plant community with dominance of perennial trees over 30 meters high with abundant vines and epiphytes. Among the most frequent species of trees are *Terminalia amazonia* ("canshán," bullywood in English), *Ceiba pentandra* ("ceiba," kapok tree), and *Swietenia macrophylla* (locally known as "caoba," mahogany). The most important fruit trees are *Spondias mombin* ("jobo"), *Orbignya cohune* palm ("corozo"), and *Brosimum alicastrum* ("Ramón"); all are very important source of food for monkeys as well as humans (CONANP 2006). In higher ground, pine and oak are the forest composition. The communities of Naha and Metzabok are located in Ocosingo municipality. They were both decreed areas for the protection of flora and fauna in 1998, due to the presence of numerous species of flora and fauna cataloged as "endemic," "rare," "threatened," or "in danger of extinction."

The location of the community of Naha is on the 17° 04′ 53" N and 91° 04′ 09" W coordinates. According to INEGI (2010), the population is 198 inhabitants belonging to 46 families. Although the traditional language is Lacandon Maya, currently, Spanish is the second used language of importance due to the need to establish communication with outsiders. The community of Metzabok is located on the 17°08′36″-17°04′53" N latitude and 91°34′42″-91°40′09" W longitude. INEGI (2010) reports 96 inhabitants grouped in 20 families. As is the case in Naha, the traditional language is Lacandon Maya, and Spanish is a currently important second language.

3.3 Monkeys in the Ancient Maya

The importance of monkeys in Mesoamerica has been made evident by numerous works, specifically in the Mayan culture, such as those presenting research by Nájera (2013; 2012; 2000) about the rich and complex symbolism of these primates in different times and places of Mayan history. Recently, Rice and South (2015) have analyzed the images of monkeys in Mayan ceramic from the Classic period. In the next section, we will present a brief summary of some remarkable aspects about the role monkeys had in pre-Hispanic Mayan cosmovision, mainly through the artistic manifestations of the Late Classic period, but also early Colonial documents have some references. This will help shed a clear light on the multiple meanings and symbolic values monkeys had in ancient times. Furthermore, it will help give a deeper context to the information gathered from contemporary Lacandon people and illustrate the cultural changes and continuities primate symbology has experienced across particular moments in history.

3.3.1 The Origin of Monkeys

Popol Vuh is one of the most important mythological lore and written historical sources of the Mayan culture. It comprises rich mythological, ritual, narrative, and historical information about highland Mayan peoples, particularly from the K'iche' people (Craveri 2013; Tedlock 1993; Recinos 1953). Because the Spanish conquistadores burned most of the codices (written Mayan documents) at the time of the conquest of Mexico, they realized that they were losing valuable information. Therefore, some of them were transcribed. We know this document was transcribed between the years 1554 and 1558 in what is currently the Quiché Department of Guatemala by descendants of three K'iche' lineages that adapted histories from a pre-Columbian ancient book originally written pictographically and hierographically which was later written in Latin (Sam 2008; Tedlock 1993). The excerpt about the creation of the world is the most famous versions of the origin of the Earth and its inhabitants in Mesoamerica. This myth narrates how creator deities originate the sky, earth, mountains, valleys with their vegetation, and water bodies. Afterward, it tells of how they created the wild animals, who were keepers of the mountains; but since these creatures could not speak nor pronounce the names of their creators, the gods dictated that their destiny would be being eaten. Then, they created the first humans, who were made from earth and mud; these creations spoke senselessly, so the gods unmade them (Sam 2008). For this reason, the creator gods decided to call upon *Xpiyakok* (deity of the sunrise) and *Ixmukane* (deity of nightfall), who knew the count of the days. Using maize (*Zea mays*) seeds and pito coral tree called locally as "colorín" (*Erythrina berteroana*) for divination, they resolved that people should be made out of wood. However, these new beings, even though they were humanlike and could talk, did not have a spirit nor heart. They could not remember their creators, and they did not worship them, so they were harshly punished by storms that flooded the world, with bats to decapitate them and jaguars that ate them. Even domestic utensils such as pots, hotplates, and grinding stones would attack them (Sam 2008). In this way, some of the wooden people from this former creation perished, but as the *Popol Vuh* makes clear, these humans became monkeys:

> This was the dispersion of the human work, the human creation. People fell, defeated. The mouths and faces of all were torn and broken. And before it was said [that] the monkeys in today's forests were their descendants. They were left as proof, because the builder and sculptor used only wood to make their flesh. This is why monkeys look like people: they are proof of a prior human making, a human creation of mere dolls and mere carved out of wood. (Tedlock 1993)

Across the Mayan region, stories about the origin of monkeys express similar divine punishments to a transgression, but they are always originated from pre-human beings (Guerrero 2015; Nájera 2013; Shaw 1972). Another passage of the *Popol Vuh* expresses this transformation of human beings into monkeys. According to this mythical passage, *Xpiyakok* and *Ixmukane* conceived *Wuqub Junajpu* and *Jun Junajpu*. The second had two sons called *Jun Batz'* and *Jun Chowen* (Sam 2008). The names of these characters are composed of a numeral and a noun, so that *Jun* is the number one while *Batz'* is the name of the howling monkey (*Alouatta pigra*) and *Chowen* is the name of the spider monkey (*Ateles geoffroyi*). Thus, their names translate as "One Howler Monkey" and "One Spider Monkey," respectively (Nájera 2013). It is interesting to point out that *Jun Batz'* and *Jun Chowen*, after mistreating their siblings, *Junaipu* and *Ixbalamke*, the hero twins of *Popol Vuh*, received as a punishment to become monkeys after they put their sashes back in place: "but in that instant these [sashes] became their tails and their appearance became that of monkeys. They immediately climbed up to the trees on the small mountains and the great mountains; they entered the forests screaming and swinging between the branches of the trees" (Sam 2008:84). The fact that *Jun Batz'* and *Jun Chowen* are described as wise ones, flutists, singers, writers, and painters in the myth is noteworthy; they were, in short, great artists (Sam 2008; Morales 2001; Tedlock 1993). These qualities were clearly associated with monkeys in the Mayan culture during the Classic period, as they were too in general across Mesoamerica (Nájera 2000).

3.3.2 *The Monkey: Lord of Writing*

In some richly painted vases from different regions of the southern lowlands of Mexico that are dated from the Late Classic (550–600 to 900–950 A.C.), images of monkeys associated with writing instruments may be found (Nájera 2013). Most of these vases are part of particular collections and lacking an archaeological context, although their place of production has been inferred from the ceramic style and chemical analyses. This has allowed the identification of the main pictorial styles in them and their relation to particular regions (Reents-Budet et al. 1994). Such is the case of the monkey illustrated in the K626 vase (Kerr 1998; Robiscek and Hales 1981, Fig. 53a). This illustration is identified as a spider monkey (*Ateles geoffroyi*) due to its morphological features (Fig. 3.2). The scene depicted in this vase as part of a topical set expressed in several ceramic containers, refers to the mythical time in which the God of Corn receives its attire after being reborn from the Underworld (Quenon and Le Fort 1997; Robiscek and Hales 1981; Coe 1973). There, the spider monkey wears a headdress with a series of sheets or leaves knotted to the center by a ribbon, although, in other examples, the same headdress is described as tied by rods or feathers, which is particularly indicative of scribes, since it confers the title of *aj k'uhun*, "he/she of the sacred books," as part of royalty, that is to say, a

Fig. 3.2 Detail of spider monkey (*Ateles geoffroyi*) in the role of scribe. From vase K626. (Drawing by Marisa Ordaz, based on Kerr (1998: 30). Digitalization by Eduardo Ordaz)

high-ranking character (Halperin 2014; Nájera 2013; Jackson and Stuart 2001; Kerr 1998). We know scribes carried out several tasks related to courtesan life (Lacadena 1996), since they constituted the administrative staff of Mayan governments from the Classic period (Beliaev 2011). Furthermore, the monkey wears a beaded necklace that occupies its entire back and has a cut seashell, which symbolizes the Underworld. Its left arm is stretched out in front of it with the palm of the hand facing upward, indicating an object that represents a book (see Fig. 3.2). Another archaeological piece in which a monkey is represented carrying a book is bowl K954. The fact that this monkey is posed over an aquatic band is interesting; this band can be recognized for the interweaved lines and the presence of chiton mollusks or sea cockroaches; these elements in conjunction symbolize the subterranean world, a dark and damp space, home to ancestors and gods, as well as the place of origin of lives and knowledge.

In other examples, the scribe [writers and accountants] monkey offers some present or tribute to a Lord. Such is the case of polychrome K5744 vase, in which the monkey holds *tamales* on a plate (Kerr 1997). While in another vase from Chocholá (K8740), the monkey holds the sign for *wa tamal* in its hand. The association between the monkey and the Maize God is also due to the fact that this deity is patron of the arts, such as sculpture and writing, according to hieroglyphic and iconographic data. Because of this, it has also been directly associated with the Mexican god Xochipilli, deity of the arts in the Postclassic period (Braakhuis 2009). Occasionally, scribes are human beings with a monkey head, perhaps representing the howler monkey (*Alouatta pigra*). This can be observed in K1225 vase (Robiscek and Hales 1981); in it, two such characters write in books wrapped in jaguar skin, and they carry painted in several parts of their body the sign for *akb'al* which is related to darkness, night, and so the Underworld. This makes the role of monkeys as intermediaries between different areas of the cosmos clear (Nájera 2013; Kidder 2009), and this is further corroborated with their presence as harmful *nagual* (*wahyis, the powerful supernatural spirits*) in different scenes.

3.3.3 Monkeys as Wahyis (Powerful Supernatural Spirits)

After deciphering T539 hieroglyph as the logogram *WAY*, many aspects of the beings described by this word in Mayan ceramics from the Late Classic have been made possible (Freidel et al. 1999; Grube and Nahm 1994; Houston and Stuart 1989). The suggestion of linguistic elements in this hieroglyph that indicated its nature of non-possessed noun allowed the establishment of the term *wahyis* as a constitutive or, put otherwise, inseparable part of the person (Velásquez 2009; Zender 2004). *Wahyis* from the Mayan Classic period are considered animated entities or the family spirits that only rulers or other important persons could manipulate, either through specific rituals, by consuming psychoactive substances, or in the dream state. This is indicated by the root *way*, which in many Mayan languages refers to dreams and the act of sleeping (Moreno 2011; Velásquez 2009). People who could control these spirits were called *wahyaw*, meaning "lord who calls the *wahyis*"; these people were able to send these spirits to provoke particular diseases to others, and it has even been suggested that the *wahyis* are the physical representations of certain ailments, represented frequently as hybrid and supernatural animals (Moreno 2011, Helmke and Nielsen 2009). Among the *wahyis* represented in painted Mayan ceramic are monkeys.

A good example of the presence of monkeys as *wahyis* is found in the codex-style vase K1203 (Robiscek and Hales 1981. 31) in which a primate lying down and carrying a cloth across its shoulders and tied in the front appears (Fig. 3.3). It holds in its right arm a plate on which a hand, an eyeball, and a bone can be appreciated (Calvin 1997). These elements have been identified to be human body parts that are devoured by the *wahyis* when they make a person ill in the dream world (Velásquez 2009). The attire we describe is one of the particular identifying features of *wahyis* which is possibly associated with sacrifices (Moreno 2011).

A remarkable anatomical feature of the monkey image in this vase is the presence of deer ears and antlers on its head. The combination of these two animals as *wahyis* is not strange; it is possible to find it in at least nine other vases in which monkeys, particularly spider monkey (*Ateles geoffroyi*), have deer ears and antlers or at least one of these features (these vases are catalogued by Kerr as K1181, K1809, K2010, K3038, K3060, K4920, K7152, K7993, and K8733). Additionally, there are other cases in which the *wahyis* have deer bodies and monkey tails; such

Fig. 3.3 Monkey as *wahyis* with deer ears and antlers. (Drawing by Marisa Ordaz, based on Robiscek and Hales (1981: 31). Digitalization by Eduardo Ordaz)

is the case of vases K3061, K3459, and K2023 (Helmke and Nielsen 2009). Through the analysis of Mayan ritual-medical texts from Colonial times, such as the *Ritual de los Bacabes*, the monkey-deer *wahyis* are the personifications of cramps (Moreno 2011; Helmke and Nielsen 2009).

The names of some of the *wahyis re*presented as monkeys have also been deciphered; an example of this is *yuch max* "flea-ridden spider monkey" which is represented in K1211 vase and *k'ihn bo'lay batz'* "hot howling monkey-jaguar" from K1743 vase (Sheseña 2010). An interesting case is one type of *wahyis* directly associated with the toponym of the archaeological site Caracol in Belize; this one has the name *xukub chih maax* "monkey with deer antlers" (Luin and Matteo 2010). The animal holds on one of its hands a cacao-chocolate (*Theobroma cacao*) fruit. This plant is associated in Mayan cosmovision to the Underworld.

3.3.4 Monkey-Cacao (Chocolate) Associations

Far from few images in Mayan art from the Classic period, monkeys are shown carrying cacao fruits or other direct relationships to this plant (Nájera 2012). Some of these examples are found in the following vases: K4599, K4691, K6312, K8234, and K8357. A special case is a howler monkey (*Alouatta pigra*) illustrated in K5070 vase (Fig. 3.4a). It has been identified as a *wahyis* thanks to the hieroglyphic text that accompanies it and the attire around the animals' neck; also, it is the only being of this type that holds a cacao fruit (Nájera 2013). It is possible that a parallel image to this is the one found in K1789 vase (Kerr 1998). In it are three sitting monkeys,

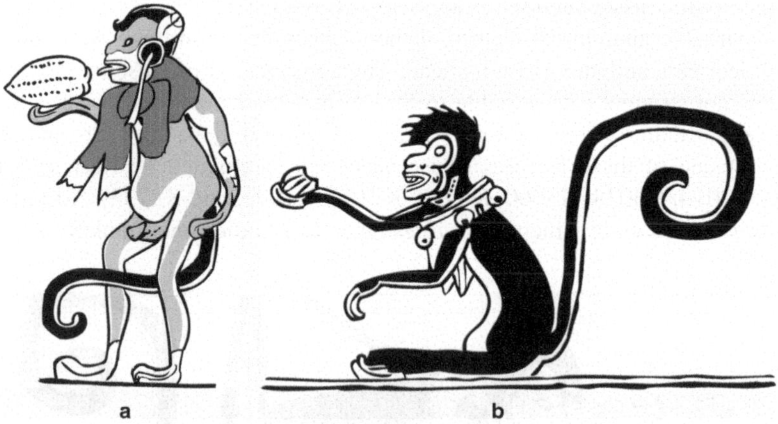

Fig. 3.4 The link between monkeys and cacao. (**a**) Detail from vase K5070 with howling monkey (*Alouatta pigra*) as *wahyis*, holding a cacao fruit. (Drawing by Marisa Ordaz, based on Grube and Nahm (1994: 700)). Digitalization by Eduardo Ordaz. (**b**) Monkey holding cacao fruit, detail from vase K1789. (Drawing by Marisa Ordaz, based on Kerr (1998: 112). Digitalization by Eduardo Ordaz)

all with their right arms spread out showing a kind of fruit and wearing an element from which three eyes hang around their shoulders (Fig. 3.4b). These elements have been interpreted as the motive "eyes of death." They also wear two cacao cobs as pectorals (Nájera 2012). The fruits can be identified as cacao (*Theobroma cacao*) due to the lengthwise lines they have and the concave shape of their top. It is interesting that, in spite of the explanations for the relationship between monkeys and cacao remarking that it is due to the fact that "the fruit was one of the little mammal's preferred foods, making it a seed disperser" (Nájera 2012:154), there is little evidence that monkey species in the Neotropics actually feed frequently from this fruit since the hard shell covering the seeds requires sharp teeth like the one of rodents like squirrels and agouties have to open these seed. Monkeys lack the dentition to open the cacao fruits. This inconsistence is shown in works of ecology, diet, and feeding habits of these primates and related species (Di Fiore et al. 2008; Roosmalen and Klein 1988; Roosmalen 1985). It is known that monkeys prefer fleshy, easily penetrable fruits (Di Fiore et al. 2008); considering this, cacao presents a considerable challenge for consumption for these animals.

Thus, it is possible that the monkey-cacao relationship shown in ancient Mayan art has a lot more to do with the cosmic role associated with this plant as well as the monkey as an intermediary that was capable to take this valuable fruit to different planes as a drink of the deities or combining psychoactive substances for ritual use. This would always take into consideration the symbolism of the cacao tree that makes it closely related to dampness, darkness, and cold, prominent features of the night, caves, and the fertility of the feminine (Nájera 2012, Martin 2006).

3.3.5 Primate Representations in the Archaeological Record

The Lacandon rainforest has several sites in which numerous cave paintings have been discovered, mostly from pre-Columbian origin (Palka 2005). For over a century, researchers have reported the existence of caves, cliffs, and rock shelters with paintings or engravings with various motifs (Thompson et al. 2005, Pincemin 1999, Tozzer 1907), although they have only been systematically studied in the past two decades (Lozada and Núñez 2014, Sánchez 2005, Pincemin 1999). In addition to the rich images of stars and handprints, animals have a significant presence in the rock art found in Lacandon territory (Pincemin 1999). The monkey is a major element in one of the places with most rock representations in the region of the Metzabok lake.

Metzabok lake is surrounded by several cliffs with cave paintings. In the middle of the lake is a rocky front known by the Lacandons as Tzibajná (or Tsibaná according to some authors), which could be translated as "house with writing" or "written house," in reference to the enormous number of images recorded on the rock (Lozada and Núñez 2014, Sánchez 2005). There is a painting representing a monkey which, according to Lozada and Núñez (2014) and Sánchez (2005), is a howler monkey, although its morphological characteristics, such as the posture of its

Fig. 3.5 Painting of a possibly spider monkey from approximately Classic or Late Postclassic Mesoamerican period found on the Metzabok lagoon, Chiapas, Mexico. (Photograph by Fernando Guerrero Martínez)

extremities and its slender body, suggest that it is a spider monkey (*Ateles geoffroyi*), painted red (Fig. 3.5). The drawing of the monkey is associated with other iconographic elements such as negative handprints, as well as anthropomorphic and quadruped figures also outlined in red, probably painted during the Mesoamerican Classic or Late Postclassic period (Lozada and Núñez 2014, Pincemin 1999).

According to the testimonials compiled by Lozada and Núñez (2014) among the Lacandons of Metzabok, the origin of these cave paintings is attributed to the gods, particularly *Hach Ak Yum*, one of the main deities, and Tsibaná, said to live in Metzabok lake. Both gods made the drawings on the rocks after the sky darkened during an eclipse, and year after year, they painted different figures (Lozada and Núñez 2014). However, the Lacandons of Metzabok believe that another god was responsible for leaving the figure of the spider monkey in the rock shelter, because, according to the story of an old Lacandon obtained by Lozada and Núñez (2014), Mensabak is said to have been the character who, in mythical times, created mankind out of mud and then painted it black: "Then he made animals, the maax (monkey), leaving the black color of his hair and then the monkey fled to the mountain, he quickly fled to the tree, he fled and we did not kill him... then he brought him to life, as he brought people to life ... the monkey is the work of Mensabak... he painted it to show that he was the one who created him" (Lozada and Núñez 2014: 508–509).

It is interesting that the image of the monkey is closely linked to the deities among the Lacandons and that their presence continues to have special meaning for present-day inhabitants. This archaeological evidence shows a certain continuity with respect to the importance of the monkey among contemporary Lacandons, despite the passage of time.

3.4 Current Local Knowledge and Cultural Significance of Non-human Primates

In Lacandon Maya, *Ateles geoffroyi* is named *ma'ax*, and *Alouatta pigra* is named *b'atz'*. Both names are non-analyzable, that is to say, they have non-decipherable etymologies. In both cases, the local taxa correspond directly to taxonomic species in western science. Berlin et al. (1973) call this correspondence one-to-one relationship, that is to say a single local generic taxon corresponds to a species in Linnaean systematics. On the other hand, for both species, the fact that their names are simple and non-analyzable indicates their degree of cultural significance. According to Turner (1988), the level of importance of a group of organisms can be accomplished through linguistic analysis and outlines of local taxonomy and classifications. According to this proposal, the most culturally significant species have the features we mention above.

In a study published for these groups, Garcia del Valle et al. (2015) point out that *Alouatta pigra* was mentioned through the free-listing technique by 40% of the Lacandon Maya that were interviewed. Similarly, *Ateles geoffroyi* was mentioned by 33% of the interviewed. Based on the work by Garcia del Valle et al. (2015), it is possible to reanalyze the data to show how, among Lacandon people, *Alouatta pigra* and *Ateles geoffroyi* occupy the 9th and 12th places in the valuation of mammals with the highest cultural significance, respectively. This includes both communities, and it refers to their placing among 35 mammal taxa. This is defined by what is known as informant consensus indexes. These indexes are defined as the degree of agreement among different interviewed persons when referring to a particular resource (Tardío and Pardo de Santayana 2008; Albuquerque et al. 2006). The most used indicator for this is frequency of mention (Alonso-Aguilar et al. 2014; Weller and Romney 1988). The elements that obtain the greatest frequency of mention in interviewed are assumed to be the most culturally significant for the studied population (Hilgert 2007; Thompson and Juan 2006). Both primates are found below the paca, locally known as "tepezcuintle" (*Cuniculus paca*); the peccary, known as "puerco de monte" (Family Tayassuidae); or the white-tailed deer, known as "venado cola blanca" (*Odocoileus virginianus*), in the significant species listings. These animals have a frequency of mention above 60%. All these species are conceived to be important because they are food and medicine (García del Valle et al. 2015).

While indexes based on frequency of mention have proved to be a precise tool to evaluate the level of cultural significance of these kinds of organisms, they do not provide much information of the reasons behind such an importance (Garibay-Orijel et al. 2007). As is mentioned above, Lacandons have used several species to solve their needs across their history, giving them a tangible value related to use. Furthermore, they have been appointed an intangible value by involving them in narrations that show the cosmovision of these people.

Currently, Lacandons from Naha and Metzabok consider both non-human primate species important because they are directly used in three cultural significance categories: as food (species that are used or have been used as food), as medicine (species that have useful ailment-combating properties), and as pets (species that are kept in the house, either in the yard in confinement or as company or decorative elements).

With regard to the "edible" category, both species of primates were consumed until about 2010. March (1987) points out that both *Alouatta pigra* and *Ateles geoffroyi* were the most hunted species for meat and that the craniums were occasionally kept as children's toys. In ancient times, the Lacandons hunted these animals with bow and arrows made from the wood of sapodilla, locally known as "chicozapote" (*Manilkara zapota*), and reed grass with points of serrated harpoon-shaped wood, which prevented the monkeys from removing them with their own hands. However, currently, Lacandons explain that these species are no longer consumed at all. This phenomenon is explained by different causes. In the words of the interviewed people, Lacandon women nowadays do not want to season or cook monkey meat because they associate them with small human children. Another reason is their preference of paca/"tepezcuintle" meat, which they find more agreeable and consequently favor above primate meat. Along with this, for many years now, the National Commission of Protected Natural Areas (Comisión Nacional de Áreas Naturales Protegidas, CONANP) together with other government offices in charge of environmental and indigenous development issues have implemented numerous programs aiming for the preservation of biodiversity and the eradication of poverty through monetary compensations in exchange for rainforest conservation activities, which in turn protect the associated fauna. Lastly, the growing tourism in Lacandon communities makes people strive to protect charismatic fauna and their natural environment, which are some of the main attractions for visitors who generate income. For all these reasons, and particularly the last two, monkey populations are not currently object to hunt. They are worth a lot more alive than they are as food.

Along with its former use as food, *Alouatta pigra* is also known for its medicinal properties to combat disease. Fifteen percent of the interviewed Lacandons stated that the meat of howling monkeys cooked as a soup is a great remedy for whooping cough, a highly contagious respiratory disease that is dangerous to these human populations. Furthermore, the hyoid bone is said to combat other respiratory diseases. Worldwide, many non-human primate species are hunted for medicinal products derived from them (Alves et al. 2010). For example, there are registers of the use of *Ateles geoffroyi* in the town of Catemaco, Veracruz state, Mexico, to cure rheumatism events by using the fried fat obtained from the monkey applied directly on the patient's body (Morales-Mávil and Villa-Cañedo 1998).

The last category of cultural significance is the use of primates as pets. Around 10% of the interviewed Lacandons mentioned this practice to be one of the reasons why these primate species, and particularly the spider monkey, are important. *Ateles geoffroyi* is the species most frequently used as pet because it is more tolerant to captive life. Contrastingly, *Alouatta pigra* is much harder to keep as a pet, since it has very high mortality rates in captivity. Duarte-Quiroga and Estrada (2003) reported that, in Mexico City, 67% of the monkeys used as pets are indeed *Ateles geoffroyi.* This may be due to the fact that spider monkeys can feed from the same fruit that humans consume unlike howler monkeys, which have a leaf-based diet.

Along with the tangible value non-human primate species hold for Lacandon people, these species also have an intangible value. This is clear by the role they have in Lacandon narrative that shows elements linked to the cosmovision of this people, that is to say, the logic and order of elements within their cosmos. With regard to this, the relationship between Lacandons and non-human primates has very particular features. Each species is profoundly related to the origins of people and their beliefs. Howling monkeys or *b'atz'* are related in origin to porcupines or *Kix pach* (*Sphiggurus mexicanus*). Lacandons conceive that, when this animal becomes old, it turns into a howler monkey. Spider monkeys or *ma'ax*, on the other hand, have a kinship, visualized in dreams, in which "one represents another." It is said that when Lacandon Maya dream of spider monkeys, it means they are part of their lineage or *onen*, and they consequently must carry their name. *Onen* is a term found exclusively among the Lacandons. According to Bruce (1975), the *onen* shares traits with totems as well as with *nagual* and *tonal*, although it also has singular features. Names in Naha still have their *onen* name along with their first names (e.g., a Lacandon name is *Bor Ma'ax*). Women only recognize belonging to the lineage but do not have their *onen* name. Currently, people from the *ma'ax* lineage are considered loud, aggressive, and even erotic. On the other hand, in Metzabok, both men and women recognize the lineage they belong to, but it has ceased to be commonly used.

One of the stories most told by Lacandons in Naha and Metzabok speaks of historical times when hunting was a common subsistence activity. Regulation of hunting was achieved through belief that if the number of hunted animals became excessive, the animal itself would come for the hunter and take him to be reprehended. So tells the following narration: "My dad told me that once upon a time a hunter went to kill a monkey, but his arrow missed and the monkey fled; the same thing happened with a peccary, by not casting the arrow the right way the animal fled and died in a cave. The man continued to hunt for monkeys and peccaries until suddenly he saw many monkeys coming towards him. They took him to live with them, he lived as a monkey, ate the same things they did, and they even provided him with females to reproduce with so that there would be more monkeys. The man did not know what to do. One day, they took him to gather fruit which stung his hand; he said it hurt. Then one of the monkeys took the thorn out and told him: 'see, you feel pain same as we do'. After a few more years living with the monkeys, the hunter was allowed to return to his family and he was asked to explain to them and to the rest of the people about his experience, so that monkeys would no longer be hunted." This narrative lets us realize that, regardless of hunt, regulations did exist to prevent an excessive use and hunt of monkeys among the Lacandon Maya.

3.5 Conclusions

With the information presented in this chapter, it let us conclude that species of non-human primates were considered a valuable edible resource in the diet of the Lacandonians not many years ago. In spite of traditional regulations expressed in narrations, field observation indicates there are few individuals of these species in Lacandon territory. However, the species ceased to be threatened because, for the Lacandon people, their significance in the generation of revenues for the community through different channels has become their most valuable feature. On the one hand, they get government subsidy programs for the conservation of ecosystems that helped the shift to stop hunting monkeys. On the other hand, monkeys bring charismatic status to tourists who visit these communities. Because of this, nowadays, as is stated by Harris (1989), these species have a much greater value for the benefits they provide when they are alive, than the ones they could provide once they are hunted and possibly gone forever. This is, without a doubt, a hopeful thought for the conservation of non-human primate species in the Lacandon rainforest.

Acknowledgments We thank the graduate program in Biological Sciences in Universidad Nacional Autónoma de México. We thank Marisa Ordaz Velázquez for her translation to English and style revision of the manuscript, as well as for the drawings of the figures we present. We thank Eduardo Ordaz Velázquez for the digitalization and edition of the images.

References

Albuquerque UP, Lucena RFP, Monteiro JM, Florentino ATN, Almeida C (2006) Evaluating two quantitative ethnobotanical techniques. Ethnobot Res Appl 4:51–60

Alonso-Aguilar LE, Montoya A, Kong A, Estrada-Torres A, Garibay-Orijel R (2014) The cultural significance of wild mushrooms in San Mateo Huexoyucan, Tlaxcala, Mexico. J Ethnobiol Ethnomed 10:27

Alves RRN, Souto WMS, Barboza RRD (2010) Primates in traditional folk medicine: a world overview. Mammal Rev 40(2):155–180

Baer P, Merrifield WR (1972) Los lacandones de México. Dos estudios. Instituto Nacional Indigenista, Ciudad de Mexico

Beliaev D (2011) Classic Lowland Maya (AD 250-900). In: Bondarenko DM, Korotayev AV (eds) Civilizational models of politogenesis, 2nd edn. Lambert Academic Publishing, Saarbrücken, pp 104–126

Berkes F (1999) Sacred ecology: traditional ecological knowledge and resource management. Taylor and Francis, Philadelphia
Berlin B, Breedlove D, Raven P (1973) General principles of classification and nomenclature in folk biology. Am Anthopol 75:214–242
Boremanse D (1984) Mitología y organización social entre los "lacandones" (*Hach winik*) de la selva chiapaneca. Estudios de Cultura Maya 15:225–249
Boremanse D (1986) Contes et mythologie des Indiens Lacandons. Contribution à l'étude de la tradition orale maya. Éditions L'Harmattan, Paris
Braakhuis HEM (2009) The tonsured maize god and Chicome-Xochitl as maize bringers and cultural heroes: a Gulf Coast perspective. Wayeb Notes 32:1–38
Bruce R (1975) Lacandon dream symbolism. Ediciones Euroamericanas, Ciudad de México
Bruner E, Cucina A (2005) Alouatta, Ateles, and the ancient Mesoamerican cultures. J Anthropol Sci 83:111–117
Calvin I (1997) Where the *wayob* live: a further examination of Classic Maya supernatural. In: Kerr J, Kerr B (eds) The Maya vase book volume 5. A corpus of rollout photographs of Maya vases. Kerr Associates, New York, pp 868–883
Cano EJ, Eroza E, Mariaca R (2009) *Tu chien k'an*: un recorrido por la cosmovisión de los lacandones del Norte desde las mordeduras de serpiente. Universidad Intercultural de Chiapas, Sociedad Latinoamericana de Etnobiología, El Colegio de la Frontera Sur, San Cristóbal de Las Casas, Chiapas
Castillo-Campos G, Narave H (1992) Contribución al conocimiento de la vegetación de la Reserva de la Biosfera Montes Azules, Selva Lacandona, Chiapas, México. In: Vázquez-Sánchez MA, Ramos MA (eds) Reserva de la Biosfera Montes Azules, Selva Lacandona: Investigación para su conservación. Publ Esp Ecosfera, San Cristóbal de Las Casas, pp 51–85
Ceballos G, Ehrlich PR (2002) Mammal population losses and the extinction crisis. Science 296:904–907
Coe MD (1973) The Maya scribe and his world. Club Grolier, New York
CONANP (2006) Programa de conservación y manejo Área de Protección de Flora y Fauna Nahá. Comisión Nacional de Áreas Naturales Protegidas, México D.F.
Craveri ME (2013) Popol Vuh. Herramientas para una lectura crítica del texto k'iche. Instituto de Investigaciones Filológicas, Universidad Nacional Autónoma de México, Ciudad de México
De Vos J (1980) La Paz de Dios y del Rey. La conquista de la Selva Lacandona (1525–1821). Gobierno del Estado de Chiapas, Tuxtla Gutiérrez, Chiapas
Di Fiore A, Link A, Dew JL (2008) Diets of wild spider monkeys. In: Campbell CJ (ed) Spider monkeys: behavior, ecology and evolution of the genus *Ateles*. Cambridge University Press, Cambridge, pp 81–137
Dirzo R (1992) Diversidad florística y estado de conservación de las selvas tropicales de México. In: Sarukhán J, Dirzo R (eds) México ante los retos de la biodiversidad. Conabio, Ciudad de México, pp 283–290
Duarte- Quiroga A, Estrada A (2003) Primates as pets in Mexico city: an assessment of the species involved, source of origin and general aspects of treatment. Am J Primatol 61:53–60
Eroza E (2006) Lacandones. Comisión Nacional para el Desarrollo de los Pueblos Indígenas, Ciudad de México
Estrela AR (2009) Etnoprimatología y su aplicación en los planes de conservación de la Caatinga brasilera. In: Costa-Neto E, Santos-Fita D, Vargas-Clavijo M (eds) Manual de Etnozoología. Tundra, Valencia, pp 242–249
Freidel D, Schele L, Parker J (1999) El cosmos maya. Tres mil años por la senda de los chamanes. Fondo de Cultura Económica, Ciudad de México
García del Valle Y, Naranjo EJ, Caballero J, Martorell C, Ruan-Soto F, Enríquez PL (2015) Cultural significance of wild mammals in mayan and mestizo communities of the Lacandon rainforest, Chiapas, Mexico. J Ethnobiol Ethnomed 11:36
García-Gil JG, Lugo J (1992) Las formas del relieve y los tipos de vegetación en la selva lacandona. In: Vázquez-Sánchez MA, Ramos MA (eds) Reserva de la Biosfera Montes Azules,

Selva Lacandona: Investigación para su conservación. San Cristóbal de Las Casas, Publ Esp Ecosfera, pp 51–85
Garibay-Orijel R, Caballero J, Estrada-Torres A, Cifuentes J (2007) Understanding cultural significance, the edible mushrooms case. J Ethnobiol Ethnomed 3(4):1–18
Grube N, Nahm W (1994) A census of Xibalba: a complete inventory of way characters on Maya ceramics. In: Kerr J, Kerr B (eds) The Maya vase book, volume 4. A corpus of rollout photographs of Maya vases. Allen Press Inc, New York, pp 683–715
Guerrero F (2015) Concepciones sobre los animales en grupos mayas contemporáneos. Pueblos y Fronteras 10(20):6–43
Halperin C (2014) Maya figurines. Intersections between state and household. University of Texas Press, Austin
Harris M (1989) Bueno para comer. Editorial alianza, Ciudad de México
Helmke C, Nielsen J (2009) Hidden identity and power in ancient Mesoamerica: supernatural Alter Egos as personified diseases. Acta Am 17(2):49–98
Hilgert N (2007) La Etnobotánica como herramienta para el estudio de los sistemas de clasificación tradicionales. In: Contreras-Ramos A, Cuevas Cardona C, Goyenenchea I, Iturbe U (eds) La Sistemática, base para el conocimiento de la biodiversidad. Universidad Autónoma del Estado de Hidalgo, Pachuca, pp 103–112
Houston S, Stuart D (1989) The *way* glyph: evidence for "co-essences" among the classic Maya. Research reports on ancient Maya writing 30. Center for Maya Research, Washington, DC
INEGI (2010) Censo de población y vivienda. INEGI, Aguascalientes
Jackson S, Stuart D (2001) The Aj K'uhun title: deciphering a classic Maya term of rank. Anc Mesoam 12(2):217–228
Kerr J (1997) The Maya vase book, volume 5. A corpus of rollout photographs of Maya vases. Kerr Associates, New York
Kerr J (1998) The Maya vase book, volume 1. A corpus of rollout photographs of Maya vases. Kerr Associates, electronic edition, New York
Kidder BB (2009) Maya scribes who would be kings: shamanism, the underworld, and artistic production in the late classic period. Master of Arts dissertation. Texas State University, San Marcos
Lacadena A (1996) A new proposal for the transcription of the a-k'u-na/a-k'u-HUN-na title. Mayab (10):46–49
Lee PC, Priston NE (2005) Human attitudes to primates: perceptions of pests, conflict and consequences for primate conservation. In: Paterson JD, Wallis J (eds) Commensalism and conflict: the human-primate interface. American Society of Primatology, Norman, pp 1–23
Lozada J, Núñez R (2014) Representaciones rupestres en la Laguna de Metzabok, Chiapas. Del trabajo arqueológico a la indagación etnográfica. In: Arroyo B, Méndez L, Rojas A (eds) XXVII Simposio de Investigaciones Arqueológicas en Guatemala, vol 2013. Museo Nacional de Arqueología y Etnología, Guatemala, pp 505–516
Luin C, Matteo S (2010) Notas sobre algunos textos jeroglíficos en colecciones privadas. In: Arroyo B, Linares A, Paiz L (eds) XXIII Simposio de Investigaciones Arqueológicas en Guatemala, vol 2009. Ciudad de Guatemala, Museo Nacional de Arqueología y Etnología, pp 1217–1231
March I (1987) Los lacandones de México y su relación con los mamíferos silvestres: un estudio etnozoologico. Biótica 12(1):43–56
Martin S (2006) Cacao in ancient Maya religion: first fruit from the maize tree and other tales from the underworld. In: McNeil C (ed) Chocolate in Mesoamerica. A cultural history of Cacao. University Press of Florida, Gainesville, pp 154–183
Morales A (2001) La creación de imágenes en la cultura maya. Estudios Mesoamericanos (3–4):111–119
Morales-Mávil J, Villa Cañedo J (1998) Notas sobre el uso de la fauna silvestre en Catemaco, Veracruz, México. Acta Zoológica Mexicana (nueva serie) 73:127–143
Moreno D (2011) Los espíritus del sueño. *Wahyis* y enfermedad entre los mayas del periodo Clásico. Bachelor's dissertation. Escuela Nacional de Antropología e Historia, Ciudad de México

Nájera MI (2000) Cambios y permanencias en la religión maya a través del análisis del significado de la figura simbólica del mono. Estudios Mesoamericanos (2):49–56

Nájera MI (2012) El mono y el cacao: la búsqueda de un mito a través de los relieves del Grupo de la Serie Inicial de Chichén Itzá. Estudios de Cultura Maya 39:133–172

Nájera MI (2013) Un acercamiento al simbolismo del simio entre los grupos mayas. In: Millones L, López A (eds) Fauna fantástica de Mesoamérica y los Andes. Instituto de Investigaciones Antropológicas, Universidad Nacional Autónoma de México, Ciuadad de México, pp 211–252

Ojasti J (1993) Utilización de la fauna silvestre en América Latina, situación y perspectiva para un manejo sostenible. Guía FAO Conservación N° 25, Roma

Palka J (2005) Arte rupestre indígena y lugares sagrados Mayas Lacandones en las Tierras Bajas de Chiapas. Bolom 2:27–38

Pennington TD, Sarukhan J (1998) Árboles tropicales de México. Universidad Nacional Autónoma de México, Fondo de Cultura Económica, Ciudad de México

Pérez-Gil R, Jaramillo F, Muñiz A, Torres M (1995) Importancia económica de los vertebrados silvestres de México. Editorial Marsa, Ciudad de México

Pincemin S (1999) De manos y soles. Estudio de la gráfica rupestre en Chiapas. Universidad de Ciencias y Artes del Estado de Chiapas, Tuxtla Gutiérrez

Quenon M, Le Fort G (1997) Rebirth and resurrection in maize god iconography. In: Kerr J, Kerr B (eds) The Maya vase book, volume 5. A corpus of rollout photographs of Maya vases. Kerr Associates, New York, pp 884–902

Quintana-Ascencio P, Ramírez-Marcial N, González-Espinosa M (1990) El medio natural de la región de Bonampak, Selva Lacandona, Chiapas. CIES, San Cristóbal de las Casas, Chiapas

Recinos A (1953) Popol Vuh. Las antiguas historias del Quiché. Fondo de Cultura Económica, Ciudad de México

Reents-Budet DJ, Ball JW, Bishop RL, Fields VM, MacLeod B (1994) Painting the Maya universe: Royal ceramics of the classic period. Duke University Press, London

Rice PM, South KE (2015) Revisiting monkeys on pots: a contextual consideration of primate imagery on classic Lowland Maya Pottery. Anc Mesoam 26(2):274–294

Robiscek F, Hales DM (1981) The Maya book of the dead. The ceramic codex. The corpus of codex style ceramics of the late classic period. University of Virginia Art Museum, Charlottesville

Rocha LC, Fortes V (2015) Perceptions and attitudes of rural residents towards capuchin monkeys, in the area of influence of the Dona Francisca hydroelectric power plant, South Brazil. Ambiente & Sociedade 18(4):19–34

Roosmalen MGM (1985) Habitat preferences, diet, feeding strategy and social organization of the black spider monkey (*Ateles paniscus paniscus* Linnaeus 1758) in Surinam. Acta Amazon 15(3–4):6–238

Roosmalen MGM, Klein LL (1988) The spider monkeys, genus *Ateles*. In: Mittermeier RA, Rylands AB, Coimbra-Filho AF, Fonseca GAB (eds) Ecology and behaviour of neotropical primates, volume II. World Wildlife Fund, Washington, DC, pp 455–537

Ruan-Soto F, Mariaca R, Cifuentes J, Limón F, Pérez-Ramírez L, Sierra-Galván S (2007) Nomenclatura, clasificación y percepciones locales acerca de los hongos en dos comunidades de la selva lacandona, Chiapas, México. Etnobiologia 5:1–20

Ruan-Soto F, Caballero J, Martorell C, Cifuentes J, González-Esquinca AR, Garibay-Orijel R (2013) Evaluation of the degree of mycophilia-mycophobia among Highland and lowland inhabitants from Chiapas, Mexico. J Ethnobiol Ethnomed 9:36

Sam LE (2008) Popol Wuj. Cholsamaj, Ciudad de Guatemala

Sánchez F (2005) Arte Rupestre de Metzabok. Una descripción preliminar. Bolom 2:61–89

Santos-Fita D, Costa-Neto E, Cano-Contreras E (2009) El quehacer de la etnozoología. In: Costa-Neto E, Santos-Fita D, Vargas-Clavijo M (eds) Manual de Etnozoología. Tundra, Valencia

Shaw M (1972) Según nuestros antepasados... textos folklóricos de Guatemala y Honduras. Instituto Lingüístico de Verano en Centro América, Ciudad de Guatemala

Sheseña A (2010) Los nombres de los naguales en la escritura jeroglífica maya: Religión y lingüística a través de la onomástica. J Mesoam Lang Linguist 2:1):1–1)30

Tardío J, Pardo de Santayana M (2008) Cultural important indices: a comparative analysis based on the useful wild plants of southern Cantabria (northern Spain). Econo Bot 62(1):24–39

Tedlock D (1993) Popol Vuh. El libro maya del albor de la vida y las glorias de dioses y reyes. Editorial Diana, Ciudad de México

Thompson E, Juan Z (2006) Comparative cultural salience: measuring using free list data. Field Methods 18(4):398–412

Thompson C, Merino G, Camacho G (2005) La exploración de las cuevas de la laguna de Metzaboc. Bolom 2:41–59

Toledo V, Ortiz-Espejel B, Cortés L, Moguel P, Ordoñez M (2003) The multiple use of tropical forests by indigenous peoples in Mexico: a case of adaptive management. Conserv Ecol 7(3):9

Tozzer A (1907) A comparative study of the Mayas and the Lacandones. Archaeological Institute of America, New York

Turner NJ (1988) The importance of a rose: evaluating the cultural significance of plants in Thompson and Lillooet interior Salish. Am Anthropol 90:272–290

Velásquez E (2009) Los vasos de la entidad política de 'Ik': una aproximación histórico-artística. Estudios sobre las entidades anímicas y el lenguaje gestual y corporal en el arte maya clásico. Ph.D. dissertation. Universidad Nacional Autónoma de México, Ciudad de México

Weller SC, Romney AK (1988) Systematic data collection. Sage Publications, Newbury Park

Zender M (2004) On the morphology of intimate possession in Maya languages and classic Mayan glyphic nouns. In: Wichmann S (ed) The linguistics of Maya writing. The University of Utah Press, Salt Lake City, pp 195–209

Chapter 4
Representation and Signification of Primates in Maya-Q'eqchi' Cosmovision and Implications for Their Conservation in Northwestern Guatemala

Marleny Rosales-Meda and María Susana Hermes

4.1 Introduction

Indigenous peoples from different cultures worldwide have established historical relationships with primates that have special representations and meanings (Ellwanger et al. 2015; Fuentes 2012; Riley 2010; Lee and Priston 2005). The pre-Columbian Mayan civilization reached its greatest development and splendor in the territory of Guatemala and southeastern Mexico (Sharer 2003). Currently, most Guatemalans descend from Mayans. There are 22 Mayan ethnic groups that share a similar cosmovision, with nature as the main cornerstone (García et al. 2009; UNESCO 1996). Mayan cosmovision is the set of values, beliefs, and ways of analyzing and interpreting the reality and the universe from the existence and collective experience of the ancient and current Mayan people (Matul and Cabrera 2007a). All the elements that exist in nature and the universe have life in the Mayan cosmology that are intrinsically linked and complement each other (García et al. 2009). Each natural element (water, air, soil, plants, animals) is integrated and represented in the cosmovision, practices, customs, and everyday life of Mayan people (Matul and Cabrera 2007a, b). Primates, in particular, have acquired special meanings and important roles that have prevailed over time (Baker 2013).

The pre-Hispanic Mayan civilization lived alongside primates of three taxonomic genera (*Alouatta palliata*, *Alouatta pigra*, *Ateles geoffroyi*, *Cebus capucinus*), and their close links are widely represented in archaeological records and ancient written and graphical sources (Rice and South 2015; Baker 2013; Fuentes 2012). The *Popol Wuj* ("Council Book" or "Community Book" in Maya-K'iche' language) is the most important ancestral Mayan book. According to Sam (2008), its narratives make a transition between the mythology and history of Mayan people

M. Rosales-Meda (✉) · M. S. Hermes
Organización para la Conservación de la Naturaleza y Desarrollo Comunitario (ORCONDECO), Alta Verapaz, Guatemala
e-mail: mrosalesmeda@orcondeco.org

B. Urbani, M. Lizarralde (eds.), *Neotropical Ethnoprimatology*, Ethnobiology,
https://doi.org/10.1007/978-3-030-27504-4_4

that explains the origin of the world, civilizations, and several natural phenomena, where primates have special roles and meanings as human ancestors. The manuscript shares how earth first appeared between the waters, and then, the Creators felt the need to be called by their names and be worshipped. For this, they created animals that populated all mountains and valleys. However, animals could not speak to name and adore them, so they proceeded to create people to fulfill this purpose. There were three instances of human creations (made from clay, wood, or corn), but only the last, made from corn, could acknowledge and worship their Creators (Sam 2008).

During the second creation, the one that would have given birth to "wood humans," the resemblance between monkeys and people is explained by stating that primates are thought to be the ancestors or "elder brothers" of "corn humans": "*Jun Junajpu* fathered two sons: *Jun B'atz'* and *Jun Chowen.* They were great sages, they had great knowledge, they were fortune-tellers on the face of the earth, of good character and manners. They taught the arts of being flutists, singers, writers, painters, sculptors of jade and silver…" (Sam 2008).

In this narrative, the relationship of the brothers' names with primates stands out: *Jun* means one or first; *B'atz'* means howler monkey in almost all Mayan languages, and *Chowen* also means monkey, as well as artist and craftsman but in Ch'ol Mayan or Yucateco language (Akkeren 2012; Sam 2008). The origin and special meaning of primates in the Mayan cosmovision from ancestral to modern times are that of "the fathers" of all arts and are widely represented in dances, music, paintings, and writings, among others (Fig. 4.1a, c).

Many archaeological records also show that primates were important characters within Mayan mythology and society. Epigraphists and iconographers claim that howler monkeys predominate in the portrayals of monkey-man scribes in Late Classic funerary pottery that represent *Jun B'atz'* and *Jun Chowen* (Rice and South 2015). Numerous examples of the representation of monkeys also appear in altars and carvings of bone, shells, and jade. Several characters with primate features are represented in pots as artists that carve, write, or paint. Their features resemble spider monkeys and howler monkeys. Figure 4.2 shows different pre-Hispanic archaeological pieces (vessels, figurines, axe, and stamp with carved or engraved drawings) that have primates as their main image.

The *Popol Wuj* also tells how *Jun B'atz'* and *Jun Chowen* are human relatives who became monkeys; in the beginning, they were created as people, but their laziness, disobedience, and envy toward their other twin brothers (*Junajpu* and *Ixbalamke*) led to their conversion to monkeys or "habitants of the trees": "…Go to catch birds -said *Junajpu and Ixbalamke*- and when *Jun B'atz' and Jun Chowen* climbed the tree, it began to grow and they could no longer descend", "…their brothers shouted: Untie the girdle of your pants, bind your lower belly and stretch them as if they were tails behind you! And when they did, the girdles became tails and they took on the appearance of spider monkeys. Immediately they went over trees and into the woods screaming and swinging among the branches" (Sam 2008). Then the twins' grandmother tried to make them human again, but when they returned and listened to music, they began dancing; this caused the grandmother to laugh out loud, but it scared *Jun B'atz'* and *Jun Chowen* and made them return to the

Fig. 4.1 (**a**) Giant kite showing an indigenous girl hugging a traditional monkey mask in Sumpango, Sacatepequez, Guatemala. These kites are specially made to fly during the "Day of the Dead" traditional celebration and honor communication between modern Mayan people and their ancestors. (**b**) Monkey character performing the "Dance of the Spider Monkeys" during a special Q'eqchi' Mayan festivity in Coban, Alta Verapaz, Guatemala. (**c**) Two traditional primate garments used in Q'eqchi' Mayan dances, which remind and honor brothers *Jun B'atz'* and *Jun Chowen* from the *Popol Wuj*. (**d**) Primate dancer characters during performance of *Xajleb' Kej* (Dance of the Deer) at a Q'eqchi' Mayan community

forests where they still remain. This narrative explicitly expresses the close kinship relationship between humans and primates in Mayan cosmovision, because people made of wood became monkeys (*k'oy*), and their descendants are howler and spider monkeys that currently live in forests (Fig. 4.3).

Generally, primate conservation efforts worldwide have focused on carrying out demographic, ethological, and ecological research without considering the beliefs, perceptions, tolerance, and use that local people have toward these species (Burton and Carroll 2005; Lee and Priston 2005). Guatemala is no exception, and primate conservation has focused on strict protection efforts inside legally protected areas without considering the sociocultural context in which they are immersed. There are no studies about the perception of primates among present-day Maya or on their relationships with primates that also address the implications and contributions to

Fig. 4.2 (**a**) Pre-Columbian Mayan vessels and figurine (**b**) with spider monkey representations, Late Classic Period (600–900 CE), Petén, Guatemala. (**c**) Basalt image of spider monkey sitting on a bench and primate-shaped axe (**d**) used in traditional rituals associated with the Mayan ball game, Classic Period (250–900 CE), Central Highlands of Guatemala. (**e**) Zoomorphic primate stamp used for impressions on surfaces, body decoration, commerce, and social distinction symbols, Classic Period (250–900 CE), Central Highlands of Guatemala (La Ruta Maya Conservation Foundation 2013a, b)

Fig. 4.3 Two adult spider monkey males (*Ateles geoffroyi vellerosus*) resting at the Temple of the Great Jaguar in the Mayan city of Tikal, Petén, Guatemala

the conservation of these species and their habitat. However, some researchers have recently documented that human communities living in close proximity to primates establish strong biological, ecological, and cultural links with them (Kansky et al. 2016; Ellwanger et al. 2015; Campbell-Smith et al. 2010; Nijman and Nekaris 2010). These studies also show that people's beliefs and attitudes toward primates have a considerable impact (positive or negative) on their survival and conservation (Rocha and Fortes 2015; Costa et al. 2013; Nungshi and Radhakrishna 2013; Khatun et al. 2012; Chauhan and Pirta 2010; Hill and Webber 2010).

Currently, three species of nonhuman primates (*Alouatta pigra*, *Alouatta palliata palliata*, *Ateles geoffroyi vellerosus*) coexist in protected areas and territories of many Mayan communities in northern Guatemala (Rosales-Meda and Hermes 2018; CONAP 2008). Most of these communities are Q'eqchi' Mayan, and their territory overlaps with the most biodiversity-rich and high conservation priority tropical rainforests of the country. This chapter addresses the representation and signification of primates in Q'eqchi' Mayan culture and cosmovision and its positive implications for the conservation of *A. pigra* and *A. geoffroyi vellerosus* in northwestern Guatemala.

4.2 Study Area

The Ancestral Rainforest Landscape (ARL) is comprised of tropical forests, mountains, wetlands, agroecosystems, and human communities (mainly Q'eqchi' Mayan) of four departments of Guatemala (northern portions of the departments of Quiche and Alta Verapaz, southern portion of Peten, and the whole department of Izabal) (Rosales-Meda and Hermes 2018; Fig. 4.4). At a biogeographical level, this region

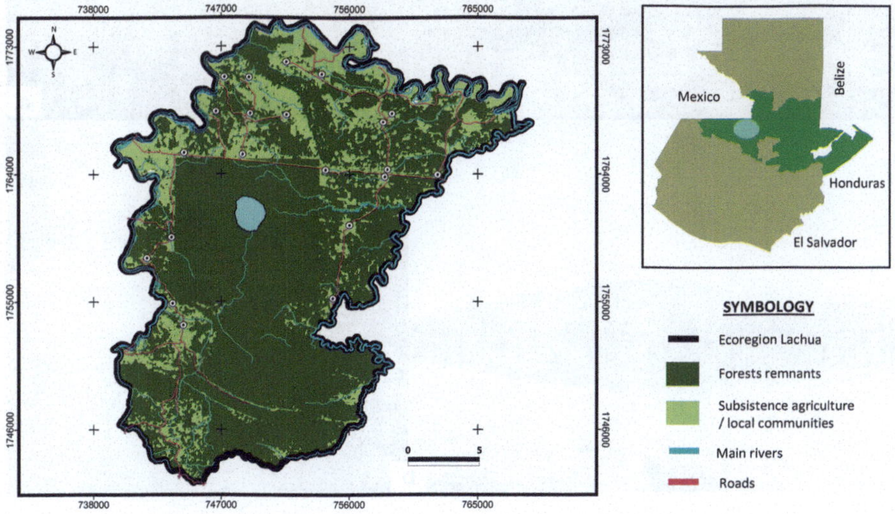

Fig. 4.4 Geographic location of the Ramsar Wetland Ecoregion Lachua in the context of the Ancestral Rainforest Landscape, northwestern Guatemala

played a crucial role as a Pleistocene refuge to maintain the country's existing biodiversity and now has one of the highest levels of endemism and species diversity (Rosales-Meda et al. 2010; Gerald and Leyden 2007; Knapp and Davidse 2007). Within the Q'eqchi' Mayan cultural-spiritual level, the distinctive hills and valleys of this landscape have a fundamental representation within their current and historical cosmovision, because their *Tzuul Taq'a* supreme deity ("hill-valley") lives and is honored in these forested mountains (Haste and De Ceuster 2001; Fig. 4.5a). Historically, the region was also inhabited by the ancient Mayan civilization, and it is common to find archaeological sites, mounds, clay pots, and other ceramic pieces from its legacy, as well as distinctive family surnames of Mayan ancestral lineage in local communities (Akkeren 2012; Rosales-Meda 2003).

The Ecoregion Lachua (535 km^2) is a Ramsar wetland located in the ARL's northwest and has been our location of work since 2002. This high-biodiversity landscape includes the Laguna Lachua National Park (143 km^2), 49 Q'eqchi' Mayan communities, and indigenous forest reserves (INAB/UICN 2010; Fig. 4.4). The region was heavily affected by Guatemala's 36-year civil war, and most local people

Fig. 4.5 (**a**) Panoramic view of sacred highlands and valleys of the Ancestral Rainforest Landscape, Guatemala. (**b**) Forested mountains and wetlands in Ecoregion Lachua, Guatemala

are victims who survived this conflict and their descendants. Despite this war, communities have resisted the historical processes of colonization and war by keeping their language, culture, and traditions alive (Haste and De Ceuster 2001). Ancestral Mayan authorities are being reconstituted after a long period of postwar absence, and elders are again occupying political positions, as they did in the past. The community economy is based on subsistence agriculture (mainly corn, *Zea mays*, and beans, *Phaseolus vulgaris*), the production of forest-shaded cardamom (*Elettaria cardamomum*), and cacao (*Theobroma cacao*) crops and seasonal fruits (INAB/UICN 2010). These activities are complemented by the use and consumption of some timber and non-timber forest products (Rosales-Meda and Hermes 2010).

Recently, the construction of mega highways, agroindustry expansion, and proliferation of evangelical churches have generated a heterogeneous and complex socioeconomic context. However, many important traditional elements persist in the representation and appropriation of their territory. These are expressed, claimed, and safeguarded through sociocultural practices and interactions with nature (Rosales-Meda and Hermes 2010; Haste and De Ceuster 2001). Nature plays an essential role in guiding individual and community life, because the conception and way of understanding the world and the divine is intimately related to all of nature's elements (Rosales-Meda and Hermes 2013). Throughout the landscape, there are many mountains, caves, primary and secondary forest remnants, rivers, and lagoons that are sacred sites for local communities (Fig. 4.5a, b). In them, distinctive ceremonies, rituals, and festivities of Q'eqchi' Mayan cosmovision are frequently practiced in which each representative natural element is acknowledged, thanked, and honored. These traditional spiritual practices also constitute powerful symbols of cultural resistance, social unity, and ancestral communion. Some ceremonies include a certain degree of syncretism with the catholic religion, but the influence of evangelical religion is increasing and imposing prohibitions to these important cultural and spiritual practices. Transculturation, youth migration to cities and the United States, and megaprojects (oil palm, livestock, oil extraction, large hydroelectric dams) also represent strong threats for traditional livelihoods, culture, and cosmovision of local communities and their biodiversity.

4.3 Primates and Our Participatory Conservation Work

In 2002, we began demographic, ecological, behavioral, and ethnobiological investigations of both threatened primates that coexist with Q'eqchi' Mayan communities in the Ecoregion Lachua landscape: *Alouatta pigra* (black howler monkey, Mayan monkey, saraguate, *B'atz'* in local language) and *Ateles geoffroyi vellerosus* (spider monkey, mico, *Max* in local language) (Hermes and Rosales-Meda 2011; Rosales-Meda et al. 2008; Rosales-Meda 2003). The black howler monkey is an endemic species of northern Guatemala, Belize, and the Yucatan Peninsula in Mexico. Individuals have large bodies and a mainly folivorous diet, which causes their metabolism and movement to be slow and sporadic. They also have an anatomical peculiarity, an elongated hyoid bone, that acts as a voice resonator and amplifier, allowing them to emit strong and characteristic vocalizations to communicate (Mittermeier 2013; Fig. 4.7a). The spider monkey is a more sensitive and vulnerable species because its diet consists almost exclusively of fruit. They have thin bodies with a bulging abdomen and long limbs and a fast metabolism. They brachiate moving with great agility and grace traveling great distances through the forest in search of food (Mittermeier 2013; Fig. 4.8a).

As part of our long-term work, we have carried out 231 semi-structured and mutually agreed-upon interviews to deeply understand local people's relationships with primates, their ecological-cultural links, perceptions, attitudes, and their implications for conservation. Interviews were conducted during three different periods: 2002–2005 (n = 104), 2010–2012 (n = 45), and 2013–2016 (n = 82). Interviews considered "Elements of the Code of Ethics and Guidelines for Anthropological Research" (Laird and Posey 2002), which are based on the principles of research ethics, equity, respect, and prior informed consent and guided by participatory research-action processes (Ander-Egg 2003). In Q'eqchi' Mayan culture, men have more knowledge about forest and wildlife, because they work in the field, plant, hunt, and make decisions regarding the use and management of land. For this reason, interviewees were adult men (ages 24–87) from 27 local communities, including farmers, park rangers, elders, community leaders, ancestral authorities, and spiritual guides.

During these 15 years, we have also shared casual conversations with our Q'eqchi' Mayan neighbors and friends, participated in traditional ceremonies and celebrations, and held private consultations about pre-Hispanic Mayan history. This has allowed us to obtain a more holistic appreciation of what primates mean and represent to the communities with whom we live and work. We have overlapped and complemented this knowledge with results of our long-term ecological research with primates to evaluate how the population distribution and conservation status of both species is related to the perception, culture, and cosmovision of Q'eqchi' Mayan communities.

4.4 Representation and Signification of Primates in Maya-Q'eqchi' Culture and Cosmovision

4.4.1 *Origin of Primates*

To explain the origin of primates, 89% of people interviewed (n = 206) quoted the *Popol Wuj*. Their narratives had some variants in the names of the protagonists and/or characters, because they were partially forgotten or mixed by previous generations by the influence of Spanish culture, war, or religions. All had a strong link with Mayan cosmovision recognizing primates as ancestors and close relatives. Most interviewees agreed that primates are the older siblings of people and the first men. Primates are also seen as reminder for people to show good behaviors in life so as to not receive punishment by the Creator and Maker (*Tzuul Taq'a*, God, *Yhave*, or other names that are now conferred to the universal Creator energy) (Cao, Macz, Pop, Seel, Xe and Xo pers comm. 2012).

4.4.2 Family Lineages, Surnames, and Proper Names

Currently, surnames such as *B'atz'* and *Chowen* (and some of their variants) are maintained in Q'eqchi' Mayan families from the north, center, and southeast regions of Alta Verapaz. Both surnames come from ancient Mayan royal lineages that have a long history and relationship with trade and wealth since pre-Hispanic times and are also remembered as lineages of influential families in early colonial times (Akkeren 2012). The meaning and importance of the surname *Chowen* are explained by analyzing the name of Cancuen archaeological site located in northern Alta Verapaz. This was a center of craftsmen who carved jade, and it is believed that its name derives from *Kan Chowen* (*kan* means sky) or "monkey of the sky." To this day, it is common to find archaeological pieces with representations of spider and howler monkeys in these regions (Akkeren 2012). In our study area and neighboring regions, the surname *Max* inspired by spider monkeys is also common. Likewise, families respectful of Q'eqchi' Mayan cosmovision give their sons the name *B'atz'* in honor of primates and the *nahual* that represents them in nature.

4.4.3 Sacred Mayan Calendar and the Nahual B'atz'

Q'eqchi' Mayan elders, spiritual guides, and ancestral authorities told us about the special spiritual representation and signification of primates by making deep connections with their cosmovision to explain how they are present in the sacred energies and cycles of nature that guide people's lives at individual, family, and community levels.

The Mayan calendar called *Cholq'ij* (sacred calendar) is used by Mayan people to keep track of time, making a close link between human life and nature (Sac 2007). This cyclical calendar has 260 days with 13 periods of 20 days (*nahuales*, which means "energies" in Q'eqchi' Mayan) each, which indicate what type of energy prevails in each day and what actions are favorable to focus on and carry out with success (García et al. 2009). The *nahual B'atz'* marks the beginning of the 20-day count and is the energy that represents and honors primates in general (Barrios 2004). Elders explain that the energy of this day favors the beginning of any process or plan to be successful and enduring. "As spiritual guides, on a B'atz' day, we perform ceremonies so that a marriage, the construction of a house, or a new business is good and long-lasting…also when assuming the position of being a spiritual guide or an authority, we seek this special energy to do it in the best way possible…" (Chocooj pers. comm.)." "…Because the *B'atz'* day is the first, is the oldest

Fig. 4.6 (**a**) Traditional Maya-Q'eqchi' ceremony called *Mayejak* where each of the 20 *nahuales* or *sacred energies* is acknowledged, thanked, besought, and honored. (**b**) Young self-taught Q'eqchi' Mayan painter born with the energy of *nahual B'atz'* sharing the process of creating his artwork inspired on the region's nature and wildlife. (**c**) Q'eqchi' Mayan elder and artist sharing an ancestral monkey story while showing a monkey's head made from coconut

(referring to the elder brothers mentioned in the *Popol Wuj)*, is the beginning, the authority, and command…on this day, we encounter the best harmony between the positive and negative…" (Xo pers. comm.). "…On the *B'atz'* day, we can also ask that our family or community problems are solved with good communication, for our crop's well-being, and the protection of all artists" (Toc pers. comm.; Fig. 4.6a). "In Q'eqchi' Mayan cosmovision, monkeys are our strength, our support, the common well-being…in a ceremony, their spirit visits us through the sacred fire and express for good or bad the things we think, feel, or do in our lives; that is why, in our days, we have a great responsibility to take care of them and to thank for the life of monkeys because of their broad meaning in our lives; they are our grandfathers and grandmothers" (Chocooj pers. comm.). Elders also shared that "through the birth date of each person, their talents and abilities can be known" (Xo pers. comm.). People born in the *B'atz'* day and *energy* have special talents and skills as "communicators and artists; they stand out in music, singing, dancing, painting, sculpture, literature, pottery, and/or weaving; they can also be good birth attendants, inventors, and nature lovers" (Toc pers. comm.; Fig. 4.6b).

4.4.4 Art

On commemorative days, celebrations, and festivities, local communities perform and enjoy the most common and important ceremonial dance of Q'eqchi' Mayan culture which is called *Xajleb' Kej* (deer dance). This traditional dance has been known since pre-Hispanic times and is currently presented with some adaptations about the Spanish invasion. This dance is very popular and revalorizes the harmonious relationship between people and wild animals. Spider monkeys have a special role recognized as playful characters that cause joy among observers and hark back to the story of the twin brothers' conversion into primates (García et al. 2009; Fig. 4.1d). In Alta Verapaz, another traditional dance called "Dance of the Monkeys" ("Danza de los monos y los micos" in Spanish) is performed at the beginning of August (Fig. 4.1b). This dance has been documented since 1872 in the town of San Pedro Carchá, where the surname *Chowen* is very common (Akkeren 2012). In our study area, artists also often draw, carve, and sculpt images with primate representations for special occasions and decoration (Fig. 4.6c).

4.4.5 Language

The word *b'atz'* is embedded as a code in the verbal communication of Q'eqchi' Mayan communities from the ARL. It is a keyword of everyday language whose symbolism and meaning is directly related to playing, fun, joy, and creativity, which is of high relevance for community interaction, social cohesion, and involvement. Examples of some words of daily and frequent use that derive from it are ***b'atz'**uul* (toy), ***b'atz'**unel* (player), ***b'atz'**unk* (play), ***b'atz'**unlenk* (play around), and ***b'atz'**unleb'aal* (place to play), among others (Misti and Xol pers. comm.).

4.4.6 Uses and Attitudes

Most people have very positive perceptions and attitudes toward both primates and consider them good animals that provide direct benefits to their lives and lands. In the interviews, 94% ($n = 217$) of the people expressed that howler monkeys "called the rain" (meaning that they make rain fall) with their vocalizations. Elders and spiritual guides explain that "*b'atz'* are very special animals because they can see when our lands are too dry and begin to pray to Tzuul Taq'a; then the rain falls which is good for us all..." (Jucub pers. comm.). "Howlers are important because they have the right word to invoke the Creator and Maker for rain, and their requests are answered; they are our witnesses before *Tzuul Taq'a.* That is why, we must be grateful for their existence, and it is our responsibility as humans to respect them so that they always live" (Toc pers. comm.). These wise phrases summarize and reflect most interviewees' perception of howler monkeys as communicators and intermediaries needed for *Tzuul Taq'a* to send rain to their lands, because in many occasions he does not listen to people due to their faults, but monkeys are indeed heard, and thanks to them, people can receive the benefits of rainfall in their crops (Fig. 4.7c).

Other uses and benefits also frequently reported for howlers are they inform the hours of day through their vocalizations, they "decorate" (embellish) their lands, they help in pruning and sprouting of trees that provide shade to their crops, and they can be attractive for tourism. Seventy-one percent ($n = 164$) stated they feel "happy" because the howler monkeys use and move thru the trees of their forested

Fig. 4.7 (**a**) Adult male howler monkey (*Alouatta pigra*) feeding in community lands, Ecoregion Lachua, Guatemala. (**b**) Adult female and baby howler monkeys (*Alouatta pigra*) feeding in secondary-growth community forest, Ecoregion Lachua, Guatemala. (**c**) Adult male howler monkey (*Alouatta pigra)* vocalizing and "calling the rain" as perceived by local people in community lands, Ecoregion Lachua, Guatemala. (**d**) Adult female and baby howler monkey (*Alouatta pigra*) feeding in community lands, Ecoregion Lachua, Guatemala. (**e**) Adult male howler monkey (*Alouatta pigra)* feeding in secondary-growth community forest in the Ecoregion Lachua, Guatemala

lands with cardamom and cacao crops. Forty-three percent (n = 99) of the people interviewed indicated that they have voluntarily left fragments of secondary forest on their lands so that howlers can live and reproduce (Fig. 4.7b, d).

Ninety-one percent (n = 210) of the interviewed people expressed that benefits they receive from spider monkeys (*max*) are the "fun," excitement, and entertainment they feel with their acting that seems human. "Spider monkeys are like us; that's why they deserve our respect…they help and care for each other and heal their bodies with tree leaves, as persons do" (Xo pers. comm.). "I like it a lot when I see spider monkeys moving through my land; they are funny, and I laugh seeing them; they do the same things we do…they are very intelligent because they are our older brothers" (Caal pers. comm.). Other uses and benefits reported for spider monkeys are that they "decorate" (embellish) the forest, spread the seeds of trees so that the forest stays alive, and may also be attractive for future tourism (Fig. 4.8b, c).

Ninety-eight percent (n = 226) of the people interviewed indicated that they have no problem with either primate species being on their land, and they think that it is good to take care of them because "they cause no harm." This contrasts with the perception they have about other wild animals such as collared peccary, raccoon, coati, and parrots that feed from their crops causing economic damage; wild cats, tayra, and opossums that kill farmyard animals; and snakes that can harm people.

Fig. 4.8 (**a**) Adult male spider monkey (*Ateles geoffroyi vellerosus*) in sacred forested mountains, Ecoregion Lachua, Guatemala. (**b**) Adult spider monkeys playfully interacting and showing similar human behaviors as perceived by local people, Ecoregion Lachua, Guatemala. (**c**) Spider monkey (*Ateles geoffroyi vellerosus*) making gestures that facilitate anthropogenic relations, feelings, and empathy toward primates, Ecoregion Lachua, Guatemala

4.5 Implications for Primate and Habitat Conservation

From 2002 to date, our participatory conservation work and long-term living and sharing with Q'eqchi' Mayan communities have allowed us to deeply understand how their culture and cosmovision has different codes, symbolisms, spiritual interactions, and sociocultural practices that have directly contributed to primate conservation over time. Both primates are sacred and respected as ancestors and close relatives of people. They are powerful symbols of cultural resistance and identity whose physical existence honors their Mayan origin and allows them to maintain close bonds with their creators and ancestors. The *Nahual* and *B'atz'* energy in the sacred Mayan calendar allows the ancestral connection and guidance with several fundamental aspects of their modern individual and collective lives. Also, people born on *B'atz'* day have special life purposes that are guided and protected by the energy of primates who are their guardians in nature. This spiritual bond makes them guardians of their counterparts, and therefore, they have the responsibility to protect them and ensure that they are not harmed, which fosters primate survival from generation to generation. Local people have a strong identification and empathy toward these species, recognizing and appreciating in them feelings, behaviors, and faculties of great resemblance and relevance for human life (Figs. 4.7 and 4.8). Particularly, howler monkeys are perceived as communicators, intermediaries, and necessary mediators between *Tzuul Taq'a* and people to send the rain needed for their crops to thrive, which means they play an essential role in the subsistence and finances of families and communities. Spider monkeys are strongly linked to artistic, ludic, and recreational processes that are very important for the expression and revalorization of the Mayan cosmovision, traditional coexistence, and Q'eqchi' Mayan community's interactions and engagement.

Local people's representation and signification of primates and their perception as harmless animals jointly represent cultural values and ethical codes embedded in the

Q'eqchi' Mayan collective unconscious that have been crucial for primate survival, limiting their hunting and favoring habitat conservation throughout the landscape. Through our participatory ecological monitoring and community-based wildlife management programs, we have extensively documented how all of this translates into key values, attitudes, and actions that synergistically support, promote, and directly benefit the primate and habitat conservation at the community and regional level.

In 15 years, we have not registered hunting incidents of primates for food or medicine, due to the anthropomorphic perception that local people have about them. Frequently, we have registered the presence of *A. pigra* troops (including females with offspring) feeding on trees available in home and community plots without any aggression or discomfort from family owners (Fig. 4.7). On the contrary, local people say they are happy that monkeys come so close to their homes. The capture of monkeys for illegal trade and pets is not common in our study area. From 2002 to date, we heard of only two incidents of death of *A. pigra* males that were recorded after teenagers wounded them with slingshots. Only five individuals (three infants and one adult female of *A. pigra*; one adult female of *A. g. vellerosus*) were registered as pets in homes and were cared for and fed as if they were human.

The sacred value of the highlands in Q'eqchi' Mayan cosmovision has allowed extensive forest cover to be maintained in community-owned highlands within agroecosystems. This directly benefits habitat and connectivity available for both primate species and is particularly important for spider monkeys due to the high diversity and abundance of fruit resources that exist in these mountains. Many local people have voluntarily protected and allowed growth of secondary forest remnants of different sizes and shapes in their lands so that primates can live, reproduce, and travel. We have verified that they make agricultural decisions regarding which areas to use for slash-and-burn agriculture by considering which secondary forests remnants are inhabited by primate troops, often deciding not to cut them down. They have also planted native fruit trees (Mayan nut, figs, sapote, tamarind, and avocado, among others) so primates may have more food and remain in their lands. We have documented the presence, abundance, reproduction, and use of resources available for both species in these locations and habitats near human communities. In them, most individuals apparently have normal appearance and healthy physical condition (Figs. 4.7 and 4.8).

We consider that ancestral links and values along with the practical benefits perceived from primates have been key to motivate and favor the positive attitudes that prevail toward these species and that habitat protection and enrichment efforts voluntarily emerge, are maintained, and are multiplied by local people.

4.6 Final Considerations

In Guatemala, as in most of Latin America, the lack of knowledge and awareness about the representation and signification of primates in the cosmovision of indigenous peoples is widespread among the scientific community and decision-makers.

Their contributions to the conservation of these endangered species and habitats have been underestimated and excluded from most education, conservation, and management policies and strategies at national and regional levels. These represent serious weaknesses in addressing primate conservation, limiting the genuine motivation and positive involvement of many human populations that coexist with these species in biodiverse tropical landscapes.

Our experience in Q'eqchi' Mayan ethnoprimatology and primate conservation was key aspects that inspired us to implement a pioneer Environmental-Cultural Popular Education Program. The program intimately links scientific-biological knowledge of our academic profession with Q'eqchi' Mayan environmental wisdom and ancestral values (Rosales-Meda and Hermes 2013). Since 2008, this program is carried out with children, teenagers, and adults to raise awareness about the holistic importance of nature with the goal that ancestral knowledge, values, and perceptions favorable for conservation are not lost, become revalued, and continue to be practiced by the new generations. This innovative process has been very successful to actively involve local people in the conservation of endangered wildlife and habitats with cultural relevance. Since 2012, the program has also allowed us to analyze and address new political and economic corporate interests that pose serious threats to biodiversity and traditional livelihoods. Through cultural revalorization, it has been possible to establish community and regional strategies and efforts to discourage the sale of indigenous lands and its deforestation to prevent the establishment of monocultures of oil palm and livestock pastures, illegal wildlife trade, and transculturation of local youth.

Considering the great risks and threats currently faced by primates and their habitats, it is important that ethnoprimatology transcends the simple documentation of the traditional knowledge and uses of species, to now help in elucidating ancestral codes, values, and links that act as motors and allies of conservation in the collective unconscious of indigenous people from their wisdom and cosmovision. We hope that the experience shared through this chapter will inspire and motivate other researchers and decision-makers to engage in participatory conservation practices and learn from indigenous peoples' cosmovision to address the conservation of endangered species and landscapes from a more holistic, inclusive, and effective perspective.

Acknowledgments We thank our partners and collaborators, the US Fish and Wildlife Service; Whitley Fund for Nature; Conservation, Food & Health Foundation; and Idea Wild, for the long-term support to carry out and strengthen our community-based environmental education, cultural revalorization, ecological monitoring, and wildlife conservation programs. We especially thank our Q'eqchi' Mayan friends, teachers, guides, and neighbors for their trust, wisdom, training, and community living during all these years and for the unconditional support provided to the conservation and management processes we have jointly created for the present and future well-being of communities and Mother Nature in the Ancestral Rainforest Landscape of Guatemala. We are grateful to Ani Youatt Cuevas and Andrea Grosse for the helpful review provided for this manuscript.

References

Akkeren RV (2012) Xib'alb'a y el Nacimiento del Nuevo Sol: Una Visión Posclásica del Colapso Maya. Editorial Piedra Santa, Guatemala. 252 pp

Ander-Egg E (2003) Repensando la Investigación-Acción Participativa. Editorial Lumen-Humanitas, Argentina. 157 pp

Baker M (2013) Revisiting Capuchin Monkeys (*Cebus capucinus*) and the Ancient Maya. Rev Primatol. https://doi.org/10.4000/primatologie.1683

Barrios C (2004) Ch'umilal Wuj: El libro del destino. Cholsamaj, Guatemala. 328 pp

Burton F, Carroll A (2005) By-product mutualism: conservation implications among monkeys, figs, humans, and their domesticants in Honduras. In: Patterson JD, Wallis J (eds) Commensalism and conflict: the human-primate interface. American Society of Primatologists, Norman, pp 24–39

Campbell-Smith G, Simanjorang HVP, Leader-Williams N, Linkie M (2010) Local attitudes and perceptions toward crop-raiding by orangutans (*Pongo abelii*) and other nonhuman primates in Northern Sumatra, Indonesia. Am J Primatol 72:886–876

Chauhan A, Pirta RS (2010) Public opinion regarding human-monkey conflict in Shimla, Himachal Pradesh. J Hum Ecol 30:105–109

CONAP (2008) Guatemala y su biodiversidad: Un enfoque histórico, cultural, biológico y económico. Consejo Nacional de Áreas Protegidas, Guatemala. 650 p

Costa S, Casanova C, Sousa C, Lee P (2013) The good, the bad, and the ugly: Perceptions of wildlife in Tombali (Guinea-Bissau, West Africa). J Primatol 2(1):1–7

Ellwanger A, Riley EP, Niu K, Tan CL (2015) Local people's knowledge and attitudes matter for the future conservation of the endangered Guizhou Snub-Nosed monkey (*Rhinopithecus brelichi*) in Fanjingshan National Nature Reserve, China. Int J Primatol 36:33–54

Fuentes A (2012) Ethnoprimatology and the Anthropology of the Human-Primate Interface. Annu Rev Anthropol 41:101–117

García A, Curruchiche G, Taquirá S (2009) Ruxe'el Mayab' K'aslemäl: Raíz y Espíritu del Conocimiento Maya. URL/PROEIMCA/DIGEBI/CNEM, Guatemala. 238 pp

Gerald I, Leyden B (2007) La vegetación en Guatemala durante el Pleistoceno Terminal y Holoceno. In: Cano E (ed) Biodiversidad de Guatemala, vol I. UVG, Guatemala, pp 15–24

Haste I, De Ceuster P (2001) Cosmovisión y Espiritualidad en la Agricultura Q'eqchi'. Textos Ak'utan No. 18. Centro Bartolomé de las Casas, Alta Verapaz, Guatemala, p 120

Hermes MS, Rosales-Meda M (2011) Monitoreo Ecológico Participativo de la Riqueza y Distribución Espacial de Mamíferos Amenazados y en Peligro de Extinción en el Parque Nacional Laguna Lachuá y Remanentes Boscosos de su Área de Influencia. Alta Verapaz. USAC-PUIRNA, Guatemala. 66 pp

Hill CM, Webber AD (2010) Perceptions of nonhuman primates in human-wildlife conflicts scenarios. Am J Primatol 72(10):919–924

INAB/UICN (2010) Plan de ordenamiento territorial de la Ecoregión Lachuá, vol 103. INAB/UICN/SIAFSA, Guatemala

Kansky R, Kidd M, Knight AT (2016) A wildlife tolerance model and case study for understanding human wildlife conflicts. Biol Conserv 201:137–145

Khatun UH, Ahsan MF, Røskaft E (2012) Attitudes of the local community towards the conservation of the common langur (*Semnopithecus entellus*) in Keshabpur, Bangladesh. Int J Biodiversity Conserv 4:385–399

Knapp S, Davidse G (2007) Flora of Guatemala revisited. In: Cano E (ed) Biodiversidad de Guatemala, vol I. UVG, Guatemala, pp 25–48

La Ruta Maya Conservation Foundation (2013a) Archeological exhibition: the world of the Mayan artist. Antigua Guatemala, Guatemala

La Ruta Maya Conservation Foundation (2013b) Archeological exhibition: Flora and Fauna of the Mayan World, Guatemala

Lee PC, Priston NEC (2005) Human attitudes to primates: perceptions of pests, conflict and consequences for primate conservation. In: Patterson JD, Wallis J (eds) Commensalism and conflict: the human-primate interface. American Society of Primatologists, Norman, pp 1–23

Laird S, Posey A (2002) Professional society standards for biodiversity research: Codes of ethics and research guidelines. In: Laird S (ed). Biodiversity and Traditional Knowledge: Equitable Partnerships in Practice. Earthscan Publications. London, UK, pp 16-38

Matul D, Cabrera E (2007a) La Cosmovisión Maya, vol I. Liga Maya de Guatemala, Guatemala. 352 pp

Matul D, Cabrera E (2007b) La Cosmovisión Maya, vol II. Liga Maya de Guatemala, Guatemala. 425 pp

Mittermeier R, Rylands A, Wilson D (eds) (2013) Handbook of the mammals of the world, vol 3. Primates. Lynx Ediciones, Barcelona, España. 95 pp

Nijman V, Nekaris KAI (2010) Effects of deforestation on attitudes and levels of tolerance towards commensal primates (Cercopithecidae) in Sri Lanka. Int J Pest Manag 56:153–158

Nungshi S, Radhakrishna S (2013) Attitudes towards primates and primate Conservation in Manipur, northeast India. Asian Primates J 3(1):29–36

Rice PM, South KE (2015) Revisiting monkeys on pots: A contextual consideration of primate imagery on classic lowland Maya pottery. Anc Mesoam 26(2):275–294

Riley EP (2010) The importance of human–macaque folklore for conservation in Lore Lindu National Park, Sulawesi, Indonesia. Oryx 44(2):235–240

Rocha LC, Fortes VB (2015) Perceptions and attitudes of rural residents towards capuchin monkeys in the area of influence of the Doña Francisca hydroelectric power plant, south Brazil. Ambient Soc 18(4):19–34

Rosales-Meda M (2003) Abundancia, distribución y composición de tropas del mono aullador negro (*Alouatta pigra*) en diferentes remanentes de bosque de la Ecoregión Lachuá. Alta Verapaz. Tesis de Licenciatura en Biología. Universidad de San Carlos., Guatemala, p 101

Rosales-Meda M, Hermes MS (2010) Capítulo 12: Avances en la validación de una normativa cinegética comunitaria en localidades Maya-Q'eqchi' aledañas al Parque Nacional Laguna Lachuá, Guatemala. In: Guerra-Roa M, Calmé S, Gallina S, Naranjo E (eds) Uso y Manejo de Fauna Silvestre en el norte de Mesoamérica, México, SEV-INECOL-ECOSUR. pp 351–380

Rosales-Meda M, Hermes MS (2013) Rax Nawom: La Naturaleza Sagrada a través de la Educación Popular Ambiental en Guatemala. In: Hernández B, De la Rosa R (eds) Un tejido de muchos puntos: Compilación sobre educación popular ambiental. Editorial Caminos, Cuba, pp 187–202

Rosales-Meda M, Hermes MS (2018) Estado de conservación de Alouatta pigra y Ateles geoffroyi en el Paisaje de los Bosques Lluviosos Ancestrales, Guatemala. In: Urbani B, Kowalewski M, Cunha RGT, De la Torre S, Cortés-Ortiz L (eds) La primatología en Latinoamérica 2 – A primatologia na America Latina 2. Volume II Costa Rica-Venezuela. Instituto Venezolano de Investigaciones Científicas (IVIC), Caracas, Venezuela, pp 481–492

Rosales-Meda M, Estrada A, López J (2008) Demographic survey of black howler (*Alouatta pigra*) population in the Lachuá Eco-región, Alta Verapaz, Guatemala. Am J Primatol 70:231–237

Rosales-Meda M, Hermes MS, Morales J (2010) Capítulo 1: Caracterización de la cacería de subsistencia en comunidades Maya-Q'eqchi' del área de influencia del Parque Nacional Laguna Lachuá, Guatemala. In: Guerra-Roa M, Calmé S, Gallina S, Naranjo E (eds) Uso y Manejo de Fauna Silvestre en el norte de Mesoamérica, México, SEV-INECOL-ECOSUR. pp 25–52

Sac A (2007) El Calendario Sagrado Maya, Método para el Cómputo del Tiempo. URL, Quetzaltenango. Universidad Rafael Landivar (URL). 50 pp

Sam LE (2008) Popol Wuj: Traducción al español y notas de Sam Colop, vol 224. Editorial Cholsamaj, Guatemala

Sharer R (2003) La Civilización Maya. Fondo de Cultura Económica, México. 882 pp

UNESCO (1996) Acuerdos de Paz en Guatemala. Colegio de Abogados, Guatemala. 264 pp

Part II
South America

Chapter 5
Ethnoprimatology of the Tikuna in the Southern Colombian Amazon

Angela M. Maldonado and Siân Waters

5.1 Introduction

In recent years, scholars studying primates have realized that most primate populations are influenced by people's activities making the animals difficult to study in isolation (Fuentes and Wolf 2002; Riley 2013). Anthropogenic activities which threaten wild primates include, but are not restricted to, overhunting, habitat destruction, retaliatory killing for agricultural crop foraging, collection for the primate pet trade (Cowlishaw and Dunbar 2000), and trade for biomedical research (Maldonado et al. 2009).

Human–primate interactions are influenced by diverse cultural, ecological, and other components which may be unique to geographical regions (Loudon et al. 2006). To deal with complex relationships between humans and nonhuman primates, interdisciplinary research methods are used in, what is commonly referred to as, ethnoprimatology (Fuentes and Hockings 2010; Loudon et al. 2006; Parathian and Maldonado 2010). Ethnoprimatologists study the diverse relationships between nonhuman and human primates often with the goal of furthering conservation aims and use a mixed-methods approach in ecological and ethnographic data collection (Fuentes and Hockings 2010; Papworth et al. 2013). Human–primate interactions occur in all tropical forests but are having an increasingly negative effect on primates (Peres and Michalski 2006; Sponsel 1997). Ethnoprimatologists working in the Neotropics investigate what indigenous groups know about primates and sustainable hunting and how they view and categorize the species they hunt for food (Endo et al. 2009; Papworth et al. 2013; Parathian and Maldonado 2010; Stafford et al. 2016).

A. M. Maldonado (✉)
Fundacion Entropika, Leticia, Colombia
e-mail: amaldonado@entropika.org

S. Waters
Department of Anthropology, Durham University, Durham, UK

B. Urbani, M. Lizarralde (eds.), *Neotropical Ethnoprimatology*, Ethnobiology,
https://doi.org/10.1007/978-3-030-27504-4_5

Traditional indigenes managed their game populations using local knowledge and cultural and social taboos to ensure sustainability (Silvius 2004). However, since colonization, the lives of many indigenous peoples have undergone substantial changes that have important ramifications for primate conservation (Silvius 2004). Early colonizers such as missionaries and government agents removed many indigenes from their nomadic lifestyles to settle them in permanent communities. The settlers' exploitation of natural resources from land that indigenes regarded as theirs not only caused extreme conflict but also encouraged indigenes to provide commodities for the settlers moving in to exploit forest resources such as timber (Nimuendaju 1952; Porro 1996; Stearman 1984; Stearman 2000; Zarate 2008). This early contact with western markets drove Amerindians to transform their environment. As a result, indigenous people modified the structure and composition of game species all over the Amazon basin for subsistence and commercial purposes (Terborgh 1999).

Stearman (2000) outlines how Amerindians' social change and modernization have had a detrimental effect on the sustainability of their hunting, namely, sedentarism, population growth, market involvement, and technological enhancements. Sedentarism plays a crucial role in the localized depletion of wildlife. For instance, nomadic tribes are now confined to settlements and are consequently heavy consumers of resources in nearby forests. Their hunting of large, long-lived primate taxa such as atelines is contributing to the drive toward these species' local extinction (Fragoso 1991; Peres 1991). Inhabitants of the Siona–Secoya horticultural villages in Ecuador had to relocate owing to the depletion of game species (Vickers 1983), and there were intra-village conflicts caused by meat scarcity and internal sociopolitical issues among the Yanomamö of Venezuela and Brazil, resulting in the fissioning and relocation of indigenous villages (Good 1987).

The results presented in this chapter form part of long-term research conducted in two Tikuna communities, Mocagua and San Martin, overlapping the Amacayacu National Park, located at the southern part of the Colombian Amazonian trapezium (see Fig. 5.1). These villages provide a comparison of two communities undergoing different rates of environmental and cultural change. Mocagua's strategic geographical location on the Amazon River facilitates access by water, thus decreasing transport costs for tourists and researchers. This, in turn, provides more income for this community. San Martin is located on the Amacayacu River, a tributary of the Amazon, and with the associated higher fuel costs and limited transport availability, access is often difficult, resulting in fewer benefits from tourism and research. The distance from the Amazon also played an important role in the historical acculturation of these communities. San Martin conserves its Tikuna language, while only Mocaguan elders still speak it. Additionally, hunting practices are more traditional in San Martin, where more people rely on hunting. Mocagua has been involved in a Humboldt's woolly monkey (*Lagothrix lagothricha lagothricha*) conservation initiative since 2003, where research and primate tourism are generating alternative income and improving local perceptions of this species for its ecological and economic services (Maldonado and Waters 2017).

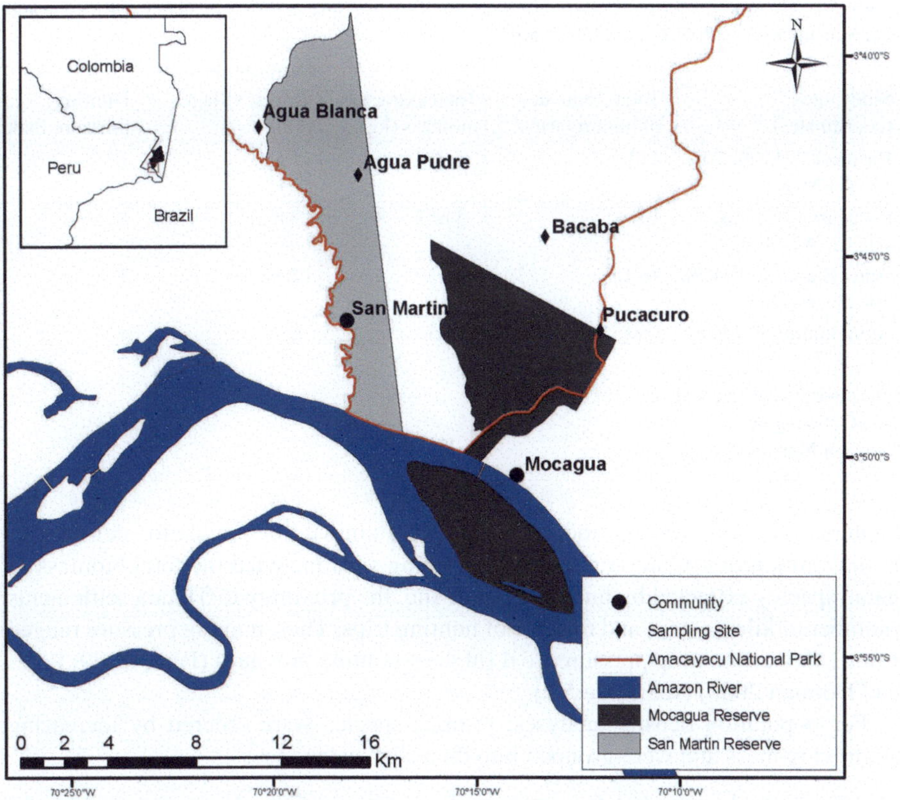

Fig. 5.1 Location of Mocagua and San Martin indigenous territories and the four sampling sites (Agua Blanca, Agua Pudre, Bacaba, and Pucacuro). Source: Maldonado (2012:25)

Qualitative hunting data collection was carried out in three different research periods. In a pilot period covering 25 days of fieldwork in August–September 2003, Tikuna hunters were interviewed, and local agreements and research permits were secured for the implementation of an investigation focusing on Tikuna hunting. An implementation period was carried out from 2005 to 2009. Monthly visits of 5–7 days were conducted in each community for ethnographic data collection by a multidisciplinary research team composed of primatologists, sociologists, anthropologists, biologists, and wildlife veterinarians. The last research period covered 32 days from March to December 2013. During this period, data collection focused on locals' views of the potential depletion of wildlife and the future of hunting in Mocagua and San Martin. A total of 46 hunters and their families were involved in this research. In addition, a randomly selected sample of community members from different gender/age classes attended 12 workshops and 28 community meetings organized by the research team.

A quantitative harvest assessment was conducted in Mocagua and San Martin in order to determine hunting sustainability, to quantify the total biomass extracted by

Table 5.1 Quantitative criteria used to rank the hunting pressure at different sites in Mocagua and San Martin. Source: Maldonado (2012:50)

Study site (coordinates)	Total frequency of hunting trips[1]	Total extracted biomass (kg)	Distance from nearest village (km)	Hunting pressure rank
Bacaba (3° 45′ S, 70° 13′ W)-MOC[a]	113	2,957	11.6	1
Pucacuro (3° 47′ S, 70° 12′ W)-MOC	165	3,657	7.8	2
Agua Blanca (3° 41′ S, 70° 20′ W)-SM[b]	180	6,139	12.5	3
Agua Pudre (3° 43′ S, 70° 18′ W)- SM	369	13,956	6.7	4

[1]During the study period
[a]*MOC* Mocagua
[b]*SM* San Martin

hunters over a 48-month period, as well as ascertain wildlife population densities in both communities. Criteria for ranking hunting sites included the total biomass of game species extracted by hunters at each site, the proximity to Tikuna settlements (number of kilometers), and number of hunting trips. Thus, hunting pressure ranged from 1 (lowest hunting pressure) to 4 (highest hunting pressure) (Peres 1999; Peres and Dolman 2000) (see Table 5.1).

For population density analyses, primate species were ordered by increasing adult body mass and were grouped into three size categories:

(i) Small body size (<1.5 kg): pygmy marmoset (*Cebuella pygmaea*), black-mantled marmoset (*Leontocebus nigricollis nigricollis*), night monkeys (*Aotus* spp.), Ecuadorian squirrel monkey (*Saimiri cassiquiarensis macrodon*), and yellow-handed titi monkey (*Cheracebus lucifer*)
(ii) Medium body size (1.5–4.0 kg): saki monkey (*Pithecia milleri*) and white-fronted capuchin (*Cebus albifrons*)
(iii) Large body size (>4 kg): Colombian howler monkey (*Alouatta seniculus*) (hereafter howler monkey) and Humboldt's woolly monkey (*Lagothrix lagothricha lagothricha*) (hereafter woolly monkey) (Peres and Dolman 2000)

5.2 The Tikuna

Earlier Tikuna were well known as nomadic hunters and gatherers, who specialized in *terra firme* habitats. They occupied the inland forests north of the Amazon River (Rianio 2003). The Tikuna's access to flooded forest (*várzea*) and the islands of the Amazon River was restricted by their neighbor and enemy, the Omagua, the largest indigenous tribe in the area (Acuña 1986; Franco 2006). As a result of early contact by Catholic missionaries in the late 1600s, a smallpox epidemic drastically reduced

Omagua numbers (Nimuendaju 1952). The Omagua, weakened by disease, could not resist the Tikuna's movement into their territory closer to the Amazon (Grohs 1974). In 1768, after the eviction of the missionaries, the Tikuna were recognized as expert hunters and fishermen (Nimuendaju 1952; Porro 1996). In the 1950s, semi-nomadic Tikuna groups living in traditional community houses near Matamata creek and the Amacayacu River were settled by Jesuit missionaries. These missionaries established the first Catholic boarding school in the area, where children were taken from their families and forbidden to speak their native language (Monica Vasquez, personal communication 2009). The villages of Mocagua and San Martin were not formally established until the 1960s (Franco 2006).

During the early 1900s, the Tikuna in Colombia underwent profound changes in their lifestyle due to their involvement in extractive economies such as rubber exploitation, the animal skin trade, illegal red cedar (*Cedrela odorata*) timber extraction, and the massive exploitation of different tortoise species for international markets (Franco 2006; Rianio 2003). During the1960s, the high levels of wildlife trade (mainly primates for the US biomedical research market) drastically affected the Tikuna's traditional use of resources, and commercial hunting was the main source of income in the area (Franco 2006). In the 1980s, the boom of coca impacted on Tikuna society as it became affluent. Traditional activities such as agriculture were abandoned, and alcoholism, prostitution, scarcity of cultivated food, and trafficking of cocaine (Franco 2006; Rianio 2003) were prevalent. As a result of their participation in extractive economies, along with the loss of cultural beliefs and taboos, Tikuna people became disconnected from nature. This disconnection has, in turn, tainted Tikuna views of conservation projects and hindered recent attempts toward cooperation with governmental and non-governmental organizations (NGOs) regarding management and sustainable use of natural resources.

Nowadays, the Tikuna indigenous group is widely distributed along the Amazon River in Peru, Colombia, and Brazil, with an estimated population of 40,000 people, and is one of the largest indigenous groups in the region (Franco 2006; Lopez 2000; Maldonado 2012). Other minority ethnic groups such as the Cocama, Yagua, and Huitoto also share territories with the Tikuna in the study area (Rianio 2003). In the Colombian Amazon, the Tikuna population (~7100 inhabitants) represents only 1.3% of the Colombian population (DANE 2005). Tikuna subsistence relies mainly on low-level slash-and-burn agriculture, hunting, fishing, gathering, and trade. Tourism also provides another source of income (Maldonado and Waters 2017). Some Tikuna are still involved in drug trafficking and illegal logging (Rianio 2003; Zarate 2008).

5.3 Traditional Hunting by the Tikuna

In the past, Tikuna social organization was composed of clans named after terrestrial animals, birds, and plant species (Lopez 2000, 2002). Clans were patriarchal and decisions made by elders (*abuelos*) and *Payés*. *Payés* were Tikuna shamans and spiritual and political authorities. Community decisions were made following advice from the *Payés* and the elders. Payés were in charge of mediating between the supernatural world, natural resources, and people (Campos-Rozo 1987) and communicated to the entire community. Nowadays, the main sociopolitical authorities are elected officers of *curaca* (headman/chief) and the *cabildo*. The elections are much influenced by the number of relatives a candidate has in the community rather than his/her leadership skills for the post (Maldonado 2012). As a consequence, current Tikuna communities lack experienced leaders to represent the community's interests. Therefore, corrupt *curacas* are common in the region. This lack of political cohesion may be related to the loss of the spiritual authorities (the *Payés*) as well.

In common with other Amazonian tribes, the Tikuna consider hunting to be one of the most important and respected activities in society, not only for the intrinsic relationship between the hunter, the *Payé*, and the spirits of the forest but also for the provision of meat for the community (Campos-Rozo 1987). This agrees with Stearman's (2000) work stating that the Yuquí and Sirionó indigenous groups in Bolivia accrue prestige through hunting and the provision of meat and not from gathering plants or planting crops which are activities that present little risk, require little skill, and are rarely considered a scarce resource. Hunting brings status to a man among his peers, and these social rewards extend to his close relatives (Stearman 1987, 1989, 1990).

5.4 *Shamans* (Payés)

As with most Amerindian groups, the Tikuna had a close relationship with nature with resource management controlled by the spiritual authorities. These authorities were mainly composed of the *Payés* and elders with their extensive knowledge of nature, such as the location of key resources, game migration and feeding habits, and seasonality of forest resources in their indigenous territory (Campos-Rozo 1987). The role of the shaman in most hunter-gatherer Amerindian tribes was also to provide spiritual protection. They interceded between the social/human world and the unpredictable world of the supernatural (Brightman 2007; Lee and Daly 1999a, b). *Payés* were also the mediators of the spiritual relationship between humans and the "owners of the game" or the "gamekeeper." The gamekeepers are mythological representations of people, animals, or plants who rule, manage, and organize the use of specific natural resources (Campos-Rozo 1987). *Payés* could perform specific rituals, including asking the gamekeeper for permission to hunt a specific animal species, to protect that species from evil spirits, or to protect game from being hunted by other communities or tribes (Brightman 2007). They also performed rituals to protect hunters embarking on hunting treks, to assist them in finding prey, and to protect them from disease. The *Payé*'s relationship with nature included the ability to transform himself into an animal (shape shifting)—the shape of a jaguar (*Panthera onca*) being the most common (Guenther 1999; Reichel-Dolmatoff 1997) (Humberto Gregorio and Leonel Panduro personal communication 2009). Thus, community *Payés* and elders supported and controlled hunting to ensure hunters were successful.

5.4.1 Hunting Taboos

The Tikuna had several ways of ensuring hunting was sustainable. For example, some species of wildlife were subject to hunting taboos, and hunting such species was forbidden by the *Payé* (Campos-Rozo 1987). Most taboos were related to the Tikuna belief that animals have a benign or malevolent spirit or that the animal might have the spirit of a *Payé* (Cardoso de Oliveira 1983). For instance, tapirs (*Tapirus* sp.), hummingbirds (*Trochilidae* sp.), and toucans (*Ramphastos* sp.) are possessed by benign spirits, while night monkeys (*Aotus* sp.), deer (*Mazama* spp.), pacas (*Cuniculus paca*), yellow-footed tortoise (*Chelonoidis denticulata*), jaguars, all snake species, owls, and most raptors have malignant spirits or might be *Payés*. Hunting bans were imposed by the *Payés* during certain months of the year for other common hunting prey to ensure that these species were not overhunted.

Disobeying a hunting restriction or ban made by the *Payé* could bring bad luck during hunting, the disappearance or scarcity of preferred game species, sadness, disease and even the death of the hunter and his family, and the entire community cursed (Reichel-Dolmatoff 1997; H. Gregorio; M del Aguila; L. Panduro, personal communication 2009). For the Tikuna, Tukano, and Yukuna Colombian indigenous groups, hunting of game species such as primates, tapirs, deer, peccaries (*Tayassu pecari*), and curassows (*Crax* sp.) was severely curtailed (Reichel-Dolmatoff 1996; van der Hammen 1992). In contrast, other game species such as pacas, agoutis, and armadillos were not subject to hunting taboos (Reichel-Dolmatoff 1996; van der Hammen 1992). As hunters are predisposed to opportunistic encounters with animals and unusual situations, Tikuna and other Colombian Amazonian societies believed hunters were the community members who had more chance of meeting the gamekeepers, which had important ecological implications (Reichel-Dolmatoff 1996). The gamekeeper protects his territory from overhunting and any other form of depletion (Maldonado 2012; Reichel-Dolmatoff 1996). Therefore, an encounter with him is dangerous, often ending with a punishment manifesting itself as an illness. This encounter usually affects people who are aware of ecological constraints, people who are actively involved in environmental damage, and people who have consciously or unknowingly violated community norms (Reichel-Dolmatoff 1997).

5.4.2 Sacred Areas

For most Amazonian tribes, salt licks represented one of the most sacred environmental areas of the forest (Reichel-Dolmatoff 1997). For the Tikuna, salt licks were the sacred place where the *Payés* met the gamekeeper personified by animals such as howler monkeys, tapirs, jaguars, deer, and macaws (*Ara* sp.) (Campos-Rozo 1987; A. Vasquez, personal communication 2009) The animals and *Payés* gathered at salt licks to talk and to receive instructions about hunting restrictions and bans. Food restrictions relating to species visiting the salt licks were communicated to the *Payé* by one of the species at the party. A party ensued at the end of the meeting where everyone got drunk (Campos-Rozo 1987; Azulay Vasquez, Leonel Panduro and Mamerto del Aguila personal communication). There are several Tikuna tales about the frequent use of salt licks as meeting locations where important decisions were made by the *Payé* regarding wildlife utilization (Maldonado 2012; Azulay Vasquez, Leonel Panduro and Mamerto del Aguila personal communication 2009). The hunting bans applied by Tikuna at salt licks may have been related to their knowledge of seasonal game migration. For instance, during the dry season, game species rely on the nutrients found in salt licks for long periods of time (Lozano 2004). Thus, source–sink dynamics of game populations were understood and relayed in Tikuna cosmological politics, playing an important role in wildlife conservation. Currently, Tikuna use of salt licks as a place to hunt tapir is very common. During the dry season, hunters establish camps to wait for the animals at night. In Mocagua and San Martin, the Bacaba creek area represents one of the most frequently visited hunting sites, where 30% of the respondents stated that they visited this site at least once a month favoring it as place to hunt large prey (Maldonado 2012).

5.4.3 *Food Restrictions*

For most of the indigenous groups distributed in the Colombian Amazon, food restrictions (better understood as abstinence, fast, or diet) were common for all members of the community (Reichel-Dolmatoff 1997; van der Hammen 1992). Reichel-Dolmatoff (1996) states that food restrictions were based on the intrinsic relationship between people and nature where humans are allied with nature, and this fact implies the observance of rules of measure. Furthermore, animals have energy that is related to the specific types of environment they live in and to the people who live in the surrounding forests. The differences in animal energies depend to a large extent on the availability and abundance of their food resources. Therefore, these considerations provide the basis for food restrictions and for certain culinary preparations (Reichel-Dolmatoff 1996). Most food restrictions were related to birth control, pregnancy, gestation, childhood, and convalescence. Other food restrictions were related to particular activities involving close contact with nature, such as hunting, gathering, cultivating, and fishing (Campos-Rozo 1987; Reichel-Dolmatoff 1997; van der Hammen 1992).

5.4.4 *Use of Primates by the Tikuna*

In common with other indigenous groups in Amazonia such as the Guaja (Cormier 2003), Tikuna traditionally kept monkeys as pets where orphaned infants, whose mothers had been killed by hunters, were reared and not traded but kept in the community (Maldonado 2012). The most common primate species reared and kept as pets by Tikuna were woolly monkey, night monkey, saki monkey, and black-mantled tamarins (Parathian and Maldonado 2010). Additionally, Parathian and Maldonado (2010) report that Tikuna used primate body parts for different purposes. They used howler monkey throat sacs as a medicinal cure for laryngitis. Woolly monkey and saki monkey skins were used to make traditional drum skins to be used for the *Pelazon* festivity which celebrates girls attaining puberty. During this celebration, woolly monkey meat was the most prized and their roasted tails viewed as a delicacy. Like other Amazonian tribes, woolly monkeys represented one of the most important primate species in Tikuna diet and were one of the most overharvested (Peres 1991). Such traditional uses are no longer practiced in most of modern Tikuna society, but some isolated communities do maintain these traditions, especially Tikuna from the Pupuña and Buenos Aires region, north of San Martin. Elderly hunters in Mocagua and San Martin revealed they hunted woolly monkeys only 3 km from the communities where settlements were established. Nowadays only experienced hunters from San Martin are successful in hunting the species. The reported densities of large-bodied primates in San Martin suggest that wild populations are depleted especially within an 8 km radius around the community (Fig. 5.2).

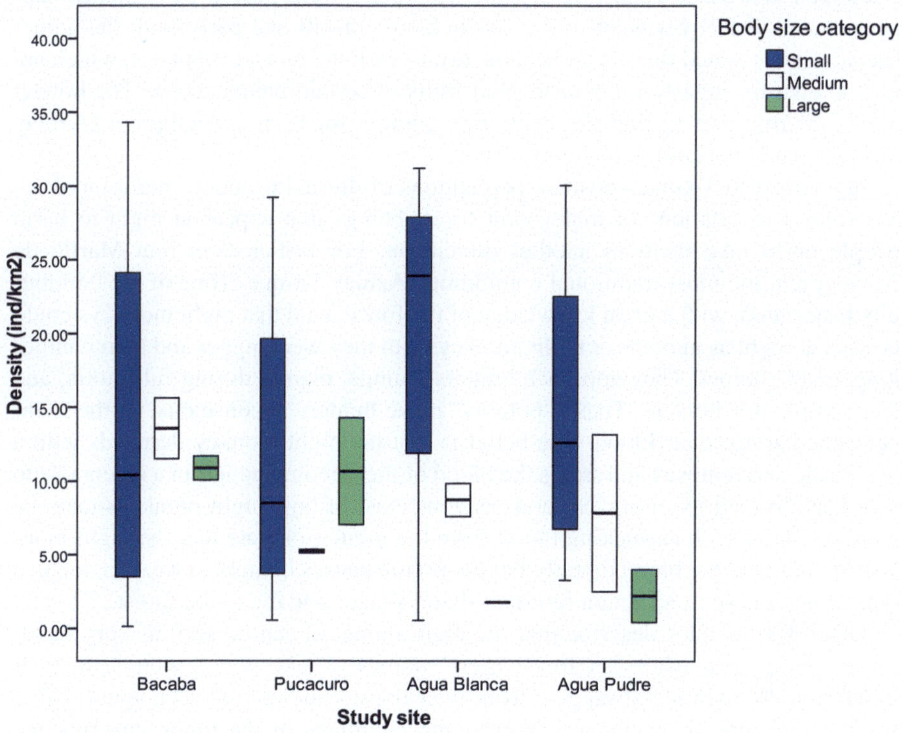

Fig. 5.2 Density of primates by size-class categories at Mocagua (Bacaba and Pucacuro) and San Martin (Agua Blanca and Agua Pudre). Source: Maldonado (2012:94)

Large-bodied primates are highly sensitive and can be the first primate species to disappear at even subsistence levels of hunting resulting in local extinctions (Laurance et al. 2006; Peres 1990, 1991; Stevenson et al. 2005). These species' vulnerability is mainly attributed to their low reproductive rates and long interbirth intervals (Di Fiore and Campbell 2007; Peres 1990). Peres (1990) states that selective hunting may affect ateline sex ratio which influences their long-term population growth.

5.5 Primates in Tikuna Folklore

In Tikuna folklore, most primates are portrayed as intelligent and jovial characters, fostering a respect for the species among local people. For instance, during the *Pelazon* festivity, a member of the community, adorned in a traditional costume made from palm fiber to represent a white-fronted capuchin monkey (*Cebus albifrons*), performs a light-hearted dance to begin the ceremony. White-fronted capuchins symbolize an important character from Tikuna folklore—a monkey who kidnaps a young girl from her community and keeps her trapped in the forest forever (Parathian and Maldonado 2010). Other important roles played by primates happened when the *Payés* went to the salt licks to consult and party with the gamekeeper. Elders stated that *Payés* became drunk with the howler monkeys, which are well known for ingesting the fermented fruits of certain palm species. The howler monkeys' role was to find the fruits ripe enough for such consumption (Azulay Vasquez, personal communication 2009).

In contrast to Tikuna's positive perceptions of diurnal primates, night monkeys are viewed as relations of malevolent forest beings that appear at night to harm people or to take them to another dimension. For instance, in San Martin de Amacayacu, the most traditional community, Azulay Vasquez (one of the community historians), with a great knowledge of the forest, said that night monkeys could be seen at night as humans or night monkeys, but they were bigger and their canines longer and sharper. They approach hunters' camps, mainly during full moon, and leave drinks for hunters. These "potions" make hunters unconscious, so the monkeys can harvest their blood. The belief is that the night monkey descends with a receptacle and removes and stores the blood of the sleeping hunter in a receptacle to take back to feed its group. Several versions exist of how night monkeys take the hunters' blood such as sucking blood from the jugular or state that the night monkeys do not take the blood directly but the potion causes hunters to weaken due to a type of anemia so they cannot return to their villages and die in the forest.

Other Tikuna tales describe that the night monkeys can be seen as very handsome, young men who visit forest users' homes to take young women to their dimension. Women just disappear from their houses and are not seen again. Other versions describe the night monkeys as the guardians of the moon ensuring the moon is never stolen. This suggests the animals have an important role for believers of this myth as the moon influences women's fertility and determines the best sowing and harvesting dates. In the past, the *Payés* preferred to gather with the gamekeepers during full moon, trusting night monkeys to be vigilant and take care of the beings involved in the gathering. These myths explain why, for traditional Tikuna, consuming night monkeys was taboo.

5.6 Primate Hunting by Modern Tikuna

The trade in primate meat finances commodities such as medicine, school supplies, and clothing, and meat is also given to the local Catholic boarding school to pay school fees and is mainly consumed by staff members, rather than the children. In addition, meat is sold in order to pay the additional costs incurred during hunting treks (Maldonado 2012). In both communities, primate meat is mainly for consumption by the hunter's family with the rest sold within the community. Hunters affirmed that they only trade in bushmeat at the beginning of the academic year when they need cash to buy school uniforms, etc. However, data collected through participant observation in Mocagua confirmed that meat was also sold to Macedonia, the nearest Tikuna community, for Catholic religious festivities (Maldonado 2012). In Mocagua, wild meat and locally made alcohol (*masato*) are often offered during the *mingas*, where a family invites community members to participate voluntarily in the clearing of land for agriculture, house construction, garden cultivation, etc. Large numbers of people attend *mingas*, when meat is offered (Humberto Gregorio and Azulay Vasquez personal communication 2009).

Atelines, mainly woolly monkeys and howler monkeys, were heavily targeted during the early 1900s as they were used as bait for hunting big cats for the skin trade. Since the early 1980s, Tikuna people at the tri-border area of Brazil–Colombia–Peru have been involved in the trapping of live night monkeys for malaria research in Colombia. This has caused a long-term impact on wild night monkey populations and their ecosystem, owing to invasive trapping methods and associated high deforestation rates (Maldonado and Peck 2014).

A total of 2101 prey items were taken by Mocagua and San Martin hunters, corresponding to 49 species of vertebrates, with a total extraction of some 26,700 kg of game meat. In rank order of preference, the primate species reported as consumed by Tikuna people were night monkey, howler monkey, and woolly monkey (see Table 5.2). These data clearly illustrate how taboos against consumption of primates can be eroded in indigenous societies, with night monkey meat now commonly consumed in the study communities. Such changes in behavior may be a reaction to the depletion of favored prey species such as woolly monkeys as the elders from San Martin acknowledged that the decreased consumption of woolly monkeys was due to the difficulty involved in locating them, rather than a lack of intention to hunt them (Maldonado 2012).

Table 5.2 Primates harvested at Mocagua and San Martin from February 2005 to February 2009, Amacayacu National Park

					Mocagua		San Martin	
Species	Common name	MBW (kg) (±SD)	Total harvest (ind.)	Total extracted (kg)	No. of harvested	Harvest (kg)	No. of harvested	Harvest (kg)
Aotus spp.	Night monkey	1.5	22	33	10	15	12	18
Alouatta seniculus	Howler monkey	6	18	108	9	54	9	54
Cheracebus lucifer	Yellow-handed titi monkey	2.2	12	26	7	15	5	11
Lagothrix lagothricha lagothricha	Woolly monkey	9.6	12	115	1	10	11	106
Saimiri cassiquiarensis macrodon	Ecuadorian squirrel monkey	1.4	11	15	5	7	6	8
Cebus albifrons	White-fronted capuchin	4.5	7	32	3	14	4	18
Pithecia milleri	Saki monkey	2.8	6	17	2	6	4	11
Leontocebus nigricollis nigricollis	Black-mantled tamarin	0.6	6	4	1	0.6	5	3
			94	*350*	*38*	*121.6*	*56*	*229*

5.7 Perceptions of Hunting Today

Hunters from Mocagua and San Martin provided different explanations regarding the decrease in game species (Mann-Whitney U test; $U = 171$, $z = -2.41$, $p = 0.016$). However, respondents agreed that there are significant changes in hunting today. These changes were (i) because animals were scarce and they had to walk long distances to hunt medium- and large-bodied prey ($n = 34$; 74%); (ii) hunters believed that the disappearance of the woolly monkey, the white-lipped peccary, the tapir, and the collared peccary close to their villages was evidence of game depletion ($n = 12$; 26%). Half the respondents ($n = 22$; 48%) stated that the reduction in preferred game species was the result of overhunting, due to an increased human population ($n = 9$; 20%). Other factors influencing the decrease in wildlife reported by respondents ($n = 15$; 32%) were as follows: (i) the use of western hunting weapons, (ii) *Payés* closing the pathways that wildlife takes to communities, (iii) the presence of white people in the forest, (iv) the noise of saws and shotguns, (v) commercial hunting to pay for children's education, and (vi) hunters targeting large prey and driving them away.

Some people had difficulty in understanding wildlife depletion is due to human influence preferring traditional beliefs. For example, some participants' comments during workshops in San Martin highlighted local people's long-held belief that some animals are immortal (e.g., giant armadillo, *Priodontes maximus*) or are incarnations of the gamekeeper. For instance, an elderly woman and her husband (who was the most respected hunter in the community) stated: "Some animals like the woolly monkey, the jaguar, the tapir and the giant armadillo, they never die, [and] they live forever, or only die if the jaguar, boa or people kill them. However, if the jaguar is not killed by people, they go back to their (supernatural) world" (Monica Vasquez and Humberto Gregorio, personal communication 2009). An elder in San Martin stated: "Most of the monkeys, but especially the large ones like the woolly and the howler monkeys, are like people, [and] they reproduce all the time, when they want" (Azulay Vasquez, personal communication 2009). In the workshops, the hunters stated their belief that woolly and howler monkeys had a life span of 100 years, while night monkeys lived up to 60 years. They also believed that these monkeys reproduced every year and that females became pregnant at 1 year old (Maldonado 2012). These hunters' local ecological knowledge of night monkeys conflicts with conventional scientific research which suggests that the Azara's night monkey (*A. azarae*) is more than 4 years old at first reproduction (Huck and Rotundo 2011), while *A. azarae* females and males in non-hunted wild populations have a mean life expectancy of 6.6 years and 6.7 years, respectively (Larson et al. 2016).

Respondents from both communities reported different preferences for animal species kept as pets (Mann–Whitney U test; $U = 105.5$, $z = -3.63$, $p = <0.001$). In Mocagua, the most common species kept as pets were the paca (27%), the acouchi (*Myoprocta* sp., 10%), and the white-fronted capuchin (10%). Most of the hunters (58%) in San Martin stated that they did not keep wild animals as pets, but in the past, the most common primates kept as pets were woolly monkeys (33%) and white-fronted capuchins (4%) (Maldonado 2012). Parathian and Maldonado (2010) interviewed other community members who talked about having owned or knowing someone who owned woolly, howler, night, and saki monkeys as pets. During the study period, black-mantled tamarins were by far the most frequently captured species kept as pets in San Martin.

5.8 Primate Watching as Alternative Income

As mentioned above, the Tikuna of Mocagua have been involved in woolly monkey conservation projects since 2003. The Woolly Monkey Project (2005–2010) determined the population status of primates and other game species to define the sustainability of hunting. This resulted in the implementation of a hunting ban which is still in force today. In addition, a primate rescue center was established by a local NGO, where visitors have the opportunity to see rescued monkeys, bringing alternative income to local people and giving them status as a community supportive of primate conservation (Maldonado and Waters 2017).

5.9 Discussion and Conclusion

It is clear that contact with missionaries drastically disrupted the Tikuna's seminomadic lifestyle, forcing them to relinquish their hunter-gatherer lifestyle and causing the erosion of food taboos and restrictions intrinsically linked to their religion, traditional knowledge, and resource management practices. This phenomenon has been widely reported for other Amerindian tribes in the Amazon basin (Brightman 2007; Good 1987; Grohs 1974; Stearman and Redford 1992). The almost total loss of traditional management practices by Tikuna in the southern Colombian Amazon is another example of the inevitable consequences of settler incursion and indigenous participation in extractive economies (Redford et al. 1995). For centuries, Tikuna in the area have tolerated the depletion of their resources and are now actively involved in market economies, where commercial extraction of resources is culturally accepted. Other factors such as population growth, lack of governance and local organization, and the need to formalize land tenure agreements between Tikuna communities also affect their use of resources. It is clear that the Tikuna need access to cash in order to satisfy needs that were never part of their traditional lifestyle in the past, such as formal education, transportation, and access to Western commodities (communication, clothing, recreation).

Experienced hunters from San Martin and Mocagua stated that hunting night monkeys was taboo for traditional Tikuna. Today, for the Tikuna, and for other Amazonian groups and *caboclos*, night monkeys are not consumed owing to the strong and disagreeable odor and taste from their subcaudal gland (Aquino et al. 2009). However, young and inexpert hunters ($\leq$30 years old; $n = 6$; 13%) do not believe, or are unaware of, taboos so hunt night monkeys due to the lack of larger-bodied prey close to communities (Maldonado 2012). The young hunters' behavior demonstrates how increasing erosion of Tikuna taboos and food restrictions leads to the consumption of primates that were not exploited previously, and this, along with the inclusion of undesirable animal species in the Tikuna diet, is expanding game depletion. Young hunters still respect elder hunters to an extent so will not go hunting alone if an elder hunter (relative) has a dream related to snakes or jaguars. The younger hunters accept this as a sign of bad luck meaning they may become lost while out hunting. Older hunters believe that the loss of the *Payés* and their spiritual connection with nature (including the gamekeeper) is one of today's triggers for the depletion of preferred game species because the traditional ways of controlling hunting have disappeared.

All male hunters stated that they always hunted with shotguns, while the two female hunters hunted with machetes and dogs. In both communities, hunters always carry a shotgun and a machete (41%) on hunting trips, while five hunters from Mocagua affirmed they also take a bow (11%), and 20% of the hunters hunt with dogs. Only one respondent, an elderly hunter from San Martin, had a blowpipe but lacked the curare (poison) to use it. The only Tikuna people preparing curare in the area are a couple of elders from the north. Hunters undertake trips on foot in the majority of the cases (52%) or on foot and canoe during the rainy season (41%). Only three elderly hunters preferred the canoe for transportation, mainly because they felt too old to walk (7%).

Social restrictions on some hunting practices endure. For example, during a community meeting in 2013, Mocaguan local authorities voiced concern about younger hunters' use of an unusual hunting tool. The *trampero* is a handmade snare that ties a wire to a shotgun and is activated when an animal or person steps on the wire, triggering the shotgun. Local people thought this was a very dangerous and lazy way to hunt as it did not demand any knowledge or hunting skill and could kill another hunter, and the prey could decompose if its owner did not visit the snare in a timely manner. The elders described the trampero as a shameful way to hunt (Leonel Panduro, personal communication 2009). We reported this practice to the Amacayacu National Park, and the Mocaguan and Park authorities forbade its use.

Parathian and Maldonado (2010) reported that during women's group discussions, the majority of participants suggested that primate meat was still considered an important source of protein for the community. They went on to say that people's diets were changing in accordance with resource availability and the Mocaguan community's decision to protect woolly monkeys made hunting that species socially unacceptable. Kinkajous (*Potos flavus*), three-toed sloths (*Bradypus variegatus*), and two-toed sloths (*Choloepus didactylus*) were appearing in dietary records, and yet, during group discussions with village elders in both communities, participants suggested these meats were traditionally considered taboo. Such decisions may be the result of one or more of the following: a decrease in available preferred prey base, social pressure from the community to refrain from hunting woolly monkeys, and/or a lessening of taboos for the consumption of kinkajous and sloths.

The elders' belief that large primates reproduce "when they want" is commonly held among local people. These local beliefs assist us in understanding local people's skepticism regarding research suggesting that several game species might become locally extinct if hunting continues unsustainably. This situation illustrates a disconnect between local people and conservation scientists regarding wildlife management that has been observed elsewhere (Dowsley and Wenzel 2008; Kreye et al. 2017).

Mocaguan informants, hunters, and other community members believed that hunting would soon die out as older, expert hunters died and younger community members were disinterested in the practice (Bonilla 2014). Even skilful young hunters (19–32 years of age) now work at the Amacayacu National Park, monitoring natural resource extraction, illustrating the paradox of young people getting education and a job, but leaving their traditional culture. This raises concern regarding the disappearance of cultural and social bonds that hunting and meat sharing among the Tikuna represent.

5.9.1 Wildlife Tourism and the Future of Primates

Since the creation of the Amacayacu National Park (ANP) in 1987, the Colombian Park System has been searching for ways to integrate conservation and sustainable use of natural resources that respect traditional Tikuna culture. The Colombian Park

System created the Special Management Regime (REM) of natural resources for overlapping areas between the park and indigenous territories. It is also working toward integrating indigenous and government legislation. In 2005, ANP implemented the "Gavilan Tatao" Tourism program that not only provided training for local people in tourism but also contained a patrolling component where the legal/illegal use of resources was monitored. Today Mocagua and San Martin offer tourism activities at the family level as part of the REM. In 2007, Maldonado and collaborators created an NGO (Entropika Foundation), which provided data on game populations for the implementation of hunting bans and restrictions as part of the REM. Since 2016, ANP and Entropika, with the financial support of the European Union, have worked together in a capacity-building project to improve governability and to establish local tourism initiatives through training and the provision of basic materials and equipment for local people, replicating successful outcomes from the Peruvian side of the Amazon River. The medium-term goal is to provide the technical capacity for local people to meet sustainable tourism standards regulated at the national level by the Tourism and Commerce Ministry and the Colombian Park System Unit.

The 10-year follow up of the hunting ban for woolly monkeys currently applied by Mocagua is intrinsically related to the monetary return that primate tourism represents for 40% of the community, along with the fame they gained from protecting this species. Moreover, our census database suggests a positive correlation between biomass and hunting restrictions supported by sustainable incomes. Primate biomass was significantly higher in Mocagua (398 kg/km^2) than in San Martin (199 kg/km^2), where hunting restrictions were not strictly applied.

However, more positively, primate watching brings welcome tourist revenue and helps local people understand that having robust primate populations brings more community benefits long term (Maldonado and Waters 2017). Younger Tikuna in Mocagua now perceive monkeys as important for the ecosystem (Parathian and Maldonado 2010). Shifting indigenous people's perception of primates from providers of meat or as pets to the animals as beneficial to the ecosystem and providers of tourism revenue may be an effective way to protect such animals. Whether the Tikuna perceive primates as food, pets, or providers of tourism revenue, the animals will continue to occupy an important part in Tikuna life and culture.

Acknowledgments We profoundly thank the Mocagua and San Martin Tikuna community members for their hospitality and continuous collaboration. Special thanks to the Panduro and del Aguila families, Leonel Panduro, Humberto and Miguel Gregorio, Azulay and Monica Vasquez, Arturo Naranjo, and Loyda and Maria Angel from San Martin. Thanks to the staff of the Amacayacu National Park for their continuous support. We thank Bernardo Urbani and Manuel Lizarralde for their comments which much improved this chapter. This study was funded by Rufford Small Grants, Rainforest Concern, the Holly Hill Trust, the Whitley Fund for Nature, ORSAS Scholarship (the United Kingdom), Russell E. Train Fellowship (WWF), and International Primate Protection League (IPPL) (the United States). Research permits to conduct this study were granted by the Colombian Park System and previous consultation processes carried out and approved by the Colombian Ministry of Interior.

References

Acuña C (1986) Nuevo descubrimiento del gran Río de las Amazonas. In: Figueroa F (ed) Informes de Jesuitas en el Amazonas, 1660–1684. Monumenta amazónica. IIAPCETA, Iquitos, pp 35–101

Aquino R, Terrones W, Navarro R, Terrones C, Cornejo FM (2009) Caza y estado de conservación de primates en la cuenca del río Itaya, Loreto. Perú Rev Peru Biol 15:33–39

Bonilla TA (2014) Usos, prácticas e ideologías socio-culturales de la cacería de dos comunidades Tikuna, ubicadas en la Amazonía Colombiana. Universidad Nacional de Colombia

Brightman M (2007) Amerindian Leadership in Guianese Amazonia. PhD thesis, Cambridge University

Campos-Rozo C (1987) Aspectos etnozoologicos relacionados con la actividad de la caza de los indigenas Ticuna, San Martin de Amacayacu (Amazonas). Universidad Javeriana. Tesis para optar al titulo de Biologo

Cardoso de Oliveira R (1983) Tempo Brasileiro. Edições Universidade Federal do Ceará. Fortaleza, Río de Janeiro

Cormier LA (2003) Kinship with monkeys. Columbia University Press, New York

Cowlishaw G, Dunbar R (2000) Primate conservation biology. University of Chicago Press, Chicago

DANE (2005) Proyecciones de Población, Grupos Etnicos. Base de Datos. Departamento Administrativo Nacional de Estadisitca, Bogota

Di Fiore A, Campbell CJ (2007) The atelines: variation in ecology, behavior and social organisation. In: Campbell CJ, Fuentes A, MacKinnon KC, Panger M, Bearder SK (eds) Primates in perspective. Oxford University Press, New York, Oxford, pp 155–185

Dowsley M, Wenzel G (2008) "The Time of the Most Polar Bears": a co-management conflict in Nunavut. Arctic 61:177–189

Endo W et al. (2009) Game vertebrate densities in hunted and non-hunted sites in Manu National Park, Peru. Biotropica

Fragoso JMV (1991) The effect of hunting on tapirs in Belize. In: Robinson JG, Redford KH (eds) Neotropical wildlife use and conservation. The University of Chicago Press, Chicago and London, pp 154–162

Franco R (2006) El Proceso del REM en el Sector Sur del Parque Amacayacu. UAESPNN, Bogota

Fuentes A, Hockings KJ (2010) The ethnoprimatological approach in primatology. Am J Primatol 72:841–847. https://doi.org/10.1002/ajp.20844

Fuentes A, Wolf L (2002) Primates face to face: conservation implications of human-primate interconnections. Cambridge University Press, Cambridge

Good KR (1987) Limiting factors in Amazonian ecology. In: Harris M, Ross EB (eds) Food and evolution towards a theory of human food habits. Temple University Press, Philadelphia, pp 407–421

Grohs W (1974) Los indios del alto Amazonas del siglo XVI al XVIII, poblaciones y migraciones en la antigua provincia de Maynas. Bonner Amerikanistiche Studies 2, Bonn

Guenther M (1999) From totemism to shamanism: hunter-gatherer contributions to world mythology and spirituality. In: Lee RB, Daly R (eds) The Cambridge encyclopedia of hunters and gatherers. Cambridge University Press, Cambridge, pp 426–433

Huck M, Rotundo MF-D (2011) Growth and development in wild owl monkeys (*Aotus azarae*) of Argentina. Int J Primatol 32:1133–1152

Kreye MM, Pienaar EF, Adams AE (2017) The role of community identity in cattlemen response to Florida panther recovery efforts. Soc Nat Resour 30:79–94

Larson SM, Colchero F, Jones OR, Williams L, Fernandez-Duque F (2016) Age and sex-specific mortality of wild and captive populations of a monogamous pair-bonded primate (*Aotus azarae*). Am J Primatol 78:315–325

Laurance WF, Peres CA, Jansen PA, D'Croz L (2006) Emerging threats to tropical forests: what we know and what we don't know. In: Laurance WF, Peres CA (eds) Emerging threats to tropical forests. University of Chicago Press, Chicago, pp 437–462

Lee RB, Daly R (eds) (1999a) The Cambridge encyclopedia of hunters and gatherers. Cambridge University Press, Cambridge
Lee RB, Daly R (1999b) Introduction: foragers and others. In: Lee RB, Daly R (eds) The Cambridge encyclopedia of hunters and gatherers. Cambridge University Press, Cambridge, pp 1–19
Lopez CL (2000) Ticunas brasileros, colombianos y peruanos: Etnicidad y nacionalidad en la región de fronteras del alto Amazonas/ Solimões. PhD Thesis, Universidad de Brasília - UnB
Lopez CL (2002) Los Ticuna frente a los procesos de nacionalizacion en la frontera entre Brasil, Colombia y Peru. Rev Colom Antropol 38:77–104
Loudon JE, Howells ME, Fuentes A (2006) The importance of integrative anthropology: a preliminary investigation employing primatological and cultural anthropological data collection methods in assessing human-monkey coexistence in Bali, Indonesia. Ecol Environ Anthropol 2:2–12
Lozano CM (2004) Efectos de la accion humana sobre la frecuencia de uso de los salados por las dantas (*Tapirus terrestris*) en el sureste del trapecio Amazonico colombiano. Maestria en Estudios Amazonicos, Universidad Nacional de Colombia
Maldonado AM (2012) Hunting by Tikunas in the Southern Colombian Amazon. Assessing the impact of subsistence hunting by Tikunas on game species in Amacayacu National Park, Colombian Amazon. LAP Lambert Academic Publishing GmbH & Co. KG., Saarbrücken
Maldonado AM, Peck MR (2014) Research and *in situ* conservation of Owl monkeys enhances environmental law enforcement at the Colombian-Peruvian border. Am J Primatol 76:658–669. https://doi.org/10.1002/ajp.22260
Maldonado AM, Nijman V, Bearder SK (2009) Trade in night monkeys *Aotus* spp, in the Brazil-Colombia-Peru tri-border area: international wildlife trade regulations are ineffectively enforced. Endanger Species Res 9:143–149
Maldonado AM, Waters S (2017) Primate Trade (Neotopics). In: Fuentes A. Ed: The International Encyclopedia of Primatology. New Jersey: John Wiley & Sons. https://doi.org/10.1002/9781119179313.wbprim0393
Nimuendaju C (1952) The Tukuna vol XLV. Publications in American archaeology and ethnology. University of California Press, Berkeley, Los Angeles
Papworth S, Milner-Gulland EJ, Slocombe K (2013) The natural place to begin: the ethnoprimatology of the Waorani. Am J Primatol 75:1117–1128
Parathian HE, Maldonado AM (2010) Human-nonhuman primate interactions amongst Tikuna people: perceptions and local initiatives for resource management in Amacayacu in the Columbian Amazon. Am J Primatol 71:1–11
Peres CA (1990) Effects of hunting on western Amazonian primate communities. Biol Conserv 54:47–59
Peres CA (1991) Humboldt's woolly monkeys decimated by hunting in Amazonia. Oryx 25:89–95
Peres CA (1999) Effects of subsistence hunting and forest types on the structure of Amazonian primate communities. In: Fleagle J, Janson C, Reed K (eds) Primate communities. Cambridge University Press, Cambridge, pp 268–283
Peres CA, Dolman PM (2000) Density compensation in neotropical primate communities: evidence from 56 hunted and nonhunted Amazonian forests of varying productivity. Oecologia 122:175–189
Peres CA, Michalski F (2006) Synergistic effects of habitat disturbance and hunting in Amazonian forest fragments. In: Laurance WF, Peres CA (eds) Emerging threats to tropical forests. The University of Chicago Press, Chicago and London, pp 105–126
Porro A (1996) O povo das águas: Ensaios de etno-história Amazônica. Editorial Vozes, Edusp, Petropolis, Vozes
Redford KH, Godshalk R, Asher K (1995) What about the wild animals? Wild animal species in community forestry in the tropics. Community Forestry Note 13. Food and Agricultural Organisation of the United Nations, Rome
Reichel-Dolmatoff G (1996) The forest within. The world-view of the Tukano Amazonian Indians. Themis Books Ltd., Totnes

Reichel-Dolmatoff G (1997) Rainforest shamans. Essays on the Tukano Indians of the Northwest Amazon. Themis Books Ltd., Devon
Rianio E (2003) Organizando su espacio, construyendo su territorio: transformaciones de los asentamientos Ticunas en la ribera del Amazonas colombiano. Universidad Nacional de Colombia, Unibiblos, Leticia
Riley EP (2013) Contemporary primatology in anthropology: beyond the epistemological abyss. Am Anthropol 115:411–422. https://doi.org/10.1111/aman.12025
Silvius KM (2004) Bridging the gap between western scientific and traditional indigenous wildlife management. The Xavante of Rio Das Mortes Indigenous Reserve, Mato Grosso, Brazil. In: Silvius KM, Bodmer RE, Fragoso JMV (eds) People in Nature Wildlife Conservation in South and Central America. Columbia University Press, New York, pp 37–49
Sponsel LE (1997) The human niche in Amazonia: explorations in ethnoprimatology. In: Kinzey WG (ed) New World Primates: ecology, evolution and behaviour. Aldine de Gruyter, New York, pp 143–165
Stafford CA, Alarcon-Valenzuela J, Patino J, Preziosi RF, Sellers WI (2016) Know your monkey: identifying primate conservation challenges in an indigenous Kichwa community using an ethnoprimatological approach. Folia Primatol 87:31–47. https://doi.org/10.1159/000444414
Stearman AM (1984) The Yuquí connection: another look at Sioronó deculturation. Am Anthropol 86:630–650
Stearman AM (1987) No longer nomads: the Sirionó revisited. Hamilton Press, Lanham
Stearman AM (1989) Yuquí: Forest nomads in a changing world. Holt, Rinehart and Winston, New York
Stearman AM (1990) The effects of settler incursion on fish and game resources of the Yuquí, a native Amazonian society of eastern Bolivia. Hum Organ 49:373–385
Stearman AM (2000) A pound of flesh: social change and modernisation as factors in hunting sustainability among Neotropical indigenous societies. In: Robinson JG, Bennett EL (eds) Hunting for sustainability in tropical forests. Columbia University Press, New York, pp 233–250
Stearman AM, Redford KH (1992) Commercial hunting by subsistence hunters - Siriono Indians and Paraguayan Caiman in lowland Bolivia. Hum Organ 51:235–244
Stevenson PR, Link A, Ramirez BH (2005) Frugivory and seed fate in Bursera inversa (Burseraceae) at Tinigua Park, Colombia: implications for primate conservation. Biotropica 37:431–438
Terborgh J (1999) Requiem for nature. Island Press, Washington, D.C.
van der Hammen MC (1992) El Manejo del Mundo. Naturaleza y sociedad entre los Yukuna de la Amazonia colombiana. Estudios en la Amazonia colombiana IV. Tropenbos, Colombia, Bogota
Vickers WT (1983) The territorial dimensions of Siona-Secoya and Encabellado adaptation. In: Hames R, Vickers WT (eds) Adaptive responses of native Amazonians. Academic Press, New York, pp 451–478
Zarate CG (2008) Silvicolas, siringueros y agentes estatales: el surgimiento de una sociedad transfronteriza en la Amazonia de Brasil, Peru y Colombia 1880–1932. Editorial Universidad Nacional de Colombia. Instituto Amazonico de Investigaciones (IMANI), Leticia

Chapter 6
Frugivorous Monkeys Feeding in a Tropical Rainforest: Barí Ethnobotanical Ethnoprimatology in Venezuela

Manuel Lizarralde

6.1 Introduction

The Barí are an Amerindian group of approximately 4000 people living on the southwestern side of the Lake Maracaibo region of northwestern South America, on both sides of the Venezuela–Colombia border (Fig. 6.1). Their language belongs to the Chibcha linguistic family. This society practices subsistence, swidden horticulture complemented by fishing, hunting, and gathering of forest products. The Barí way of living is mostly associated with Amazonian cultures. They depend on manioc, supplemented with bananas and plantains as the main starches, and bocachico fish and monkeys as main sources of protein, in addition to pacas, peccaries, tapir, curassows, and river turtles as seasonal sources (Beckerman 1975, 1983; Beckerman and Lizarralde 2013).

Their environment is classified as a hyper-humid, tall tropical forest (Pittier 1948; Huber and Alarcón 1988; Aymard 2011). Based on its biodiversity of both fauna and flora, this region was denominated "Refugio del Catatumbo" by Steyermark (1982) and like the Southern Central American wet forests shares floristic similarities with the Amazonian and the Guayana Shield lowlands (Steyermark 1982; Gentry 1990; Aymard 2011).

However, Barí forest biodiversity is not as high as that found in the Amazon in places such as Yasuni Park in Ecuador (Finer et al. 2009) or Manu in Peru (Terborgh 1999). Ecological and cultural aspects of Barí ethnoprimatological work have been published in two book chapters (Lizarralde 2004, 2019). This book chapter focuses on combining ethnobotanical and ethnoprimatological information to detail the holistic understanding that the Barí have of their natural resources.

M. Lizarralde (✉)
Department of Botany and Environmental Studies Program, Connecticut College, New London, CT, USA
e-mail: mliz@conncoll.edu

B. Urbani, M. Lizarralde (eds.), *Neotropical Ethnoprimatology*, Ethnobiology,
https://doi.org/10.1007/978-3-030-27504-4_6

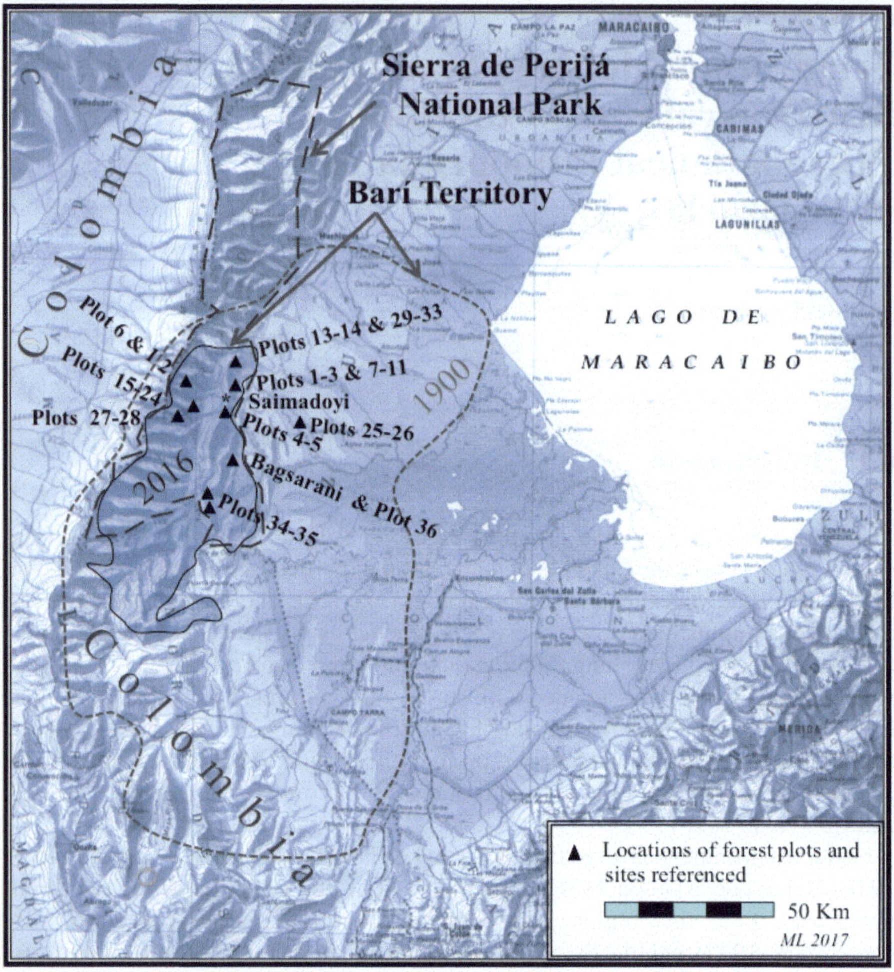

Fig. 6.1 Barí Territory in the 1900s and 2016 as well as locations of forest plots censused in Venezuela

A word for primate does not exist in the Barí language. However, when asked in Spanish about all of their monkeys, they provide a list of six different animals. Among them are four species found in their forest which are considered the proper primates: (1) white-bellied or variegated spider monkey (*Ateles hybridus*, I. Geoffroy Saint-Hilaire, 1829, *su'shà* in Barí (the apostrophe (') in the Barí words is to represent a glottal occlusion sound similar to a combined "g" and "k" sound), average weight 9.5 kg; see Fig. 6.2), (2) red howler monkey (*Alouatta seniculus*, Linnaeus, 1766, *borò* or *kámashkò'da* in Barí (the two names are dialectical variations in the Barí language), average weight 6.0–8.5 kg; see Fig. 6.3a), (3) Sierra de Perijá white-fronted capuchin (*Cebus leucocephalus*, Gray, 1865, *barashì* in Barí, weight 1.0–3.5 kg), and (4) gray-handed night or owl monkey (*Aotus griseimembra*, Elliot,

Fig. 6.2 Spider monkey pet (immature) in June of 1999 near the Barí community of Bagsarani

1912, *kogchi'bà* or *bá'bora* in Barí, weight 0.8–1.3 kg; see Fig. 6.3b [Linares 1998]). The Barí also include under the Spanish gloss of "monos" (monkeys), or related to them, two other species of mammals: kinkajou (*Potos flavus*, Schreber, 1774, called *bíshwì* in Barí) and eastern lowland olingo (*Bassaricyon alleni*, Thomas, 1880, called *bo'sábù* in Barí), both associated with night monkeys because of their nocturnal arboreal habitat and/or prehensile tails (Lizarralde 2002).

Primates have always played a significant role in Barí culture. This importance can be observed in their mythology where monkeys have much more complex and rich mythological stories than any other animals, including fish. Even though the work of Stephen Beckerman (1983) in the 1970s and 1980s shows that fish were even the most important source of protein, monkeys likely surpassed them in pre-contact times (Lizarralde 2019). According to Beckerman (1975), by weight, monkeys accounted for 5.6% of the consumed proteins and 25% of hunted animals. This figure seems to be relatively low, especially in contrast to the great role monkeys play within the Barí culture. The consumption of monkeys must have been much higher in the past in order for them to figure so prominently in their mythological stories (Lizarralde 2019). Before the 1920s, Barí territory was much larger and supported a fraction of the population density that it has in recent times (Lizarralde 1991; Lizarralde and Lizarralde 2018). Their traditional hunting did not likely have a negative impact on communities of forest animals in the early 1900s as it has had,

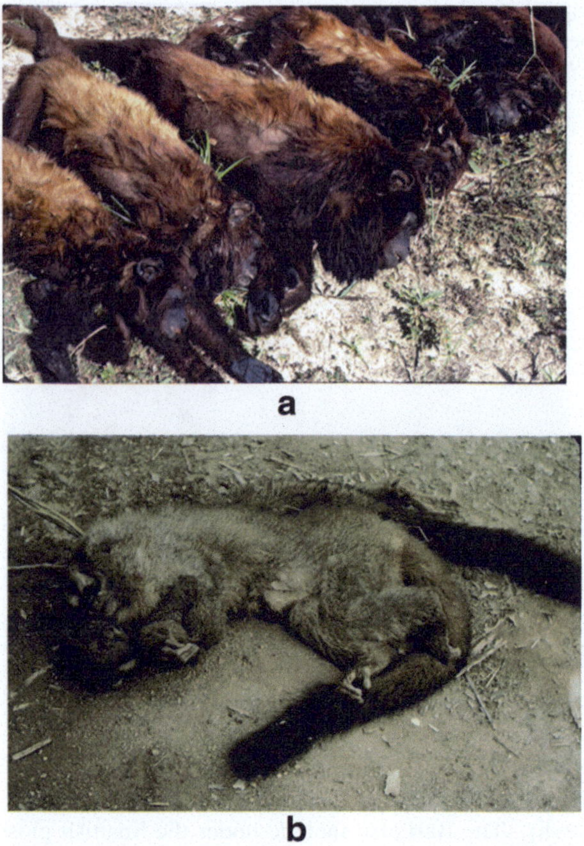

Fig. 6.3 A group of red howler monkeys (**a**, above) and two gray-handed night monkeys (**b**, below) hunted in 1988 near the Catatumbo River

for instance, in Amazonian Peru due to the recent increase in population density of the Matsigenka (Ohl-Schacherer et al. 2007).

The Barí people have a detailed knowledge of their flora and fauna due to their intimate relationship with their environment and hunting activity (Lizarralde 1997). They recognize close to a thousand plant taxa and more than 350 animals. Also, they have developed a rich understanding of the behavior and diet of the animals they hunt, which enables them to predict the movement and location of these animals. While walking in the forest, the Barí are constantly examining fresh fruits found on the ground in order to determine which animals are eating them, for they can recognize the biting patterns of most taxa. To anticipate the movement of these animals, the Barí hunters check which foods are available, and their seasonalities can help to predict the movements of hunted forest animals.

Besides providing a food source, the other important role monkeys play in indigenous cultures is as pets, which is also common in many other indigenous cultures in the Amazon (Cormier 2003; Cormier and Urbani 2008; Lizarralde 2002; Shepard 2002; Stafford et al. 2016). Cormier (2003: 115) has observed that the Guajá people of Brazil have many pet monkeys, sometimes outnumbering the human inhabitants at some hearths. The Barí definitely enjoy having monkeys as pets, and children continually ask their parents to bring them home from hunting expeditions. The most common pet monkeys among the Barí are spider and night monkeys, while capuchin monkeys are rare. Howler monkeys have never been recorded as pets because they are harder to feed according to the Barí. Monkey pets play a complex role of socialization and enculturation of children to develop a hunter's ability to discern vocalizations and scents of monkeys while in the forest (Cormier 2003).

The term *ethnoprimatology* was first coined by Leslie Sponsel (1997) as the subdiscipline that studies indigenous people's relationship with their nonhuman primate populations. For the New World, there are relatively limited ethnoprimatological studies (e.g., Cormier 2003, 2004; Cormier and Urbani 2008; Lizarralde 2002, 2019; Parathian and Maldonado 2010; Shepard 2002; Stafford et al. 2016). However, with the exception of Cormier (2003, 2004), which provided brief information on the number of useful plants ($N = 275$) and the percentage (65.2%) of those feeding monkeys, there have been no studies in which ethnoprimatology and ethnobotany overlap in the present. This chapter combines these two subdisciplines for the case of the Barí, providing not only the number and percentage of trees that provide food for primates but also details about tree species demographics and the proportion of the forest providing food for monkeys. This is a new ethnobotanical perspective not applied to the present in ethnoprimatological studies with the exception of the work of L. Cormier (2004).

The theory behind this work is that Barí people have complex and detailed knowledge of their flora, like many indigenous people in the world (Lizarralde 1997, 2004). However, this knowledge is quite variable in a given population, and it is important to recognize and address this variability. Indigenous societies do have their experts who maintain complex and rich information about their flora and fauna and then share this information with community members. Ethnobotanical work completed by Brent Berlin (1992), Berlin et al. (1974), William Balée (2013), and Eglee and Stanford Zent (2002) has provided detailed examples of this extensive local knowledge by various indigenous people in the New World. The work of Shepard et al. (2001) has demonstrated that indigenous people such as the Matsigenka of Peru also have rich local ecological knowledge of their forest, which can provide excellent examples of animals' relations to their food sources. Berlin's proposal that *saliency* (e.g., abundance, size of tree, outstanding characteristics such as a big colorful flower or fruit) would make a tree better known and more utilized (Berlin 1992) has directly been observed in the case of the Barí knowledge of food-providing trees for monkeys.

Another important theoretical aspect of this chapter is *historical ecology*, since people and animals make changes in their environment and adapt to it (Balée 2013). Cormier (2003: 157) states, "historical ecological perspective takes into account the mutual influence of culture change and environmental change over time." This perspective contrasts with the notion of an "ethnographic present" and the belief that cultures are unchanged and traditions are being maintained unchanged by millennia. In fact, culture and the environment are constantly changing, perhaps slower in indigenous societies where technology does not evolve as fast as it is in developed nations, but still, changes do occur. Understanding these changes can only be accomplished by taking into account historical events and framing them with people's notions of adaptation and evolution of behaviors, similar to the way the work of Fikret Berkes has expanded our cognition of the Cree people in Canada (2012). According to Berkes, the Cree have been able to manage their fish and game and adapt to changes of their faunal population without detrimental effect. Drawing on these different theoretical approaches, I am elaborating below on the Barí case.

6.2 Primate Frugivory

The coevolution of angiosperms and frugivores started 80 million years ago, and the diversity of seed and fruit sizes and types peaked between 55 and 50 million years B.C.E. (Eriksson 2016). This evolutionary process developed as a mutualistic interaction, with fruit trees providing nutritive resources to frugivores (mostly birds and mammals) and frugivores offering seed dispersal services to plants (Eriksson 2016; Herrera 2002; Jordano 2000). The relationship between fruit trees and frugivores is often diffuse, with most frugivores taking advantage of various species of fruiting trees at any given time and vice versa (Lambert and Garber 1998). The larger the animal is, like a spider monkey (Fig. 6.4) or a tapir, the greater the number of seeds of different species of plants they can consume and disperse.

Trees with very large fruit (12–20 cm diameter) are rare in the Barí forest and are species that were once dispersed by megafauna that became extinct with the arrival of humans in the region at the end of the Pleistocene (Guimarães et al. 2008). These

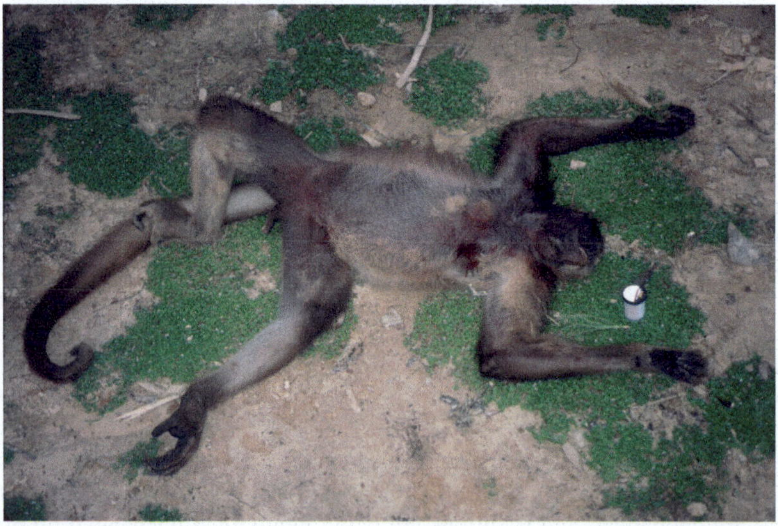

Fig. 6.4 A hunted female spider monkey (mature) in June of 1999 near the Barí community of Bagsarani

tree species have likely survived due to their dispersal by humans and/or agoutis and pacas. For example, in the Barí territory, we have at least three species of trees that were food for extinct megafauna (Gentry 1974, Janzen 1982): algarrobo or copal (*bwai boj'bá* in Barí, *Hymenaea courbaril*), the tree gourd (*shiima* in Barí and *tapara* in Venezuelan Spanish, *Crescentia cujete*), and the cannonball (*kóba* in Barí and *coco-de-mono, taparo chuco, muco* in Venezuelan Spanish, *Couroupita guianensis*). These trees produce thick 10–20 cm pods that were broken open by megafauna such as native horses, gomphotheres, and ground sloths (Janzen and Martin 1982; Janzen 1983). However, these trees are very rare in the Barí forests, with very few specimens recorded. For example, the cannonball tree has only been recorded once in all my forest plots; the copal has only been observed one time, a few kilometers northeast of Saimadoyi; and only a few tree gourds have been found at old Barí longhouse sites. Despite the literature suggesting that cannonball fruits can be dispersed by peccaries (Prance and Mori 1978) and copal by agoutis (Hallwachs 1986, Asquith et al. 1999), this does not seem to be the case in the Barí territory, as there appears to be little regeneration of these species. The rarity of these uncommon trees with their extralarge fruits could help to explain the abundance of other trees with medium and large fleshy fruits since the latter are dispersed by primates (Terborgh 1992).

Since the 1970s, many studies have provided detailed information on the topic of fruit trees and frugivores. For example, in the Peruvian Amazon, Terborgh (1986) recorded that 80% of mammalian and avian biomass is frugivorous. In the Eastern Amazon, Balée (1987) observed that 86% of 138 species of trees produced fruits eaten by hunted animals in Tembé territory. Cormier (2004) reported that in the Guajá territory, 90 species of plants were consumed by spider monkeys, 88 for capuchin monkeys, and 74 for black bearded saki monkey. In another study in Barro Colorado Island in Panama, out of the 291 species of trees, 78% produced fleshy fruits (Howe 1984). According to various other studies (Fleming et al. 1987; Jordano 2000, Smythe 1986), 70–94% of the Neotropical trees produce fleshy fruits in the low rainforest. The amount of fruits produced by central Amazonian forest trees is quite high and ranges from 0.85 to 1.3 metric tons per hectare annually (Fittkau and Klinge 1973). Therefore, multiple research studies confirm that lowland Neotropical forest trees provide an important source of food for birds and mammals and in significant amounts that facilitated their coevolution (Eriksson 2016; Fleming et al. 1987; Jordano 2000; Lambert and Garber 1998; Terborgh 1983).

The size of the fruit is an important consideration in their selection by monkeys according to Terborgh (1983: 75) and Eriksson (2016). Larger species of monkeys have preferences for larger fruits (>3 cm), while smaller monkeys focus on smaller fruits (<1 cm). Most of the fruits consumed by primates are medium- (1–3 cm) to large-size (>3 cm) seed drupes or berries. In the New World, primates consume mostly fruits that are "medium (44%) to large (45%) sized drupes (36.3%) and berries (22.2%) that contain seeds of between 0.5 and 2 cm (45.6%) in maximum dimension" (Lambert and Garber 1998: 16).

According to Lambert and Garber (1998), New World monkeys have the tendency of swallowing bigger seeds than Old World monkeys. Capuchin monkeys can swallow seeds as big as 1.8 cm and spider monkeys as big as 3 cm (Rowell and Mitchell 1991; van Roosmalen 1985, cited in Lambert and Garber 1998: 16–17). This is important because New World monkeys can disperse seeds from a greater number of species of trees, especially ones that are large seeded and are therefore not effectively dispersed by smaller-bodied frugivores.

Various studies in the New World detail the diversity of fruiting tree species consumed by primate communities. One of the most extensive studies of monkey diets in the New World is based in Manu Park, Peru (Terborgh 1983). This study recorded that 170 species of plants and 55 botanical families provided food for the primates studied, which represent about 20% of the total flora of Manu (1983: 62). Also in Manu, Lambert and Garber (1998: 16) indicated "153 plant species from 37 families in [their] sampled of the most common fruits eaten by the 14 platyrrhine taxa." According to Link and Di Fiore (2006), spider monkeys in Ecuador consume 152 species of fruit and disperse 133 of these. All of these studies support primates having a quite diversified diet in different regions.

According to the Barí, monkeys feed on many species of trees, and their forest has plenty of food for them. This view is quite common in the literature about frugivory and primate diet (e.g., Eriksson 2016; Milton 1978, 1980, 1982; Milton et al. 2005; Stevenson et al. 2015; Terborgh 1983). For example, Lambert and Garber (1998: 21) offer a compelling observation: "Given the tens of millions of fruits eaten by primates each year and the hundreds of millions of seeds transported by primates and other animals, dispersers can have a collective ecological effect on the local history of the forest and the present day distribution of trees."

In conclusion, primates are not passive agents in their forest but actively influencing the distribution and composition of tree species in their forest because fruit trees rely on seed dispersers to move their seeds away from the maternal tree, thereby increasing the seeds' chances of survival. The Janzen–Connell effect shows the importance of frugivores for maintaining ecological health of forests (Terborgh et al. 2008). This movement of seeds also influences genetic mixing and gene flow within and between populations of trees (Wilson and Traveset 2000). Therefore, anthropogenic changes to forest, such as defaunation and fragmentation, can affect these processes and therefore the regeneration of forest trees (Neuschulz et al. 2016).

6.3 Methodology

Data for this chapter were collected over 34 months of fieldwork starting in 1988 and ending in 2002 as well as many more months of analyzing it and writing about the Barí ethnoecology. The focus of the research was ethnobotanical looking at not only knowledge variation among individual Barí people but also data on species of trees edible for different animals, and their relative densities were recorded. While collecting information about their flora, the Barí eagerly volunteered information about their fauna and their diet, especially in regard to those that they hunted. In multiple interviews, the Barí were able to consistently indicate the types of fruits or plants that were consumed by different kind of animals, including primates.

I marked off 36 small 30 × 50 m forest plots (including two smaller plots of 20 × 50 m which were added for logistical purposes), covering 5.4 hectares of forest, in order to learn about the variation of the Barí knowledge of trees and to capture a representative census of the forest composition of species (see Fig. 6.1 for the map with locations of places mentioned in the text). I chose this size for the plots because it fits well on an 8.5″ × 11″ piece of paper at the scale of 1:200, facilitating the ease of manipulating this map while presenting a questionnaire for interview informants. From these plots, a list of Barí tree taxa was compiled. The trees censused were equal or bigger than 10 cm dbh (diameter at breast height, 130 cm), which is a standard measurement botanists use for collecting tree species information in forest plots. For this study, 20 adult Barí (7 women and 13 men) were interviewed for each of those forest plots to document their consensus and variation regarding the ethnotaxa. Six Barí men who were good hunters then went over the list of 227 species of trees and confirmed which species provided food for primates and other animals.

For the identification of these species, 398 botanical vouchers were collected between 1988 and 1995, most of them fertile and containing either flower or fruit, if not both. Six sets of each of the vouchers were deposited in the Victor Manuel Ovalles Herbarium (MYF) located in the Pharmacy School at the Universidad Central de Venezuela in Caracas. Three hundred and two of the vouchers were identified by a number of taxonomists who visited the V. M. Ovalles Herbarium as well as by Dr. Stephen Tillet, the director at that time. A set of approximately 100 vouchers, out of the 302, were identified by Dr. Paul Berry, Gerardo Aymard, and other taxonomists in 1996 at the herbarium of the Missouri Botanical Garden. In the last 4 years, a Venezuelan taxonomist, Jose Ramón Grande Allende, also worked on 48 vouchers that were not identified or that were misidentified. Ninety-six vouchers have not been identified, as some of these vouchers may have been discarded in the V. M. Ovalles Herbarium collection due to poor preservation which resulted in the destruction of identifiable features. Scientific names were corrected and updated by G. Aymard in 2018. Originally more vouchers were planned to be collected with the goal of collecting from one and two thousand of them. Also, due to a strong case of hepatitis, the author had to leave the field earlier and was not able to collect more later since the two Colombian guerrilla groups (FARC and ELN) entered the area of research, making this research rather too risky to continue.

The current Barí territory covers 231,000 hectares that are still mostly forested. It is important to indicate that the 5.4 hectares of forest censused consisted of mostly primary forests with a few sections of secondary forests. Not only the forests in this study are focused mostly around the village of Saimadoyi (22 plots), but also a dozen plots extend from the Santa Rosa River on the north to Río de Oro on the south, and two plots on the mid-altitude section of the Sierra de Perijá were added to try to capture a greater number of tree species. Two additional plots (#25 and #26) are by the Ariquaisa River east and outside the Barí reservation (see Fig. 6.1). These last two plots are particularly interesting since they represent what was the largest type of forest originally found in the 1900s Barí territory which originally had a large population of monkeys but which today has been mostly deforested by cattle ranchers.

6.4 Number of Forest Trees That Produce Food for Primates

In tropical rainforests, we commonly find that species of trees have populations that have extreme numbers ranging from many (potentially hundreds in some cases) to very few individuals, with some being extremely rare with only one or few individuals known by the local population in the region (Terborgh 1992; Gentry 1996; Lizarralde 1997; Brent Berlin 1987: pers. comm.; Glenn Shepard Jr. 1996: pers. comm.; Zizka et al. 2018). The contribution of rare species to the tropical diversity has been recognized (Wills et al. 2006; Kenfack et al. 2007); however, their spatial distribution remains poorly understood (Zizka et al. 2018). The latter recent work on rarity in the Neotropics identified 26,315 species for Amazonia, of these 10,080 species as putatively rare within this region (Zizka et al. 2018). Inside Amazonia most collections of rare species were in the sub-Andean region and on the Guayana Shield and in few areas scattered across the study area. The authors also found that rare species are homogeneously distributed through most parts of the lowland Neotropics and Amazonia but more concentrated in highlands, with no clear disjunction patterns within lowland areas. These results suggest that a considerable proportion of rare plant species has surprisingly large distribution ranges (e.g., *Peridiscus lucidus*; see Aymard and Arellano 2018) and that collections of rare species across most of the lowland Neotropics, and in particular in Amazonia, show no clear directionality.

These sparse recordings were also observed in the Barí territory with almost half of the species represented only once or twice in the plots censused (Lizarralde 1997). Therefore, it is also relevant to include the number of individual trees since this information offers the demographics and densities of these species, and it indicates the relative amount of food available for the region. The important information missing here is the basal size of the trees that is key to relative abundance of fruits (which will be provided in the near future). Even though the Barí mentioned hog plum (*baróo* in Barí, *Jobo* in Spanish, *Spondias mombin*, Anacardiaceae) or breadnut tree (*barúu* in Barí; *Ramón*, *Charo*, or *guaimaro* in Spanish; and Mayan nut in English, *Brosimum alicastrum*, Moraceae) as important for their population of monkeys, having the number of individual trees representing different species might provide a more accurate perspective of which tree species are more important primate food in the Barí forests (instead of only listing species without their number of individual trees representing their species). Further in this chapter, the most important botanical families of trees and species are ranked.

Out of all the trees censused ($N = 3664$) in the 5.4 hectares of forest, 2476 individual trees (67.8%) provide food for monkeys. Among the 28 botanical families that provide food for monkeys, the top 10 represent 57.5% of all the trees that were

Table 6.1 The 10 most important botanical families in this study providing food for monkeys

		Individual trees	
Botanical family	No. of species	Number	Percentage
Arecaceae (Palmae)	8	758	20.69
Fabaceae (Leguminosae)	13	390	10.64
Moraceae	14	228	6.22
Sapotaceae	4	193	5.27
Burseraceae	3	168	4.59
Lecythidaceae	5	97	2.65
Annonaceae	5	91	2.48
Bignoniaceae	1	87	2.37
Combretaceae	1	45	1.23
Anacardiaceae	2	43	1.17
Total	**57**	**2100**	**57.3%**

censused in the forest plots in the Sierra de Perijá (see Table 6.1). They also represent 86.2% of all the trees that provide food for monkeys.

These leading botanical families are consistent with other regional studies (Russo et al. 2005). Terborgh (1983) also pointed out that the most important botanical families that provide food for the monkeys in Manu Park (Peru) are Moraceae, Annonaceae, Palmae, and Fabaceae. In another study, Stevenson (2004: 277) stated that the most important botanical families of fruit for wooly monkeys were Moraceae, Lecythidaceae, Anacardiaceae, Mimosaceae (in the Fabaceae), Urticaceae, Burseraceae, and Sapotaceae. Also in Manu park, Lambert and Garber (1998: 16) pointed out that 53% of these trees were represented by five botanical families: "Moraceae (23 species, including 12 *Ficus* species), Leguminosae (16 species), Sapotaceae (8 species), Palmae (8 species), and Annonaceae (6 species)." These findings indicate the Barí forest is not unique or exceptional in terms of the diversity of species feeding its primates.

6.5 Species of Trees that Produce Food for Primates

Looking at the number of species of trees based on this research, mostly around the village of Saimadoyi (but also around Bachichida, Kumanda, Bokshí, and Bagsarani), we note that 102 species of trees provide food for many animals, including monkeys (see Table 6.2 for the complete list and their details). Therefore, 44.9% of the 227 species of trees censused are food sources for primates.

The 18 most important species of trees noted below represent 46% of all the trees included in this study that provide foods for monkeys (see Table 6.3). This statistic, however, does not represent the preference of these fruits by monkeys but only the most abundant of all fruits known to feed them. Among the most important tree species in this list of trees, the Barí indicated that hog plum (*Spondias mombin*,

Table 6.2 List of trees in the Barí territory that provide food for monkeys

Botanical family	Scientific name	Native name (Spanish or English names)	Density (no. individual trees recorded)	Botanical voucher number	Fruit characteristics and edibility
Anacardiaceae	*Spondias globose*	*Íshiraberi* (ciruelo pequeño)	0.03% (1)	ML307 (it is in MO as S. mombin)	Yellow fruit (2–3 cm drupe) edible for many animals including monkeys and people
Anacardiaceae	*Spondias mombin*	*Baróo* (jobo or hog-plum)	1.15% (42) higher at 700–1200 msnm of the Sierra of Perijá	ML155 and ML319	Yellow fruit (3–4 cm long and 2–2.5 cm diameter drupe) edible for many animals including spider, capuchin, and howler monkeys
Annonaceae	*Annona mucosa*	*Dabaikaa* (anoncillo)	0.08% (3)	ML179	Yellow fruit (8–10 cm) edible for many animals
Annonaceae	*Duguetia lucida*	*Bichirabú* (yaya, anoncillo, yarayara amarillo)	0.19% (7)	ML201	Red fruit (6–8 cm) edible for many animals including spider and howler monkeys
Annonaceae	*Guatteria amplifolia*	Chirabuu babai (yaya “negra”)	0.06% (2)	–	Orange sweet fruit (1–2 cm) for many animals including spider, capuchin, and howler monkeys
Annonaceae	*Guatteria ucayalina*	*Tairù*	0.68% (25)	ML229	Bitter fruit edible for many animals including capuchin and spider monkeys
Annonaceae	*Oxandra venezuelana*	*Chirabuu* (yaya)	1.47% (54)	ML219	Orange fruit (1 cm drupe) edible for many animals
Apocynaceae	*Couma macrocarpa*	*Aino’bà* (perrillo, milk tree)	0.03% (1)	ML188	Yellow sweet fruit (4–5 cm berry) consumed by all monkeys, many animals, and people
Apocynaceae	*Mesechites trifida*	*Bo’baa*	0.11% (4)	ML258	Yellow fruit (15–40 cm long and 0.5 cm diameter) edible for many animals including spider and howler monkeys
Araliaceae	*Schefflera morototoni*	*O’boo* (yagrumo macho, orumo)	0.22% (8)	ML181	Yellow small fruit (8 mm) edible by some animals

Arecaceae (Palmae)	*Attalea butyracea*	*Ara'ta* (corozo, wine palm, American oil palm)	2.35% (86)	–	Orange/yellow fruit edible (3–4.5 cm diameter and 5–8 cm long) for many animals including capuchin monkeys and humans
Arecaceae (Palmae)	*Bactris gasipaes* var. *chichagui* type 2	*Téchi* (macanilla, wild peach palm, chontaduro, wild pejibaye)	0.82% (30)	ML090	Orange-red fruit (1.0 cm diameter and 1.5 cm long) edible for many animals and humans
Arecaceae (Palmae)	*Bactris major* var. *major*	*Kari'bai* (macanilla pequeña, prickly palm)	2.81% (103)	ML128	Red fruit edible (1.5–3.5 cm diameter and 2.5–4.5 cm long) for many animals and humans
Arecaceae (Palmae)	*Chamaedorea* cf. *pauciflora*	*Burubüü* (suma-yuca, sangapilla)	0.19% (7)	ML008	Fruit edible (0.5–0.8 cm diameter and 1.2–1.5 cm long) for some animals
Arecaceae (Palmae)	*Euterpe oleracea*	*Arihbei* (açai palm, palmicha)	1.39% (51)	ML319	Black fruit edible (1–1.8 cm diameter) for capuchin and spider monkeys as well as humans
Arecaceae (Palmae)	*Oenocarpus bataua* var. *bataua*	*Arúu* (mapora grande, seje, patauá, batauá, aricagua, milpesos, trupa)	4.12% (151)	ML326	Black fruit (2.2–2.5 cm diameter and 2.5–4.5 cm long) edible for many animals including howler, capuchin, and spider monkeys as well as humans
Arecaceae (Palmae)	*Oenocarpus mapora*	*Keki* (mapora mediana, bamboo palm, bacaba, sinamillo)	8.57% (314)	ML320	Purple fruit (1.5–2.5 cm diameter and 2–3 cm long) edible for many animals including monkeys and humans
Arecaceae (Palmae)	*Socratea exorrhiza*	*Logsó* (palma de cacho, walking or stilt palm)	0.47% (16)	ML325	Yellowish fruit (1.5–2 cm diameter and 2.5–3.5 cm long) edible by many animals, especially *Steatornis caripensis* (oilbird) and many monkeys
Bignoniaceae	*Jacaranda copaia* subsp. *spectabilis*	*Chiri'baa* (huamansamana or pata de elefante)	2.37% (87) (secondary forest tree)	ML266	Green fruits edible (10 cm) for many animals (mostly parrots and macaws but some mammals too)
Malvaceae	*Quararibea* cf. *aristeguietae*	*Sobo'bò*	1.04% (38)	ML168	Green fruits edible for many animals including monkeys

(continued)

Table 6.2 (continued)

Botanical family	Scientific name	Native name (Spanish or English names)	Density (no. individual trees recorded)	Botanical voucher number	Fruit characteristics and edibility
Boraginaceae	*Cordia bicolor*	*Nunku'bóo* (alatrique, caujaro blando, or cautaro)	0.33% (12)	ML263	Sweet fleshy yellow/green fruits (0.8 cm berry) edible for many animals including spider monkeys
Burseraceae	*Dacryodes* sp. (cf.)	*Tootogbáa* (urá)	0.79% (29)	ML084	Yellow-orange fruits (2 cm) edible for many animals
Burseraceae	*Protium sagotianum*	*Ishku'baa* (tacamahaco)	2.95% (108)	ML101	Red very sweet fruit (4–5 cm, white flesh around seeds) appreciated by monkeys and humans
Burseraceae	*Protium tenuifolium*	*Loai shkugbá* (tacamahaco)	0.85% (31)	ML198	Red fruits edible (2 cm) for many animals including monkeys
Chrysobalanaceae	*Chrysobalanus icaco*	*Asaj'bõõgba laba'dou* (icaco, coco plum, paradise plum)	0.03% (1)	–	Yellow fruit (4 cm) is eaten by all monkeys. Domesticated plant and anthropogenically distributed in forest near villages
Chrysobalanaceae	*Licania* sp. 1	*Asaj'bõõgba* (different than *icaco*)	0.19% (7)	ML267	Yellow sweet fruit (4 cm) is like *Chrysobalanus icaco* and consumed by many animals, including monkeys
Chrysobalanaceae	*Licania* sp. 2	*Asaj'bõõ*	0.03% (1)	–	Yellow fruit (5 cm) is eaten by capuchin, howler, and spider monkeys, besides other animals
Combretaceae	*Terminalia amazonia*	*Son'bà* (palo amarillo, roble coral)	1.23% (45)	ML285	Orange fruits (1 cm) edible for many animals
Euphorbiaceae cf.	*Croton* sp. cf.	*Shumi'dae'*	0.25% (9)	ML167	Fruits (1 cm) edible for many animals
Euphorbiaceae	*Hyeronima alchorneoides*	*Yio'* (pilón)	0.22% (8)	ML289	Red-dark purple sweet fruits (2–3 cm) edible for many animals including birds and monkeys
Fabaceae/ Caesalpiniaceae	*Brownea coccinea* subsp. 1	*Shkúbabá abama* (rosa de monte, mountain rose)	0.52% (19)	ML215	Green fruits (20–25 cm long pod) edible for many animals

Fabaceae/ Caesalpiniaceae	*Brownea coccinea* subsp. 2	*Shkúbabà* (rosa de monte, palo de cruz)	6.63% (243)	ML170	Green fruits (20–25 cm long pod) edible for many animals
Fabaceae/ Caesalpiniaceae	*Dialium guianense*	*O'chiri* (cacho, velvet tamarind, parajuba)	0.46% (17)	ML223	Orange very sweet fruit (2 cm) appreciated by monkeys and humans
Fabaceae/ Caesalpiniaceae	*Dialium heaxstaminatumi*	*Ichow* (wild tamarind)	0.41% (15)	ML082	Green very sweet fruit (2 cm) highly appreciated by human, many animals, and monkeys
Fabaceae/ Caesalpiniaceae	*Peltogyne paniculate*	*Bo'shi* (nazareno, zapatero, purpleheart)	0.11% (4)	–	Green sour fruit (5 cm pod) consumed by howler monkeys and many other animals
Fabaceae/ Mimosoideae	*Inga* sp.	*Kamio'ba* (guamo de monte)	0.14% (5)	ML054	Green fruit edible (40 cm long pod) for many animals, including monkeys and people
Fabaceae/ Mimosoideae	*Inga* sp. (wild *Inga spectabilis*)	*Kããkarabá* (guamo, wild ice-cream-bean)	0.49% (18)	–	Green fruit (25–30 cm long pod) edible for many animals, including monkeys
Fabaceae/ Mimosoideae	*Inga cocleensis*	*Kamashkorou nondyiruku* ("howler monkey *Inga*")	Very low (0)[a]	ML327	Green fruit (18–22 cm long pod) edible for many animals, including monkeys and humans
Fabaceae/ Mimosoideae	*Inga scabriuscula*	*Nondyíruku* (guamo chivo, ice-cream-bean)	1.42% (52)	ML248	Green sweet fruit (40–45 cm long and 1.8 cm diameter pod) edible for many animals and humans
Fabaceae/ Mimosoideae	*Inga quaternate*	*Birichíboo abama* (guamo venezolano, Venezuelan ice-cream-beam)	0.03% (1)	ML238	Green sweet fruit (5–20 cm long and 3 cm wide pods) edible for many animals, including capuchin, howler, and spider monkeys
Fabaceae/ Mimosoideae	*Inga marginata*	*Birichíboo* (guamo)	0.16% (6)	ML249	Green sweet fruit (14 cm long and 1 cm diameter pod) edible for people, monkeys, and many other animals
Fabaceae/ Mimosoideae	*Parkia* sp. 2	*Kõ'dai abama*	Very low (0)[a]	–	Green fruit or flower edible for many animals, including monkeys
Fabaceae/ Papilionoideae	*Platypodium elegans*	*Dyiroaibakãã bãshï* (draque, graceful platypodium)	0.27% (10)	ML313	Green fruit (10 cm long samara) edible for many animals, including howler and spider monkeys; Howler monkeys also eat its leaves

(continued)

Table 6.2 (continued)

Botanical family	Scientific name	Native name (Spanish or English names)	Density (no. individual trees recorded)	Botanical voucher number	Fruit characteristics and edibility
Flacourtiaceae	*Laetia procera*	*Shirokaru* (jobo macho, cuero de rana)	0.14% (5)	ML275	Green fruit (1.5–2 cm drupe) edible for many animals including spider monkeys
Flacourtiaceae	*Lindackeria paludosa*	*Bohkãã kaa* (sarakura)	0.46% (17)	ML252	Orange fruit (0.8 cm drupe) edible for many animals
Indeterminate	Indeterminate	*A'doukàru*	0.55% (20)	ML355	Green sweet black fruit (2 cm) edible for many animals including all monkeys
Indeterminate	Indeterminate	*Batu'bó*	0.19% (7)	ML160	Sour fruit (3.5 cm drupe) edible for howler and spider monkeys
Indeterminate	Indeterminate	*Birinkaru*	0.11% (4)	ML351	Fruit edible for many animals
Indeterminate	Indeterminate	*Dyiroaibakããá*	0.03% (1)	–	Fruit supposedly edible for many animals
Indeterminate	Indeterminate	*Kobeekaa*	0.14% (5)	–	Fruit edible for many animals, including primates
Indeterminate	Indeterminate	*Liruu maama*	0.03% (1)	–	Fruit edible for many animals, including primates
Indeterminate	Indeterminate	*Maa*	0.22% (8)	–	Fruit edible for green parrots (*Amazona* spp.), other smaller parrots, toucan (*Ramphastos ambiguous*), kinkaju, spider monkey, and other animals
Indeterminate	Indeterminate	*San'boo*	Very low (0)[a]	–	Fruit edible for many animals
Indeterminate	Indeterminate	*Sasan'boo*	Very low (0)[a]	–	Fruit edible for many animals, including spider monkey and kinkajou
Indeterminate	Indeterminate	*Shiborokoo abama*	Very low (0)[a]	–	Fruit edible for many animals, including monkeys
Indeterminate	Indeterminate	*Tó'boo*	0.33% (12)	ML182	Fruit (1 cm) edible for many animals and humans

Indeterminate	Indeterminate	*Tu'tu'boo*	0.22% (8)	–	Fruit edible for many animals
Indeterminate	Indeterminate	*Tubi'tróo'boo*	0.33% (12)	ML093	Fruit edible for many animals
Lauraceae	Indeterminate	*Chiru'chiru'kaa*	0.06% (2)	ML328	Fruit edible for many animals (currasows, toucan, parrots, and kinkajou, among others)
Lauraceae	*Licaria* sp.	*Shi'bóo*	0.55% (20)	ML157	Green bitter fruits (4 cm) edible for many animals
Lecythidaceae	*Cariniana pyriformis*	*Bahku* (bacú, albarco, Colombian mahogany)	1.20% (44)	ML089	Brown fruit 10 cm with small seed (1 cm) like a pine nut edible for capuchin, howler, and spider monkeys
Lecythidaceae	*Gustavia* sp.	*Logsorobogboo* (chupon)	0.06% (2)	–	Fruit edible for many animals, including monkeys
Lecythidaceae	*Gustavia* cf. *hexapetala*	Logsorologsoro abama (chupon)	0.03% (1)	ML297	Green fruits (1 cm) edible for many animals, including monkeys
Lecythidaceae	*Gustavia speciose*	Logsorologsoro (chupon)	0.25% (9)	ML154	Green fruit (5 cm drupe) edible for many animals, including monkeys
Lecythidaceae	*Lecythis corrugate*	*Lugshuu*	1.12% (41)	ML257	Green fruit (5 cm drupe) edible for many animals, including paca, agouti, green parrots, macaws, bracket deer, and peccary, among others
Melastomataceae	*Bellucia* cf. *pentamera*	*Shóotu'baa*	0.76% (28)	ML196	Green fruit (4–5 cm drupe) edible for many animals
Meliaceae	*Swietenia macrophylla*	*Buyó'baa* (caobo)	0.41% (15)	ML159	Green fruit (12–15 cm) edible for some animals
Moraceae	*Brosimum alicastrum*	*Barúu* (breadnut, Mayan nut, guaimoro, Ramón)	0.46% (17)	ML158	Yellow fruit (2 cm) and seed edible (1.5 cm) for spider, capuchin, and howler monkeys, among many other animals
Moraceae	*Brosimum lactescens*	*Ko'doai* (coquino amarillo, leche dulce)	0.82% (30)	ML265	Red fruit (0.8–1.2 cm drupe) edible for many animals, including monkeys

(continued)

Table 6.2 (continued)

Botanical family	Scientific name	Native name (Spanish or English names)	Density (no. individual trees recorded)	Botanical voucher number	Fruit characteristics and edibility
Moraceae	*Ficus* sp.	*Ishiba abama* (higuerote gigante)	0.08% (3)	–	Green fruit (3 cm fig) edible for many animals, including monkeys
Moraceae	*Ficus* sp.	*Ishibaa* (higuerote)	0.27% (10)	ML197	Green fruit edible (2.5 cm fig) for many animals
Moraceae	*Ficus* sp.	*Luru'boo abama* (≠ *luru'baa*)	Very low (0)[a]	–	Fruit (2 cm fig) edible for many animals
Moraceae	*Ficus* sp.	*Luru'baa* (mapapalo)	0.06% (2)	–	Green fruit (2 cm fig) edible for many animals
Moraceae	*Ficus* sp.	*Moeshiba* (giant strangler fig)	0.14% (5)	ML150	Green fruits (2.5 cm fig) edible for many animals, including monkeys
Moraceae	*Ficus* sp.	*Luru'boo* (≠ *luru'baa*)	0.08% (3)	ML323	Green fruit (3 cm fig) edible for many animals
Moraceae	*Ficus insipida*	Ishibaugbou	Very low (0)[a]	ML034	Green fruit (2.5–3 cm fig) edible for many animals, including monkeys
Moraceae	*Helicostylis tomentosa*	*Shíndwe* (charo macho)	0.82% (30)	ML234	Yellow highly appreciated edible sweet fruit (3–5 cm diameter) for many animals, including monkeys and people
Moraceae	*Maclura tinctorial*	*Ishdãkaa* (mora)	0.14% (5)	ML114	Green fruit (1–1.5 cm) edible for many animals, including monkeys and people
Moraceae	*Maquira guianensis*	Dagyikogbaa	1.86% (68)	ML286	Red fleshy sweet fruit (0.8–1.6 cm) edible for many animals, including monkeys
Urticaceae	*Pourouma cecropiifolia*	*Chi'bidãa* (yagrumo, Amazon tree grape, uvilla)	0.44% (16)	ML232	Purple fleshy fruit (2.5–3 cm) edible for many animals, including, curassows, green parrots, kinkajou, and howler and spider monkeys
Urticaceae	*Cecropia* sp. 2	*Tu'baa* ("secondary forest cecropia")	1.06% (39)	ML180	Green fruit edible for many animals
Primulaceae	*Ardisia guianensis*	*Drã'biña* (coralberry or marlberry)	0.11% (4)	ML276	Blue-black fruit (0.4 cm berry) edible for many animals, including monkeys and people

Myristicaceae	*Virola* cf. *sebifera*	Labibúu (camaricaro, red ucuuba)	1.04% (38)	ML233	Red fruit (2 cm long and 1.5 cm diameter dehiscent capsule) edible for mostly birds but some monkeys too
Polygonaceae	*Coccoloba* sp.	A'doudakaa babai	0.03% (1)	ML254	Black fruit (0.6 cm) edible for many animals
Polygonaceae	*Triplaris caracasana*	*Chirahbáakaa* (palo de Maria, palo santo)	0.19% (7)	ML279	Red fruit (3 wing samara with a 1 cm seed) edible for many animals, including monkeys
Rubiaceae	*Genipa americana*	*Mamañio'baa* (caruto, genipapo, huito)	0.08% (3)	ML240	Green fruit (10–15 cm drupe) edible for many animals, including kinkajou and capuchin and spider monkeys
Rubiaceae	*Posoqueria latifolia*	*Ko'dobo'báa* (manzanillo, monkey apple)	0.03% (1)	ML264	Yellow sweet fruit (5 cm berry) edible for many animals, including monkeys
Sapindaceae	*Dilodendron costarricense*	*Kandya* (machirio tamarindo, forest tamarind)	0.19% (7)	ML152	Red bitter fruit (4 cm berry) edible for many animals, including capuchin, spider, and howler monkeys
Sapotaceae	Indeterminate	*Lorogbá*	1.34% (49)	ML190	Large yellow sweet fleshy fruit (6 cm) is an important food for hunted animals like paca, agouti, bracket deer, green parrots, macaws, kinkajou, and spider, capuchin and howler monkeys
Sapotaceae	*Pouteria reticulata*	*Abo'bò*	1.26% (46)	ML153	Yellow fruit (7 by 4.5 cm) edible for many mammals including all monkeys
Sapotaceae	*Pouteria* sp.	*Burúma*	2.27% (83)	ML270	Yellow fruit (6 cm drupe) edible for many animals including monkeys
Sapotaceae	*Sarcaulus* sp.	*Kwí*	0.41% (15)	ML213	Fruit (2.5–3 cm drupe) edible for many animals, including spider monkeys
Malvaceae	*Herrania albiflora*	*Birón'wã'wã* (cacao de monte, cacaito, monkey cacao)	0.14% (5)	ML019	Yellow-orange very sweet fruit (7.5–10.5 cm) edible for many animals including monkeys

(continued)

Table 6.2 (continued)

Botanical family	Scientific name	Native name (Spanish or English names)	Density (no. individual trees recorded)	Botanical voucher number	Fruit characteristics and edibility
Malvaceae	*Pterygota* colombiana	*Bo'soro'baa*	0.19% (7)	ML235	Fruit (10–15 cm) edible for many animals, including capuchin, howler, and spider monkeys as well as humans
Malvaceae	*Theobroma* sp. 1	*Daairu'bá tagtabaankorai* (green flower wild cacao)	Very low (0)[a]	–	yellow fruit (15 cm) edible for many animals
Malvaceae	*Theobroma* cacao	*Daairu'bá bashï* (white flower wild cacao)	0.06% (2)	ML321	Yellow fruit (15 cm) edible for many animals
Malvaceae	*Theobroma* sp. 3	*Daairu'bá bokimai* (red flower wild cacao)	Very low (0)[a]	–	Yellow fruit (15 cm) edible for many animals
Malvaceae	*Theobroma* sp. 4	*Daairu'bá karikanshundu* (yellow flower wild cacao)	Very low (0)[a]	–	Fruit (10 cm) edible for many animals
Ulmaceae	*Ampelocera macrocarpa*	*Luríi*	1.04% (38)	ML268	Yellow edible (1.5–2.2 cm) fruit for paca, agouti, parrots, toucan, macaws, squirrels, kinkajou, and monkeys
Lamiaceae	*Vitex divaricate*	*Maama* (totumillo, white fiddlewood)	0.25% (9)	ML381	Green fruits (0.8 cm) edible for many animals, including spider monkeys
Total: 28 families	**102 species**		**67.58% (2476)**	**79**	

[a]These trees were not plotted, but either botanical vouchers were collected or the Barí mentioned these in interviews as important

Table 6.3 The 18 most important tree species recorded in the Barí plots

Species	Family	No. of trees	Percent
Oenocarpus mapora	Arecaceae (Palmae)	314	8.6
Brownea coccinea	Caesalpiniaceae (Fabaceae)	262	6.6
Oenocarpus bataua var. *bataua*	Arecaceae (Palmae)	151	4.1
Protium sagotianum	Burseraceae	108	3.0
Bactris major var. *major*	Arecaceae (Palmae)	103	2.8
Jacaranda copaia subsp. *spectabilis*	Bignoniaceae	87	2.4
Attalea butyracea	Arecaceae (Palmae)	86	2.4
Pouteria sp.	Sapotaceae	83	2.3
Maquira guianensis	Moraceae	68	1.9
Oxandra venezuelana	Annonaceae	54	1.5
Inga scabriuscula	Mimosaceae (Fabaceae)	52	1.4
Euterpe oleracea	Arecaceae (Palmae)	51	1.4
Indeterminate (*loro'bá*)	Sapotaceae	49	1.3
Pouteria reticulata	Sapotaceae	46	1.3
Terminalia Amazonia	Combretaceae	45	1.2
Cariniana pyriformis	Lecythidaceae	44	1.2
Spondias mombin	Anacardiaceae	42	1.2
Lecythis corrugate	Lecythidaceae	41	1.1
Total		**1686**	**46.0%**

Anacardiaceae) was among the top, especially in the mid to high altitudes, between 700 and 1000 meters above sea level on the Sierra de Perijá, where, according to the Barí, the population of spider and howler monkeys is much higher than in other parts of their territory. In fact, the Barí indicated that it was quite obvious that spider monkeys feasted on these fruits because the fat in their abdomen becomes yellowish to orange during the seasons of hog plums, from November to December and June to September. The author was able to confirm this observation, while the Barí were butchering a dead monkey in June of 1999 (see Fig. 6.4).

Another important source of food is breadnut (*Brosimum alicastrum*, Moraceae), which is also a favorite fruit tree for the monkeys of other Neotropical regions (Russo et al. 2005; Stevenson 2015; Terborgh 1983). It is not surprising that hog plum and breadnut are essential fruit trees for primates and many other animals in these forests. This observation is based on the fruit mass these two trees normally produce (Stevenson et al. 2015). Even though the density of breadnut is not very high (0.46%, with 17 individual trees registered) in the plots recorded, it is much more abundant at higher altitudes between 250 and 700 meters above sea level on the Sierra de Perijá (around plot 27–28) and Serrania de Abusanqui (around Bagsarani village), where the population of spider and howler monkeys is also very high according to the Barí, which was also observed several times between 1995 and 1999.

Similarly, in Yalbac (Belize), in an area surrounded by Mayan ruins and on a large man-made mound, it was observed that the breadnut tree attracts wildlife, including spider and howler monkeys, parrots, and toucans (personal observation in mid-March of 2008, 2011, and 2013). In Perijá, the Barí indicated that the fruits of the breadnut tree are abundant in April, in the beginning of the rainy season, when avocados are also ripe (M. Lizarralde 1995: pers. obs.). In these periods, the Barí also consume a substantial amount of breadnut tree fruits too. Cormier and Urbani (2008: 381) stated that breadnut tree density is very high near Mayan ruins and it is the preferred food for spider monkeys. Terborgh (1983) also mentioned several times that *Brosimum* fruit was very popular for monkeys in Manu as well as Kinzey and Norconk (1993) for monkeys in the Venezuelan Guayana region of Guri Lake, Bolivar state, mainly *B. alicastrum* subsp. *bolivarense* and *B. guianense* (Aymard et al. 1997).

Russo et al. (2005) stated that *Brosimum* was among the most preferred genus in four regions for spider monkeys: Tinigua (Colombia), Yasuni (Ecuador), Voltzberg (Surinam), and Barro Colorado Island (Panama). Stevenson et al. (2015) listed three species of *Brosimum* (*B. guianensis*, *B. alicastrum*, and *B. lactescens*) as very common for monkeys in Tinigua. In the Barí plotted forest, we recorded 47 *Brosimum* trees (17 *B. alicastrum* and 30 *B. lactescens*).

Most tree fruits favored by primates are fleshy and juicy like hog plum or have oily flesh like American oil palm (*Attalea butyracea*). Others such as breadnut or Mayan nut (*Brosimum alicastrum*) also provide starchy seeds that attract a large number of not only monkeys but also birds and other animals. A few trees such as Bacú (*Cariniana pyriformis*) and other medium-size palm fruits provide oily seeds too. Apparently, some fruits are alternative choices if the most preferred fruits are not available in particular times of the year.

Also, the large seeded fruits of different palms, bamboo palm (*Oenocarpus mapora*) and seje or patauá palm (*Oenocarpus bataua* var. *bataua*), are other highly preferred food sources found near the communal houses or where these longhouses once stood. Anthropogenic forests were created where people discarded fruits near their villages or homes, dispersing different tree species and ultimately shaping the forest community structure after villages were abandoned (Balée 2013; Rival 2016). Rival (2016) stated that *Oenocarpus bataua* was found frequently in old encampments sites. We have observed this pattern clearly with palm species like seje-patauá and American oil palm. The Barí confirmed this, which is also evident from aerial photographs of longhouses that show higher densities of these palms in their vicinity. In fact, further to confirm this observation, plot #36 was established on a former longhouse location where spider monkey hunting was the primary source of protein. This plot had a very high density of different species of palm. In this plot, there were 41 bamboo palms (*Oenocarpus mapora*, 19.9% density out of 206 individual trees 10 cm dbh were recorded), 12 prickly palms or cubarro (*Bactris major* var. *major*, 5.8% density), and 35 seje palm (*Oenocarpus bataua* var. *bataua*, 17.0% density). Therefore, these three palm species had a higher density with 42.7% of them recorded in this forest plot. This finding could be an indication that this entire region (the Lora River basin) also has a high density of these palms because they also appear to have been the most commonly gathered wild (anthropogenic) fruits. Also, this region has a very high abundance of monkeys, especially spider monkey with troops of up to 50 individuals (Lizarralde 2002, 2019), who are at the same time dispersing these seeds. Other studies suggest that the genus *Oenocarpus* is also eaten and dispersed by monkeys (Link and Di Fiore 2006; Russo et al. 2005; Stevenson 2015).

There are two other groups of important trees belonging to the various species of legumes/ice-cream-beans and figs that play an important role in the diet of monkeys as well. For example, one of the types of wild ice-cream-bean trees is called the "howler-monkey-ice-cream-bean" (*kamashkorou nondyiruku, Inga cocleensis* subsp. *megantha*). It is not surprising that the fleshy fruits preferred by the Barí people are the same ones that monkeys also consume due to their high caloric content in the form of sugars, which is needed to sustain our large brains. According to Lambert and Garber (1998: 24), "Legume pods ... were among the top ten fruit species eaten by many New World monkeys." In Voltzberg, Surinam, *Inga* was the second most consumed fruit among spider monkeys (Russo et al. 2005). Stevenson et al. (2015) also reported five species of *Inga* to be very important to primates. This situation is similar for the Barí since they have 13 species of legume trees with edible pods making up to 380 (15.5%) of individual trees in the plots.

The use of figs for food by monkeys has been consistently cited in the literature (e.g., Díaz-Martín et al. 2014; Terborgh 1983; Lambert and Garber 1998; Stevenson et al. 2015). However, Terborgh (1983, 1986) points out that some figs are the last resort when other fruits are not available, even though figs are important choices in other studies (Lambert and Garber 1998). Stevenson et al. (2015) reported three species of figs to be high in terms of proportions of fruits handled by primates. However, these trees can produce a significant amount of fruits throughout the year, attracting an array of animals (including various species of monkeys) and therefore enhancing their dispersal (Díaz-Martín et al. 2014). For this reason, the Barí like to use fig trees as a place to hunt monkeys by placing a blind with platforms on the crown of nearby trees (Lizarralde 2019). In their forest, we have recorded 23 individual fig trees, many of which tended to be large in general, reaching a diameter of the trunk at breast high over 2 meters and a height of 30–40 meters.

According to Terborgh (1983), the most important tree genera that feed monkeys are *Ficus* (Moraceae), *Brosimum* (Moraceae), *Guatteria* (Annonaceae), *Casearia* (Salicaceae), *Inga* (Fabaceae), *Cecropia* (Urticaceae), *Celtis* (Cannabaceae), and *Cissus* (Vitaceae). The main reason is that all of these genera "contain soft pulpy material surrounding the seeds" (Terborgh 1983: 74). Another study by Russo et al. (2005: 1033) notes that "consistent preferences by spider monkeys for *Brosimum*, *Cecropia*, *Virola*, and *Ficus*. *Brosimum* may be preferred because of the generally large quantities of fruit produced and long periods of fruit availability on individual trees."

Primates also have certain preferences for fruits that have bright colors, and these preferences can vary from region to region according to Lambert and Garber (1998). The colors of choice are yellow and red, which account for 50% of fruits (and 80% including also orange and purple) in Manu (Lambert and Garber 1998). A similar pattern emerges among the Barí primates with yellow and red as their preferred fruit colors followed by orange, black, brown, and purple. Therefore, yellow, orange, or red fleshy or pulpy fruits, starchy or oily seeds, figs, and sweet pods are among the preferred fruits among primates in Perija.

Table 6.4 Color of fruits for all edible species of trees (n = 102 species/2476 individuals)

Fruit color	Individuals (n = 2476) No.	%	Species (n = 102) No.	%	All forest (N = 3664) %
Red	407	16.4	10	9.8	11.1
Yellow	345	13.9	23	22.6	9.4
Orange	251	10.1	7	6.9	6.9
Red + yellow + orange	**1003**	**40.5**	**40**	**39.2**	**27.4**
Purple	338	13.7	3	2.9	9.2
Black	207	8.4	4	3.9	5.7
All above colors	**1548**	**62.5**	**47**	**46.1**	**42.2**
Green	665	26.9	29	28.4	18.2
Brown	44	1.8	1	1	1.2
Total[a]	**2257**	**91.2**	**77**	**74.5**	**61.6**

[a]Totals exclude the lumped categories of "red+yellow+orange" and "all colors above"

Green edible fruits are the most popular color preferred by monkeys in Sierra de Perijá with 665 individual trees (26.9%) and 29 species (28.4%). The reason for it is that they are produced by two of the most important and populous families that commonly produce green fruits (Moraceae and Leguminosae). However, vibrant colors are also very popular if we combine all colors (except green and brown; see Table 6.4). These account for 62.5% (1548 individual trees) of fruits. If we include in all plots censused (N = 3664), this represents 42.2% of all trees. Unfortunately, we do not have information for 25 species and 219 (8.8%) trees that produced edible fruits for monkeys nor information for all other trees that do not produce fruit for monkeys. Like other studies (Lambert and Garber 1998), red fruits are represented with 407 (16.4%) individuals and 10 (9.8%) species, followed by yellow with 345 (13.9%) individuals and 23 (22.6%) species and orange with 251 (10.1%) species. Combining yellow, red, and orange fruits, there are 1062 (42.9%) individual trees and 40 (39.2%) species. Surely, bright color fruits attract monkeys and indicates their ripeness, but surprisingly, green fruits seem to be preferred in Perijá.

6.6 Discussion

As previously discussed, this high proportion of two-thirds of the forest trees feeding monkeys in the Sierra de Perijá is not unusual in other Amazonian forest regions. It is interesting to note that some of these fruit trees' common names are associated with primates (e.g., *Herrania albiflora* as "monkey cacao," *Posoqueria latifolia* as "monkey apple," and "coco-de-mono" for *Couroupita guianensis*). We also see the "howler-monkey-ice-cream-bean" (a translation of the Barí name *kamashkorou nondyiruku*, *Inga cocleensis*). These names clearly indicate the trees' functions of providing important primate foods.

We need to address the fact that primates are clearly dispersing seeds of the fruit trees that they feed on, especially those which have fleshy or oily mesocarps with smaller-size seeds not exceeding more than 2 or 3 cm. Apparently, spider monkeys can consume seeds as big as 3 cm (van Roosmalen 1985) and white-faced capuchins as big as 1.8 cm without not only destroying but also dispersing them (Rowell & Mitchell 1991, in Lambert and Garber 1998). This dissemination makes these trees more abundant since it is well known in the literature that primates play an important role in dispersing tropical rainforest seeds in the tropical rainforests (Eriksson 2016; Karubian et al. 2015; Stevenson et al. 2015; Terborgh 1986). It is known that white-faced capuchin monkeys disperse live seeds at an average distance of 235 m and as far as 700 m (Valenta and Fedigan 2009). Spider monkeys disperse seeds even farther (average of 443 m and as far away as 1280 m) since they cover a larger territory and move much faster than other monkeys (Link and di Fiore 2006).

Primates clearly exist in an intricate relationship with their forest. They not only are dependent on many fruit trees but also actively change the composition of their forest. As humans "anthropogenically" create "cultural forests," primates must be doing something very similar, by making their forest a "primatogenical" one. Therefore, we could argue that most of the Neotropical forests are *primatogenical* since monkeys and humans disperse fruits of most trees, but monkeys still do most of it. For example, the work of Anzures-Dadda et al. (2011) indicated that with the absence of howler monkeys in southern Mexico, the sapling densities and recruitment of *Brosimum alicastrum*, *Dialium guianense*, *Manilkara zapota*, and *Damburneya ambigens* decrease. Chaves et al. (2011: 177) state that "spider monkeys are effective seed dispersers" with most seeds (>86%) undamaged from 71 species of trees from 23 plant families in the Mexican Lacandon rainforest. Russo et al. (2005: 1033) states that "[s]pider monkeys are one of the first animals to be hunted out of forests…, and their loss is likely to have important consequences for demography, community structure, and gene flow of trees and lianas in tropical forests." These primates are the largest arboreal frugivorous species that can disperse a greater number of larger seeds, which are still viable.

In this chapter, I am not questioning the Balée (2013) concept of cultural or anthropogenic forest but wonder the role of monkeys in the reproduction of Neotropical forests. It is clear Amerindians have been responsible for the distribution of many species of trees (like *Attalea butyracea*, *Bactris gasipaes*, Brazil nuts, canon ball, copal, and tree gourd). Even monkeys help in the reproduction of these cultural forests. According to Balée (1994: 149), the Ka'apor people of Brazil reported that capuchin monkeys frequented old or abandoned gardens dispersing the seeds of *Spondias mombin* and wild cacao nearby, therefore "responsible for the high frequency and sensitivity of other species in fallows." However, monkeys seem too play a major role in the reproduction of the forest in general. For example, "Link & De Luna (2004), for example, estimated that spider monkeys could ingest up to 1352 g (mean: 289 g, N = 90) of fruits … in a single feeding bout of the palm *Oenocarpus bataua*" (Link and Di Fiore 2006: 243). Therefore, larger monkeys like spider monkeys can consume and disperse wider range of fruits that other animals cannot consume and disperse (e.g., birds, bats, agouties, and pacas), being responsible to the reproduction of more species of trees in the rainforest. Link and di Fiore (2006: 243) stated "spider monkeys and other ateline primates are significant dispersers for many species of plants throughout the Neotropics."

In the late Pleistocene, South America lost a large number of its megafauna that were seed dispersers (7 genera of large mammals that were more than a ton heavy and many more that weighted more over 44 kg, Guimarães et al. 2008). Today, the largest land mammal is the tapir that disperses the large quantities of largest seeds. But because "tapirs frequently defecate in salty lakes, a site unsuitable for successful seedling recruitment" (Donatti et al. 2007: 118), largest monkeys do play a major role dispersing many species of fruiting trees. In the late Miocene Amazon, there used to exist a 20 kg monkey related to spider monkey named *Caipora bambuiorum* found in NE Brazil. It became extinct only few thousand years ago likely due to human action of hunting (Defler 2019). This species was probably dispersing larger seeds of fruit trees, and these could explain the existence of some of the rare trees that lost their dispersers 10–12 thousands years ago.

6.7 Conclusion

Of the 227 species of trees found in the Barí territory, 102 species provide food for monkeys. The question is whether this 44.9% of trees that primates feed on is a low, normal, or high percentage. This percentage would have been higher if my plots were bigger and fewer in number since they would not include larger number of rare species that are mostly not food for monkeys. However, this is not abnormally low for other studies. There are only a few studies that provide some potential references. According to Stevenson et al. (2015: 2), primates in the Colombian forests were "responsible for 64% of the fruits manipulated across species." Stevenson's percentages are potentially higher than would be the case for all species of trees since they preselected a group of trees (73 species) that are "zoochorous ripe fruiting plants that had good crown visibility" (Stevenson et al. 2015: 3). In another project, Cormier (2003) also stated that 65.2% of the plant species known by the Guajá people of Brazil are also food for monkeys. It is clear that increased numbers of monkey species and therefore size variation will result in subsequent increases in the number of trees that feed monkeys. Perhaps also because the Barí forests have only four species of monkeys, this percentage is smaller (versus seven for the Guajá in the Brazilian Amazon). However, this could have been different if forest plots farther from the Barí village and in areas where monkey densities are higher would have been included in this research.

In the Barí territory, primates are reproducing a type of forest that can continue to provide them with food. The absence of primates could change the future and nature of this forest to one that promotes bird- or wind-dispersed seed trees. Therefore, the conservation of monkeys is key to the future of this forest, which could continue to support larger fauna. This research shows that primates are likely to be keystone species in the reproduction of two-thirds of the Barí forest. Another important point is that the field of ethnobotanical and ethnoecological ethnoprimatology work with indigenous people could provide new lights on the ecology and diet of primates in the tropical forests.

Acknowledgments I am very thankful to all the Barí people for allowing the author to conduct research and reside in their cultural and natural environment and for providing prodigious amount of information and support for this research, especially from Akirihda, Iribi, Arukbá, Sarukbá, Abuyokba, Abokoré, Abohkín, Ataktabá, Mandabó, Ashkoró, Oroksá, Elizabeth Asigbera, and Emilio Aleobaddá. The help of Andres Achirabu, David Aleobaddá, and all their family members, in particular, made this work possible. Without their generosity and kind support, this chapter could not have been written. Funding for research was provided by Connecticut College (R. F. Johnson Faculty Development Fund). I want to thank my friend Jorge Cruz, anthropologist and writer, for reviewing and correcting my abstract in Spanish, Christopher Cobalth for editorial suggestions, my former student Zoe Diaz-Martin'14 and my mentor/padrino Dr. Stephen Beckerman for making many useful suggestions and corrections. I am also very grateful to Gerardo Aymard (PORT) because he corrected the botanical and taxonomical nomenclature of trees in this chapter and Paul Berry for the identification of my vouchers. Finally, I am extremely grateful to my wife, Anne-Marie, for her meticulous editorial assistance with this chapter. I am very grateful for all the helpful comments that Leslie Sponsel and William Balée provided in this chapter to improve it and the other editor of this volume, Bernardo Urbani, for all the work involved in the production of it.

References

Anzures-Dadda A, Andresen E, Martínez ML, Manson RH (2011) Absence of howlers (*Alouatta palliata*) influences tree seedling densities in tropical rain forest fragments in southern Mexico. Int Primatol 32(3):634–651. https://doi.org/10.1007/s10764-011-9492-0

Asquith NM, Terborgh J, Arnold AE, Rivero CM (1999) The fruits the agouti ate: Hymenaea courbaril seed fate when its disperser is absent. J Trop Ecol 15:229–235

Aymard G (2011) Bosques húmedos macrotérmicos de Venezuela. Biollania (Spanish Edition) 10:33–46

Aymard G, Arellano PH (2018) First report of Peridiscaceae for the vascular flora of Colombia. Harv Pap Bot 23(1):109–121

Aymard G, Norconk M, Kinzey W (1997) Composición florística de comunidades vegetales en islas en el embalse de Guri. Río Caroní. Estado. Bolívar. Venezuela. Biollania 6 (Spanish Edition), pp 195–233

Balée W (1987) A etnobotánica quantitativa dos índios Tembé (Rio Gurupi, Pará). Bol Mus Para Emílio Goeldi, Botanica 3(1):29–50

Balée W (1994) Footprints of the forest: Ka'apor ethnobotany—the historical ecology of plant utilization by an Amazonian people. Columbia University Press, New York

Balée W (2013) Cultural Forest of the Amazon: a historical ecology of people and their landscapes. University Alabama Press, Birmingham

Beckerman S (1975) The cultural energetics of the Bari (Motilones Bravos) of Northern Colombia. Ph.D. thesis, Department of Anthropology, University of New Mexico

Beckerman S (1983) Carpe diem: an optimal foraging approach to Barí fishing and hunting. In: Hames RB, Vickers W (eds) Adaptive responses of native Amazonians. Academic Press, New York, pp 269–299

Beckerman S, Lizarralde R (2013) The ecology of the Barí: rainforest horticulturalists of South America. University of Texas Press, Austin

Berkes F (2012) Sacred ecology, 3rd edn. Editorial Routledge, New York

Berlin B (1992) Ethnobiological classification. Princeton University Press, Princeton

Berlin B, Breedlove DE, Raven PH (1974) Principles of Tzeltal plant classifications. Academic Press, New York/London

Chaves ÓM, Stoner KE, Arroyo-Rodríguez V, Estrada A (2011) Effectiveness of spider monkeys (*Ateles geoffroyi vellerosus*) as seed dispersers in continuous and fragmented rain forests in southern Mexico. Int J Primatol 32(1):177–192. https://doi.org/10.1007/s10764-010-9460-0

Cormier LA (2003) Kinship with Monkeys: the Guajá foragers of eastern Amazonia. Columbia University Press, New York

Cormier LA (2004) Monkey ethnobotany. In: Stepp JR, Wyndham FS, Zarger RK (eds) Ethnobiology and cultural diversity. Proceedings of the seventh international congress of ethnobiology. International Society of Ethnobiology, pp 313–325

Cormier L, Urbani B (2008) The ethnoprimatology of spider monkeys (Ateles spp.): from past to present. In: Campbell CJ (ed) Spider monkeys: behaviour, ecology and evolution of the Genus Ateles. Cambridge University Press, Cambridge, pp 377–403

Defler T (2019) History of terrestrial mammals in South America: how South American mammalian fauna changed from the Mesozoic to recent times. Springer, Dordrecht

Diaz-Martin Z, Swamy V, Terborgh J, Alvarez-Loayza P, Cornejo F (2014) Identifying keystone plant resources in an Amazonian forest using a long-term fruit-fall record. J Trop Ecol 30:291–301. https://doi.org/10.1017/S0266467414000248

Donatti CI, Galetti M, Pizo MA, PR Guimarães JR, Jordado P (2007) Living in the Land of Ghosts: Fruit Traits and the Importance of Large Mammals as Seed Dispersers in the Pantanal, Brazil. In AJ Dennis et al. Seed Dispersal: Theory and its Application in a Changing World. CAB International, Wallingford pp 104–123

Eriksson O (2016) Evolution of angiosperm seed disperser mutualisms: the timing of origins and their consequences for coevolutionary interactions between angiosperms and frugivores. Biol Rev 91:168–186

Finer M, Vijay V, Ponce F, Jenkins C, Kahn T (2009) Ecuador's Yasuní Biosphere Reserve: a brief modern history and conservation challenges. Environ Res Lett 4(3):034005. https://iopscience.iop.org/article/10.1088/1748-9326/4/3/034005/fulltext/. Accessed 15 Dec 2016

Fittkau EJ, Klinge H (1973) On biomass and trophic structure of the central Amazonian rain forest ecosystem. Biotropica 5:2–14

Fleming TH, Breitwisch R, Whitesides GH (1987) Patterns of tropical vertebrate frugivore diversity. Annu Rev Ecol Evol Syst 18:91–109

Gentry A (1974) Coevolutionary patterns in Central America Bignoniaceae. Ann Mo Bot Gard 61:728–759

Gentry A (1990) Floristical similarities and differences between Southern Central America and Upper and Central Amazonia. In: Gentry AH (ed) Four neotropical rainforests. Yale University Press, New Haven, pp 141–160

Gentry A (1996) A field guide to the families and genera of woody plants of Northwest South America: (Colombia, Ecuador and Peru) with supplementary notes on herbaceous taxa. Chicago University Press, Chicago

Guimarães PR Jr, Galetti M, Jordano P (2008) Seed dispersal anachronisms: rethinking the fruits extinct megafauna ate. PLoS One 3(3):1–13., e1745. https://doi.org/10.1371/journal.pone.0001745

Hallwachs W (1986) Agoutis (*Dasyprocta punctata*): the inheritors of guapinol (Hymenaea courbaril: Leguminosae). In: Estrada A, Fleming TH (eds) Frugivores and seed dispersal. W. Junk Publishers, Dordrecht, pp 285–304

Herrera CM (2002) Seed dispersal by vertebrates. In: Herrera CM, Pellmyr O (eds) Approach plant-animal interactions: an evolutionary. Oxford Press, Blackwell, pp 185–208

Howe HF (1984) Implications of seed dispersal by animals for tropical reserve management. Biol Conserv 30:261–281

Huber O, Alarcón C (1988) Mapa de vegetación de Venezuela, con base en criterios fisiográfico-florísticos. 1:2.000.000. MARNR, The Nature Conservancy, Caracas

Janzen DH (1982) Fruit traits, and seed consumption by rodents, of Crescentia alata (Bignoniaceae) in Santa Rosa National Park, Costa Rica. Am J Bot 69(8):56–57

Janzen DH (1983) Hymenaea courbaril: (Guapinol, Stinking Toe). In: Janzen DH (ed) Costa Rican natural history. University of Chicago Press, Chicago, pp 253–256

Janzen DH, Martin PS (1982) Neotropical anachronisms: the fruits the gomphotheres ate. Science 215:19–27

Jordano P (2000) Fruits and frugivory. In: Fenner M (ed) The ecology of regeneration in plant communities, 2nd edn. CAB International, Wallingford, pp 125–166

Karubian J, Ottewell K, Link A, Di Fiore A (2015) Genetic consequences of seed dispersal to sleeping trees by white-bellied spider monkeys. Acta Oecol 68:50–58. https://doi.org/10.1016/j.actao.2015.07.005

Kenfack D, Thomas DW, Chuyong G, Condit R (2007) Rarity and abundance in a diverse African forest. Biodivers Conserv 16:2045–2074

Kinzey W, Norconk MA (1993) Physical and Chemical Properties of Fruits and Seeds Eaten by Pithecia and Chiropotes in Surinam and Venezuela. Int J Primatol 14(2):207–227

Lambert JE, Garber PA (1998) Evolutionary and ecological implications of primate seed dispersal. Am J Primatol 45:9–28

Linares O (1998) Mamíferos de Venezuela: Sociedad Conservacionista Audubon de Venezuela. Caracas, Venezuela

Link A, De Luna AG (2004) The importance of Oenocarpus bataua (arecaceae) in the diet of spider monkeys at Tinigua National Park, Colombia. Folia Primatol (abstract) 75:391

Link A, Di Fiore A (2006) Seed dispersal by spider monkeys and its importance in the maintenance of neotropical rain-forest diversity. J Trop Ecol 22(3):235–246

Lizarralde R (1991) Barí settlement patterns. Hum Ecol 19(4):428–452

Lizarralde M (1997) Perception, knowledge and use of the rainforest: ethnobotany of the Barí of Venezuela. Ph.D. thesis. Anthropology Department, California University at Berkeley

Lizarralde M (2002) Ethnoecology of monkeys among the Barí of Venezuela: perception, use and conservation. In: Fuentes A, Wolfe L (eds) Primates face to face: the conservation implications of human and nonhuman primate interactions. Cambridge University Press, Cambridge, pp 85–100

Lizarralde M (2004) Indigenous knowledge and conservation of the rainforest: ethnobotany of the Barí of Venezuela. In: Carlson TJ, Maffi L (eds) Ethnobotany and conservation of biocultural diversity, Advances in economic botany, vol 15. The New York Botanical Garden Press, Bronx, pp 112–129

Lizarralde M (2019) Etnoprimatología Barí en la Sierra de Perijá del estado Zulia. In: Urbani B, Ceballos-Mago N (eds) La Primatología en Venezuela. Tomo I Academia Nacional de Ciencias Exactas, Físicas y Naturales/Universidad Simón Bolívar (Colección Conjunta Acfiman/USB), Caracas

Lizarralde M, Lizarralde R (2018) Los Barí. In: Pereira MA, Rivas P (eds) Los Aborígenes de Venezuela, Volumen 5 (Monografía 52). Fundación La Salle de Ciencias Naturales, Caracas, pp 725–886

Milton K (1978) Behavioral adaptations to leaf-eating by the mantled howler monkey (Alouatta palliata). In: Montgomery GG (ed) The ecology of arboreal folivores. Smithsonian Institution Press, Washington, D.C., pp 535–549

Milton K (1980) The foraging strategy of howler monkeys: a study in primate economics. Columbia University Press, New York

Milton K (1982) Dietary quality and demographic regulation in a howler monkey population. In: Leigh EG, Rand AS, Windson DM (eds) The ecology of a tropical forest: seasonal rhythms and long-term changes. Smithsonian Institution Press, Washington, D.C., pp 273–289

Milton K, Giacalone J, Wright SJ, Stockmayer G (2005) Do frugivore population fluctuations reflect fruit productions? Evidence from Panama. In: Dew L, Boubli JP (eds) Tropical fruits and frugivores: the search for strong interactors. Springer, Dordrecht, pp 5–36

Neuschulz EL, Mueller T, Schleuning M, Böhning-Gaese K (2016) Pollination and seed dispersal are the most threatened processes of plant regeneration. Sci Rep 6:29839. www.nature.com/scientificreports/. https://doi.org/10.1038/srep29839

Ohl-Schacherer J, Shepard GH Jr, Kaplan H, Peres CA, Levi T, Yu DW (2007) The sustainability of subsistence hunting by Matsigenka native communities in Manu National Park, Peru. Conserv Biol 21:1174–1185

Parathian HE, Maldonado AM (2010) Human-nonhuman primate interactions amongst Tikuna people: perceptions and local initiatives for resource management in Amacayacu in the Columbian Amazon. Am J Primatol 71:1–11

Pittier H (1948) Trabajos Escogidos. Ministerio de Agricultura y Cria, Caracas
Prance GT, Mori SA (1978) Observations on the fruits and seeds of neotropical Lecythidaceae. Brittonia 30:21–33
Rival L (2016) Huaorani transformations in twenty-first century ecuador: treks into the future of time. The University of Arizona Press, Tucson
Rowell TE, Mitchell BJ (1991) Comparison of seed dispersal by guenons in Kenya and capuchins in Panama. J Trop Ecol 7:269–274
Russo SE, Campbell CJ, Dew JL, Stevenson PR, Suarez SA (2005) A multi-forest comparison of dietary preferences and seed dispersal by Ateles spp. Int J Primatol 26(5):1017–1037. https://doi.org/10.1007/s10764-005-6456-2
Shepard G Jr (2002) Primates in Matsigenka subsistence and world view. In: Fuentes A, Wolfe K (eds) Primates face to face: the conservation implications of human and nonhuman primate interactions. Cambridge University Press, Cambridge, pp 101–136
Shepard G Jr, Yu D, Lizarralde M, Italiano M (2001) Rainforest habitat classification among the Matsigenka of the Peruvian Amazon. J Ethnobiol 21(1):1–38
Smythe N (1986) Competition and resource partitioning in the guild of neotropical terrestrial frugivorous mammals. Annu Rev Ecol Syst 17:169–188
Sponsel LE (1997) The human niche in Amazonia: exploration in ethnoprimatology. In: Kinzey WG (ed) New World primates: ecology, evolution, and behavior. Aldine, Chicago, pp 143–165
Stafford CA, Alarcon-Valenzuela J, Patino J, Preziosi RF, Sellers WT (2016) Know your monkey: identifying primate conservation challenges in an Indigenous Kichwa community using an ethnoprimatological approach. Folia Primatol 87(1):31–47
Stevenson PF (2004) Fruit choice by woolly monkeys in Tinigua National Park, Colombia. Int J Primat 25(2):367–381
Stevenson PR, Link A, González-Caro S, Torres-Jiménez MF (2015) Frugivory in canopy plants in a Western Amazonian forest: dispersal systems, phylogenetic ensembles and keystone plants. PLoS One 10(10):1–22., e0140751. https://doi.org/10.1371/journal.pone.0140751
Steyermark JA (1982) Relationships of some Venezuelan forest refuges with lowland tropical floras. In: Prance GT (ed) Biological diversification in the Tropics. Columbia University Press, New York, pp 182–220
Terborgh J (1983) Five New World primates: a study in comparative ecology. Princeton University Press, Princeton
Terborgh J (1986) Community aspects of frugivory in tropical forests. In: Estrada A, Fleming TH (eds) Frugivores and seed dispersal. Dr. W. Junk Publishers, Dordrecht, pp 371–384
Terborgh J (1992) Diversity and tropical rain forest. Scientific American Library, New York
Terborgh J (1999) Requiem for nature. Island Press, Washington, D.C.
Terborgh J, Nuñez-Iturri G, Pitman NCA, Cornejo Valverde FH, Alvarez P, Swamy V, Pringle EG, Timothy Paine CE (2008) Tree recruitment in an empty forest. Ecology 89(6):1757–1768
Valenta K, Fedigan L (2009) Spatial patterns of seed dispersal by White-Faced Capuchins in Costa Rica: evaluating distant-dependent seed mortality. Biotropica 42(2):223–228
van Roosmalen MGM (1985) Habitat preferences, diet, feeding strategy and social organization of the black spider monkey (Ateles paniscus paniscus Linnaeus 1758) in Surinam. Acta Amazon 15:1–238
Wills C, Harms KE, Condit R, King D, Thompson J, He F, Muller-Landau HC, Ashton P, Losos E, Comita L, Hubbell S, Lafrankie J, Bunyavejchwin S, Dattaraja HS, Davies S, Esufali S, Foster R, Gunatilleke N, Gunatilleke S, Hall P, Itoh A, John R, Kiratiprayoon S, Kassim SR, Sukumar R, Suresh SH, Fang Sun I, Tan S, Yamakura T, Zimmerman J (2006) Nonrandon processes maintain diversity in tropical forests. Science 311:527–531
Wilson MF, Traveset A (2000) The ecology of seed dispersal. In: Fenner M (ed) The ecology of regeneration in plant communities, 2nd edn. CAB International, Wallingford, pp 85–110
Zent E, Zent S (2002) Los Jodï: sabios botánicos del Amazonas venezolano. Antropologica 97–98:29–70
Zizka A, ter Steege H, Do Céo M, Pessoa R, Antonelli A (2018) Finding needles in the haystack: where to look for rare species in the American tropics. Ecography 41:321–330

Chapter 7
Memories, Monkeys, and the Mapoyo People: Rethinking Ethnoprimatology in Eco-Historical Contexts of the Middle Orinoco, Venezuela

Bernardo Urbani

7.1 Introduction

Ethnoprimatology as the study of the interface between human and nonhuman primates has been expanding theoretically and methodologically since it was incorporated as a primatological research agenda following the seminal paper written by L. Sponsel (1997) (e.g., Fuentes and Wolfe 2002; Gumert et al. 2011; Radhakrishna et al. 2013; Dore et al. 2017; and chapters therein). Today, ethnoprimatology is open to a large set of questions that range from the inquiry on the views about primates to primate conservation itself. Ethnoprimatology is in a position to integrate multiple experiences that allow an ample vision of how sociocultural interconnections are built around primates, and what defines them.

With the previous premises in mind, the objectives of this ethnoprimatological research are (a) to explore the ethnohistorical information regarding the uses of primates by past indigenous groups of the Middle Orinoco, (b) to summarize ethnoprimatological memories and current ethnoprimatological accounts of the Mapoyo people, and, (c) to provide a reflection on the primates and the Mapoyo framed within the contexts of the traditional ecological knowledge (TEK) paradigm and the historical changing *continuum*.

7.2 The Mapoyo

Henley (1983) indicates that the Mapoyo used the term *Wanai* as their ethnonym. However, as pointed out by Scaramelli and Tarble (2007), today they use *Mapoyo* as their self-denomination, as also observed when this research was conducted.

B. Urbani (✉)
Centro de Antropología, Instituto Venezolano de Investigaciones Científicas, Caracas, Venezuela
e-mail: bernardourbani@yahoo.com

B. Urbani, M. Lizarralde (eds.), *Neotropical Ethnoprimatology*, Ethnobiology,
https://doi.org/10.1007/978-3-030-27504-4_7

Mapwoi, *Mopwe*, *Mopoi*, and *Mopue* have also been reported as other designations for the Mapoyo (Gilij 1965; Henley 1983; 1975; Perera 1992; Granadillo 1997; Medina 1997; Villalón 2003, 2007a, b, Villalón and Granadillo 2000; Scaramelli and Tarble 2007; CDC 2014; Schwartz et al. 2014). Mattéi-Muller (1975, 2003) suggested that the Mapoyo is a Carib language closely related with the extinct Tamanako† and the still extant Yabarana. In the past, the Mapoyo used to have close inter-ethnical relationships with another Carib group, the Pareca†, with whom they also used to marry until the 1920s when the Pareca† apparently became extinct (Henley 1983). The Mapoyo live in the Middle Orinoco region of Venezuela between the Caripo and Villacoa Rivers—having an ancestral territory limited by the Suapure and Parguaza Rivers—near the large towns of Morichalito, Los Pijiguaos, and La Urbana in the Cedeño District of the State of Bolívar (Henley 1983; Scaramelli and Tarble 2007) (Fig. 7.1).

Using information from Gilij (1965 [1780]) and Codazzi (1841), Henley (1983) inferred that between the eighteenth and nineteenth centuries the Mapoyo population ranged between 100 and 200 persons. Today, according to the National Census, the Mapoyo population has increased in number to 365 persons by 2001 (INE 2001) and to ~400 persons in 2006 as estimated by the Mapoyo themselves (Scaramelli and Tarble 2007). By the mid-1970s, Henley (1975) reported about ten remaining Mapoyo speakers. By the late 1990s, seven speakers were alive (Granadillo 1997). At the time when the field research of this study was conducted (2011–2012), four Mapoyo speakers remained out of around 400 Mapoyo. This is to say <1% of the total population of persons self-identified as Mapoyo speaks the language. For that reason, the Mapoyo language was proposed to be declared part of the UNESCO List of Intangible Cultural Heritage, gaining that status in 2014 (CDC 2014; Schwartz et al. 2014). In spite of the imminent extinction of the Mapoyo language, this indigenous society currently employs multiple strategies to restructure their cultural frontiers and conserve their identity as Mapoyo (Scaramelli and Tarble 2006).

Mapoyo subsistence is based on fishing, hunting, fruit and honey collection, horticulture (mostly corn [*Zea mays*], sugar cane [*Saccharum* spp.], and manioc [*Manihot esculenta*]), the raising of chicken and pigs, as well as the insertion into the local market economy where other daily use items are obtained (Jerozolimski and Szeplaki 2005; Scaramelli and Tarble 2007). In addition, the Mapoyo extract the wood of the Orinoco's *alcornoque* (*Bowdichia virgilioides*) within their lands in order to obtain economic resources for buying community goods (Jerozolimski and Szeplaki 2005). Cattle ranching, road construction, and active mining activities are affecting Mapoyo territory, by limiting the access to their cultural heritage sites as well as their natural resources (Schwartz et al. 2014; B. Urbani, personal observation.).

Published ethnobiological information on the Mapoyo is limited; although an overview on resources used in the Middle Orinoco—and by the Mapoyo—was compiled by Tarble et al. (1994). Medina (1997) provided a list of names of plants of ethnobotanical significance found within the Mapoyo territory as well as others associated with their agricultural and culinary practices. In addition, Torrealba (2011) and Torrealba and Scaramelli (2018) presented a study on the extraction of

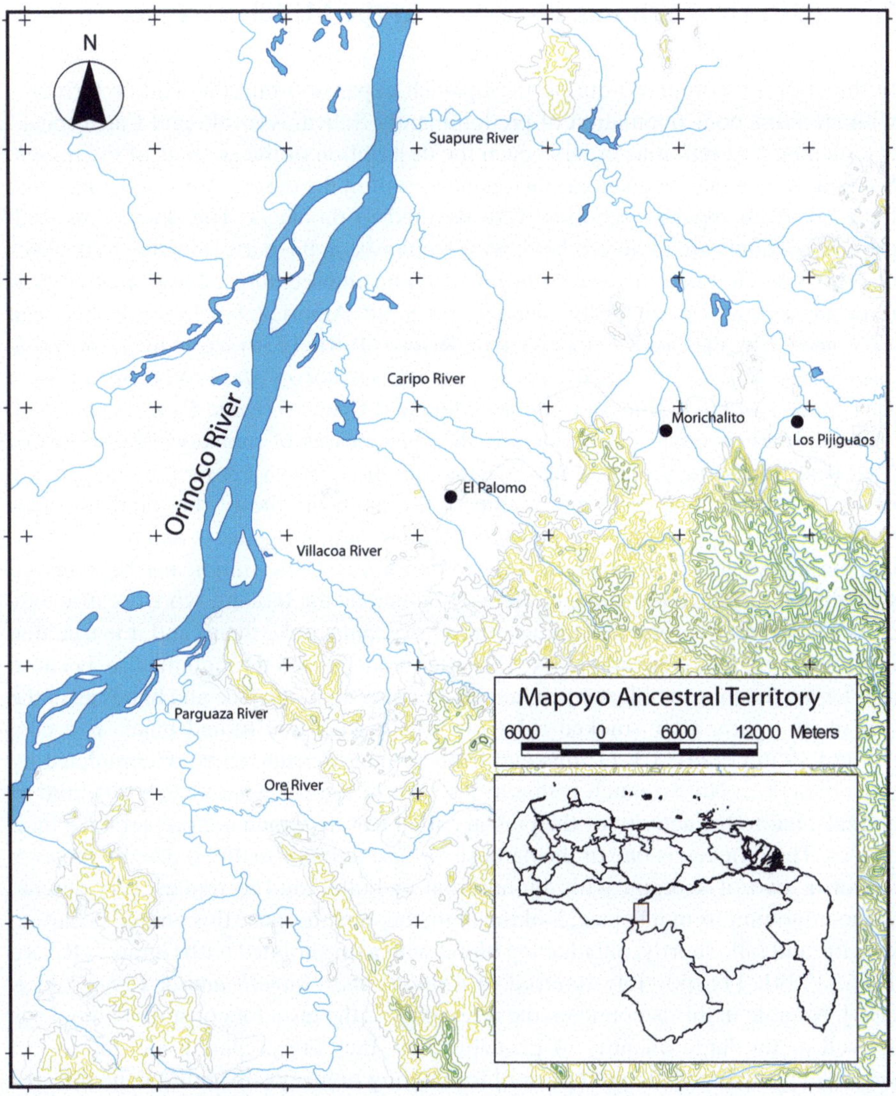

Fig. 7.1 Location of the Mapoyo ancestral territory. (Redrawn from Scaramelli and Tarble [2007] in cooperation with N. Martín and Y. Bernal. Courtesy of F. Scaramelli and K. Tarble)

tonka beans (*Dipteryx odorata*) by this indigenous people, and Brites (1994) provided a list of plants and animal parts used for medicinal purposes. An ethnozoological study was conducted by Meza (2013) and Meza and Menezes (2016) regarding the historical production of turtle oil in the region, and Medina (1997) also recorded a comprehensive repertoire of names of insects, birds, fishes and other aquatic animals, domestic and wild land animals, and body parts of vertebrates. To our knowledge, this is the first study fully devoted to any mammal order in Mapoyo lands.

7.3 Historical Ethnoprimatology in the Middle Orinoco

In the eighteenth century, during the Spanish expansion into the Middle Orinoco, with an indigenous population of predominantly Saliva, Arawak, and Carib speakers, members of religious orders began the description of the animals of the region. Sometimes primate received initial emphasis within the descriptions of the regional fauna, or even separate sections were devoted to monkeys. The documents with primate accounts are reviewed here, with emphasis on the parts referring to the uses of primates. This is a historical corpus of eighteenth-century ethnoprimatological information that is particularly abundant for South America. For example, between 1728 and 1729, the Jesuit Father Juan de Rivero (1681–1736) wrote his *Historia de las misiones de los llanos de Casanare y los ríos Orinoco y Meta.* When referring to the primates of the region between the Meta and Orinoco rivers, Rivero expressed, "Although all mounts of America are abundant in lots of monkeys, the shores of Meta and Pauto, the mount of Macaguane and the forests of Airico, could provide monkeys to many kingdoms and provinces. So many are those who see themselves through the trees, crossing and jumping by their natural liveliness and restlessness, that the arrows get tangled in the crowd (of monkeys), as they might also be observed with their figures and movements. Heavily hunted by the Indians, who, because they do not have salt to prepare them, use fire to dry and smoke them; and it is true that it is horrifying to see an animal like this after it has passed through the fire, because as it has so much resemblance to a human creature, so in the face and hands as in the rest of its figure, after smoked a lot of boys roasted and turned black as a coal (Rivero 1956 [1729]: 10–11; this and forthcoming translations were completed by the author) (...) Not so much is due to the skill of the Indians to catch this kind of animals, but to the activity of the poison called Curare which are prepared for their arrows. The monkey is one of the hardest and last animals to die; it usually receives a shot or two of shotgun with ammunition or bullet, and to remain very serene, without moving from its place, looking at the hunters; because this animal is hard to give up, and falls shortly, only having blood with a tip anointed with Curare" (Rivero 1956 [1729]: 11) [possibly referred to *Alouatta macconnelli* and/or *Cebus olivaceus*]. Notable in this account are the references to the taste for consuming monkeys as well as the large quantity of primates, how they are prepared, and the use of *curare* [a plant-based alkaloid poison] for hunting monkeys. In addition, he referred to the use of gunshots to obtain primate prey, possibly, also used by indigenous hunters at that date.

By mid-eighteenth century, in his *opera magnum*, the *Historia natural, civil y geográfica de las naciones situadas en las riveras del río Orinoco* (1749), the Jesuit missionary Joseph Gumilla (1686–1750), wrote the following in his "Chapter XIX: From their hunting parties, animals that they kill for their gift, and others that are saved for their gift,": "If someone is going have bad fortune, and they do not find wild pigs, they do not return empty to their places; because in all those forests there is abundant and multitude of Monkeys of many species, in which to choose all their tastes, and to use their harpoons. It is to be known, that each Nation of Indians likes of a kind of Monkeys, and hates the others: the Achaguas are get frantic by the Yellow Monkeys, which they call *Arabata* [Gumilla's italics]; these in the morning and in the afternoon infallibly make an intolerable noise, with echoes so low, that they cause horror. The Tunevo Indians like the Black Monkeys very much: they are very ugly and brave; and when they see people, they rage furiously to the last branches of the trees, shaking them, and scolding, with that the Hunters killed them at their wills. The Jiraras, Ayricos, Betoyes, and other Nations hate the two named species of Monkeys, and they pursue and like the White Monkeys, which are also large, no less than the yellow and black ones: their flesh is good; but no matter how much fire it is given, it is always hard; the liver of these monkeys is a gift and appreciated. (Gumilla, 1984 [1749]: 260–261) [*Alouatta macconnelli*, *Ateles belzebuth* or *Chiropotes chiropotes* and *Cebus olivaceus*, respectively] (...) As regards to the large number of several species of Monkeys or small Monkeys, all those Nations eat of them; there is no scruple; because this way, like the big Monkeys, they are only kept with wild fruits, very healthy and tasty; of which the Indians also use during their hunt; and in the journeys that the Fathers make for those and other forests, they observe the fruiting trees where Monkeys are eating, and they surely eat and keep those fruits,..." (Gumilla 1984 [1749]: 261–262). Gumilla provides a comment of the hunting practices and preferences for different species of primates of the Middle Orinoco indicating their meats to be very tasty. He also provides information on the observation of monkeys' feeding behavior and diet for the purpose of emulating feeding selection by indigenous people and Spaniards alike.

In 1759, the Franciscan friar Antonio Caulín (1719–1802) in his *Historia Choro-Graphica, Natural y Evangelica de la Nueva Andalucía*, refers to monkeys in his "Chapter VII. Of the animals, that are raised in these countries, and their properties:" "In the fertile and leafy mounts where these animals commonly inhabit, which can be counted in the class of Monkeys, of a reddish color, and the magnitude of greyhound. They have a grown beard like male goats; and their flanges are very medicinal for those who suffer from asthma, and other affections of the chest, drinking the water, which has been prepared as an infusion. There are other four or five species of Monkeys of various colors, and magnitude, all with prehensile tails, excepts little ones, who breed in the headwaters of the Orinoco River, very funny, whose tail is similar to that of the Cat; and all are sustained by wild fruits." (Caulín 1966 [1759]: 75–76) [*Alouatta macconelli*, and other Orinocan primates, including *Cebus olivaceus*]. This chronicle stands out from an ethnoprimatological perspective because it indicates the use of parts of howler monkeys for medicinal purposes (see Urbani and Cormier 2015), as well as information on their diet and morphology.

In 1763, the French cartographer S. Bellin, member of the French Maritime Academy and the Royal Society of London, referring to the monkeys of Spanish Guayana, wrote that: "There are in all Forests a prodigious amount of Monos and Micos [appears in Spanish in the original French text] of several species that the Indians eat; but each Nation has a determined taste for a certain species of Monkey, excluding all other: some prefer Yellow Monkeys, others like Black Monkeys, and others hate these two species and eat only White Monkeys, which are the same size as the yellows and the blacks: their meat is tasty, but it is always hard, no matter how much they are cooked. As for the Micos, there are also several species, which are in all these Nations, and they can be eaten without disgust, because they only eat healthy and tasty fruits" (Bellin, 1986 [1763]: 60) [*Alouatta macconnelli, Ateles belzebuth* or *Chiropotes chiropotes* and *Cebus olivaceus*, respectively]. This account highlights the differentiation and variability of primates due to the flavor of the meat according to several indigenous societies of the Venezuelan Guayana.

In 1780, Jesuit priest Filippo Salvatore Gilij (1721–1789) wrote another extensive work about the Orinoco region, the *Ensayo de Historia Americana.* In the Fifth Volume, entitled "Of the animals of the interior countries of the Orinoco," he dedicated a section on primates, "Chapter V. On the monkeys," where, when referring to the indigenous peoples and the use of primates, he commented: "... In the Orinoco they [the monkeys] are so abundant, that sometimes you see the trees full of them. They play on top of one another, they eat the fruits, and as they are seated after long eating, getting entangled in the branches with their tails, they are hanging gracefully. The monkey, as everyone knows, is a very unproductive animal, and one finds in them a kind of lust not known to the other brutes. A funny little animal in itself disgusted me for this, and I got rid of him immediately, as well as the rare ones, that bothered me with its continuous croaking. The flesh of the monkeys is not unpleasant, and the Orinoco peoples make great use of it. (Gilij 1965 [1780]: 217) [...] Pleasant and tastier than any other is the flesh of the howler monkey. (In tamanaco Aravatá, in Maipure Maravé.) The howlers have the size of an ordinary dog, but with a long beard, a reddish color and a long tail. His head produces horror, because it looks like a man with a beard. It skin is held in great esteem, and the tamanacos believe it is convenient for children to sleep on it without danger of disease" (Gilij 1965 [1780]: 217–218) [*Alouatta macconnelli*] [...] "Considering the flavor of its meats, the howler monkey is great; highly esteemed, as much by the Orinoco peoples as by the Spaniards. I have eaten it several times, but above all I found delicious [the monkeys] that I ate in the Parecas region. I wanted to cut its head and feet off and place them away from my sight. But these (if we remove the nausea they naturally give), being very fatty, are considered by the Indians as the best snack. I contented myself, without heeding the praises of the barbarians, with only the meat. I found it tender, juicy and of a rather delicate flavor. It is true that there was nothing else to take away the hunger, but I believe that a well-cooked howler monkey would be esteemed by the gluttons" (Gilij 1965 [1780]: 218) [...] "But the most beautiful monkey of the New World is, as let me stand by its pelage. It has been discovered in recent times, and to my knowledge it is found only in the Guaviare River, and perhaps also in the upper Orinoco. In eight and ten years and more, that I was there, I saw only one brought from Cabruta, I would not know if by the Spaniards or by the Indians..." (Gilij 1965: 218) [*Lagothrix lagothricha*] [...] There are undoubtedly other monkeys in the large savannas, but I did not see them. There are also small, but rare, and not esteemed except by women, which they like, as here our dogs (Gilij 1965 [1780]: 218) [*Cebus olivaceus* or *Aotus trivirgatus*]. Gilij's chronicle emphasized the preference for howler monkeys by extinct Carib groups (e.g., Pareca† and Tamanako†) providing its name: *Aravatá.* In the Middle Orinoco, at the town of Cabruta, the traffic of primates from other regions of the Orinoco is noted, in this case a wooly monkey, possibly brought by indigenous peoples or Spaniards. It also suggests the preference as pets by women for smaller monkeys, likely capuchins or night monkeys.

At the end of the eighteenth century, the observant Franciscan Ramon Bueno refers to primates in "Chapter III. The monkeys" of his book *Tratado Histórico sobre la provincia de Guayana* (1785), stating, "The monkeys that, through these vast lands and wherever one transits, find them in heaps, from branch to branch jumping, recognizing and scrutinizing everything, are composed of several classes. Some, and these are more common, they are called *machines*, whose color is between white and brown, long tail, and with it they hang down, keep and cover their heads [...] When they find a cornfield, all descend on it, leaving one sentinel in the highest tree and adjacent to the farm to give the signal to the companions shouting, if people come. They break the stalk of the corn, take out the corn and, in leaving it, they tie it with another one, and thus it hangs on the arm, leaving the other free to climb the branches, so that, as there are many who walk together, carrying at least two corn cobs each, to cause a noticeable loss. It is a matter of laughter to see them flee with the corn, jumping from branch to branch, and never letting a corn cob drop. They have another grace, and that is, going where they walk, and stopping to look at them, they straighten the hand to the ass, and ruling it, they throw dirt at us, and also break some dry branches and do the same..." (Bueno 1965 [1785]: 106) [*Cebus olivaceus*] [...] The howler monkey has the same disposition in the face, eyes, ears, hands and feet as the capuchin; but this one is saffron colored. In the throat it has a thing, similar to a water container, and, since it is so hairy, it looks like a Moorish beard. They grumble, morning and evening […] When looking at people, they usually stay much seated on the branches. They practice the same ceremony as the first ones, throwing at us the excrement with great seriousness and pause; They show us their teeth with a silent growl and they shake the branches to amaze us [...] They are easy to kill, not excusing myself as many times as I am embarked on my trips, because the Indians are very fond of eating them. They eat fruits, and do not steal corn, and like the others; and from far away they present to the sight a color the most beautiful, nevertheless of being so ugly. You cannot have them, like the others in the house for being very crybabies (Bueno 1965 [1785]: 106–107) [*Alouatta macconnelli*] [...] The capuchin [referring to *Chiropotes chiropotes*; see note d in Table 7.1] a monkey of the same color as the R(espected) F(athers) Spanish Capuchins They have the same figure. A very venerable and long beard. They do not allow dirt in it. Every time they smooth it with their hands. Its head is in such a disposition that it forms a small fringe. Eyes very typical of rational and very majestic. They are sustained by fruits, and are not found everywhere. The first we discovered were the missionaries, each in the part that interested them [each religious order], and so many other animals that were lacking in their news. The Indians call them *huicha*, because they sing like that... (Bueno 1965 [1785]: 108) [*Chiropotes chiropotes*]. Bueno's narrative spread the pan-Neotropical notion of cornfield raids by capuchin monkeys. Again, the interest in hunting howler monkeys and the difficulty in domesticating them is highlighted, but not because of their diet, but due to their vocalizations. He also reported an onomatopoeical characterization of the bearded saki. Here as well as in the other chronicles, the descriptions of the monkeys were likely provided mostly by local indigenous peoples, including the encounter with novel species of primates to European eyes.

In order to finish this part of the chapter, it is relevant to point out the existence of the Cerro Mapoyo, a granite *inselberg* containing a site of archaeoprimatological interest. This topographical prominence has historical importance, and it is currently located within the Piaroa (Wõthîhã) territory. It has a large rock art panel that includes the depiction of primates (F. Scaramelli, personal communication, 2011).

7.4 The Mapoyo and the Primates

7.4.1 Framing the Field Research

This section addresses the perception of, interaction with, and uses of primates by the Mapoyo people of the Middle Orinoco of Venezuela. This ethnoprimatological study was carried out in El Palomo, the principal Mapoyo village, and the surrounding forests in five field visits between May 2011 and July 2012. However, the author had also visited them in 1999. For this study, structured interviews as well as unstructured interviews/open conversations were conducted. In addition, direct observation, participant observation, and ethnocartography (the creation of a mental map on the distribution of primates) were also completed. Several Mapoyo adults participated in this ethnoprimatological project, especially one of the last four Mapoyo speakers alive at that time.

7.4.2 Identifying the Monkeys Within Mapoyo Lands

Traditional knowledge about primates has been forgotten trans-generationally. In this essay, an exercise was made for registering the Mapoyo TEK on the monkeys present in their territory. There are four species of primates in an environment characterized by a matrix of Orinoco savannas with *chaparro* (*Curatella americana*) and semi-deciduous gallery forests as well as evergreen forests of the Venezuelan Guiana Shield. Among the Mapoyo there is no single generic name for "monkey," although, the names of the four individual primate species have been recorded (Table 7.1). Table 7.2 presents the phonetic transcriptions of the Mapoyo names for monkeys, showing variability depending on the Mapoyo speaker.

As seen in Table 7.1, four species of monkeys, following a Linnaean classification, are also identified as primates in the Mapoyo ethnozoological classification. However, squirrels (*Sciureus aestuans*; *karirí* in Mapoyo) as well as three-toed sloths (*Bradypus trydactilus*) and kinkajous (*Potos flavus*) have been classified as

Table 7.1 Names of monkeys in the Mapoyo territory

Linnaean names	Spanish names[a]	English names	Mattéi-Muller (1975)	Medina (1997)[e]	Villalón (2003)	This study
Alouatta macconnelli	Araguato	Howler monkey	arawata	arawata	–	arabatá
Chiropotes chiropotes	Capuchino, capuchino del Orinoco, mono negro	Bearded saki	warakaru[c]	–	–	pisá
Cebus olivaceus	Mono, mono blanco, maicero	Wedge-capped capuchin	pisa[d]	pisa[d]	pissa[d]	warakarú
Aotus trivirgatus	Mono tití, cara rayada, tigrillo[b]	Night monkey	makïrikïri	–	–	makirikirí

[a]Spanish names as reported in the region
[b]The Spanish name "mono de noche [night monkey]" is often used in this part of the Middle Orinoco to refer to the kinkajou (*Potos flavus*). It is described as nocturnal, yellowish, and striped. For the purpose of this chapter, when referred to "night monkeys," it indicates *Aotus trivirgatus*
[c]Mattéi-Muller (1975) assigned the Mapoyo name *warakaru* to *Ateles belzebuth*. The author confuses the spider monkey (*Ateles belzebuth*) with the bearded saki (*Chiropotes chiropotes*) likely because both are blackish primates of the Venezuelan Guayana. The Mapoyo do not identify spider monkeys in their territory
[d]Mattéi-Muller (1975), Villalón (2007b), and likely Medina (1997) misidentified the name for *Cebus* possibly because—contrary to the rest of Latin America—in the Orinoco basin, the Spanish name "capuchino [capuchin]" refers to *Chiropotes* and not *Cebus/Sapajus*. In this chapter, when the noun "capuchin" is indicated within the text, it refers to *Cebus olivaceus*
[e]Medina (1997) provided the name "warakaru" for "mono (monkey)"

Table 7.2 Phonetic transcription of the names of monkeys in the Mapoyo territory

Mapoyo speaker	Reference	*Alouatta macconnelli*	*Chiropotes chiropotes*	*Cebus olivaceus*	*Aotus trivirgatus*
Petra Reyes†[a]	Medina (1997)[c]	'aɾaβata ~ 'aɾawata	–	pi'sa[e]	–
Bernabé Reyes†	Villalón (2007b)[c]	a'ɾa:βaˌta	–	'piṣṣa[e]	–
Jose S. Reyes[b]	This study (transcribed by F. Medina, personal communication, 2019)[d]	aɾaβa'ta [arawata]	pi'sa [pisa]	waraka'ɾu[f] [warakaru]	makiɾiki'ɾi [makirikiri]

[a]Recorded by F. Medina on March, 1996
[b]Recorded by B. Urbani on November 19, 2011
[c]Using the International Phonetic Alphabet (see IPA 2015)
[d]The phonetic transcription was kindly performed by F. Medina using the International Phonetic Alphabet (IPA 2015) as well as the phonetic proposition for the Mapoyo language made by Villalón (2007b) [showed within brackets]. This latter proposal was discussed with the Mapoyo themselves as a practical alphabet for their language. For the purpose of potential future comparisons with other Carib language phonetic studies, the last vowel of this transcription (following Villalón 2007b) must be observed as having a tilde as found in the last column of Table 7.1
[e]See note d in Table 7.1
[f]The same word transcribed as 'warakaɾu by Medina (1997).

“monkey-like” animals because they live in tree canopies and their bodies are alike. The squirrel is described as having a palm-like tail, feeding on fruit, and is not annoying. Collared tamanduas (*Tamadua tetradactila*) are characterized as being scarce and having some similarity with proper monkeys. On one occasion, it was mentioned that, apparently, a long time ago there used to be a large primate, but it is gone.

7.4.3 Remembering the Origin of Monkeys

Recalling a story told by his 1880-born grandparent, José María Sandoval, while gathering together in the preparation of *cucurito* (*Attalea maripa*), the then young J. S. Reyes (personal communication, 2011) carefully listened to what his grandfather said, “ The monkeys were people with nails like humans. The teeth, ears, and the shape of the face were of people but it became an animal because it was looking around, jumping, because it was a *safrisco*” (a noun/adjective between “nosy” and “meddling” [this is a Venezuelan Spanish word difficult to literally translate into English]). After that, continuing with the story itself—told twice to BU by his grandson, J. S. Reyes—in order to be transcribed and shared with other people. A long time ago, “during the winter (rainy season) there was not much rain. At that time, the monkey (referring to the capuchin monkey, *warakarú*) was a person who called God (*maiwaka* in Mapoyo language [Mattéi-Muller 1985]) to his Uncle, and was also the culprit for the beginning of rain. The Uncle had a bag to keep *yopo* (hallucinogenous powder prepared with *Anadenanthera peregrine*) and the plate used to inhale it, and while he was working, his nephew (the monkey) opened it. That is where the cicadas, the owners of the water, were confined. The cicadas flew out. The Uncle heard the noise they caused, which meant that it began to thunder and rain a lot. That is why the monkeys are considered to be so meddlesome, and that is also why it rains; because of the monkey. As a result of his action and being *safrisco*, the Uncle lashed him to a tree with vines. He was white because his color had been washed out by the rain and ended up sleeping in the trees. For this reason, the monkey looks like a human, but the difference is the tail. The monkeys were people.”

In another context, a second version of the narrative was also told by J. S. Reyes, saying that “one day the Creator was making a *curiara* (a canoe made from a large tree bark), while telling him (the monkey) not to open the *chacara* (a purse). God did not feel that he had to say that: do not open it. But when he was away at the mountain, he heard thunder; it was the water that came. The monkey had opened the bag, causing the downpour. God noticed that a bird’s sweat did not allow it to get wet and that is why the bird did not get wet, while he told a curassow to cut a leaf so it would not get wet, and that is why it sings when it is going to rain. But the monkey, because he was a *safrisco*, a vagabond, he did not obey; God told him to climb the vines and get wet, and so he got wet. Then, the monkey turned white, after having been black; because the water washed the monkey's color out, losing its original dark color. That is why, when water falls, one becomes ashen, like the monkey. God also told the monkey that from that moment on, he would live on the trees: eating fruits and having everything he needed so that he could stay in the trees without falling. The monkey stayed there, searching within the trees. On the other hand, when the monkey opened the *chacara*, the cicadas also came out. That is why cicadas are seen around the month of April everywhere, because of the monkey." Thus, this is also why "according to the Mapoyo, today's monkey was turned into an animal, while for the *criollos* [Creoles], the monkey appears through Noah's ark" (J. S. Reyes, personal communication, 2011). Previously, cosmological accounts involving primates were not reported (see Mattéi-Muller 1985; Villalón 2003; CDC 2014; Schwartz et al. 2014).

Because the monkeys were people in the old times, as indicated by elder Mapoyo, today it is said that if a person climbs a tree he/she will become a monkey. In this sense, during the *Semana Santa* (the Holy Week), it is suggested that it is preferable not to bathe in the rivers because one can be transformed into a fish, and not to climb trees because the person may become a monkey. In addition, children were not allowed to climb the Cerro Caripito because from there they could hear frightening sounds of cooking pots and howler monkeys (F. Scaramelli, personal communication, 2011).

7.4.4 Remembrances on Primate Ethnoecology

Feeding, behavioral, and descriptive information on the four primates present within Mapoyo territory is reported in this section. Regarding the diet of the monkeys, Table 7.3 offers a summary of particular plant species reported as consumed by three of them. Howler monkeys are known to consume leaves and some common fruits of the forest; however, no specific plant is reported as consumed by this monkey species as it is not seen in Table 7.3. In the case of bearded sakis, they are reported as consumers of immature fruits, especially *moriche* palm (*Mauritia flexuosa*). It is known that neither bearded sakis nor howler monkeys eat corn, and the second is cautious not to enter cornfields. On the contrary, capuchins are repeatedly indicated as consumers of corn, particularly *jojoto* (young corn); that is the reason they are named as *maiceros* (corn eaters). The story of corn consumption included above indicated that they took and ran away with the maize cobs slung over their shoulders. The narrative of capuchins as raiders of cornfields is common within the range of this primate group in Latin America. In addition, capuchins are known to eat nonpoisonous fruits as well as mangoes near the village of El Palomo at the time when the sympatric iguanas (*Iguana iguana*) are also consuming these fruits. Capuchins eat the eggs of yellow-spotted river turtles (*Podocnemis unifilis*) in February, during the dry season, when they dig for them in the spawning grounds. Capuchins are also attributed the consumption of spiders, worms, and orthopterans.

Night monkeys are reported to feed on ripe mangoes, seeds, palm fruits, as well as insects. As is observed in Table 7.3, the majority of plant species corresponds to palms; this might be related to the fact that they are also useful for humans. Consequently, they are reported to be associated to particular animals like primates. Another group of plants reported in Table 7.3 are found in domestic spaces.

Regarding the behavior of the monkeys in the Mapoyo territory, the capuchin is informed to be *safrisco* as well as tricky, brave, and a bully. They are also considered to be visitors of the *conucos* (slash-and-burn or swidden gardens), where they are heard vocalizing and also procuring corn as stated above. Capuchins are found in groups of 10–15 individuals. The capuchins are recognized for moving nimbly on the branches both in the higher and lower parts of the forest as well as sleeping within lianas. Another common primate from the region is the howler monkey, which is especially recognized for their characteristic vocalizations, onomatopoeically described as "oh oh oh" and "uh uh uh." It is said that those vocalizations are usually performed when it is going to rain or when showers are falling down at the time howlers are in the branches about to sleep. While navigating in Laguna del Corozo for fishing, it is common to see Orinoco river dolphins (*Inia geoffrensis*) and listen to howler monkeys. They are also recognized as primates that occupy the canopy of the forest and walk in groups of 10–20 individuals. When howler monkeys and wedge-capped capuchins are compared, it is indicated that the first can hang by their tails in the branches of the trees while the second cannot.

Night monkeys tend to inhabit the forest surrounding bodies of water such as the *morichales* (*Mauritia flexuosa* patches). The Laguna del Corozo, Karamakate, and Caripito River stand out as sites where they are found. This explains why it is difficult to observe them around the village of El Palomo. During the day, night monkeys live in holes "like parrots," being recognized as the only monkeys in the region that sleep in this way: inside old hollow of trees such as *congrio* (*Acosmium nitens*) and dry *moriche* palms (*Mauritia flexuosa*). While moving at night, they vocalize a sound that onomatopoeically resembles an "um um." These monkeys come out—and they are seen—mainly during the full moon. This monkey species has the tendency to flee the areas where fires occur, living in tall forests with many lianas. Its locomotion does not differ from that of diurnal monkeys. It is also known as the monkey that lives in smaller groups.

Bearded sakis are known to live in large groups of 10–30 individuals in the highest part of the forests. They are also known to sleep on branches with lianas. Along the Caripito River, bearded sakis have three points where they cross it, as it was also reported that "those monkeys use routes around the hill" (S. Bastidas, personal communication, 2011). Bearded sakis and capuchins are considered to be more common in tall forests. However, like the night monkeys, capuchins are also found to visit flooded habitats. Howler monkeys are known to be elsewhere within Mapoyo lands. Although capuchins and howler monkeys are more abundant along the Orinoco River, their populations were even larger when the region was less occupied by foreign people. It is suggested that in terms of abundance the order is *Cebus* → *Alouatta* → *Chiropotes* → *Aotus*.

Table 7.3 Plants used by monkeys in the Mapoyo territory[a]

Linnaean names	Mapoyo names[4]	Spanish names	English names	*Chiropotes chiropotes*	*Cebus olivaceus*	*Aotus trivirgatus*
Oenocarpus spp., Arecaceae[b]	Peri[d]	Sejé	Seje palm	–	X	–
Attalea maripa, Arecaceae[b]	Wasay[e]	Cucurito	Maripa palm	X	X	X
Mauritia flexuosa, Arecaceae[b]	Attawai[d]	Moriche	Moriche palm	X	X	X
Astrocaryum spp., Arecaceae[b]	Kupa[e]	Macana, macanilla	Macana palm	X	X	–
Inga spp., Fabaceae	–	Guama	Ice-cream-bean tree	–	X	–
Psidium spp., Myrtaceae[c]	Wanapoimë[d]	Guayabo	Guava	X	–	–
Magifera indica, Anacardiaceae[c]	Manko[d]	Mango	Mango	X	X	X
Zea mays, Poaceae[c]	Ë'nai[d]	Maíz, jojoto	Corn	–	X	–
Indetermined	Indet.	Cupata	Indet.	–	X	–

[a]As indicated in the text, there are no specific names for plants selected by Guianan howler monkeys (*Alouatta macconnelli*)
[b]Palm species
[c]Domestic plant, although mangoes and guavas might be observed outside the domestic areas
[d]from Mattéi-Muller (1975)
[e]from Medina (1997)

Ethnocartographically, a schematic mental map of the distribution of primates in the Mapoyo territory was prepared. Fig. 7.1 as a base map indicates that bearded sakis are located mostly in the tall forests in the north, near the Orinoco River, and around the Cerro Caripito, a prominent granite *inselberg* that clearly marks this indigenous land. Wedge-capped capuchin monkeys are located elsewhere; however, they tend to be found near crops in the eastern part of the Mapoyo lands and around Cerro Caripito. Night monkeys are identified as scarce, mainly around the lagoon (Laguna del Corozo, the *cienaga*). Howler monkeys are very common within the whole Mapoyo territory, as indicated above.

The Mapoyo description of the four primates relates to their phenotypic distinctions. Howlers are recognized as the largest monkeys within the Mapoyo territory, being reddish orange with yellowish overtones and a black face. They are also distinguished by a long tail and thick arms. For their part, the capuchins have a white to grayish ash color similar to "mold," also with some yellowish overtones. They present a body similar to that of a cat, with long arms like the howler monkeys, but differentiated by being thinner. The bearded sakis are blackish and have faces similar to those of the howler monkeys. They are also distinguished by their white scrotum sacks, bearding with sideburns, a long hairy tail, and intermediate arms in size compared with those of howlers and wedge-capped capuchins. Finally, night monkeys are identified by their small size but long tails, having brownish color and "painted" in the face as if they have four eyes (two ocular globes and two spots over those globes that look like eyes). This primate species is also identified behaviorally as the only nocturnal one, difficult to see during the day, and a tree hole dweller as indicated above.

7.4.5 Primate Hunting

In the past, Mapoyo "grandparents" ate the four kinds of monkeys of the region as well as sting rays (Batoidea), dwarf caimans (Crocodylidae), and electric eels (*Electrophorus electricus*). In principle, those animals were not eaten by women or children. It was also mentioned that the gold tegu (*Tupinambis teguixin*) was also an animal that should not be eaten. It was indicated that in the present, the Mapoyo do not hunt monkeys because they have a "sense" of humankind, they resemble humans, and because the Mapoyo are "already civilized" (S. Bastidas, personal communication, 2011). Currently, game animal order of preference is ranked by the tapir (*Tapirus terrestris*), deer (*Mazama* spp. and *Odocoileus virginianus*), peccaries (*Tayassu tajacu* and *T. pecari*), red-footed tortoise (*Chelonoidis carbonaria*), and agoutis (*Agouti paca* and *Dasyprocta leporina*). Chaffanjon (1889) pointed out that the Mapoyo used to hunt river dolphins (*Inia geoffrensis*) and manatees (*Trichechus manatus*). It was said that some time ago, animals like peccaries (*Tayassu* spp.) crossed through the town of El Palomo. Those animals no longer appear there, not because they are overhunted but because a national road—disruptive for animals—is nearby. This road reached the Mapoyo territory by the early 1980s as photographically presented by Henley (1983). It was indicated that they hunt agoutis when they entered *conucos* for corn. Today, the Mapoyo said that primates in the region are hunted by their neighboring indigenous *hermanos* (brothers) such as the Curripako (or Kúrrim: Arawak), Piapoco (or Tsáse: Arawak), Hiwi (or Guahibo: Guahiban), E'ñepa (or Panare: Carib), and Piaroa (or Wõthîhã; Salivan); the latter with shotguns. Present-day, Mapoyo maintain amicable relations with these indigenous neighbors (Scaramelli and Tarble 2007; B. Urbani, personal observation.). As an anecdote, a Mapoyo hunter indicated that a few years ago he went out to hunt a pair of bearded sakis with a Curripako man.

By about 1950, the Mapoyo bought blowguns from the Eñepa to hunt howler monkeys, guans (Gracidae), and curassows (*Crax* spp.). Henley (1975, 1983) also reported that the Mapoyo used to buy blowguns made of bamboo from the Pareca† (Carib). The Parecas†, who were specialists making blownguns, ended their contact with the Mapoyo in the 1920s, when allegedly they became extinct. Bueno (1965 [1785]) indicated that the Mapoyo were exceptional in their ability with blowguns as well as with arrows and spears. In El Palomo, it was commented that they preferred to use a spear called *puyón* (large point) without *curare*, as well as blowgun darts made of cotton and *cucurito* palm (*Attalea maripa*) with *curare*. The *puyón* is used today to catch large fishes. Monkeys were also hunted with bow and arrow with tips impregnated with *curare*. In the case of hunting howler monkeys, and also other monkeys, they used a measure of "two fingers" (about 3 cm) of *curare* from the tip of the arrow point. In this way they fell "light" to the ground and did not take long after "scratching" their bodies. Once hunted, primates were cleansed of "guts and hairs." Then, the monkeys were prepared in a stew; that is, boiled with *ají* (*Capsicum* spp.) and *mañoco* (cassava [*Manihot esculenta*] grits). Before conducting the search of howlers, they were previously located by vocalizations, mainly between Laguna del Corozo and Cerro Macho. In the past, when the Mapoyo went out to collect *sarrapia* (seed of *Dipteryx odorata*), they preferred to hunt peccaries and deers, but during the dry season, between January and May, they also hunted curassows and monkeys. The Mapoyo tended to hunt in parties of 4–5 people using bows and arrows; however, sometimes they would go out alone. The Mapoyo reached their hunting places by foot or by bike (Scaramelli and Tarble 2007). In the nineteenth century, the Mapoyo were also reported to use hunting *curiaras* (usually 10–15 m long dugout canoe) with 20 persons on the Caripo River; also described as an ideal place for fishing (Chaffanjon 1889).

The Mapoyo preferred howler monkeys because they provided more meat due to their larger body size, although night monkeys were considered to have tastier flavor. *Chiropotes* ranked second among the preferred monkey prey in the past. Currently, they are mostly hunted by the neighboring Piaroa people. For that reason, it is said that they are now difficult to find on the south side of the Cerro Caripito. Thus, bearded sakis are considered relatively less common today than they used to be in the past; however, it was also said that their numbers are recovering slowly because hunting has diminished. Monkey body parts are not used for medicine by the Mapoyo, nor are they used for making any kind of Mapoyo material culture. Nevertheless, Brites (1994) reported the use of the core of a tree named *palo de mono seje* (seje monkey tree). This name might have been attributed because of the use of this plant by monkeys. Unfortunately, the scientific name was not provided, but considering that the uses of the term *seje*, which refers possibly to a palm species listed in Table 7.3; that is, "seje palm (*Oenocarpus* spp.) fed on by wedge-capped monkeys." The *palo de mono seje* is prepared as a drink, which is consumed in one to three shots in order to counteract the venom of poisoned animals (Brites 1994).

7.4.6 Interacting with Monkeys as Pets

Primates are not frequently found in captivity in Mapoyo households; in fact, historically there has been a tendency to not keep monkeys as pets. Actually, while this study was conducted, only one capuchin monkey was registered in the village of El Palomo, which had a human name ("Rufino"), used with an empathic sense. This particular monkey was taken while fishing at the lagoon, as—years ago—a baby bearded saki was also found in the Caripito River. The latter was kept and later returned to the forest. Other infant capuchin monkeys have been at times maintained as pets. For instance, though, it was said that in the past, night monkeys were also preferred at home because of their docility, beauty, and for their pleasant vocalization of a tonal "woop woop."

In the forest, not as pets but as habituated wild primates, women used to see the "meek" bearded sakis while washing clothes in the Caripito River. Likewise, during fishing days it is reported that night monkeys sometimes are found observing the Mapoyo. This nocturnal monkey has also appeared looking at people in bathing pools. When howlers interact with humans in the forest, it is said that they sometimes toss excrements.

7.5 Conclusions

I would like to close this chapter by highlighting three final remarks that might be drawn from this ethnoprimatological study among the Mapoyo. First, some ethnoprimatological "universal" patterns were found when compared with other indigenous societies in Latin America, mainly from lowland South America. As compiled by Urbani (2005), howler monkeys are ranked as the preferred primate hunting game in the tropical Americas. For the Mapoyo also, howlers were the most highly selected primate game. A taboo for eating monkeys also exists in this society of the Middle Orinoco. In the past, Mapoyo women and children were prohibited from eating monkeys, as encountered in multiple indigenous groups of lowland South America (Urbani and Cormier 2015). As an Amazonian cosmological trend, it is common in native societies to transform animals into humans or vice versa (see Viveiros de Castros 1992, 1998). This was observed in a Mapoyo ancestral narrative on the creation of the monkey. The Mapoyo story on the origin of the monkey—after provoking the rain by opening God's personal bag—is parallel to that reported for other extant Carib societies of Venezuela. For example, among the Ye'kuana, *Yadaakadu*, also the nephew of the Creator, opened the bag of the latter, and the night escaped (de Civrieux 1970). In this way, the night was created. As a punishment, the Creator transformed his nephew into a monkey (de Civrieux 1970). So, the role of capuchin monkeys in the creation of relevant natural phenomena such as the rain (Mapoyo) or the night (Ye'kuana), and the transformation of humans into monkeys appears to be a pan-Carib pattern in northern South America. Variability in the uses of monkeys was also observed for the Mapoyo. For instance, they tend to avoid keeping primates as pets—contrary to what is commonly found among many South American native groups (Cormier 2006)—and to avoid using primate body parts for medicinal purposes (e.g., Alves et al. 2010).

Second, ethnobiological research in general, and ethnoprimatological in particular, becomes a challenge in indigenous societies given the continuous decline in their native language and cultural lexicon. As suggested by Zent (2009), by losing aboriginal linguistic capabilities, traditional ecological knowledge (TEK) and its transmission are in peril. Accordingly, Stringer (2016: 14) emphatically pointed out that "the current mass extinction of both languages and species has given rise to a vibrant, interdisciplinary movement with the goal of maintaining and revitalizing linguistic, cultural, and biological diversity" alike. TEKs are actually possible sources of inspiration for ecological reconstruction in changing contexts (Martin et al. 2010). In fact, by understanding TEKs, it will be possible to apply informed policies not only to ensure the survival of biodiversity, but also the future viability and vitality of indigenous societies that depend on that biodiversity. In the case of ethnoprimatological research, it has particular relevance because primates are key mammals for maintaining forest dynamics. As argued by M. Lizarralde (this volume), the forests of the tropical Americas seem to actually be *primatogenical* forests, or culturally and biologically made by both humans and nonhuman primates. Therefore, I suggest that the roles, social constructions, interactions, imaginations, perceptions, uses, and empathic reflections about primates are fundamental for understanding not only the cultural baggage surrounding primates as understood by indigenous peoples that sympatrially share spaces and landscapes with them but also for conserving the environment itself where both humans and nonhuman primates live.

Third, to end, this study stands also as a reminder of the usefulness of the comparison of the historical ethnoprimatological information and its contrast with the current primatological ethnography. For example, the fact that the reported preference for howler monkeys and its name *arabatá* by extinct Carib societies (e.g., eighteenth-century Pareca† and Tamanako†) of the Middle Orinoco that likely preceded the use of the territory (Durbin 1977, 1985) occupied by the Mapoyo today is noteworthy. The Mapoyo, as a contemporary indigenous society that also used to have howler monkeys (*arabatá*) as the preferred monkey species, reflects that beside the chronological *long durée* distance from possible ethnical predecessors (see Comaroff and Comaroff 1992), there is a symbolic synchrony across centuries as they similarly share not only the significant (the howler monkey named *arabatá*) but also, at least, one of its meanings (howlers as preferred primate game species). Even more, interesting enough, the common name for "howler monkey" in the whole territory of Venezuela is *araguato*, a word that seems—as showed in this study—to have a Carib linguistic root (see above: *arawata*, *arabatá*). Therefore, the sense of the TEK as discussed above should also be revisited as it is now in comparison with how it will, most certainly, change in the future by the current younger generation of Mapoyo. Those changes, based on the relation with the state, a reconfigured nature, and the relationships with the nearby *criollo* society and other indigenous societies within the Mapoyo territory, will likely occur and serve in mediating futures sensitivities about primates and other animals. Accordingly "as indicated by Ford (2001), at the turn of the new millennium, ethnobiological research is at a 'crossroad.' In this sense, ethnobiology is confronting multiple challenges in a fast changing world. [In this direction, B. Urbani and L. A. Cormier emphasized that] ethnoprimatology is not exempt to those challenges, in which biological, ecological, cultural, philosophical, sociopolitical, historical, religious, and even linguistic realms as well as global, national, regional, community, and family economies impact on multinational and domestic realities that modulate contemporaneous indigenous uses, interactions and perceptions of nonhuman primates" (Urbani and Cormier 2015: 275–276). This study reflects this realm, as forthcoming ethnoprimatological research will face predictably in the future.

Acknowledgments I wish to thank all Mapoyo members of El Palomo community for their hospitality and support, in particular to *capitanes* (headmen) Simón Bastidas and Argénis Bastidas. There, in El Palomo, I especially want to express my gratitude to one of the last Mapoyo speakers, José Secundino "Candecho" Reyes, dear friend and companion of long conversations about monkeys and the life in the Orinoco. To him is dedicated this essay. To Kay Tarble for the invaluable editing of the English text and comments, and to Franz Scaramelli for introducing me to the Mapoyo people, encouraging this research, and for our discussions in Caracas, Chicago, and the field. Both, Kay and Franz are my mentors in South American lowland anthropology and role models in ethnographic research—thanks a lot for the teachings. To Francia Medina (Universidad Central de Venezuela) for the transcription of the Mapoyo names of primates and sharing her knowledge of the Mapoyo language, and to Nuria Martín (Instituto Venezolano de Investigaciones Científicas) for the support in preparing the audio tracks; to N. Martín and Yheicar Bernal (Instituto Venezolano de Investigaciones Científicas) for helping with the edition of the map used in Fig. 7.1, shared by F. Scaramelli and K. Tarble. *Gracias* to Manuel Lizarralde and the external reviewer for their comments on an earlier draft of the manuscript. To the friends at Los Pijiguaos, Morichalito, and Caicara for their cooperation when this study was carried out. To the Instituto Venezolano de Investigaciones Científicas for collaborating with the fieldwork (project # 1061), and in particular to the members of the Center for Anthropology. I appreciate the support of the personnel of the libraries at the Fundación La Salle de Ciencias Naturales, Instituto Venezolano de Investigaciones Científicas, School of Anthropology of the Universidad Central de Venezuela, and the Instituto de Patrimonio Cultural. To Ana María Resnik for her support, comments, and love.

References

Alves RN, Souto Wedson MS, Barboza Raynner RD (2010) Primates in traditional folk medicine: a world overview. Mammal Rev 40:155–180

Bellin S (1986) [1763] Descripción geográfica de la Guayana. Ediciones de la Presidencia de la República, Caracas

Brites N (1994) Espacios y tiempos sagrados: tradiciones y ritos en las prácticas funerarias de los grupos Wanai y Wóthuja de sector Parguaza-Suapure, estado Bolívar. Honor Thesis. Universidad Central de Venezuela, Caracas

Bueno R (1965) [1785] Tratado Histórico sobre la provincia de Guayana. Biblioteca Academia Nacional de la Historia. Col Fuentes Historia Colonial Venezolana, Caracas

Caulín Fray A (1966) [1759] Historia de la Nueva Andalucía. Tomo I. Academia Nacional de la Historia. Fuentes para la Historia Colonial de Venezuela 81, Caracas

Chaffanjon J (1889) L'Orénoque et el Caura. Librairie Hachette et Cie., Paris

CNC (Centro de Diversidad Cultural) (2014) Mapoyo oral tradition and its symbolic reference points within their ancestral territory Venezuela (Bolivarian Republic of) inscribed in 2014 (9.COM) on the list of intangible cultural heritage in need of urgent safeguarding. UNESCO, Paris. https://ich.unesco.org/en/USL/mapoyo-oral-tradition-and-its-symbolic-reference-points-within-their-ancestral-territory-00983

Codazzi A (1841) Resumen de la geografía de Venezuela. H. Fournier y compañía, p 648

Comaroff J, Comaroff J (1992) Ethnography and the historical imagination. Westview Press, Boulder

Cormier LA (2006) A preliminary review of Neotropical primates in subsistence and symbolism of indigenous lowland South American peoples. Ecol Environ Anthropol 2:14–32

de Civrieux M (1970) Watunna, Mitología Makiritare. Monte Ávila Editores, Caracas

Dore KM, Riley EP, Fuentes A (2017) Ethnoprimatology: a practical guide to research on the human-nonhuman primate Interface. Cambridge University Press, Cambridge

Durbin M (1977) A survey of the Carib language family. In: Basso E (ed) Carib speaking Indians culture, society and language. Anthropol Papers, vol 28. University of Arizona Press, Tucson, pp 23–38

Durbin M (1985) A survey of the Carib language family. In: Manelis Klein HE, Stark LR (eds) South American Indian languages; retrospect and prospect. University of Texas Press, Austin

Ford RI (2001) Introduction: ethnobiology at the crossroads. In: Ford RI (ed) Ethnobiology and the millennium. Past promise and future prospect. The Museum of Anthropology at the University of Michigan, Ann Arbor

Fuentes A, Wolfe LD (2002) Primates face to face. Conservation implications of human-nonhuman primate interconnections. Cambridge University Press, Cambridge, NY

Gilij FS (1965) [1780] Ensayo de Historia Americana. Tomo I. Academia Nacional de la Historia, vol 71. Fuentes para la Historia Colonial de Venezuela, Caracas

Granadillo T (1997) Marcadores de personas nominales y verbales en la lengua Mapoyo. Honor Thesis. Universidad Central de Venezuela, Caracas

Gumert MD, Fuentes A, Jones-Engel L (2011) Monkeys on the edge. Ecology and management of long-tailed macaques and their interface with humans. Cambridge University Press, Cambridge

Gumilla J (1984) [1749] Historia Natural, Civil y Geográfica de las naciones situadas en las riveras del Orinoco. Tomo I (edición facsimilar) Edit. Carvajal S.A., Santander de Chilicao

Henley P (1975) Wanai: Aspectos de pasado y presente del grupo indígena Mapoyo. Antropológica 42:29–52

Henley P (1983) Los Wanai (Mapoyo). In: Coppens W (ed) Los aborígenes de Venezuela, vol II, 1st edn. Fundación La Salle de Ciencias Naturales, Caracas

INE (Instituto Nacional de Estadística) (2001) Población indígena e índices de crecimiento según entidad federal. Censos 1992 y 2001. Instituto Nacional de Estadística, Caracas

IPA (International Phonetic Association) (2015) IPA Chart. International Phonetic Association, Kiel. https://www.internationalphoneticassociation.org/sites/default/files/IPA_Kiel_2015.pdf
Jerozolimski D, Szeplaki A (2005) Somos mapoyo [video]. Fundación La Villa del Cine/ Cooperativa Estrella Films, Caracas
Martin JF, Roy ED, Diemont SAW, Ferguson BG (2010) Traditional Ecological Knowledge (TEK): ideas, inspiration, and designs for ecological engineering. Ecol Eng 36:839–849
Mattéi-Muller MC (1975) Vocabulario básico de la lengua Mapoya. Antropológica 42:57–78
Mattéi-Muller MC (1985) Fragments de mythes Wanai. Amerindia, Rev d'ethnolinguist amérind 10:127–136
Mattéi-Muller MC (2003) Pémono: Eslabón perdido entre Mapoyo y Yawarana, lenguas Caribe ergativas de la Guayana noroccidental venezolana. Amerindia 28:33–54
Medina F (1997) Una introducción a la fonética y fonología Mapoyo (Caribe). Honor Thesis. Universidad Central de Venezuela, Caracas
Meza E (2013) Aceite de tortuga: Colonialismo, agencia y producción en el Orinoco Medio durante los siglos XVIII y XX. Honor Thesis. Universidad Central de Venezuela, Caracas
Meza E, Menezes LF (2016) Agencia indígena y colonialismo: una arqueología de contacto sobre la producción de aceite de tortuga en el Orinoco Medio, Venezuela (siglos XVIII y XIX). Amazônica 7:375–402
Perera MA (1992) Los últimos Wanai (Mapoyo). Contribución al conocimiento de otro pueblo amerindio de que desaparece. Rev Españ Antrop Americana 22:139–161
Radhakrishna S, Huffman MA, Sinha A (2013) The macaque connection: cooperation and conflict between humans and macaques. Springer, New York
Rivero J (1956) [1729] Historia de las Misiones de los llanos de Casanare y los ríos Orinoco y Meta. Argra, Bogotá
Scaramelli F, Tarble K (2006) Los Wanai (Mapoyo), Notas sobre su situación actual y actualización bibliográfica. In: Coopens W (ed) Los aborígenes de Venezuela, vol II, 2nd edn. Fundación La Salle de Ciencias Naturales, Caracas
Scaramelli F, Tarble K (2007) Los Mapoyo (Mopue). In: Freire G, Tillet A (eds) Salud indígena en Venezuela. Ministerio del Poder Popular de Salud, Caracas, pp 141–186
Schwartz S, Torrealba G, Navas AM, Escudero M, Pérez J, Wierdak N (2014) Mapoyo. Lista de Patrimonio Cultural Inmaterial que requiere medidas urgentes de salvaguardia. UNESCO. Centro de Diversidad Cultural, Caracas, p 2014
Sponsel LE (1997) The human niche in Amazonia: explorations in ethnoprimatology. In: Kinzey WG (ed) New world primates. ecology, evolution, and behavior. Aldine de Gruyter, New York
Stringer D (2016) When grasshopper means lightning. Landscape Mag 5:14–19
Tarble K, Pina I, Fernández FJ (1994) Una perspectiva histórica de la percepción de recursos en el Orinoco Medio. Bol Mus Arqueol Quibor 3:155–170
Torrealba G (2011) La economía política de la sarrapia: Etnografía histórica de las actividades extractivas entre los Mapoyo del Orinoco Medio, Venezuela. Master Thesis. Instituto Venezolano de Investigaciones Científicas, Caracas
Torrealba G, Scaramelli F (2018) Las estaciones sarrapieras: los Mapoyo y las economías extractivas del Orinoco Medio, Venezuela. Bol Mus Para Emílio Goeldi Cienc Hum 13:293–314
Urbani B (2005) The targeted monkey: a re-evaluation of predation on New World primates. J Anthropol Sci 83:89–109
Urbani B, Cormier L (2015) The ethnoprimatology of the howler monkeys (Alouatta spp.): from past to present. In: Kowalewski MM, Garber PA, Cortés-Ortiz L, Urbani B, Youlatos D (eds) Howler monkeys: behavior, ecology, and conservation. Springer Press, New York
Villalón ME (2003) Mopue waimuru. Idioma mapoyo. Instituto de Patrimonio Cultural, Caracas
Villalón ME (2007a) Una caracterización del sistema vocálico de la lengua mapoyo. Antropológica 107–108:13–39
Villalón ME (2007b) Una caracterización del sistema consonántico de la lengua mapoyo y propuesta para un alfabeto práctico. Antropológica 107-108:41–70

Villalón ME, Granadillo T (2000) Los marcadores de personas de la lengua Mapoyo. Proceed 49th Int Congr Americanists, vol 1, Quito
Viveiros de Castro E (1992) From the enemy's point of view. Humanity and divinity in an Amazonian society. University of Chicago Press, Chicago
Viveiros de Castro E (1998) Cosmological deixis and Amazonian perspectivism. J Roy Anthropol Inst 4:469–488
Zent S (2009) Traditional Ecological Knowledge (TEK) and biocultural diversity: a close-up look at linkages, delearning trends and changing patterns of transmission. In: Bates P, Chiba M, Kube S, Nakashima D (eds) Learning and knowing in indigenous societies today. UNESCO, Paris

Chapter 8
Co-ecology of Jotï, Primates, and Other People: A Multispecies Ethnography in the Venezuelan Guayana

Stanford Zent and Egleé Zent

8.1 Introduction

After a modest beginning (Sponsel 1997), ethnoprimatology is now poised to occupy the vanguard of contemporary discourses charting the course of Amazonian anthropology, considering its strategic relevance for reconceptualizing the relationship between human societies and the so-called natural environment. Since the early period of European colonization until quite recently, understandings of the Amazon-Orinoco rain forest and the indigenous peoples living there were based on the Cartesian dichotomy between nature and culture. Models of Amazonian *cultural ecology* postulating stringent environmental limiting factors on indigenous cultural development assumed that sociocultural formations were obligated to adapt to an invariant nature (Steward and Faron 1959; Lathrap 1968; Meggers 1971; Gross 1975). Comparative as well as ethnographic studies adhering to the *structuralist* research program represented Amazonian mythologies, cosmologies, rituals, and social structures as being elaborate symbolic constructions based on deeper cognitive structures of an oppositional nature, especially the binary contrast of nature/culture and its permutations (Levi-Strauss 1969, 1973; Dumont 1976; Mayberry-Lewis 1979).

Dualism persisted until the 1990s when it started to break down under the weight of a growing body of evidence showing widespread anthropogenic modification of the natural environment, such as palm forests, Brazil Nut tree groves, Amazonian dark

S. Zent (✉) · E. Zent
Centro de Antropología, Instituto Venezolano de Investigaciones Cientificas, Caracas, Venezuela
e-mail: srzent@gmail.com

B. Urbani, M. Lizarralde (eds.), *Neotropical Ethnoprimatology*, Ethnobiology,
https://doi.org/10.1007/978-3-030-27504-4_8

earths,[1] and large-scale earthworks (Baleé 1989; Denevan 2001; Lehman et al. 2003; Baleé and Erickson 2006). That the footprint of ancient human societies can be detected in the biota, soils, and landforms, effectively undercuts the notion of pristine ecosystems. The paradigm of Amerindian *perspectivism* shifts the dualistic focus from existence to perception, namely, the distinction between external versus internal points of view. Seen from the outside, animals (and spirits) differ from humans (and each other) in a superficial, corporeal sense, while from an inner, immanent vantage point, they are equally (inter)subjective, social, and culture-bearing (Viveiros de Castro 1998; Lima 1999). The recently christened "*multispecies turn*" in anthropology takes this one step further by blurring the lines between *Homo sapiens* and other species at ecological as well as phenomenological levels. Humans and nonhuman species coinhabit the same spaces, and the entanglement of their minds, bodies, and agencies mutually creates the conditions for each other's existence (Kirksey and Helmreich 2010; Kohn 2013). The environment is understood as neither an inherently natural object nor a cultural by-product but rather as a dense network of relationships established among a population of intermingled and interacting people, animals, plants, rocks, artifacts, and other entities. *Ecogony* refers to a research approach that explores the manifold root causes, whether material, ideological, spiritual, or functional, which underlie particular interactions between human groups and their local environments. This approach also questions the ontological separations of society and nature, subject and object, but does so from an ethnographic rather than an a priori philosophical point of view. Thus it advocates the careful study of the environmental ethics, encompassing representations (e.g., of nature, humanity, species, predation, mutualism, transformation, interpenetration, life, death) linked to actions (e.g., subsistence, settlement, healing, socialization, ritual), of particular biocultural groups (Zent 2014a, b, c).

The utility and durability of models purporting to explain Amazonian socionatural realities will depend ultimately on their coherence with the ethnographic record. The ethnography of human-nonhuman primate relationships is crucial in this sense for the same reasons that primatology became an important branch of biological and cultural anthropology. Other primates are our closest relatives and therefore offer the optimal subjects for probing the boundaries of human exceptionalism. Ethnoprimatology was developed in part to correct the bias of excluding human contact/influence from consideration in research on primate behavior and ecology. The results of field research in different cultural and ecological contexts point to long-term coexistence and ongoing interactions between human and nonhuman primates that go beyond simple ecological exploitation (e.g., preda-

[1] Amazonian dark earth, also called *terra preta* 'black soil,' is fertile soil that was formed by mixing charcoal and other organic substances (e.g., discarded vegetal matter and plant residues; fish, animal, and human bones; excrement) with otherwise relatively infertile soil. It is believed that this class of soil is the product of human activities associated with settlement and agriculture, formed over long time periods. It has been found in numerous locations spread throughout the Amazon basin, typically in small patches and together with pottery remains. Soil scientists have determined that this category of soil really consists of several types according to their chemical and generative properties, e.g., *terra preta*, *terra mulata*, and *terra preta nova* (Baleé 2010).

tion, crop-raiding) or competition (e.g., eating same food source). Fuentes (2006) demonstrates that the adaptive outcome of this contact is a two-way street. Whereas some simian species have adapted behaviorally, physiologically, and demographically to sympatric human populations and anthropogenic landscapes, it can also be shown that numerous human groups have incorporated simians into their economic, social, and symbolic systems (Lizarralde 2002; Shepard 2002; Cormier 2003, 2006). This finding raises the questions of how shared or complementary adaptations affect human representations of nature and in turn how such representations affect patterns of engagement between people and the other primates.

Lowland South America is characterized by high endemic cultural diversity on the one hand and by high biodiversity and wide distribution of neotropical monkeys on the other. Given these conditions, one might expect to find a prolific range of socio-ecological interactions. Yet, as Cormier (2006) observes, focused research on ethnoprimatology in this area is extremely scarce, which provides one of the justifications for the present paper. The purpose here is to serve up an ethnographic account of the co-ecology of the Jotï people and the monkeys (and related species) that share their territory. The information presented here is not the product of planned research on the topic of ethnoprimatology but rather was taken, in bits and pieces, from fieldwork with other foci carried out by the authors among the Jotï over a 20-year period. As such, this report is basically descriptive, not analytical, and contains gaps that should be covered during future field trips.

8.2 Ethnographic and Ecological Context

The Jotï are a culturally, linguistically, and phenotypically distinct, indigenous people of the Venezuela Guayana. Today they have a population of approximately 1500 persons distributed among 25 communities in northeastern Amazonas State and southwestern Bolívar State. First contact between the Jotï and western society took place in 1969. At that time, they displayed a cultural pattern with the following traits: interfluvial habitat orientation, dispersed and nomadic settlement pattern, simple material technology, scarce presence of western goods, mixed horticultural-hunting-gathering subsistence economy, and band level of sociocultural integration (Jangoux 1971; Coppens and Mitrani 1974; EibelEibsfeldt 1973; Guarisma 1974). Missions were established in Jotï territory shortly thereafter, ushering in a new era of rapid demographic and cultural change. In 1971, evangelical Christians affiliated with the US-based New Tribes Mission (NTM) founded the mission of Caño Iguana, in Amazonas State. In 1983, Catholic nuns sent by the Colombian-based congregation María Inmaculada de la Beata Laura Montoya established an outpost among Eñepa (Panare) and Jotï groups at the confluence of the Kayamá and Moyá Rivers in Bolívar State. Subsequently, a large proportion of the Jotï local groups migrated toward either one of these poles of intercultural contact. In addition to their coordinated efforts to achieve the religious proselytization of the local indigenous population, the missionaries installed schools and medical dispensaries at the missions; taught the Spanish language, local language literacy, and other subjects; distributed western material goods; and introduced new forms of social organization. Although the missionaries are no longer present at either Iguana (since 2006) or Kayamá (since 2016), their educational and health facilities remain somewhat intact, and they are still the main centers of demographic gravity. About a quarter of the Jotï population have remained outside the ex-missions' sphere of influence and preserve a very traditional lifestyle. Some of these groups have migrated downriver and maintain intermittent social and economic contacts with neighboring indigenous groups, while others are extremely isolated from outsiders. Thus, while the Jotï have experienced considerable social, economic, and intellectual changes in the last four decades, these impacts are uneven across the entire population. Furthermore, basically all of the Jotï communities existing today still maintain a way of life and worldview that is closely connected with the forest environment, and their subsistence economies are for the most part independent and exhibit very few if any direct economic ties to outside markets. Thus they still share many cultural traits and similar relationships with the natural environment (Zent and Zent 2008, 2012).

The Jotï homeland is located in and around the northern sector of the Sierra Maigualida mountain range. This topographically diverse region varies in altitude from 150 to 2400 meters above sea level and is covered by dense and high forests (evergreen and semi-deciduous; basimontane, premontane, and montane), except at peak elevations (above 2000 m asl) where tepui meadows and scrub prevail. At lower altitudes, mosaics of small transitional formations of scrubland, woody, and open savannas can also be observed (Huber 1995:42). The macroclimate for most of this region is classified under the Köppen system as wet tropical with a dry season of 2 months (defined as less than 75 mm). Our own measurements of pluviosity from within the Jotï territory produced a range of 2400–2700 mm/yr, with May–August being the wettest months (>300 mm) and December–March the driest months (<100 mm).[2] Temperatures throughout the region oscillate between 30.8° (±2.4°) and 21.7° (±0.5°) C (range of 18–44 °C). Relative humidity is variable according to the season, but the average daily maximum value was 92.6 (±1)%, and the average daily minimum value was 63.7 (±11.9)%. According to our botanical surveys carried out at four sites within the Jotï territory, the forests of this region

[2] We took daily readings of rainfall using a Tru-Chek rain gauge during fieldwork carried out in four Jotï communities from 1996 to 1999 (see first project description under Methodology section). The entire period of observation covered 24 months but was not continuous: June–August 1996, January–September 1997, January–May 1998, September–December 1998, and January–February 1999. Monthly totals for each community site were extrapolated according to the levels recorded and proportion of days in the month for which readings were made. The range of pluviosity reflects the highest and lowest annual totals, as calculated by summing the site-specific monthly totals (Zent et al. 2001).

exhibit some of the highest levels of alpha (within-plot) diversity ever recorded in the Guiana shield region of South America (133–191 species/Ha). Beta diversity is also unusually high, as less than 20% of the total inventory of species collected can be found at more than one site. The primary forests usually present at least two distinct arboreal strata, canopy heights varying from 15 to 30 meters, and emergent trees up to 30–40 meters high. The most conspicuous plant families are the Burseraceae, Moraceae, and Sapotaceae, while the most species rich family is Fabaceae (38 species) (Zent and Zent 2004).

8.3 Methodology

The data on which this chapter is based come from different research projects that the authors have carried out among the Jotï during the last 20 years. The research objectives, sites, and methods of three main projects are summarized here.

1. Following initial field surveys (1994, 1996), we carried out an extensive study of Jotï ethnobotany and behavioral ecology in 1996–1999 at four sites: Caño Iguana (IG), Caño Majagua (MA), and Caño Mosquito (MO) in Amazonas State and Kayama (KA) in Bolívar State (see Fig. 8.1 Map of Jotï communities, detailed description of these settlements is found in Zent 1999; Zent and Zent 2004). Besides the standard ethnographic techniques of participant observation and key informant interviewing, data concerning knowledge and use of primates and other animals were collected in the course of plant specimen collections, structured interviews in forest plots, food resource accounting (i.e., counting and weighing of food items harvested), focal person follow observations, and casual conversations. Mammalian names and classifications in the Jotï language were recorded using the field guides by Emmons and Feer (1990) and Linares (1998) as prompts.
2. From 2001 to 2006, we worked together with the Jotï on a collaborative project dedicated to land self-demarcation and community-based mapping. All land borders and places of cultural or natural significance were mapped using GPS (>5000 data points), and geo-referenced maps were generated using GIS. In addition to the ethnocartographic data, we conducted semi-structured interviews about various topics related to conceptualizations of the earth and the cosmos, life forms and their habits (plants, animals, fungi, spirits), land use patterns, resource management practices, life histories, ethnomedical beliefs, food taboos, ritual practices, and environmental ethics. This included the recording of myths and stories in which primates and other related animals are key figures. Most of the data during this period comes from Kayamá or Caño Iguana.[3]

[3] An ancillary outcome of this project was the development of an alphabet for writing the Jotï language based on a process of community-based consensus. This alphabet is used in the present paper to write all of the words appearing in the Jotï language (see Quatra 2008 for details).

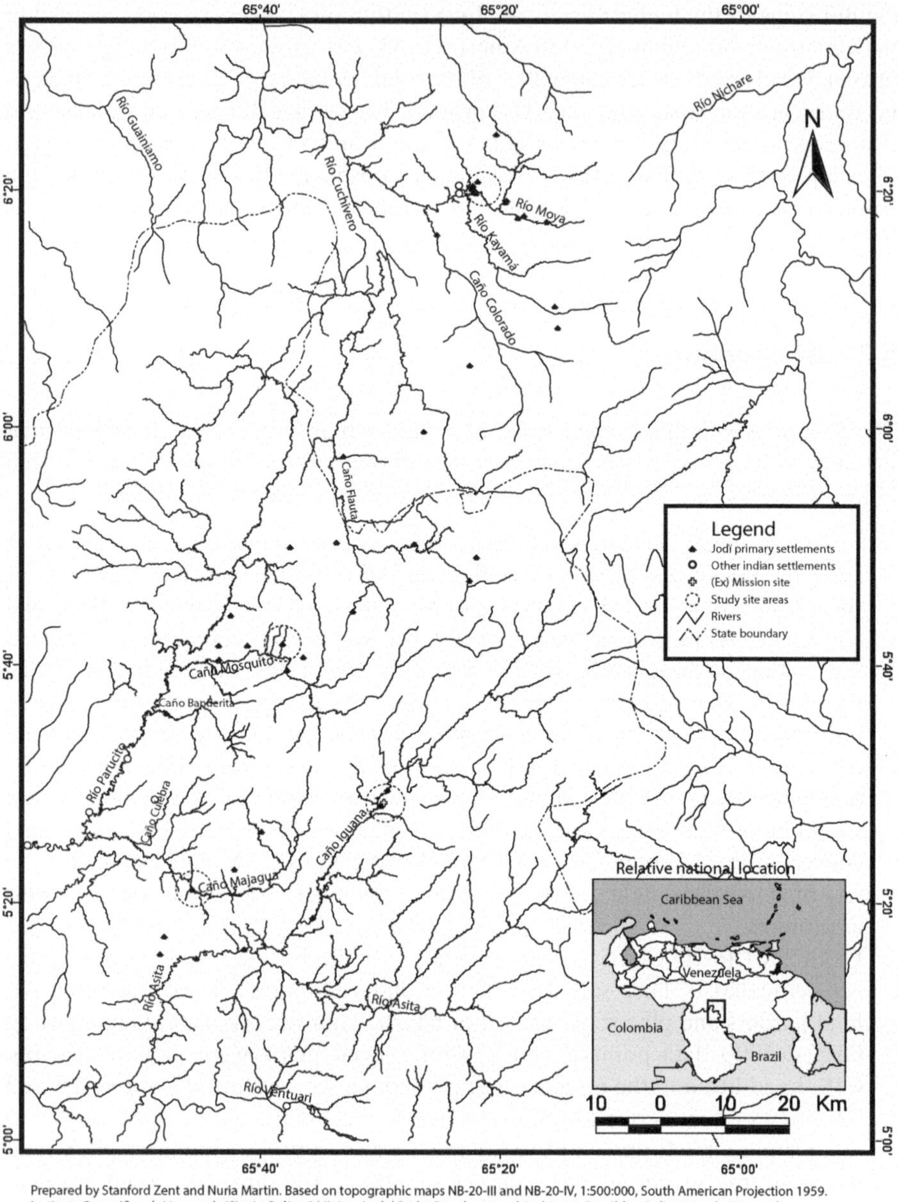

Prepared by Stanford Zent and Nuria Martin. Based on topographic maps NB-20-III and NB-20-IV, 1:500:000, South American Projection 1959. Instituto Geográfico de Venezuela "Simón Bolívar", Ministerio del Poder Popular para el Ambiente, República Bolivariana de Venezuela.

Fig. 8.1 Map of Jotï communities in the Venezuelan Guayana

3. A project entitled "VITEK: Vitality Index of Traditional Environmental Knowledge" was carried out in Kayamá and neighboring communities from 2011–2016. A big part of this project involved the creation of a bilingual computerized database (Jotï-Spanish) organized by 16 TEK-related domains: plants,

animals, interspecific relationships, ecotopes, soils, climate, landscape, agriculture, pet-keeping, hunting, fishing, collection, food preparation, ethnomedicine, crafts, and architecture. Semi-structured interviews were conducted with informants recognized by their peers to be local experts in the subject. The interviews on plants, animals, interspecific relationships, pet-keeping, hunting, and food preparation yielded specific information about different primates and related species.

8.4 Nomenclature and Classification of Primates Among the Jotï

There are six primate species that are recognized and named by the Jotï. Two other species of the Raccoon family (Procyonidae) are closely associated with them. This grouping of ethnozoological taxa is shown in Table 8.1. Most of the species listed here have more than one name in the Jotï language. The primary name refers to the default name, the one that is most widely known and spoken. Half of these constitute primary lexemes (i.e., single terms). The others are secondary lexemes, ostensibly of the productive kind, in which the first constituent specifies the modifier, *uli* 'big' or *jani* 'small', and the second constituent denotes the superordinate category, *jkwayo* or *mujkëlo*. Based on the linguistic evidence alone, one might conclude that the latter refer to folk generics and therefore the binomial expressions designate folk specifics. Other evidence, such as the results of free listing exercises and descriptive statements about the morphology and behavior of each taxon, suggest instead that all of the terms listed here, whether mono- or polylexemic, make up a contrast set of the same approximate rank, which would appear to be folk generic. However, this too would oversimplify the cognitive organization of this association of folk taxa, as discussed below (see Fig. 8.2). The secondary names are less common; they consist of nicknames, ritual names (e.g., used in mythical narratives, ceremonial performances, or shamanic recitations), or antiquated names. At least some of these are readily analyzable, such as *Abiyë jkojko maja* 'bushy bearded animal' for the bearded saki and *Duweyo jtudï* 'redheads' for the howler. The secondary names provide clues for discerning the cognitive associations between taxa, as in *Duwëwe jkwayo* 'red jkwayo' for the howler, *inï jkwayo* 'night jkwayo' for the kinkajou, and *Jtajwä nimo* 'old howler' for the olingo.

Based on the evidence at hand, there is no one-to-one categorical correspondence between the primate order as defined in western biosystematics and Jodï ethnobiological classification. Instead, the Jotï associate or link all of the monkey and procyonid species mentioned in Table 8.1 into a single category on the basis of perceived

Table 8.1 Jotï ethnoprimate nomenclature

Jo primary name	Jo secondary names	English name	Systematic name
Uli jkwayo	*Mojkajte, jkwakyolojto, uli awela*	Spider monkey	*Ateles belzebuth*
Jani jkwayo	*Jkyabo jkwayo, bulodo, bule jeno dea*	Wedge-capped capuchin monkey	*Cebus olivaceus*
Jkwaijlë	*Abiyë jkojko maja, jkwayo*	Brown bearded saki	*Chiropotes chiropotes*
Imo	*Nimo, duwëwe jkwayo, duweyo jtudï*	Red howler monkey	*Alouatta seniculus*
Jkijki		Squirrel monkey	*Saimiri sciureus*
Uli mujkëlo	*Nujtinë jlojlo, inï jkwayo*	Kinkajou	*Potos flavus*
Jkujkujtu[a], *jani mujkëlo*[b]	*Nujtinë jlojlo, waijlo mujkëlo, mujkïkï (jelë)*	Owl monkey	*Aotus trivirgatus*
Jani mujkëlo[a], *dodo jtejte*[b]	*Mujkïkï (jtu), jtajwä nimo, jwani jelë, waijlo mujkëlo*	Olingo	*Bassaricyon gabbii*

[a]Name used in northern dialect (Bolívar State)
[b]Name used in southern dialect (Amazonas State)

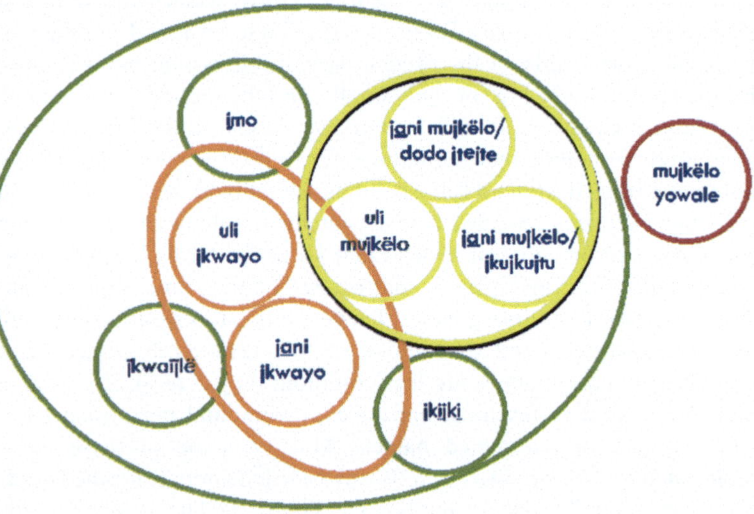

Fig. 8.2 Graphic schema of Jotï classification of the *Jkwayo–Mujkëlo* complex

shared or similar morphological, behavioral, and ecological traits.[4] It should be noted, however, that this grouping is not based on absolute criteria of category membership but instead on variable degrees of resemblance among members, thus

[4]The Jotï ethnobiological classification of primates is therefore identical to that of the Waorani of Ecuador. The latter recognize primates as a unique group but also include the non-primate kinkajou and olingo within the group of "monkeys" (Papworth et al. 2013). It is interesting to note that both groups rely heavily on hunting monkeys for meat procurement.

more like a fuzzy set. Together, these make up what has been called a covert (i.e., unnamed) category at the intermediate rank (Berlin 1992). Intermediate taxa "form natural, biologically well-founded groupings whose recognition is based on obvious patterns of similarity" and more commonly "correspond closely to portions of recognized biological families (p. 144)." In the present case, the intermediate category encompasses all members of the primate order inhabiting the Jotï territory as well as the kinkajou and olingo. The inclusion of the tree-dwelling procyonids can be explained by the observable fact that they display several traits common to the primates: quadrupedal mammals, fur-covered bodies, grasping hands, flexible feet, long tails used for grasping or balance, short snouts, exclusively arboreal habitat, and omnivores that subsist mainly on tree fruits or flowers. This hypothesis draws further support from the observation that the body and habits of the coati (*Nasua nasua*; *walijkyḛna*, *walina*, or *jtawä jkajwiyë* in Jotï), the only other member of the Procyonidae family in this area, differ from this description, and hence it is not included in the group. In free-listing exercises, it is often the case that the informant names all or most of the group members in succession.

The aforementioned category is divided into two main subcategories or clusters of inclusive taxa, which correspond to the terms *jkwayo* and *mujkëlo*, respectively. This distinction reflects the divergence between primate and procyonid, but not completely. The term *jkwayo* (sg.), or *jkwayodï* (pl.), is polysemic, and therefore its precise meaning must be inferred from the communicative context and purpose. In the first place, it may refer to the folk generic, as in the productive constructions of *ja̱ni jkwayo*, *uli jkwayo*, and *i̱nï jkwayo*. When the term *jkwayo* is expressed without a modifier, the speaker is usually referring to the focal members of the taxon, spider monkey (*uli jkwayo*) or capuchin monkey (*ja̱ni jkwayo*). Of these two, the spider monkey may be considered the more prominent one from economic and symbolic standpoints (see below), but the capuchin monkey is the more central figure in terms of social relations with other species. *I̱mo* and *jkwaijlë* are more peripheral members of the group but still members as can be judged from the secondary names, *duwëno* 'red' *jkwayo* for the howler and just plain *jkwayo* for the bearded saki. As for *jki̱jki̱*, unfortunately we do not have sufficient data at this time for a definitive placement, but one informant did use the expression *jkwayo jawabo meka̱ bada* 'same type of animal as *jkwayo*' to describe it. If we take into account all of the different applications of the lexeme *jkwayo*, it becomes apparent that the term in its maximally extended sense refers to all of the creatures mentioned here (Table 8.1). Thus *jkwayo* functions as a sort of pseudo-label for the entire grouping even though it is not explicitly used as such.

The subcategory of *mujkëlo* (pl. *mujkëlodï*) subsumes *uli mujkëlo* (kinkajou), *dodo jtejte/ja̱ni mujkëlo* (olingo), and *jkujkujtu/ja̱ni mujkëlo* (owl monkey), all of which are nocturnal species.[5] However, the distinction is somewhat fuzzy since

[5] Besides all of them being nocturnal, apparently their habitats overlap as well. Emmons and Feer (1990) report that the kinkajou, olingo, and owl monkey have been observed on the same trees at the same time.

the kinkajou is also called the *inï jkwayo* 'night monkey' and the olingo is nicknamed *jtawä nimo* 'old howler'. When asked to describe what a *mujkëlo* looks like, a common first response is that it is *jkwayo-bo* 'like a monkey'. *Uli mujkëlo* seems to occupy a focal position within the class because it is the first member listed during controlled elicitation exercises and can be named by the single lexeme *mujkëlo*, the unmarked form, whereas the others cannot. It also provides the strongest connection to the *jkwayo* cluster since it is the referential prototype of *inï jkwayo* 'night monkey'. The other two taxa are labeled by marked forms, and their physical presence is much rarer throughout the Jodï habitat. Our Jodï collaborators stated in fact that *jkijki* (squirrel monkey) is absent or extremely sparse on the Bolívar side of their homeland, and not so rare on the Amazonas side. This rarity probably leads to their being less well-known, and one consequence is that we are able to detect a certain ambiguity or inconsistency with regard to which names are applied to which species. Thus in the northern dialect (Kayamá area), the binomial *jani* ('small') *mujkëlo* is applied to the olingo, whereas in the southern dialect (Caños Mosquito, Iguana, and Majagua), the same name refers to the owl monkey.

A picture recognition and naming exercise was conducted during a visit to Kayamá in July 2017 in order to confirm the categorical distinctions and names given above. A total of ten persons from the community, five men and five women, were shown drawings[6] of the primate and procyonid species mentioned here and asked to provide the names. Ten out of ten respondents agreed on the identification and naming of *jani jkwayo* (*Cebus*), *uli jkwayo* (*Ateles*), *imo* (*Alouatta*), and *jkwaijlë* (*Chiropotes*). Six out of ten agreed on *jkujkujtu* (*Aotus*) and *uli mujkëlo* (*Potos*). Three out of ten named *jani mujkëlo* (*Bassaricyon*), and no one was familiar with *Saimiri*. Based on this exercise, then, the members of the *jkwayo* cluster appear to be more salient for people of this community. The lesser salience, or agreement, in regard to the *mujkëlo* cluster may explain part of the observed nomenclatural variance and overlap among its members. The fact that *Saimiri* went unrecognized seems to confirm its total absence in this area.

Associated in a peripheral sense with the entire *jkwayo-mujkëlo* taxonomic cluster is the mouse opossum, *Micoureus demerarae* or *Marmosa murina* or both. It is named *mujkëlo yowale*, which translates to *mujkëlo* 'opossum'. While this productive name clearly situates the organism within the *yowale* generic taxon, it also expresses an identity or affinity with the *mujkëlodï*. This is probably motivated by the fact that its niche and food habits overlap to some extent with the other members of this cohort.

The totality of evidence reviewed here suggests that the *jkwayo-mujkëlo* grouping is organized more along the lines of a fuzzy set with graded category membership, a finding consistent with the intermediate categories found in other ethnobiological classification systems (Berlin 1992:24). A graphic representation of this categorical structure, including relative affinities among members, is depicted in Fig. 8.2.

[6]The drawings used for this exercise were reproductions from Linares (1998).

8.5 Kinship Between Monkeys

The existence of a covert taxon encompassing all primates is bolstered by the widespread belief that there is kinship among monkeys. This belief seems to be based less on perceived morphological similarities and more on interspecific social relationships. The most commonly stated relation is that the capuchin (*jani jkwayo*) and the saki (*jkwaijlë*) are *jluwëna*. The *jluwëna* is one of the most prominent nonconsanguineal relationships recognized in the Jotï kinship system and is based on the principles of affinity, reciprocity, generation, and heritability. It is essentially an affinal relationship in the sense that the preferred marriage is with the daughter of a man's *jluwëna*. For this reason, the Jodï often translate *jluwëna* as *suegro* 'father-in-law' in Spanish, but the semantic correspondence is only tangential. It is reciprocal in that it entails a dyadic exchange relationship between two persons (males) who call each other *jluwëna*. Besides giving or exchanging women, *jluwënadï* (plural) also provide other trade goods, food, labor (e.g., to cut trees for a new garden), and shelter (one sleeps in the house of *jluwëna* during visits to other settlements) to each other. It is heritable to the extent that a person recognizes as *jluwëna* the son of someone who their father called *jluwëna*. However, *jluwënadï* are not just inherited but may also be created by establishing and sustaining an exchange relationship. It is generational because the son (and father) of *jluwëna* is called *ji* and, in turn, the son of *ji* is called *jluwëna*, and so on. In general, it is a relationship characterized by closeness, friendship, positive affect, cooperation, and support. The perception that the capuchin and the saki are *jluwëna* most likely reflects the notion of nonantagonistic coexistence or even symbiosis between them. Thus it is often noted that these two species sometimes travel together or that their chance encounters, especially at a common food source, are characterized by an initial episode of simulated fighting which turns into general merriment and harmonious feeding for all.

The kinship of spider monkey, *uli jkwayo*, and howler monkey, *imo*, constitutes another salient family tie. Their relationship is likened to men who call each other *ji*. The category of *ji* is also a fundamentally affinal, rather than consanguineal, connection, but it differs fundamentally from *jluwëna* in the sense of social distance. Whereas a person considers his *jluwëna* to be a close friend, one's *ji* is not especially friendly and can even be considered a potential rival or enemy. It also alternates in generation with *jluwëna*, such that the son or father of a person's *jluwëna* is their *ji* and vice versa. If *jluwëna* approximates the category of father-in-law, then *ji* is close to being brother-in-law, where the latter is not by rule or preference a reliable giver of women but in actual practice may do so. The application of such terminology to describe spider monkey-howler monkey relations, and by way of contrast to the relationship between capuchin and saki, is probably intended to spotlight the social distance between them. Another term which has been used to describe this relation is *jlewadï*, which signals dislike and avoidance. While both species are invariably considered to be gregarious among members of its own kind, they are also aloof when it comes to interacting with other species. It is said that the spider monkey neither aggresses nor socializes with other species. The howler monkey, by contrast, is more territorial and hence defensive. If a troop of howlers cross paths with some spider monkeys, they may take to "scolding" or "yelling" at their cross-species rivals, ostensibly for trespassing their travel routes, but then one or both withdraw. They are never seen feeding in one spot at the same time, as is claimed for capuchin and saki monkeys.

The relationship of the spider monkey and the capuchin monkey, while acknowledged to be somewhat distant, is nevertheless seen as brotherhood, in which the former is the younger brother and the latter is the elder brother. Morphologically, such designations are consistent with the Jotï belief that the second-born son will be taller than the first-born son. From the standpoint of spider monkey, capuchin monkey is his *jnë* 'older brother/relative', while the latter refers to the former as his *nawï* 'younger brother/relative'. The relationship between brothers is a close and cooperative one, especially when they are younger and live under the same roof or close by, and thus it may be assumed that the Jotï regard spider monkey and capuchin monkey to be especially close. However, observations of the two species in the field, as expressed by the Jotï themselves, do not substantiate the hypothesis of close interaction or joint participation in activities. Even though they display similar movements and feeding habits, they are not observed traveling or feeding together. We postulate that the association is derived from myth, where both species are prominent actors in a primordial time period when animals were invested with bodies and behaviors more like humans. In this alternate time-space, there are multiple episodes in which spider monkey and capuchin monkey can be found participating in activities together, like planting cultivated plants in gap gardens (see Sect. 8.8).

The only other putative kinship relation involving primates that we are aware of involves owl monkey (*jkujkujtu*), which is considered to be the *nawï* 'younger male relative' of both spider and capuchin monkeys. The category of *nawï* is semantically expansive and contextual in the sense that it may be used to refer to a diverse range of relatives, including son and younger brother, but there are alternative terms for naming these. The most common reciprocal of *nawï* is *jlae* which is the formal form of 'father' but is also used to specify any male relative of an older generation. In general, the relationship is not particularly close, and we think that this better explains why it is used to represent the relationship between the owl monkey and the more focal members of the category. It is worth noting that the owl monkey is nocturnal, whereas the others mentioned here are diurnal, and therefore there really is no basis for direct interaction. Furthermore, its body size is smaller than most of the others.

8.6 Economic Significance

The cognitive salience of the *jkwayo-mujkëlo* complex is undoubtedly motivated, at least in part, by their economic significance. All of the taxa mentioned here are prey species hunted for food. A couple of these are also exploited for bones which are sharpened into multipurpose needles and worn as ornaments (usually inserted into the pierced earlobe). The blowgun, loaded with curare-tipped darts, is the weapon of choice for hunting arboreal game. This type of hunting is usually carried out by solitary hunters or small hunting parties (two to three persons).

For diurnal game species, the hunting party sets off at dawn and walks along footpaths aptly named *jkwayo mana* 'monkey trails' looking for sign of game such as discarded food remains or feces or listening for their distinctive calls or movements. To improve their chances of encounter, the hunter(s) will rub on their hands or other body parts a perfume-like substance called *jkwayo jlebona dekawa*. This substance is extracted from small sacks (scent glands?) found in the lower neck-upper chest region of hunted primate specimens. The logic behind this practice is that conspecifics of the animal from which the perfume was drawn are attracted to the hunter because he carries the smell of the animal on him. The inconspicuous fungi called (*uli*) *jkwayo wa̠ña* (unidentified specimen, collection number ZHMO-1008) is another magical hunting substance used specifically for hunting monkeys (Zent et al. 2004; see Sect. 8.10). When the nearby presence of monkeys is suspected, especially spider, capuchin, and bearded saki monkeys, the hunter pinpoints their precise location by making specific calls or whistles that they are inclined to respond to. The spider monkey responds to an imitation of its call, while the capuchin monkey responds to an intermittent, monotone whistle. Another tactic is to go to certain fruiting trees where the monkeys are expected or to a spot along the trail that they are known to use and wait behind a blind for them to come. After the precise location of the prey has been determined, the trick is to get close enough for a clear shot while remaining unseen and unheard. If one of the prey sentinels spots the hunter(s), the troop will immediately bolt and then must be chased, sometimes over long distances. The spider monkey, in particular, is quite fast and predisposed to flee up to 2–3 kilometers without stopping. The howler monkey, by contrast, is adept at hiding, and the hunter may have to climb the tree to locate it, at which time it may also decide to run.

The nocturnal species – kinkajou, olingo, and owl monkey – may be hunted during the day if the hunters happen to find their tree dens. Upon spotting a promising hole up in a tree, a hunter will bang on the tree trunk below with a machete to arouse the occupant from its slumber. Meanwhile, the hunter's companion takes aim at the entrance with his blowgun. When the startled animal sticks its head out of the hole to see what the ruckus is about, the shooter fires a shot. An alternative method is to climb the tree and seize the prey by hand. Less commonly, these species are hunted when they go to feed on the flower nectar of *Caryocar microcarpum*. The hunter goes to flowering trees of the species at dusk and waits for the animals to come.

Box 8.1 provides a personal recollection of a hunting outing in which squirrel monkeys (*jki̠jki̠dï*) were successfully hunted. Several aspects of this hunting experience are worth calling attention to because they give insight into the typically Jotï strategy of hunting monkeys and other game. The first is that it involves preparation, beginning with the women singing before dawn. The Jotï often sing before going out to forage, naming different animals that they hope to encounter. It is believed that this will influence the animal's *jkyo aemo* 'spirit master' to send one of their charges in the hunter's direction (see Sect. 8.9). Hunting songs may be sung by anyone from the house, including women, who may not take an active part in the hunt itself. Thus hunting is very much a collective enterprise involving the efforts of more people than just the hunters who actually go out looking for game to shoot.

Secondly, the hunters set off with both blowgun and lance in hand and thus were prepared to pursue a large variety of terrestrial or arboreal game that they might encounter. However, they also had a specific game type in mind – *jkwayodï* 'monkeys' – and went to a specific place to find it. To find the monkeys, the hunters went directly to its food source, a tree that they knew is eaten by monkeys and is fruiting at that moment in time. Thus an obviously important part of Jotï ethnoprimatological natural history is a knowledge of the different fruit- and flower-bearing tree species – phyto-indicators (Table 8.2) – eaten by different primate species. After the hunt was over, Jkamilo and Elijke recited an extensive list of common plants whose leaves, fruits, or flowers are eaten by the squirrel monkey (Table 8.3).

In the event described here, the initial target tree did not bear game so the hunters moved on and eventually found the squirrel monkeys at another fruiting individual of the same species. In addition to their knowledge of the appropriate phyto-indicators, the Jotï use additional information such as the territorial ranges, travel routes, and nesting habits of the primate species. The last half of the hunting involved lots of walking and observing but no more kills. Just as valuable as the actual game capture, however, was the gathering of information about monkey traces and movements that can be used in future outings. Such details emerge when hunters recall the circumstances of the hunt to an audience back home. For example, our notes of other squirrel monkey hunting events at the same community include the following

Table 8.2 Phyto-indicators of squirrel monkey (*jkijki*)

Jo name	Scientific name
Ulu	*Attalea maripa* (Aubl.) Mart.
Muli	*Socratea exorrhiza* (Mart.) H.Wendl.
Jani bade	*Oenocarpus bacaba* Mart.
Jkolöwa ji	*Attalea* sp.
Ikyeka mau	*Protium* sp.
Jkwayajkiabo jtawï	*Bathysa bathysoides* (Steyerm.) Delprete
Jkwayo wejkao	*Pourouma* sp.
Jkalimane	*Couma macrocarpa* Barb. Rodr.
Jani iyë jyëï	*Trichilia pallida* Sw.
Wejtolo	*Cecropia* sp.
Imo jtabali	*Ceiba* sp.
Jkwayo jtu jtawï	*Clavija lancifolia* Desf.
Jtelela	*Ficus* sp.
Jkuwëjte jani wejkao	*Pourouma* sp.
Jkwayo oneka dodo	*Enterolobium schomburgkii* Bentham
Uli aejkwa jyëï	*Virola surinamensis* (Rob.) Warburg
Weya jyëï	*Hirtela* spp.
Jkwaijlë jtawï	*Licania apetala* (Meyer) Fritsch
Wanejko jyëï	*Chrysophyllum argenteum* Jacquin
Jedä jkwayajkiabo	*Bathysa* sp.
Walijkyëna jtawï	*Erythroxylum* sp.
Jkawïle jtawï	*Erisma uncinatum* Warm.

Table 8.3 Rank order biomass of hunted animals

English name	Latin name	% Wt.	Rank
Long-haired spider monkey	*Ateles belzebuth*	23.03%	1
Tapir	*Tapirus terrestris*	20.31%	2
White-lipped peccary	*Tayassu pecari*	13.15%	3
Paca	*Agouti paca*	5.96%	4
Red howler monkey	*Alouatta seniculus*	4.36%	5
Black curassow	*Crax alector*	4.34%	6
Blue-throated guan	*Pipile pipile*	4.22%	7
Red-rumped agouti	*Dasyprocta leporina*	3.94%	8
Wedge-capped capuchin monkey	*Cebus olivaceus*	2.99%	9
Brown-bearded saki	*Chiropotes chiropotes*	2.94%	10
Smooth-fronted caiman	*Paleosuchus trigonatus*	2.11%	11
Red-and-green macaw	*Ara chloroptera*	1.39%	12
North Amazon red squirrel	*Sciurus igniventris*	1.16%	13
Great long-nosed armadillo	*Dasypus kappleri*	1.08%	14
Southern tamandua	*Tamandua tetradactyla*	0.92%	15
Spix's guan	*Penelope jacquacu*	0.82%	16
Collared peccary	*Tayassu tajacu*	0.74%	17
Gray-winged trumpeter	*Psophia crepitans*	0.74%	18
South American coati	*Nasua nasua*	0.68%	19
Lesser razor-billed curassow	*Mitu tomentosa*	0.44%	20
Others		4.68%	

contextual information: a squirrel monkey was shot while eating *waiyo walema jyëï* (*Helicostylis tomentosa* (Poepp. & Endlicher) Rusby) in the upper Majagua, one-day's walk from Jkawale community (Jkayupare July 17, 1997); two squirrel monkeys were bagged while climbing over *jedä malawa ibuju* (*Abuta rufescens* Aublet), a vine that was draped over a tree of *ejko luwe jyëï* (*Inga* sp.) bearing fruits (Baijkyo January 24, 1999).

Finally, we can also note the agency of the squirrel monkeys in response to being hunted. When the hunters were spotted, a cry of danger rang out, thus manifesting the importance of in-group communication for their survival, and the troop attempted to flee. Thus they kept their losses to three members. Their extreme wariness probably reflects their "acculturation" to humans and their penchant for attacking them.

Box 8.1 Personal Recollection of Hunting Squirrel Monkeys

A party of three, Jkamilo, Elijke, and I, set off to hunt at 5 a.m. Beautiful singing duets had started well into the night, probably about 3 a.m. in the middle of the dry season at Jkawale community, located on the left bank of a tributary of the Parucito River. The piercing voice of the women cut the night silence. The five houses of the community were next to each other. Therefore it was easy waking up for the beautiful but incomprehensible music. The dim forest

(continued)

night resounded sonorously, magnifying the human tunes over the trees. While painting their arrow tips with black curare, the men hummed tones. Elijke and Jkamilo quickly left their houses, each one carrying a blowgun and a lance, and I hastened to follow. We all got into the community au-jtawï [lit. water tree, translation: 'canoe'] and quietly paddled up the Majagua River for over an hour. A crescent moon drew the profile of the forest. It reflected the trail left by our vessel over the soft waters. Lazily, the seasonally flooded forest awoke with a bright sun while we disembarked on the other side of the Majagua. Walking fast, Jkamilo led the party toward a patch of jani waiño trees (Pseudolmedia laevis (R. & P.) Macbride). *As usually happens, Jkamilo and Elijke were expecting that some monkeys would congregate around the trees to eat. The daily forest creatures started to awaken with the morning as well as the humidity and heat. The two hunters walked quietly but steady, looking up and leaving a distance of about 10 meters between them. Their tracks were almost imperceptible to me. They produced no sound. Almost 2 hours later, we arrived at a very noisy patch of dense canopied forest. The red fruits of many trees of jani waiño hung heavily from the branches where what seemed to be over 2 dozen jkijki moved quickly and recklessly at least 15 meters above the ground. Once they recognized our presence, they shrieked painfully, alerting the troop of danger. Holding their blowguns and quivers, Jkamilo and Elijke moved hastily but attentively, producing no sounds unfamiliar to the forest that would disperse the prey any more. Silently, they pointed to different individuals who, smelling danger, scampered in various different directions. This strategy proved effective since each hunter could follow just a few jkijki at the most. Two, three … seven, many arrows one after another were rapidly shot. Most of the little animals broke away in boisterous commotion, while at least four to five were sieged meters away. It was not easy to see what was happening, and the hunters asked me to stay motionless in one corner. The whizzing of flying darts was constantly breaking through the background forest noise, hitting diverse surfaces, leaves, wood, bodies,* etc., *falling back down to the ground. After 10 minutes, no noise nearby could be associated with a human hunter. Forty minutes elapsed when the steps of Elijke followed by Jkamilo were clearly heard. They had shot three jkijki prey that amounted to almost 3 kilograms of meat. Additionally, Elijke was holding a tiny infant jkijki that was alive and produced what seemed like scared sharp cries. With leaves of A. maripa palm, Elijke wove a small basket to carry the baby monkey, and we continued our walk through the forest. After a few more hours of trekking on jkwayo mana 'monkey trails' and searching for signs of game activity that could be used as input for the next hunting trip, we returned to the community. Along the trail, Jkamilo pointed to a Jkaile tree* (*Micropholis egensis* (A.DC.) Pierre), *which had been the nesting place of a male howler monkey that he shot last dry season, adding that the fruit of this tree was the animal's preferred food* (Egleé Zent field notes, *Jkawale balo*, December 24, 1998).

Although all of the species mentioned here are deemed edible, such designation is not without restrictions. The capuchin, howler, and spider monkeys can be eaten by young and old alike but should be purified by *yu dekawa* ‘medicine’, which may consist of verbal supplications addressed to the animals’ respective spirit masters (see Sect. 8.10). The bearded saki is edible for everyone except for toddlers (<3 years old) who would become dizzy and cry if they were to eat it. Taxa belong to the *mujkëlo* sub-cluster are more dangerous. The kinkajou and olingo can be eaten by adults and adolescents after purification to neutralize any toxic effects, but never by children who can become very ill if they do so. Young women not far removed from first menstruation are also prohibited from eating or preparing this type of meat. The owl monkey is safely eaten only by elders (i.e., past reproductive age).

Food resource harvest data gives us a good idea of the considerable importance of primates in the Jotï diet. Data were collected at four Jotï communities, Caño Mosquito (MO), Caño Majagua (MA), Caño Iguana (IG), and Kayamá (KA), during the 1997–1999 field season. In sampling periods varying from 2 weeks to several months, the name and weight of all food types, whether of plant or animal origin, brought back to selected households were recorded in all of the communities mentioned. This data gives us an approximate idea of the relative composition of the Jotï diet, the dietary contribution made by primates-procyonids species, and variations of hunting levels or pressure by community. Animal species make up 20% of the total food biomass recorded, of which 11% are hunted, 5% are collected, and 4% are fish or aquatic dwellers. The relative contribution per major animal class by weight is as follows: mammals 46%, fish 23%, insects 21%, birds 8%, and reptiles 2% (Fig. 8.3). It is likely that a substantial portion of protein and fat requirements are satisfied by animal sources and the majority of these are obtained by hunting. Of the top ten species of animals hunted by the Jotï in terms of raw weight contribution, four are primates (Table 8.3): the spider monkey (*Ateles belzebuth*) in the top spot (23.03%), the howler monkey (*Alouatta seniculus*) in fifth place (4.3%), the capuchin monkey (Cebus olivaceus) (2.99%) at number nine, and the brown-bearded saki (*Chiropotes chiropotes*) coming in at tenth place (2.94%). Thus roughly a third of all meat consumed by the Jotï is monkey meat. Perhaps this explains why the phrase *jkwayo(nï) ju jtoba ibï dekae* ‘intending to hunting monkey’, or variations thereof, is frequently used to express the notion of hunting in general. Another remarkable result is the dominant contribution of the spider monkey, even over larger-bodied animals such as the tapir and peccaries.

The breakdown of hunting yields per community is shown in Table 8.4. Here we see that the proportion of the total food supply, in terms of gross weight, derived from game animals ranges from 9% in Kayamá, the largest Jotï community, to 30% at Caño Mosquito. Of that amount, mammals comprise approximately 40–60% of the total raw weight, being highest at Caño Iguana (61%). Primates make up anywhere from 9% (Iguana) to 89% (Mosquito) of the mammalian total. The spider monkey (*A. belzebuth*) accounts for 34% (Kayamá) to 75% (Mosquito) of the total of primates hunted. The contribution of the capuchin is highest (31%) at Kayamá and lowest (2%) at Majagua. The howler is most prominent (22%) at Majagua and Kayamá communities and least prominent (5%) at Mosquito. Saki representation is

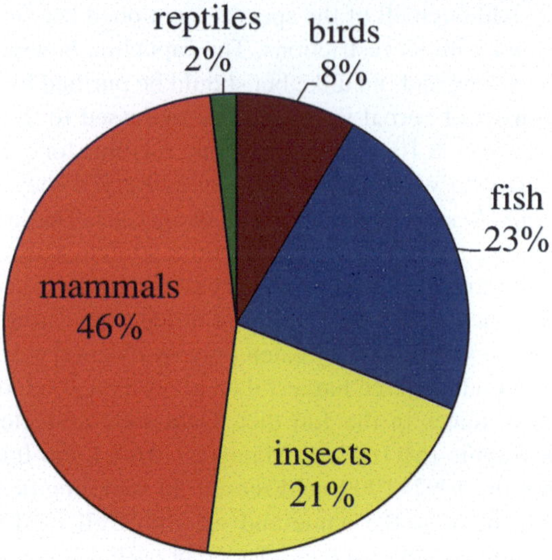

Fig. 8.3 Contribution of animal class by weight

Table 8.4 Dietary contributions of primates-procyonids by community

Dietary contributions (% biomass)	MO	MA	IG	KA
Undomesticated animals	30%	27%	16%	9%
Mammals	44%	41%	61%	43%
Primates	89%	54%	9%	13%
Ateles belzebuth	75%	70%	62%	34%
Cebus olivaceus	11%	2%	4%	31%
Chiropotes chiropotes	8%	5%	19%	13%
Alouatta seniculus	5%	22%	11%	22%
Saimiri sciureus	0	1.3%	0	0
Aotus trivirgatus	0	0.3%	0	0
Potos flavus	0	0	4%	0

highest at Iguana (19%) and lowest (5%) at Majagua. Considering the essentially opportunistic nature of Jotï hunting, we are inclined to assume that the intercommunity variations recorded here reflect differences in the availability of this game per area. However, we have no way of knowing whether such differences may be the result of natural or human drivers. What is clear, however, is that the representation of primates is significantly higher among the smaller, more nomadic and less acculturated communities (89% at MO, 54% at MA) and much lower among the large, sedentary, (ex)mission communities (9% at IG, 13% at KA). This constitutes strong evidence that the traditional reliance of the Jotï on primates for food is severely impacted by the shift from nomadic to sedentary settlements. Obviously, primates are very vulnerable to overhunting when the local population grows to several hundred people, and they basically remain in the same place for long periods lasting three to four decades.

8.7 Social Significance

It can be said that monkeys as well as other animals have a social status in Jotï society to the extent that they are raised as pets and exhibit social behaviors in the "wild" that are recognized as being similar to human beings. The Jotï are prolific pet-keepers and consider all of the species of the *jkwayo-mujkëlo* group to be suitable pets. However, the species most appreciated for this are the spider monkey and the capuchin monkey. Since animals are not bred in captivity, they must be taken in the forest, usually during hunting outings. The process begins by capturing an infant, which is accomplished by killing the mother. We have seen this being done for both species mentioned above, and there is no need to seize or secure the infant monkey because it will cling to the lifeless body of its dead mother for as long as it takes to transport it back home and appears to do what can only be described as weeping. Once back home and the parent's carcass disposed of, it is then confined by tying a fastened chord around its leg. It may be given a bath, or smoke is blown over it to calm it down. The first food given is masticated banana or papaya. The diet can be expanded after a few days to include masticated maize or sweet potato or whatever food is on hand. Once it has become accustomed to its owner, it will be taken off the leash and allowed to run free inside the house, spending most of its time up in the rafters. Monkeys are raised until adulthood but are not appreciated very much after that point. They may be killed if they become especially aggressive – the capuchin is a notorious biter – but, different from game birds that are also raised, are normally not eaten. But more frequently, they escape or choose to leave the house although they may come back home occasionally to visit their former owner (who will then feed them).

Among Jodï groups in direct contact with other indigenous group, like the communities we studied in the Caños Mosquito and Majagua, pet monkeys have become valuable trade goods. This has spurred selective hunting of females carrying infants. In one 4-week period of observation recorded at Caño Mosquito in 1998, we observed the capture of five infant spider monkeys. Not one of them survived more than a couple of days after being captured, even before embarking on the trip downriver to the other Indian villages, but we were told of more successful capture events. We can only imagine the negative consequences of this practice for the maintenance of the monkey populations in this area.

Jotï conceptions of the social distance/closeness of the *jkwayo-mujkëlo* group members to people can be detected from the natural history descriptions provided for each one (Table 8.4). Like people, monkeys live in small to large social groups, follow group leaders, and rely on verbal communication among themselves. Within the larger co-resident group, they may form smaller family groups around a pair-bond. Only the spider monkey is described as having "group marriage" (i.e., multiple partners for both males and females). The capuchin monkey stands out for its intelligence but also for its aggressiveness, even to the point of using sticks as weapons against intraspecific and interspecific rivals. The howler monkey may fight with individuals from rival groups but not among members of the same group. Meanwhile the spider monkey does not display aggressive behavior at all. The saki monkey is noted for its ceremonial feasts, when different local groups come together on the ground, and for its amiable attitude toward other monkeys. The *mujkëlodï* live in somewhat smaller social groups and form very stable pair-bonds. Most of the species described here have a single offspring at a time, the main exception being the kinkajou which may give birth to twins. The kinkajou and howler monkey have seasonally fixed reproductive cycles, mating in dry season and giving birth at the beginning of wet season, while all of the others can mate and breed at any time.

Ecologically, all of the animals described here are primary forest inhabitants, making their homes in the upper strata of the largest trees and rarely touching the ground (Table 8.5). They are invariably described as favoring mountain forests far from any human settlement. The capuchin, howler, and owl monkeys are considered to be more sedentary or confined to a fixed territory, while the spider monkey and saki are seen as the most nomadic. The favorite foods of the monkeys in most cases double as highly valued edible fruits for humans, so it is safe to say there is considerable dietary overlap. Fruit of the *ulu* palm (*Attalea maripa*), in particular, stands out as a major food of human as well as nonhuman primates in this area, thus indicating some degree of competition. Other fruits of importance for both groups include *jlude* (*Dacryodes* spp.), *jkaile* (*Micropholis egensis*), *luwe* (*Inga* spp.), *wajlikye* (*Ecclinusa guianensis*), and *jkalimane* (*Couma macrocarpa*).

8.8 Monkeys in Jotï Cosmology

Primates are conspicuous figures in Jotï myth and ritual. Members of the *jkwayo* fraction, especially spider monkey, capuchin monkey, howler monkey, and saki monkey, played generative roles in the primeval time-space. Several mythical narratives recall their input in past events which explain certain aspects of the contemporary world order. In this section, we describe two examples.

The origin of the Jotï staple foods and other economic products – mostly cultivated plants and game animals – is recounted in the tale of the food tree, *jkwë jtawï jkajka*. This story is a culture-specific variation of a region-wide myth that identifies a primordial tree as the main source of food that people have and tells how it was chopped down to gain access to its nutritious bounty. Neighboring groups like the Eñepa, Piaroa, Yekuana, and Pumé have a similar myth (de Civrieux 1970; Boglar 1978; Wilbert and Simoneau 1990; Mattei-Müller 1992).

According to the Jotï variant, crops came to light in the biosphere at the same time as animals and, by inference, modern-day humans. Paraphrasing the story, it begins by reminding us that a band of animal-people once lived in the land of the setting sun (i.e., west or downriver). Much like Jotï bands today, this one consisted of different animal-persons, each with their own distinct personality. Besides *Imo* (howler monkey), *Uli jkwayo* (spider monkey), *Jkwaijlë* (saki monkey), and *Jani jkwayo* (capuchin monkey), there were *Inï ajkuli* (paca, *Agouti paca*), *Yowä* (tapir, *Tapirus terrestris*), *Jkwajtïbo* (deer, *Mazama americana*), *Jkyado ajkuli* (agouti,

Table 8.5 Natural history descriptions of *jkwayo-mujkëlo* group members[a]

Taxa	Distinctive morphology	Social habits	Reproductive habits	Preferred foods	Common habitats	Interspecific associations
Uli jkwayo (*Ateles belzebuth*)	*Ja uli ja meka* 'large body'; *Jwï walejte meka jawabo mejwa jkuwë kyabo meka* 'black backside, white below chest'; *Enena bukë baï jwai meka* 'prehensile tail'; *ujku balejkï meka* 'sharp teeth'	*Lowalibï ide mekadï* 'do not fight among themselves'; *Aejadï jodena mee dekadï bada* 'live in large multi-male, multi-female groups'; *Aejadï adï, jawabo aejadï manimo mekadï* 'polygamous'; *Ae ñaña meka, uli dawo meka* 'troop leader is elder male'	*Bëña mamona jawa bujto ibï deka bada* 'offspring are born at any time of year'	*Jtawï ade jkwaï deka* 'eats tree fruits', *ulu* (*Attalea maripa*), *luwe* (*Inga* sp.), *jlude* (*Dacryodes* spp.), *jkaile* (*Micropholis egensis*)	*Inëbei jtudöna, inëwa ajkudänï inëmawa jkyodänï mee dekadï* 'inhabits high forest, hilltops, or mountain slopes'	*Wëdeka jade* 'not friendly with any other species'; *Imo lëbuwa deka* 'howler is distant, potentially friendly relative'
Jani jkwayo (*Cebus olivaceus*)	*Uli ja meka* 'large body'; *Duwëjka manï jwalejte jkï meka bada, jkaladena jkyabo jawabo jkyajka ïö meka* 'some are brown-black, others have white-gray hair'; *Mëjnadä cinco meka jawa bada ña bëkya* 'have five legs in all'	*Jodena meejka jadï bada* 'live together in groups'; *Ae ñaña meka* 'there is a troop leader'; *Namana jawa jobe deka bada* 'call to others in their group'; *Abaladïnï jelënï jwï deka jawa jkwaïde mowaibï* 'use sticks to attack rivals in order to defend food source'	*Ñajadowa jkibï meka bada* 'have one offspring at a time'	*Jtawï ade jkwaï deka* 'eats tree fruits', *ulu* (*Attalea maripa*), *awajto* (*Attalea macrolepis*), *luwe* (*Inga* sp.), *wajlikye* (*Ecclinusa guianensis*), *mau* (*Protium* spp.)	*Jtau jkuwë, inëwanï mee dekadï* 'live in montane forest'; *Namana jkyo ñaña ma, jawabo jaede bada jani jkwayo jkyo* 'some groups are territorial, others are not'	*Jkwaijlënï jawa jluwëna, jawa ijtekï dekadï* 'friend and close relative of saki, they share food'; *Walijkyëna wëidëu deka* 'enemy of coati'

(continued)

Table 8.5 (continued)

Taxa	Distinctive morphology	Social habits	Reproductive habits	Preferred foods	Common habitats	Interspecific associations
Imo (*Alouatta seniculus*)	*Jtaikye uli ja meka* 'very large body'; *Bëjkya duwëwe manï, ikyu jwïnï kyabo meka* 'red all over, males have white patch on rear end'; *Ae manï abiyë jkojko meka bada* 'older individuals have bushy beard'; *Enena bukë baï jwaï meka* 'prehensile tail'; *Jkwabo jawa neinï jlebona meka* 'has fragrant scent glands on neck'; *Bale jku meka* 'has dark, sharpened teeth'	*Aejadï jodena mekadï* 'live in large groups'; *Ae ñaña mali deka bada* 'there is a troop leader who is male'; *Abaladï lïkë lowalibï, neinamajadï lowalibï ide mekadï* 'they fight with other troops, but do not fight among themselves'; *Abajladï jobe dona aku deka bada* 'communicate with each other'	*Aujka malijka mekadï* 'monogamous pair-bonds'; *Jtuwö bae jaï jkyewayakï, ojku bae lïdï meka* 'mate at beginning of dry season, give birth at beginning of rainy season'; *Ekyo jkwa ini ñajadowa meka bada* 'give birth to one offspring'	*Jtawï ade jkwaï deka* 'eats tree fruits', *jkaile* (*Micropholis egensis*), *luwe* (*Inga* spp.), *wajlikye* (*Ecclinusa guianensis*), *jkalimane* (*Couma macrocarpa*), *ulu* (*Attalea maripa*), *mau* (*Protium* spp.); *wenewene bu* (*Brownea coccinea*) flower; *Jtido aiye jkwaï deka* 'eats young leaves', *jtabali* (*Ceba pentandra*), *jkaile* (*Micropholis egensis*), *jtuku aiye* 'leafy plants'	*Inëwanï, jkaile jtunï ebolo jkwa jwejwe meka bada* 'lives in mountain, top of *jkaile* tree, makes nest in liana tangle'	*Wëdeka jade* 'has no companions'; *Uli jkwayo lëbuwa deka* 'spider monkey is distant, potentially friendly relative'; *Lowalibï ide meka, wai deka* 'avoids, does not fight, spider monkey'
Jkwaijlë (*Chiropotes chiropotes*)	*Walejte manï, jwï kyabo meka* 'negro, con espalda blanco'; *Abiye jkojko* 'bushy beard'; *Jtu jkalado meka* 'skull parted in the middle'; *Enena buka mïkë nadabe deka u dowaki dïmë* 'non-prehensile tail'	*Aejadï jodena mee dekadï* 'vive en grupos grandes'; *Ae ñaña uli ja, aka u deka* 'there is a troop leader, the largest one, who walks in the lead'; *Aujka malijka mekadï* 'monogamous pair-bonds'; *Nenï maluwe maï dekadï* 'they have ceremonial feasts on the ground when many gather together and mate'	*Bëña lïdï deka aejadï baebï jto ibï deka bada* 'mate all the time and give birth at any time'	*Jtawï ade jkwaï deka* 'eats tree fruits', *jkaile* (*Micropholis egensis*), *luwe* (*Inga* spp.), *wajlikye* (*Ecclinusa guianensis*), *jkalimane* (*Couma macrocarpa*), *jtijti* (?); *Ajkukë* 'spiders', *Añu janï* 'seed grubs'	*Jkyonï inëwa jku jwëla, inëwa jwïnï; inëba majae* 'lives in montane forests'; *jkaile jtau majae mee deka bada* 'lives in *Micropholis egensis* tree'; *Bëña u deka* 'not territorial, always on the move'	*Jani jkwayo jawa jluwëna, jodenï u lïdï* 'friend and close relative of capuchin monkey, they live and move together'

Jkujkujtu (*Aotus trivirgatus*)	*Mejwa duwëjka, jwïjta kyajte* 'reddish chest, gray-whitish back'; *Ïö jtejte jkwayo bou meka* 'has tufted hair, similar to spider monkey'; *Enena bukë ïö jkwayo bou mawa* 'tail hair similar to spider monkey'; *Uli búle dale* 'large eyes'; *Bulë minë jkalijkakï meko bada* 'has stripes above eyes'; *Enena bukë baï jwaï ide meka* 'non-prehensile tail'	*Jawa aujka jawabo inijka mee deka* 'lives in monogamous family group (mate and children)'	*Bëma lïdï dekö* 'gives birth at any time of year'	*Jtau ade jkwaï dïmë* 'eats tree fruits', *luwë* '*Inga* spp.', *jkaile* '*Micropholis egensis*', *wejkao* '*Pourouma* spp.', *ulu* '*Attalea maripa*', *jawa bëjkya* 'all kinds'; *jkawale bu* '*Caryocar macrocarpa* flower'	*Inëwadä jtau jkuwë majaena mee deka* 'lives in montane forests'; *jtau jkojko jkwa, ebolo jtejte mee deka* 'inside tree hollow or nest made of vines'; *Ulu jtunï abu meka* 'sleeps in top of ulu palm'; *Namana jkyo ñaña ma* 'territorial'	*Uli mujkëlo wëdeka, jawaña jodena u dekadï* 'kinkajou is its companion, sometimes they travel together'

(continued)

Table 8.5 (continued)

Taxa	Distinctive morphology	Social habits	Reproductive habits	Preferred foods	Common habitats	Interspecific associations
Uli mujkëlo (*Potos flavus*)	*Ïö duwëwe meka* 'reddish hair'; *Najne jedodo jawabo ïö manï jtaikye jtamïwa meka bada* 'thin skin, thick hair'; *Jelë bawa manï enena bukë yoo meka* 'long, round, pole-like tail'; *Enena bukë baï jwai meka* 'prehensile tail'; *Ïnëjae manï jwalejte bawa mekae ena jkwayo ïnëjaeda meka bada* 'dark meat, like *jkwayo* monkey'	*Ña jadowa jawabo aujka mee deka bada* 'lives in monogamous family group or solitary'	*Ini ña jadowa, jawabo wejkyadï inimo meka* 'have one or two offspring'; *Ojku bae lïdï meka* 'gives birth at beginning of rainy season'	*Jtau ade jkwaï dïmë* 'eats tree fruits', *luwë* '*Inga* spp.', *jkaile* '*Micropholis egensis*', *awajto* '*Attalea macrolepis*', *ulu* '*Attalea maripa*', *wejtolo* '*Cecropia* spp.', *delela* '*Ficus* sp.'	*Jtau mebï jle deka bada, jtawï jkojko jkwa mee deka* 'lives in tree hole, high up in tree'; *Ñaña namana jkyo manï bë jawabo bekya jkyoka balebï deka bada* 'some are territorial, others are nomadic'	*Jani mujkëlo jawabo jkujkujtu wëdakadï bada* 'associates with olingo and owl monkey'

Dodo jteite (*Bassaricyon gabbii*)	*Mujkëlo bou maja ña jani ja* 'similar to kinkajou but smaller'; *Duwëwe mejkwa duwëjka* 'hair color is white-gray, reddish-yellow chest'; *Jedodo jkenowano, jawabo ïö meka bada* 'soft skin and hair'; *Búle jkalijka ja*; 'has marks or rings around eyes'; *Ïnë jae jtaikye ñajae da manï kyabo mekae bada* 'white meat, not palatable'; *Enena bukë jtejte* 'tufted skin and tail'	*Cinco jadï jodena mee dekadï* 'lives in small groups of five'; *Jawa auka mee deka bada* 'monogamous'	*Jawa ininï jkojko jkwakï lïdïnï deka bada* 'gives birth in tree hole'; *Ñajadoa meka bada* 'has one offspring'	*Jtau ade jkwaï dïmë* 'eats tree fruits', *luwe* '*Inga* spp.', *jkalimane* '*Couma macrocarpa*', *ulu* '*Attalea maripa*', *wejtolo* '*Cecropia* spp.'; *Jkawale bu nai deka* '*Caryocar macrocarpa* flower'	*Jtawï jkojko jkwa abu deka* 'sleeps in tree hole'; *Jtau jtunë balebï madëkö bada* 'lives in tree canopy'; *Namana jkyo jaede ja bada,bëña bïkë balebï jto ibï waibï kyewayakï deka bada* 'not territorial'	*Wë deka jade* 'has no companions'
Jkijki (*Saimiri sciureus*)		*Aejadï jodena mee dekadï bada* 'live in large groups'; *Jtidoja lïkï,abajlau ode jyu jtï dekau, jawa ama jau de mona* 'female nurses other infant that is not her offspring if the real mother is absent'		*Jtau ade jkwaï dïmë* 'eats tree fruits', *ulu* '*Attalea maripa*', *muli ji* '*Socratea exorrhiza*', *jani bade* '*Oenocarpus bacaba*', *waiño* '*Pseudolmedia spp.*', *wejtolo* '*Cecropia spp.*'		

[a]Sources: Jairo Mölö (02/12/2014), Alberto Melomaja (07/10/2014), Ofelia Konojto (07/19/2014), Clemencia Uriah (07/12–13/2014), Jeremi Jkailebo (07/12/2014), Nestor Uriña (07/13/2014), Rosa Jono (07/19/2014), Noe Jono (07/22/2014, 07/10/2014)

Dasyprocta leporina), *Uli jkali* (northern Amazon red squirrel, *Sciurus igniventris*), *Uli jñome* (porcupine, *Coendou prehensilis*), *Ja̲ni o̲jko̲* (nine-banded long-nosed armadillo, *Dasypus novemcinctus*), and *Uli o̲jko̲* (great long-nosed armadillo, *Dasypus kappleri*), among others. According to one version, these primordial people were simply bored and decided to leave their territory and walk toward the land of the rising sun (i.e., east or upriver). Another version states that they left and went east in search of peace because they were constantly under attack and in danger of being vanquished by another group, the *Jkaliwekï* (Jotï name for Panare/Eñepa), who possessed more potent curare poison. This event explains why the ancestors of contemporary Jotï became nomadic. So they set off walking toward the east, eating wild tree fruits along the way, and stopping only to rest at night along the trail. By the afternoon of the second day, they stopped to rest again when spider monkey heard a *jkilëka* (orange-cheeked parrot, *Pyrilia barrabandi*), chirping and singing, and decided to go and investigate. When he approached it, the parrot flew away, and at that very moment, it also transformed permanently into an animal, thus inaugurating a series of multiple metamorphosis that would soon follow. Then spider monkey noticed before him stood a huge tree, which was *jkyo jkawïle* (*Erisma uncinatum* Warm.), and on the ground around the tree was a wide assortment of edible fruits and vegetables. Not wanting to share his delicious discovery, spider monkey kept it a secret, but he did let on that this was a good place to settle down. The others agreed, and so they built two longhouses there, one for the families headed by spider monkey and the other for the followers of howler monkey. The secret got out when paca found a maize grain under spider monkey's hammock and ate it. This led to a fight between the two, and spider monkey was forced to tell his companions about the food tree. Different episodes that occur in the myth explain particular morphological traits of certain species. For instance, the distinctive opening under the cheek pouch of the paca is a memento of the fight its ancestor had with spider monkey. The dark crown of the capuchin monkey is the consequence of its ancestor putting his hands on his head after eating charcoal-covered, roasted yam. All of the animal-people then went to see the tree and decided to cut it down to get at all the food hanging from the branches. Spider monkey, howler monkey, and saki monkey all took turns chopping the tree with a big ax made from a toucan beak. Meanwhile, tapir and deer dumped the leftover wood chips into the river, upon which they turned into the different types of fish and aquatic creatures. When the enormous trunk of *jkwë jtawï jka̲jka̲* fell, it liberated the rich store of food it contained, such as *ja̲ni jwa̲ne* (small yam variety), *uli jwa̲ne* (large yam variety), *jtamu* (maize), *jwalulë* (plantain), *jedala* (banana), *jkalala* (sugarcane), *iyë* (sweet potato), *më* (yautia), *nana ju* (pineapple), *ale* (manioc), and other crops like *jkawai* (tobacco), newa̲ (cotton), and *jkulilu* (annatto). Tired from the work but pleased with the result, everyone sat down on the trunk and began to converse animatedly. Spider monkey thought to himself, "they will want to change their voices, walking sticks and instruments." Then he said: "I am going to sing." He stood up and walked alongside the fallen trunk. When he reached the branches, he turned into the spider monkey with a long and handsome tail. He asked one and all "how are you going to sing to each other?" Each person answered "I will sing this way" and proceeded to sing just like they do today. Following spider monkey's example, everyone stood up and walked in procession along the trunk, and when they reached the branches, they too transformed into animals. There were many different types of mammals, birds, and fish. From that day forward, the animals have maintained their present-day *habitus* (sensu Biourdieu, as historically and culturally contingent modes of being and acting) and went off to live in their respective habitats: mountain, savanna, river, or elsewhere.[7]

Besides explaining where cultivated food plants came from, there are additional creation theories encoded in the narrative: what the origin of biodiversity is, how animals got their shapes and sounds, why they went their separate ways, when did humans differentiate from nonhuman animals, why trees constitute quintessential sources of food, and why slash-and-burn cultivation is a staple of the Jotï economy. In more abstract terms, it describes the creation of a definite socio-biological order from an unstable and amorphous state and thereby provides an explanation for the present state of things (cf. Wright 1993–1994:39–41; Naddaf 2005:4). Like other Amerindian groups such as the Hohodene, a chaotic time-space is followed by a cycle in which first are created the "conditions of social reproduction" and second "the details by which life and culture are sustained and renewed (Wright 1993–1994:40)." The inquisitive, part-selfish, part-cooperative disposition of some of the key personalities in the story also offers a reading onto the human ethos today (cf. Naddaf 2005:38). The fact that monkeys stand out as some of the leading actors in this all too "human" drama is probably no accident (Sec. 8.9). Spider monkey is unquestionably the lead protagonist who, abetted by his foils capuchin monkey and howler monkey among others, kick-starts the epic transformation of the world to its present biodiverse state. If primates had a hand in bringing about the modern world, as the food tree myth asserts, then their anthropogonic role should also be acknowledged.

The broader meaning of the food tree tale as a multilayered creation myth is apparent if we consider what it has to say about the dawn of modern humankind. The transformation of the ancient animal-persons into animals as they are today implies as corollary the emergence of human-persons as a perceptually distinct species apart from the animals (i.e., as ego to the animals' alter). The fact that *nï jotï* 'true people' separated from animal-people concurrently with the discovery of cultivated plants points to the importance of horticulture to the Jotï sense of self and subsistence (Zent and Zent 2012). Another important point is that biological speciation – the present-day essences and forms of the characters in the story – is not seen as an external creation or imposition or accident but instead as an act of individual volition. By their own wish, each character in the myth turned into their present-day nonhuman *habitus* and proceeded to diverge into separate habitats. Thus it differs from other Amerindian traditions in which a divine creator, trickster, or cultural hero is responsible for such transformations (Alvarsson 1997; García Tomás 1994; Civrieux 1980). This also means that humanity is not merely an inherited condition but also a choice, one that must be reaffirmed in the present by behaving in appropriate ways. The *nï jawa* 'real foods' eaten by *nï Jotï* are those acquired from the primordial food tree, most of which are garden crops, while the converted animals

[7]The *habitus* is located on the subject's body and consists of "bundles of affects," that is to say, the set of capacities and behaviors typical of a being (Viveiros de Castro 1998:478).

are today's prey and predators that went on to adopt their own species-specific food-ways. The diaspora of animals into their specific niche spaces insinuates a cooperative arrangement of ecological diversification, while the adoption of distinctive sounds alludes to voluntary linguistic diversification.

The interspecific differences expressed here refer only to perceived physical bodies and behaviors. The question remains to what extent are mental characteristics the same or different? The key actors appearing in the myth are given animal names but are depicted as acting like people, hence our use of the term "animal-persons." By this we mean that not only did they take on a human appearance but also that they were once endowed with the same capacities of agency, intention, and sociality as humans. This viewpoint appears to be consistent with the more generalized Amerindian philosophy of the ontological equivalence of people, animals, plants, and other sensate beings (cf. Viveiros de Castro 1986, 1992, 1998). The extent to which there is equivalence in the present is addressed in the following section.

In the second narrative, of which we only have a fragment, we learn that the focal *jkwayodï* are responsible for the spread and diversification (into different kinds of folk generics and specifics) of the original crops. The smart monkey-people, capuchin monkey and spider monkey, are said to have taken the seeds of the plants eaten by people and dropped them along the riverbanks (*jedä o̲neka majae*) and onto small clearings where trees had fallen down (*jtawï lai jwi de jae*). This primordial event sets the precedence for one of the distinctive features of the Jotï horticultural system, namely, the cultivation of natural forest gaps (Zent and Zent 2012). This practice may have been more widespread before the easy acquisition of steel tools and may represent an adaptation to a dispersed, nomadic settlement pattern, but it is still practiced today. Additionally, spider monkey is credited with being the one who planted the first yams at the large rapids on the Kayamá River below the ex-mission site (Ivan Juae, November 2011).

This folktale enhances the creative status of monkeys as conscious managers of landscape modification and the inventors of a successful food production strategy. In this case, they serve as teachers or models for people to emulate.

8.9 Monkeys as Paradigmatic Persons

While the prominence of monkeys in Jotï myths points to their status as paradigmatic persons in the imaginary past, there are other beliefs that reinforce the view that they exhibit many human traits in the present. Like people, monkeys and many other animals (e.g., mammals, birds, some arthropods) live in houses, are grouped into families, go out hunting, raise gardens, paint their bodies, and have ceremonial feasts, among other customs. In short, they are cultured and goal-driven and therefore symmetric to people (Zent 2013a). However, just as all people do not possess the full complement of humanly attributes, not all animals (and plants) are considered to be complete persons. To make up for their deficiencies, they have *jkyo a̲e̲mo(dï)* and spiritual selves which are always associated with personhood to varying degree. J*kyo a̲e̲mo* are the equivalents of spirit guardians or masters of the animals abundantly reported in the Amerindian literature (Reichel-Dolmatoff 1971; Overing and Kaplan 1988; Århem 1996; Whitehead 2002; Cormier 2003). They are typically ascribed the qualities of being powerful, polymorphic, semidivine, hypostatic, and protective. A common depiction is that of a family group consisting of parents and their children who take care of (e.g., feed, shelter, defend) all members of their generic class. They are able to conglomerate all of their kind and regulate the flow of their populations across different areas (Zent 2006, 2009). They are also the main regulators or intermediaries of predatory events. This means that hunters must interact with the *jkyo a̱emo* (hereafter referred to as "master") to transact the safe capture of their animals. The masters of animals coinhabit the world of people, living and moving among them, but they are invisible to the uninitiated eye.

An animal's master possesses certain trees or fungi on a spiritual level that are, in turn, tied to the material incarnation of these organisms. For example, *uli jkwayo ae* 'spider monkey's master' is the spiritual owner of the plants bearing his name, *uli jkwayo jtawï* [*Leonia* sp.] and *jkwayo jtu jelë* [*Heliconia* sp.]. The bark and leaves of the first one are considered to be effective inductors for hunting spider monkeys (Alberto Melomaja, January 2002). He is also associated with the round, reddish mushroom called *uli jkwayo ñajki̱no* [*Lycoperdon* sp.]. Males and females wash themselves with a liquid substance extracted from this mushroom to invigorate their hunting skills. They can also insert a few drops of the liquid in their nostrils to restore connections with various masters that were lost by certain transgressions (Jailo Mölö, November 2011).

The masters of monkeys are considered to be particularly effective intermediaries of human-animal-plant interactions for producing good results (e.g., hunting success, good health, no accidents). This prominence is no doubt related to the close resemblance of monkeys to people in morphology and behavior. In similar fashion, the masters of capuchin monkey, spider monkey, howler monkey, and saki monkey are perceived as being *more alike* to people than those of the other animals in an imaginary alterity space between entities. Some Jotï recognize kinship relations with monkeys, claiming that capuchin monkey, spider monkey, howler monkey, and saki monkey are their *ñajti jluwëna* 'truly good *jluwëna*' (Ijtö April, 2002), referring to the respective masters rather than the physical organism itself. As mentioned earlier, the *jluwëna* is the most cooperative and affective nonconsanguineal relationship in the Jotï kinship system (Sect. 8.5). The perceived closeness of primates to people is also reflected in their pet-keeping habits. The capuchin and spider monkeys are among the most preferred animals as pets, and it is stated that raising them actually turns them into "a total person," based on the logic that the interchange of foods and other substances, words, touches, and essences leads to the consubstantiation of entities (Zent 2008). In second place, the howler monkey is mentioned as being amenable to being brought up alongside people and capable of affective and sensitive behavior toward those around him/her. The saki is considered to be less adaptable to the human domestic space, but also at the same time "like people" (Gerardo Liye, October 2016).

The typical abodes or houses of the masters correspond to hills, mountains, and other landforms that appear at a distance to have similar shapes to the houses that people make. These topographical objects are very pervasive throughout the Jotï landscape and are important reference points in their ethnocartography. A common formula for constructing Jotï toponyms is by naming a place for a plant or animal that is particularly abundant or somehow associated with that location. Such associations are often based on the belief that the masters of the named species reside in the vicinity. The main primate taxa – *uli jkwayo* (spider monkey), *ja̲ni jkwayo* (capuchin monkey), *i̲mo* (howler monkey), and *jkwaijlë* (saki monkey) – appear in ~5% of the toponyms in which animals are named, thus providing some measure of the importance and proliferation of the primate masters as coinhabitants of the territory (Zent and Zent 2012). For example, *uli jkwayo i̲nëwa* 'spider monkey mountain' is a somewhat common toponym that designates several mountains known for their resident populations of spider monkey and where the species' master dwells underneath the ground. The summit of one of these mountains, in particular, located in the upper reaches of the *duwëwe jedä* 'red river', was a favorite site for realizing the culturally important nasal septum-perforation rite of passage (for details of the ritual see Zent 2006) before the establishment of the missions far to the north and south (Jtukyabolae, June 2003). Another example is the place called *jkwaijlë lajo* 'bearded saki monkey rapids'. It marks the entrance of the saki master's home and is a sacred site where people can transform into monkeys and monkeys into people (Alberto Melomaja, January 2002). *Jkwayo jedä* 'monkey river' is the name given to the nearby Caura River, one of the largest in the region. In this case, the reference is not based on monkey population density or their masters but the connotation that this river, like the monkey, is big, strong, and influential, from the Jotï perspective a keystone species.

The association of primates and "strength" is expressed in other ways as well. Certain rituals involve burning leaves or burying something at the base of a hardwood tree. The *jkwayo jtawï* is one of the trees most sought after for this purpose (Maliela Yalúa, December 2004). The liana *jkwayo i̲nimo ju* (Prionostemma sp.) 'monkey children vine' is considered to produce sturdy stems which are good for lashing together logs and poles for construction purposes. This is the vine that fastens and keeps erect the tree trunks that hold up the middle layer of the cosmos. The divine entity known as *Jkyo ae* 'Forest Master' is a generalized force in the living world that takes different forms, one of which is the thunder. As thunder, it may also transform himself temporarily into *judeko jyëï* (*Symphonia globulifera* L.f.) and then call himself *jkwaijlë* 'saki monkey', whose coal black fur matches the color of the peraman wax produced from the resin of this tree. By using the monkey-person's name, he is able to conceal his true identity as a powerful glue capable of snatching unsuspecting beings and leave them fastened to the tree (Noe Jonö, December 2004). Another story about the Forest Master tells of his (spiritual) pet, *mujkëlo* 'kinkajou'. Forest Master places *mujkëlo* inside a gourd at daybreak, where it stays during the day, and lets it out at nightfall to walk outside (Alvaro Ulijtujtea, April 2002). The associated *mujkëlo jtawï* 'kinkajou tree' is considered an indicator of good soils to open a new garden (Magalita Ñejtojkuamaja, November 2011).

As in other cultures, the combination of human and primate attributes provides models for the imagination of monster-like creatures among the Jotï. One of these is the creature called *uli jkwayo jawabo*, which is described as a sort of spider monkey, but much bigger and hairier. It is reputed to be a big predator, and its real name should not be uttered out loud because if it hears you it will be provoked to attack (see Zent 2013b). The most fearsome animal-person-spirit hybrid entity, however, is the fabled sasquatch- or mapinguari-like creature known as *Bulu ja*. The Jotï describe it as the biggest monkey alive, and its phenotype is described as a mix of howler and spider monkey attributes: bearded, reddish, hairy, muscular, with prominent sharp teeth, long arms and legs, longer tail, and with big inverted paws (Alvaro Ulijtujtea, April 27, 2002). It is considered to be extremely dangerous and fond of feeding on human prey. Although it does not transmit diseases or abduct people, *Bulu ja* is the leading predator of pregnant women and bearded men. It is able to track people by their smell and kills them with the ax-like lip of its mouth. It will devour all that it is able to trap and can eat up to hundreds in a short amount of time (Dino Jkailebo, October 15, 2005). There is no way to kill or eliminate this formidable foe draped in black and white hair. It inhabits rocky mountain shelters, and therefore to avoid an attack, one should not mention its name when walking near to such places (Iné Baiyeja June 11, 2005).

8.10 Primates and Eco-cosmological Regulation

In the previous section, we touched on the metaphysical life of monkeys in the Jotï universe. Another aspect of this life is their role in the regulation of proper behaviors with respect to trophic relationships and subsistence-related pursuits in general. Like the Makuna and other Amazonian peoples, the Jotï subscribe to a model of socio-ecological relatedness among distinct species based on the moral principles of predation and exchange (cf. Århem 1996). For the Jotï, predation is permitted, even facilitated, by the masters of animals, plants, and fungi so long as the predator reciprocates the gift. Reciprocity is achieved by following strict codes of conduct during the realization of subsistence acts. These may include showing proper respect (i.e., not playing around, not uttering a demeaning comment), not vocalizing the names of certain entities, reciting *yu* blessings when appropriate, not killing too many individuals at a time, not harvesting too much of one thing, following certain procedures in the handling of body parts, observing food taboos, holding ceremonial feasts, inviting *jkyo aemo* to attend the feasts, and many others. Violations of cultural rules set into motion a disconnection of the violator with other beings in the universe, especially those directly involved in the transgression. The effect is disrelatedness and solitude, an intensely undesirable state that should be avoided because it is tantamount to being amoral and asocial, which is to say inhuman. The loss of one's connectivity results in the loss of effectivity to hunt, gather, garden, and carry out just about any productive activity. More severe transgressions make the person vulnerable to becoming the object of predation, where predator and prey undergo role reversal. Sickness is a common outcome of a predatory attack, for example, a sudden illness or an unusual malady. A clear symptom of predation is the radical change of a person's normal mannerisms, which is indicative of a loss of spiritual essence. The *aemo* of animals, plants, and fungi, as well as other spiritual beings, are the agents most responsible for such reverse predation which takes place on an immaterial plane of existence but can have effects on the material body. Århem (1996) refers to this phenomenon as mystical predation and argues that it is not very different from predation in a biological sense.

Primates are sometimes mentioned as tangible prey that are capable of becoming intangible predators if proper rules of socio-ecological engagement are broken. Along with *uli yewi* 'jaguar', spider monkey is one of the figures most often associated with mystical predation and, in particular, is responsible for enforcing rules associated with hunting monkeys and their consumption. Consequently, prey species like spider monkey, howler monkey, and bearded saki monkey are all considered food taboo during the couvade and infant stage lest the spider monkey master takes the baby to his abode and turns him/her into one of his own (Noe Jono, December 2004). The small child risks a similar fate if the first time that he/she eats spider monkey meat the *yu* blessing is not performed (Maliela Yalúa, October 2005). All types of *jkwayo-mujkëlo* meat are regarded as potentially "dangerous" for children if not purified by reciting *yu* and blowing a compound mixture of masticated leaves over the meat to be consumed (Benito Nodï, October 2005). The preferred leaves for this purpose come precisely from those trees, vines, or palms named for the taxon in question. For instance, if the child will be given kinkajou for the first time, the *yu* blessing must contain *mujkëlo jtawï* leaves and so on.

Predators are often considered to be clever, cunning, capable, vigorous, generous, and trickster-like. Perhaps for this reason, the teeth and skull fragments of the capuchin and spider monkeys are favorite adornments worn to keep away bad or harmful entities. The power of intelligence associated with the artifact is equated with protection. The Jotï recognize monkeys as being among the smartest animal, the most intelligent being capuchin monkey, which is more common in the Kayamá area (Tito Jonö, June 2005), and squirrel monkey, which is found in the Caño Iguana region (Gerardo Liye, October 2016). The skull of the capuchin monkey is highly valued and worn on the necklace during the realization of many subsistence activities, including hunting trips. This species is known for its artfulness and ability to avoid capture, even employing deceit to disorient the hunter as to the specific location of troop co-members (Kamilo, December 1998).

One of the most frequently cited offenses associated to flawed hunting is the inappropriate manipulation or rupture of the animal's *wa̱ña* (Zent 2005). The *wa̱ña* consists of the gall bladder and the bilic secretions that it contains. According to some mythological versions, spider monkey invented the *wa̱ña* in order to avoid that he and his kind would ever go extinct. It was then passed along to other animals but only to those who were formerly incarnated persons. Thus a select group of animals possess the *wa̱ña* today. Like curare, it is a very bitter, powerful substance and must be handled with great care. During the process of butchering and cleaning the animal carcass, the *wa̱ña* must be removed without damage and properly disposed of (see Zent 2005:51–53 for a more detailed description). If this procedure is not followed, it can trigger severe punishment, such as mystical predation or excommunication by the game masters. As with most other organic entities in Jotï eco-cosmology, the *wa̱ña* is not merely a material object but conveys a spiritual force that connects a metaphysical network of different plants, animals, fungi, and people. The network becomes visible when we consider the remedial treatments that must be applied when it is not handled correctly. According to which animal's *waña* is involved, the treatment may encompass particular parts (bark, leaves) of particular plants or particular parts (hymenia, pilei, inner liquid) of particular mushrooms. Proper handling of the *waña* guarantees not only sustained reproduction of the animal population and therefore hunting success but also open communication channels with hypostatic entities of the cosmos.

Besides the *waña*, spider monkey is assigned credit for other inventions, some of which he discarded and later converted into natural elements found in the world today. One of them is a finger-shaped fungus that sprouts on rotten logs or damp earth that is called *uli jkwayo waña* (*ñajkino*) or *jkyo waña ñajkino*. This fungus is conceptualized as a lively, sentient creature that has the power to restore a fallen hunter's ability to obtain game and maintain animal populations at the same time. Consistent with the logic of the doctrine of signatures, the shape of *jkyo waña*, the fungus, resembles *waña*, the animal organ. A drop or two of *jkyo waña* liquid is inserted into the hunter's nostril to ensure that the animal he shoots actually dies or can be located. This, in turn, re-establishes the connection between all hunters, all animals of the kind killed, and the animal's master.

8.11 Conclusions

This paper has attempted to demonstrate the importance of nonhuman primates for the cognitive, economic, social, cosmological, and philosophical life of the Jotï. Monkeys and monkey-like procyonids comprise a perceptually salient, albeit formally unnamed, intermediate-ranked category within the Jotï ethnobiological classification system. Monkeys are hunted frequently and are a major source of meat, accounting for about a third of the total weight of hunted animals. However, it was also observed that the amounts of monkey meat procured in the larger and more sedentary communities are significantly lower than in the more traditional nomadic settlements, thus indicating that populations of these animals have been impacted as a result of changes in the settlement pattern since contact with westerners. The fact that overnight hunting trips and more extended trekking expeditions are sometimes explained as *jkwayonï jujtoba ibï wai dekae* 'gone monkey-hunting' suggests a connection between the hunting of this type of game and the traditional nomadic settlement pattern. All of the monkeys found in this area are deemed edible and relatively safe, but the meat must first be purified with *yu* blessing, and consumption is restricted only for young children. The kinkajou, olingo, and owl monkey are also eaten but potentially more dangerous; therefore more restrictions apply, and they are rarely hunted. The social value of monkeys can be inferred from the practice of capturing and raising them as pets. The process of domesticating the animals turns them into *nï jodï jawabo* 'like complete people', and they are looked upon as family members. Raising monkeys as pets affords people an intimate exposure with their habits, a knowledge that can be put to use for hunting purposes. Moreover, people's observations of monkeys' intra- and interspecific patterns of social behavior provide opportunities to reflect upon their own behavior.

Turning to a consideration of the representation of primates in the expressive culture of the Jotï, we can appreciate that their value is much more than just something good to eat. Here we explored their representations as human-like actors in myth, as models of personhood, and as spiritual brokers in eco-cosmological philosophy. In the food tree myth, monkeys are portrayed as key protagonists in momentous events leading to the creation of the modern world, including the discovery of cultivated plant foods and other useful products; the emergence of diverse biological organisms, including humans, from the primordial band of animal-persons; and the adaptive radiation of species into different ecological niches. Meanwhile, the narrative telling about the dissemination of economic plants in forest gaps traces the origin of a food production technology, not to mention the anthropogenic landscape, back to monkeys. Primates and other biological entities (animals, plants, fungi) are depicted by the Jotï as thinking and acting like people, in other words possessing personhood to the extent that they have conscious, cultural, social, and spiritual attributes. Monkeys, in particular, are considered to be more like people (i.e., possess more personhood attributes) than other animals. Similar to other Amerindians, the Jotï seem to hold them as simultaneously paradigms of otherness and of sameness whose contacts and interactions with people, of both material and immaterial kinds, provide a figurative illustration of the polyvalent integration of multiple species in the day-to-day life of human societies and ecologies (cf. Cormier 2003; Erikson 2000). Monkeys are symbolic participants in notions of eco-cosmology, whose agency helps to ensure proper rules of conduct in interspecific relationships, especially in regard to upholding the moral principle of predation with reciprocity. This in turn maintains balance among all elements in the cosmic food web (cf. Århem 1996).

The spider monkey (*uli jkwayo*), in particular, stands out for playing a central role in Jotï hunting behavior, mythology, cosmogony, and the spiritual regulation of predation. Spider monkey-person is the main protagonist in more than one mythological account. He is the main axeman responsible for chopping down the primordial food tree. He seduced humans to sing animal songs which then induced the transfiguration of people to animals. He created the *wa̱ña* as a physical organ that functions on a metaphysical plane to reaffirm the connectivity among different living entities. Of all the animals, and all the monkeys, the spider monkey is the one perceived as being the closest to humans in appearance and action. He is represented as a being too human to be just an animal and too animal to be only human. Clever and mischievous, crafty inventive, congenial and useful, the spider monkey embodies a recurrent human-animal motif in the multilayered universe of the Jotï. While lacking a rigid morality or purpose, the mythological accounts of spider monkey reflect the flexible and fluid daily happenings of Jotï hunters. One of the lasting legacies of spider monkey is to serve as an example of the connection between agency and transformation, which might be considered one of the pillars of the Jotï philosophy of life. In the food tree myth, spider monkey decides to discover the food tree, to cut it down, to adopt its own language (animal-specific call), and to take on a new body, and other animals followed suit. But the example is pertinent not just for animals, since plants, both wild and cultivated, are also often represented by the Jotï as once being "plant-persons." In some cases, they purposely changed to provide for the necessities of contemporary human-persons, but on the condition that they retain their spirit masters to protect them and ensure that respectful treatment and reciprocity is fulfilled.

It is hard to imagine what the Jotï would look like today if primates were removed from the picture. Harder still to imagine how monkey populations would be affected by the withdrawal of humans. The zone of connection uniting people and primates is sufficiently complex to realize that monkeys are not simply material resources to be exploited nor symbolic foils but are extended participants in the human drama of cultural adaptation and reproduction. The question of the significance of anthropogenic habitats and human hunting pressure for monkey populations in the Circum-Magualida is still in need of investigation. Recognition of interspecific interdependence has implications for ecocentric conservation. In order to safeguard cultural traditions, it will be necessary to preserve, among other things, the conditions that propitiate the continuation, although not necessarily the replication, of relationships between humans and their nonhuman primate counterparts. Furthermore, if the ultimate lesson of Jotï ethnoprimatology is that humanity, defined as an intellectual, social, and spiritual condition, is not exclusive of one species (cf. Viveiros de Castro 1996:12), then biodiversity conservation becomes a basic human right.

References

Alvarsson J (1997) Trickster Versus Culture Hero -The Roles of Thokwjwaj and 'Ahutsetajwaj in 'Weenhayek Mythology. Acta Americana 5(2):89–102

Århem K (1996) The cosmic food web: human-nature relatedness in the Northwest Amazon. In: Descola P, Pálsson G (eds) Nature and society: anthropological perspectives. Routledge, London, pp 185–204

Baleé W (1989) The culture of Amazonian forests. In: Posey DA, Balée W (eds) Resource management in Amazonia: indigenous and folk strategies, Advances in economic botany, vol 7. New York Botanical Garden, New York, pp 1–21

Baleé W (2010) Amazonian dark earths. Tipiti 8(1):Article 3

Baleé W, Erickson C (eds) (2006) Time and complexity in historical ecology. Columbia University Press, New York

Berlin B (1992) Ethnobiological classification: principles of categorisation of plants and animals in traditional societies. Princeton University Press, New Jersey

Boglar L (1978) Cuentos y mitos de los Piaroa. Universidad Católica Andrés Bello, Instituto de Investigaciones Históricas, Centro de Lenguas Indígenas, Caracas

Civrieux M de (1980) Watunna: An Orinoco Creation Cycle. Austin, University of Texas Press

Coppens W, Mitrani P (1974) Les Indiens Hoti: compte rendu de missions. L'Homme XIV(3-4):131–142

Cormier L (2003) Kinship with monkeys: the Guajá foragers of eastern Amazonia. Columbia University Press, New York

Cormier L (2006) A preliminary review of neotropical primates in the subsistence and symbolism of indigenous lowland South American peoples. Ecol Environ Anthropol, Paper 4. http://digitalcommons.unl.edu/icwdmeea/4

de Civrieux M (1970) Watunna: mitología makiritare. Monte Avila Editores, Caracas
Denevan WM (2001) Cultivated landscapes of native Amazonia and the Andes. Oxford University Press, Oxford, UK
Dumont JP (1976) Under the rainbow: nature and supernature among the Panare. University of Texas Press, Austin
EibelEibsfeldt I (1973) Die Waruwádu (Yuwana), ein kurzlich entdeckter, noch unerforschter Indianerstamm Venezuelas. Anthropos 68(1–2):137–144
Emmons L, Feer F (1990) Neotropical rainforest mammals, 1st edn. University of Chicago Press, Chicago
Erikson P (2000) The social significance of pet-keeping among Amazonian Indians. In: Poberseck P, Serpell J (eds), Companion Animals and Us. Cambridge University Press, Cambridge
Fuentes A (2006) Human-nonhuman primate interconnections and their relevance to anthropology. Ecol Environ Anthropol. Paper 4. http://digitalcommons.unl.edu/icwdmeea/4
García Tomás MD (1994) Buscando Nuestras Raices. Cumpanamá, Cosmovisión Chayahuita. Tomos 3–4. Lima: Centro Amazónico de Antropología y Aplicación Práctica. Tomo 3:453 p
Gross D (1975) Protein capture and cultural development in the Amazon basin. Am Anthropol 77:526–549
Guarisma V (1974) Los Hoti: introducción etno-lingüística. Tesis de Licenciatura. Caracas: Universidad Central de Venezuela
Huber O (1995) Geographical and physical features. In: Berry P, Holst B, Yatskievych K (eds) Flora of the venezuelan Guayana, vol 1: Introduction. Missouri Botanical Garden, St. Louis, pp 1–61
Jangoux J (1971) Observations on the Hoti Indians of Venezuela. Unpublished manuscript
Kirksey SE, Helmreich S (2010) The emergence of multi-species ethnography. Cult Anthropol 25(4):545–576
Kohn E (2013) How forests think: toward an anthropology beyond the human. University of California Press, Berkeley
Lathrap D (1968) The hunting economies of the tropical forest zone of South America: an attempt at historical perspective. In: Lee RB, Devore I (eds) Man the hunter. Aldine, Chicago, pp 23–29
Lehman J, Kern DC, Glaser B, Woods WI (eds) (2003) Amazonian dark earths: origin, properties, management. Springer, Dordrecht
Levi-Strauss C (1969) The raw and the cooked, Mythologiques, vol 1. The University of Chicago Press, Chicago
Levi-Strauss C (1973) From honey to ashes: an introduction to the science of mythology, Mythologiques, vol 2. Harper & Row, New York
Lima TS (1999) The two and its many: reflections on perspectivism in a Tupi cosmology. Ethnos 64(1):107–131
Linares O (1998) Los mamíferos de Venezuela. Sociedad Conservacionista Audubon de Venezuela, Caracas
Lizarralde M (2002) Ethnoecology of monkeys among the Barí of Venezuela: perception, use, and conservation. In: Fuentes A, Wolfe LD (eds) Primates face to face: the conservation implications of human-nonhuman primate interconnections. Cambridge University Press, Cambridge, pp 85–100
Mattei-Müller MC (1992) Yoroko: confidencias de un Chamán Panare. Armitano Editores, Caracas
Mayberry-Lewis D (ed) (1979) Dialectical societies: the Ge and Bororo of central Brazil. Harvard University Press, Cambridge
Meggers BJ (1971) Amazonia: man and culture in a counterfeit paradise. Aldine, Chicago
Naddaf G (2005) The Greek Concept of Nature. Albany: State University of New York Press
Overing J, Kaplan MR (1988) Los Wóthuha (Piaroa). In: Lizot J (ed) Los Aborígenes de Venezuela. Volumen III. Etnología Contemporánea II. Fundación La Salle, Caracas, pp 307–411
Papworth S, Milner-Gulland EJ, Slocombe K (2013) The natural place to begin: the ethnoprimatology of the Waorani. Am J Primatol 75(11):1117–1128
Quatra MM (2008) Bajkewa Jkwïkïdëwa-Jya. In: Diccionario Básico Castellano-Jodï. Ediciones IVIC, Caracas

Reichel-Dolmatoff G (1971) Amazonian Cosmos – The Sexual and Religious Symbolism of the Tukano Indians. The University of Chicago Press, Chicago
Shepard GH Jr (2002) Primates in Matsigenka subsistence and world view. In: Fuentes A, Wolfe LD (eds) Primates face to face: the conservation implications of human-nonhuman primate interconnections. Cambridge University Press, Cambridge, pp 101–136
Sponsel LE (1997) The human niche in Amazonia: explorations in ethnoprimatology. In: Kinzey WG (ed) New world primates: ecology, evolution, behavior. Aldine De Gruyter, New York, pp 143–165
Steward JH, Faron LB (1959) Native peoples of South America. McGraw-Hill, New York
Viveiros de Castro E (1986) Araweté: os deuses canibais. Jorge Zahar/anpocs, Rio de Janeiro
Viveiros de Castro E (1992) From the enemy's point of view: humanity and divinity in an Amazonian society. University of Chicago Press, Chicago
Viveiros de Castro E (1996) Os pronomes cosmológicos e o perspectivismo ameríndio. Mana 2(2):115–144
Viveiros de Castro E (1998) Cosmological deixis and Amerindian perspectivism. J R Anthropol Inst 4(3):469–488
Whitehead N (2002) Dark Shamans. Kanaimá and the Poetics of Violent Death. Duke University Press, Duke
Wilbert J, Simoneau K (eds) (1990) Folk literature of the Yaruro Indians. Latin American Center Publications, University of California, Los Angeles
Zent EL (1999) Etnobotánica Hotï: explorando las interacciones entre la flora y el ser humano del Amazonas Venezolano. PhD Dissertation, University of Georgia, Athens
Zent EL (2005) The hunter-self: perforations, prescriptions and primordial beings among the Hodï, Venezuelan Guayana. Tipiti 3(1):35–76
Zent EL (2006) Noções de Corporalidade e Pessoa entre os Jodï. Mana 12(2):359–388
Zent EL (2008) Interpenetración de Esencias: la fabricación de cuerpos entre los Jotï, Guayana venezolana. Antropológica LII 110:89–122
Zent S (2009) Estudio Piloto del VICAT: Índice de la Vitalidad de los Conocimientos Ambientales Tradicionales, Proyecto de investigación. Laboratorio de Ecología Humana, Instituto Venezolano de Investigaciones Científicas (IVIC), Caracas
Zent EL (2013a) Jodï Ecogony, Venezuelan Amazon. Environ Res Lett 8:015008. http://stacks.iop.org/1748-9326/8/015008
Zent EL (2013b) Contrapunteo de sonidos y silencios: germinadores de vida entre los Jodï, Venezuela. Copérnico VIII 18:32–42. https://doi.org/10.13140/2.1.4668.9923. http://copernico.uneg.edu.ve/numeros/c18/c18_art04.pdf
Zent EL (2014a) Ecogonía I. Desovillando la noción de naturaleza en la tradición occidental. Etnoecológica 10(6):1–13
Zent EL (2014b) Ecogonía II. Visiones alternativas de la biosfera en la América Indígena ¿Utopía o continuum de una noción vital? Etnoecológica 10(7):1–21
Zent EL (2014c) Ecogonía III. Jkyo jkwainï: la filosofía del cuidado de la vida de los jotï del Amazonas venezolano. Etnoecológica 10(8):1–28
Zent EL, Zent S (2004) Floristic composition of four forest plots: Sierra Maigualida, Venezuelan Guayana. Biodivers Conserv 13:2453–2483
Zent S, Zent EL (2008) Los Jodï. In: Perera MA (ed) Los aborígenes de Venezuela. Ediciones IVIC, Monte Avila Editores, ICAS, Fundación La Salle, Caracas, pp 499–570
Zent S, Zent EL (2012) Jodï horticultural belief, knowledge and practice: incipient or integral cultivation? Boletim du Museu Paraense Emilio Goeldi 7(2):293–338
Zent S, Zent EL, Marius L (2001) Informe Final del Proyecto 'Etnobotánica Cuantitativa de los Indígenas Hotï de la Región Circum-Maigualida, Estados Amazonas y Bolívar, Venezuela', Research Project Report prepared for Consejo Nacional de Investigaciones Científicas y Tecnológicas (CONICIT), Ministerio de Ciencia y Tecnología, Republica Bolivariana de Venezuela
Zent EL, Zent S, Iturriaga T (2004) Knowledge and use of Fungi by a Mycophile society from the Venezuelan Amazon. Econ Bot 58(2):214–226

Chapter 9
Primates in the Lives of the Yanomami People of Brazil and Venezuela

Jean P. Boubli, Bernardo Urbani, Hortensia Caballero-Arias, Glenn H. Shepard Jr, and Manuel Lizarralde

9.1 Introduction

9.1.1 Preliminary Remarks

The Yanomami are an indigenous people who have largely maintained their traditional way of life and cultural patterns despite ongoing cultural transformations experienced in recent decades due to gold mining, epidemic diseases, missionary activity, contact with the market economy, and national indigenous policies. Among the tropical rainforest peoples in South America, they inhabit one of the largest indigenous reservations, a binational territory of nearly 180,000 km^2 located on the Brazil–Venezuela border. They live in relatively small autonomous villages called *shabono* and their subsistence activities depend on horticulture, hunting, fishing, and gathering. The Yanomami rely on forest animals as their main source of protein, primates being among the most intensely targeted animals on hunting expeditions. Primates also play an important role in Yanomami material culture and mythology. This chapter provides an overview of the role of primates in Yanomami culture, including (a) a comprehensive review of the available literature for information

J. P. Boubli (✉)
School of Environment and Life Sciences, University of Salford, Salford, UK
e-mail: j.p.boubli@salford.ac.uk

B. Urbani · H. Caballero-Arias
Centro de Antropología, Instituto Venezolano de Investigaciones Científicas, Caracas, Venezuela

G. H. Shepard Jr
Coordenação de Ciências Humanas, Museu Paraense Emílio Goeldi, Belém do Pará, Brazil

M. Lizarralde
Department of Botany and Environmental Studies Program, Connecticut College, New London, CT, USA

B. Urbani, M. Lizarralde (eds.), *Neotropical Ethnoprimatology*, Ethnobiology,
https://doi.org/10.1007/978-3-030-27504-4_9

relevant to the classification, use and perception of primates; (b) original information from the villages of Maturacá in Brazil (J.P. Boubli [JPB]) and the Upper Orinoco of Venezuela (H. Caballero-Arias [HCA]); and (c) a summary on the current conservation status of primates in their territory. Since most of the available information had to be gleaned from anecdotes dispersed in literature sources on various other subjects, this chapter could be considered the first consolidated study of Yanomami ethnoprimatology to date.

9.1.2 The Yanomami

The Yanomami are one of Amazon's largest and most traditional indigenous societies, occupying an extensive territory in the Orinoco-Amazon interfluvial region between Brazil and Venezuela (Fig. 9.1). Yanomami is the generic term that identifies an entire linguistic family also known as Yanomama, made up of four linguistic groups: Yanomami (Yanomamɨ) located mainly in Venezuela; Yanomae (Yanomam, Yanomami), mostly found in Brazil; Sanema (Sanumá), the northernmost group situated between Venezuela and Brazil; and Shirian (Yanam, Ninam) a smaller population located in the northeast area in both sides of the border (Fig. 9.1). The Yanomami language family has no known relatives and is considered linguistically

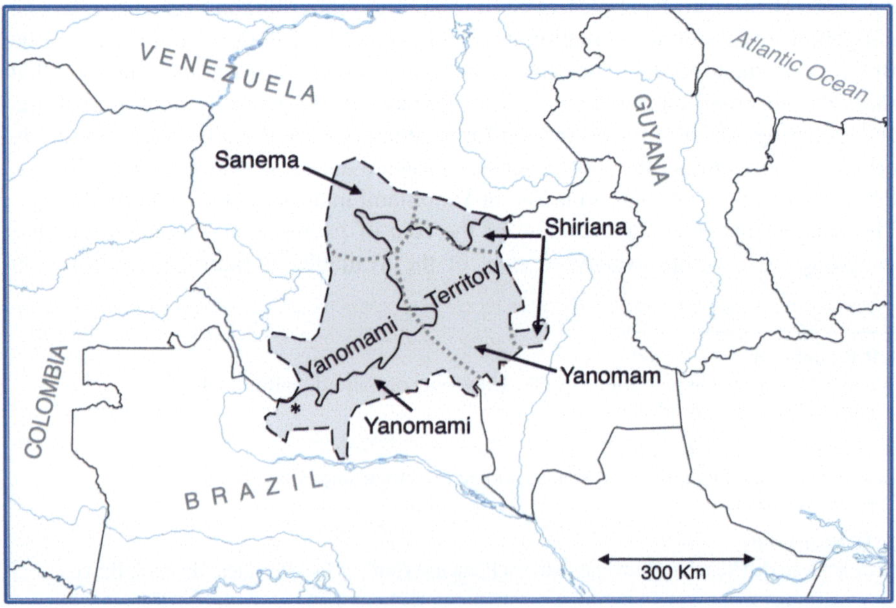

Fig. 9.1 Yanomami territory in Brazil and Venezuela. Dashed lines indicate the approximate borders between the different Yanomami dialect groups. Asterisk indicates the location of the Yanomami village of Maturacá

isolated, making it difficult to trace the ancient origins of the group (Albert 1989; Migliazza 1972; Smole 1976). Migliazza (1982) suggested a distant affiliation between Yanomami and the Panoan language family based on apparent cognates; however, a more recent review of Panoan linguistics by Fleck (2013: 24) considers this hypothesis "less plausible."

The Yanomami territory is centered on the Parima mountain range, the natural divide between Brazil and Venezuela. The area is known biogeographically as the Pantepui and it is characterized by the presence of large sandstone and granite mountains of altitudes from 800 to 3014 meters above sea level that form the watershed between the Orinoco and Amazon basins. Rivers draining the mountainous region are small, fast flowing, and very difficult to navigate due to numerous rapids and waterfalls. This region remains one of the most remote and ecologically diverse in South America.

The Yanomami population is over 35,000. The latest official censuses counted 24,603 Yanomami in Brazil (ISA 2017), and 11,431 in Venezuela (INE 2011). In Brazil, 228 villages have been recorded, distributed over a 96,650 km^2 of demarcated territory, shared between the Brazilian states of Amazonas and Roraima (ISA 2006). This territory constitutes a protected area currently identified as *Terra Indígena Yanomami* (Yanomami Indigenous Land), which was approved by a presidential decree on May 25, 1992. In Venezuela, there are over 220 Yanomami villages in Amazonas state within the Alto Orinoco-Casiquiare Biosphere Reserve with a total area of 82,662 km^2. This protected area, together with the Parima-Tapirapecó National Park located inside the biosphere reserve, was decreed on June 5th, 1991.

The Yanomami were made famous by Napoleon Chagnon's notorious ethnography *Yanomamo, The Fierce People* (1968) and an accompanying series of films in collaboration with Timothy Asch, notably *The Ax Fight* (1975). Chagnon's research and films focused especially on traditional warfare patterns, making the Yanomami out as an exceptionally violent indigenous society. Chagnon has been accused by several anthropologists of overstating the role of warfare, particularly revenge-based homicide, in Yanomami culture (Davis 1977; Albert 1989; Ramos 1987; Ferguson 1995; Sponsel 1998).

During the late 1980s and 1990s, the Yanomami received significant media attention as a result of the gold rush on the Brazil–Venezuela border, culminating in the *Haximu* massacre in 1993, in which some 16 Yanomami were killed by illegal gold miners in Yanomami territory in Brazil near the Venezuelan border. The Yanomami once again attracted significant international media attention in 2000 with the publication of Patrick Tierney's controversial book, *Darkness in El Dorado*, which accused Chagnon and other scientists working among the Yanomami of numerous unethical practices. However, the most serious allegations of genocidal medical experimentation were debunked (American Anthropological Association 2002; Gregor and Gross 2004). *The Falling Sky* (2014), Davi Kopenawa's magnificent first-person autobiography, translated and edited by anthropologist Bruce Albert, charts the history of Yanomami contact with the outside and paints a philosophical vision of Yanomami shamanism as an ongoing struggle to protect their land, their culture, and the entire world from an apocalyptic cataclysm. From fierce primitive tribe to victims of Western greed and scientific hubris, to shaman-warriors protecting the earth from calamity, the Yanomami have maintained a special fascination within and beyond anthropology for more than half a century.

Prior to the arrival of missionaries in the middle of the twentieth century, the Yanomami lived in large communal houses (*shabono* of 50–150 people) and maintained a semi-nomadic (or semi-sedentary) lifestyle, moving to new areas in decade-long cycles (Chagnon 1968; Milliken and Albert 1997). Today, some Yanomami live in individual houses with their extended families in settled, permanent villages, while others maintain the more traditional village structure, with seasonal migrations between different *shabonos* (Albert and Le Tourneau 2007). The Yanomami cultivate manioc and plantains, among other crops, in small family gardens as well as peach palms in old garden sites around their villages (Milliken et al. 1999). The Yanomami are patrilocal and patrilineal with men living temporarily with their wives' families as bride-service for a few years. Their kinship system is Dravidian with a high preference for cross-cousin marriage within the same village/shabono (Lizot 1988). Their political organization is relatively egalitarian, with one "headman" (or more than one when the community is large and has multiple patrilineal groups) who leads the community less through fiat than by fostering consensus. Headmen tend to be older but still active men who are respected by the community for their leadership and guidance. Shamans are also important members of the community performing frequent rituals involving the consumption of psychoactive *Virola* and *Anadenanthera* snuffs ("*paricá*") to heal individuals and protect the community from evil *hekura* spirits (Lizot 1973; Milliken et al. 1999).

In most Yanomami communities, hunting of forest animals is the major source of dietary protein (see Fig. 9.2a, b); however, in communities located along larger rivers, canoes and fishing gear have been introduced through contact with neighboring indigenous or riverine populations (Boubli 1997; Menezes 2010). Hunting among the Yanomami is carried out exclusively by men (Valero 1984). Yanomami men hunt mostly during the day, since nighttime ventures into the forest are proscribed due to fear of attack by *hekura* spirits (Boubli, personal observation). This salient cultural element is reflected in Yanomami zoological classification, which divides all arboreal animals (including monkeys) into two main groups, diurnal (*paso bi*) and nocturnal (*haso bi*). Nocturnal animals (*haso bi*) include night monkeys as well as olingo, kinkajou, porcupine, and opossum. With increasing contact with outside traders, the Yanomami acquired flashlights and shotguns, which has made it easier to venture into the night. The only night hunting observed by one of the authors in Maturaca (JPB) was paca hunting (*Cuniculus paca*), which was done from a canoe floating down the river and looking for the typical reflective eye shine of this large rodent. No night treks in the forest were ever witnessed (Boubli, personal observation).

Traditionally, the Yanomami hunted with arrows shot from unusually long (by Amazonian standards) palm or hardwood bows. Over the past few decades, however, firearm use has become widespread among most Yanomami hunters (Fig. 9.2a), though many retain their skills with the bow and arrow. The Yanomami start hunting in their early teens and quickly become remarkable animal trackers. They practice

Fig. 9.2 (**a**) Yanomami hunter with his shotgun used to kill a large peccary in Pico da Neblina mountain. (**b**) A black curassow (*Crax alector*) hunted by a Yanomami hunter in Pico da Neblina National Park, Brazil. (**c**) Yanomami men returning from a collective hunting expedition bringing large bundles of smoked meat to be shared in the village of Ariabú and Maturacá in Pico da Neblina National Park, Brazil. (**d**) Neblina black uakari monkey (*Cacajao hosomi*) hunted by a Yanomami man along the Iá River in Pico da Neblina National Park, Brazil. (**e**) Yanomami children in the Yanomami village of Purima, Mavaca, Alto Orinoco, 2005. (Photographs **a**–**d**, **f**: Jean P. Boubli; Photograph **e**: Hortensia Caballero)

two kinds of hunting, solo hunting on shorter outings close to the village (*rami*), and collective hunting treks (*heniyomou*) mostly for ritual feasts (Fig. 9.2c). In both, they search for a variety of animals, but preferentially hunt large mammals and birds (Fig. 9.2a, b). Primates are among their preferred prey, due to the diurnal habit of most species, relatively large body size, and conspicuous social nature (Fig. 9.2d). In this chapter, we discuss the importance of primates to the Yanomami people with a focus on the village of Maturacá in Brazil.

9.2 Primates in Yanomami Culture: A Review

Most of the relevant literature on Yanomami interactions with primates was drawn from research carried out in Venezuela (see Table 9.1). A few additional observations were gleaned from studies carried out in Brazil (Becher 1974; Saffirio and

Scaglion 1982; Milliken et al. 1999; Albert and Milliken 2009). Comparable information on primate hunting, folk-taxonomy, mythology, and cultural practices among the related Sanema people (e.g., Taylor 1974; Colchester 1981, 1982) are omitted in this review, since the Sanema are geographically, linguistically, and culturally distinct from the main Yanomami subgroups (Yanomaï, Yanam, and Ninam) discussed here.

The Yanomami recognize eight taxa of monkeys, lumping two species of uakari under *hõsõmɨ* (Fig. 9.2d). Ten primate species are reported from the Yanomami trans-national territory (cf. Boubli 2006), with fairly consistent names reported in multiple studies (see Table 9.1). It is not clear to what extent the listed variations in terminology are due to dialect differences, as opposed to inconsistent orthographical conventions across the studies.

9.2.1 Monkey Pets

The Yanomami (Lizot 2004), like many other Amazonian peoples (Cormier 2003, 2006; Lizarralde 2002, 2019; Shepard 2002), tend to take monkeys as pets, typically recovering surviving baby monkeys after adult individuals have been killed during a hunt (Fig. 9.2e) (Baker 1953). Infant monkeys can be breastfed by women (Smole 1976) until they wean (Baker 1953). In this sense, monkey pets kept by the Yanomami take on a degree of human kinship and, therefore, are never eaten by their keepers (see Cormier 2003 for human–primate kinship concepts among the Guajá). In the Upper Orinoco, the Yanomami mostly call their pets by their species names, e.g., *hima* (dog), *hoashi* (capuchin monkey), etc. (original observation by HCA). According to Lizot (2004), the vocative *thãri* refers to animal companion, and *thãriyë yaro* is translated as "my animal companion," however, this expression seems to be used more in mythical language.

The Brazilian Yanomami have been observed to use pigments from *Bixa orellana* (red), *Genipa americana* (black), and *Picramnia spruceana* (purple) to paint monkeys and other pets, much as humans are decorated with body paint (Fuentes 1980). In the Upper Orinoco region, one of the authors (HCA) has observed that monkey pets in a *shabono* are adorned with cotton "down" in preparation for feasts, just like people. The Yanomami often keep several species as pets, including howler monkeys, spider monkeys (Fig. 9.2e), capuchins, and night monkeys. Animal pets, including monkeys, are not typically killed or consumed, but they are punished if they misbehave. When pet monkeys die, they are burned outside the *shabono* and their bones are buried (Valero 1984), a practice distinctive from Yanomami funerary practices for humans (Lizot 1988), but still including cremation and special treatment of the bones as a core element.

Mattei-Muller (2007) provides a photographic record of *Aotus trivirgatus*, *Saimiri cassiquiarensis*, and *Cebus olivaceus* pets among the Yanomami. Caballero-Arias (2011) reports a juvenile spider monkey as a child's pet; Cocco (1987) reports a capuchin monkey (*hoaxi*) being kept by a girl in Iyëwei-teri; Steinvorth de Goetz

Table 9.1 Yanomami primate terminology from published studies and personal observations

				Venezuela				Brazil		
Linnaean taxonomy[a]	English common name	Spanish (Venezuela)	Portuguese (Brazil)	Grossa (1975)	Lizot (1975a, 2004)	Finkers (1986)	Mattei-Muller (2007)	Boubli, this study	B. Albert, pers. comm.	H. Ramirez, pers. comm.
Alouatta macconnelli[b]	Red howler monkey	Araguato. Mono aullador	Guariba	*iro*	*iro*	*iro, weyurasi* (="large ♂")	*weyurasi, iro*	*iro*	*iro a* / pl. *iro kɨkɨ, iro pë*	*Iro*
Ateles belzebuth[c]	Spider monkey	Marimona, mono araña	Coatá	*pasho*	*pasho*	*pasho*	*pasho*	*pasho*	*pasho a* / pl. *pasho kɨkɨ, pasho pë*	*pasho*
Cebus albifrons[g]	White-fronted capuchin monkey	Mono capuchino, mono blanco, mono cariblanco	Macaco prego, Cairara	*hoachi*	*hoashi*	*hoashi*	*yawerewë, hoashi, yarimɨ*	*hoashi*	not observed	*hoashi, yawerewë, yarimɨ*
Cebus olivaceus[d]	Wedge-capped capuchin monkey	Mono capuchino común	Cairara	*hoachi*	*hoashi*	*hoashi*	*yawerewë, hoashi, yarimɨ*	not observed	*yarima a* / pl. *yarima kɨkɨ, yarima pë*	Not observed
Saimiri cassiquarensis[g] *(ex S. sciureus)*	Squirrel monkey	Mono titi, mono ardilla	Macaco-de-cheiro	*culisi, curisi, echëechem* (="little monkey")	*kurisi, ëshëëshëmɨ*	*kurisi, ëshëëshëmi*	*kurisi, ëshëëshëmɨ*	*ëshëëshëmi*	*kusi si* / pl. *kusi sikɨ, kusi sipë*	*ëshëëshëmi, kurisi*
Aotus trivirgatus	Night owl monkey	Mono de noche	Macaco-da-noite	*Cucumü*	*makurutamɨ mo, kuukuumɨ*	*makurutami, kuukuumɨ, kukuumi*	*makukushimɨ (mo), makurutamɨ, kuukuumɨ*	*makurutami, kuukuumɨ, kukuumi*	*kuukuu moxi* / pl. *kuukuu moxipë*	*kuukuumi, makurutami*

(continued)

Table 9.1 (continued)

				Venezuela				Brazil		
Linnaean taxonomy[a]	English common name	Spanish (Venezuela)	Portuguese (Brazil)	Grossa (1975)	Lizot (1975a, 2004)	Finkers (1986)	Mattei-Muller (2007)	Boubli, this study	B. Albert, pers. comm.	H. Ramirez, pers. comm.
Cheracebus lugens	Black titi	Viudita	Zogue-zogue	*hoquepocomii*	*yõkõyõkõmɨ* (="a monkey species" (Lizot 1975a, b)), *hõkõhõkõmɨ*	*hõkõhõkõmi*	*yõkõyõkõmɨ*, *hõkõhõkõmɨ*	*hõkõhõkõmi*	*yõkoxi a* / pl. *yõkoxi pë*	*hõkõhõkõmɨ*, *yõkõyõkõmɨ*
Chiropotes israelita[g] *(ex C. satanas*[e]*)*	Bearded saki	Mono capuchino del Orinoco, mono barbudo	Cuxiú	*wisha*	*wisha*	*wisha*	*wisha*	*wisha*	*wĩxa a* / pl. *wĩxa kɨkɨ*, *wĩxa pë*	*wisha*, *wɨsha*
Cacajao ayresi[f]	Black-headed / Aracá uakari	Mono chucuto	Bicó	–	*hõsõmɨ*	*hõsõmi*	*hõsõmɨ*	*hõsõmi*	Not observed	*hõsõmɨ*
Cacajao hosomi[f]	Black-headed / Aracá uakari	Mono chucuto	Bicó	–	*hõsõmɨ*	*hõsõmi*	*hõsõmɨ*	Not observed	Not observed	Not observed

Abbreviations: Bra.: Brazil; Ven.: Venezuela

[a]Following the latest classification of primates from the Venezuelan Guayana (Urbani and Portillo-Quintero 2018)

[b]iro (Bra: Saffirio and Scaglion 1982, Ven: Migliazza 1972), ilo (Ven: Migliazza 1972)

[c]paxo (Bra: Saffirio and Scaglion 1982, Bra: Milliken et al. 1999), pasció (Ven: Biocca 1966), pasho (Ven: Fuentes 1980, Ven: Eguillor-García 1984), basho (Ven: Chagnon 1992), paso (Ven: Migliazza 1972), pašo (Ven: Migliazza 1972)

[d]yarim (Bra: Saffirio and Scaglion 1982; referred by the authors as *Cebus* [*Sapajus*] *apella*), hoashi (Ven: Fuentes 1980), hoaxi (Ven: Cocco (1987), howashi (Ven: Chagnon 1992)

[e]wixa (Bra: Saffirio and Scaglion 1982; misidentified as *Lagothrix* sp.), wisha (Ven: Fuentes 1980), wishia (Ven: Biocca 1965)

[f]Lizot (1975a, b) does not list a Yanomami name for *Cacajao*, the reason of this absence seems to be explained as Lizot (2004:110) later indicated that "[this monkey] is less frequent in the region inhabited by the central Yanomami," where Lizot conducted most of his field research, as did D. Grossa

(1968) recorded a capuchin kept as a pet by a Yanomami girl in the Upper Orinoco; and Herzog-Schröder (1999b) recorded a man on a canoe with a capuchin pet. Grossa (1975) indicates that Yanomami girls care for capuchin monkeys as if they were babies, matching a similar observation made by Becher (1974) in a Brazilian *shabono*.

9.2.2 Primate Hunting

The Yanomami have been observed to hunt all primate species found in their region (Viguera 1968, Chagnon 1992, Boubli 1997). Large-bodied social monkeys are the preferred prey species because they are relatively easy to locate as they move together in groups; however, most species are also quick, agile, and acrobatic and often escape into high branches or hilly forests (Biocca 1966). The Yanomami also consider monkey meat to be "good tasting" (Biocca 1966). So deeply associated is primate consumption with the Yanomami that non-indigenous peoples once referred to them as "monkey eaters": *guaharibos, guaicas, guaribas,* and *guaribas blancos* in Spanish colonial documents going back to the eighteenth century (Caballero-Arias 2014), or *guajaribos, uaharibos, uaribas,* and *uajaribos* (Koch-Grünberg 1924; de Barandiarán 1965) in later Brazilian sources. These varied terms derive from the Tupi-Guarani word *guaríba* or *uaríva* for "howler monkey" (*Alouatta* spp.) (Koch-Grünberg 1924). Yanomami are sometimes willing to venture dangerously close to enemy territory in search of their monkey prey. Valero (1984) reports an episode that nearly resulted in warfare when hunters from two enemy communities encountered one another in the forest while hunting monkeys. She also mentions that at a funeral she observed, monkey and curassow meat were reserved only for direct relatives of the deceased, while large prey (tapir and peccary) was served to other participants involved in mixing the plantain beer with the ashes of the dead (Valero 1984).

The Yanomami of the Upper Orinoco have a particular preference for spider monkey (Eguillor-García 1984; Valero; 1984; Cocco 1987), the largest neotropical monkey species (Emmons 1990). Several authors note that the Yanomami consider spider monkey to be particularly delicious (Steinvorth de Goetz 1968; Grossa 1975; Smole 1976), with a taste said to be "similar to agouti" (Smole 1976). At a feast in Mahekodoteri (Platanal) observed by Chagnon (1992), the Yanomami hosts were particularly proud of the 17 spider monkeys (*basho/pasho*) they had hunted. Finkers (1986) witnessed four spider monkeys and two howlers served at a three-day feast at a Yanomami community. Hames (1979), in a study of hunting among Yanomami living in a Ye`kuana on the upper Orinoco of Venezuela, recorded 18 spider monkeys among the 20 primate individuals hunted during an observation period of 216 days. Still, primates represented only about 3% of the total body weight of terrestrial animals hunted, with tapirs and peccaries accounting for most of the meat (Hames 1979). Likewise, Saffirio and Scaglion (1982) recorded capuchins (*Cebus olivaceus*), titi monkeys (*Cheracebus lugens*, misidentified as "*Lagothrix* sp."), howlers (*Alouatta macconnelli*), and spider monkeys (*Ateles belzebuth*) in a five-month study of Yanomami hunting on the Catrimani river in Brazil. They noted that spider monkey was the second most preferred prey species, after white-lipped peccary (see also Urbani 2005). Despite resource pressure and the necessity of new hunting strategies in an "acculturated" village setting, they recorded 152 kg of primates hunted, compared to only 38.5 kg for the "unacculturated" village (Saffirio and Scaglion 1982).

As a strategy for hunting primates, the Yanomami "use an ingenious and simple system: they flush a group of monkeys, they start shouting, 'Oh! Oh! Oh!,' the others 'Ih! ih! ih!.' The monkeys get frightened and remain immobilized in terror, giving time for the hunter [to carefully aim and] to shoot" (Cocco 1987: 194, authors' translation). When hunting capuchin monkeys, the Yanomami have been observed to use dogs (Cocco 1987). A favorite arrow tip for hunting capuchins, spider monkeys, and howlers among the Venezuelan Yanomami is the *pei-namo*, made of the wood of the light palm tree *Iriartella* sp. (*yoroama*) coated with curare (Cocco 1987, Signi-Sánchez and Morales-Mago 2008). This arrow tip has 4 or 5 notches that make the tip break off inside the animal's body so it cannot be removed (Cocco 1987). The *mamokori*, also a curare-coated point, is often used for hunting primates (Good 1989). In addition, Good (1989) recorded the use of the barbed-bone *u namo* point to hunt monkeys and other arboreal animals and birds. Herzog-Schröder (1999a) reported the use of the *huso mamo* spear with a narrow, notched palm tip coated in curare, causing the monkey to become immobilized and to fall quickly. Similarly, Chagnon (1968) observed the Yanomami making large quantities of palm-tipped arrows (about 40 cm long) with lateral cuts so that the curare-coated arrow tip breaks off inside the animal. Similarly, Boubli (this study), also recorded *Iriartea* sp. Palm arrow tips with perpendicular cuts coated with curare, primarily used for hunting monkeys (Fig. 9.3).

Thus, the Yanomami appear universally to prefer curare-tipped arrows when hunting monkeys (Baker 1953; Grossa 1975; Lizot 2004) since the poison relaxes the animals' muscles, releasing their grip from tree branches so they fall from the forest canopy as they die (Chagnon 1968; Cocco 1987). However, if the poison is not well prepared, the monkey may remain high up in the canopy, gripping a branch in *rigor mortis* (Biocca 1965). Yanomami shamans have been observed applying curare to arrows while reciting the names of the animal species that will be hunted in order to ensure an effective outcome for the hunt (Finkers 1986). According to Lizot (1988), arrowheads with curare are reserved for use only in wars and for hunting spider monkeys. Biocca (1965) indicated that bearded saki monkeys should be hunted with arrows coated in old curare; if freshly made curare were used when hunting this animal, the hunter's remaining arrowheads would rot with mold (Biocca 1965). Fuentes (1980) recorded the Venezuelan Yanomami making an arrow with the bony tip of a stingray tail (order Myliobatiformes), used on rare occasions to hunt smaller vertebrates, including the bearded saki. Ethnobotanical studies among the Yanomami of Brazil recorded arrow tips especially for primate hunting made from the palm trees *Jessenia bataua* and *Iriartella setigera*, coated with the resin of *Virola elongata* (Milliken et al. 1999, Albert and Milliken 2009). Arrows meant for

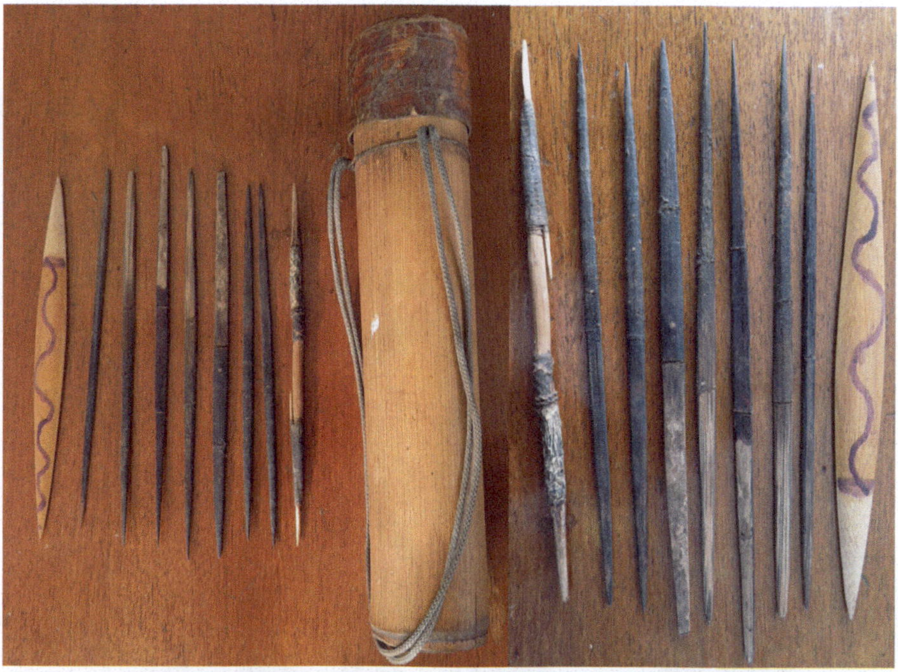

Fig. 9.3 Arrow tips made by Yanomami hunters from Marari village, Brazil. The picture on the left shows an arrow-tip set and their quiver made of bamboo with a deer skin cap. The picture on the right is a close-up of the arrow tips showing the characteristic perpendicular cuts on the *Iriartera sp*. Curare-coated tips meant to allow for easy breaking inside their primate targets (the 7 tips in the middle). On the outside left (bamboo tip) and right (bone harppon-like tip) are arrow tips meant for large terrestrial prey. (Photograph: Jean P. Boubli)

other types of prey, in particular large-bodied terrestrial animals, have tips made from monkey bone: *ẽthẽri,* as described below or bamboo (Figs. 9.3 and 9.4a).

Finkers (1986) includes visual documentation showing a group of Yanomami men carrying bundles of spider monkeys on tumplines, their foreheads bearing the weight, from the hunting site to a camp in the forest for cooking. When preparing monkey meat, the Yanomami first pass it over a fire to singe off the hair (Grossa 1975). Only then are the animals gutted and butchered (Finkers 1986). Men and women work together in forest cooking camps, wrapping monkey meat in leaves or tying it with thin vines to roast over a fire, or boiling it in pots (Cocco 1987). Large monkeys, and spider monkeys especially, are tied into a characteristic "seated" or "fetal" position with vegetable fibers for roasting (Fig. 9.2f) (Smole 1976; Grossa 1975; Steinvorth de Goetz 1968). Monkey brain is considered a delicacy and is highly esteemed (Chagnon 1968). The hunter typically distributes the cooked meat among those present in the cooking camp, giving a few small pieces to those who participated in the hunting expedition but reserving the majority of the meat for the feast in the *shabono* (Finkers 1986). After butchering and roasting in such forest

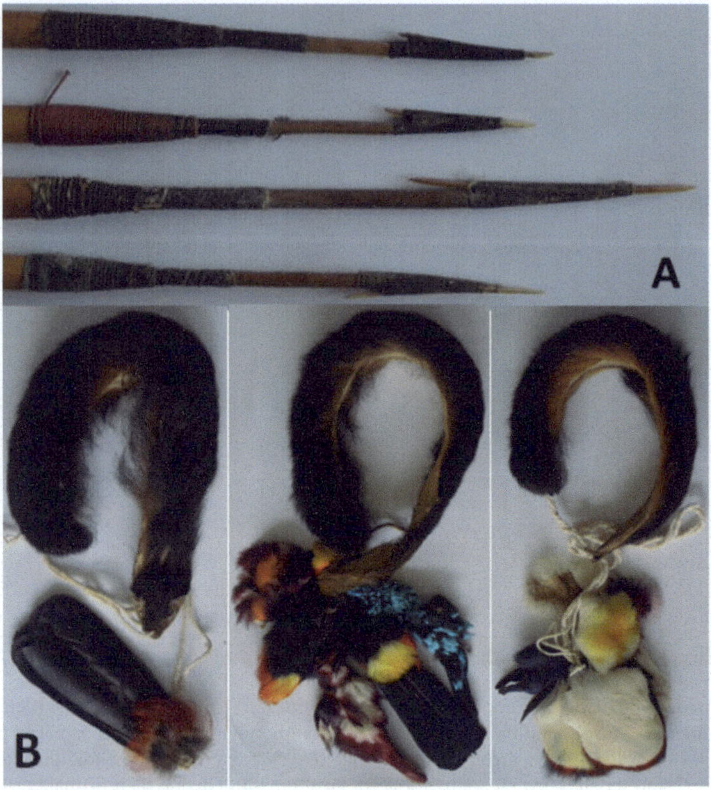

Fig. 9.4 (**a**) Yanomami arrow tips made with monkey bones from the Ocamo area in Venezuela. Lengths from the end of the rolled thread to the arrow tip, from top to bottom: 18.5 cm, 18 cm, 24 cm, 18.5 cm (coll. Bernardo Urbani); (**b**) Yanomami bracelets made with monkey skins. Lengths of the bracelets, right to left: 43 cm (associated bird feathers: *Ramphastos vitellinus/tucanus*), 44 cm (associated bird feathers: *Cotinga cayana* [turquoise/black], *Xipholena punicea* [purple/white], *Ramphastos tucanus* [black/yellow]), 29 cm (associated bird feathers: *Cyanerpes* sp. [blue/black], *Ramphastos vitellinus/tucanus* [white/red/yellow]) (coll. FLSCN)

camps, prepared monkey meat is carried back to the central community (Grossa 1975) (Fig. 9.2c). The monkey meat must be well roasted to the point that it looks almost burned; otherwise, the Yanomami will not eat it (HCA). Surplus smoked monkey meat is considered to be an especially prized item to be left hanging in the *shabono* (Smole 1976).

The extensive literature on the Yanomami (mostly in Venezuela) mention a number of food taboos and other cultural practices related to primates. These taboos and practices vary from one Yanomami locality to another, as well as between different age sets. For example, Lizot (1988) reported that adolescents and young adults of both sexes (approximately 11–25 years old) were prohibited from eating bearded sakis and howler monkeys in that study region. Finkers (1986) reported that children of roughly 9–14 years of age should not eat night owl monkeys or titi monkeys.

In three Yanomami communities located in the Mavaca river basin, Finkers (1986) reported that adolescent girls and adult women between roughly 12 and 40 years of age tend to avoid eating monkeys altogether. Both Eguillor García (1984) and Finkers (1986) observed that pregnant women avoided eating spider monkey's meat, since it was said to spoil their breast milk. Food taboos also apply to Yanomami house pets, for example, dogs are not allowed to eat howler monkey meat lest they become lazy, blind, and infected with botfly larvae (Finkers 1986). If someone steps on a spider monkey skeleton (*pasho ishi*) found lying in the forest, they will become frail and sick (Lizot 2004). The bones and remains of monkeys and other animals are typically thrown into the fire hearth. To discard of animal bones carelessly makes a hunter lose his hunting abilities, becoming *sina*, a bad or unlucky hunter (Lizot 1992). The ashes of the spider monkey pubis or hip bone (*pasho ishiki*) are said to be mixed with food or placed on a person's head as a kind of curse, causing the person to fall to the ground with powerful cramps and wide-open eyes (echoing the symptoms of tetanus) until they die (Lizot 2004).

The practice of using animal terms to create personal names and group names among the Yanomami (Lizot 1973) also appears associated with some instances of primate food taboos. Terms for primates including howler monkeys (*iro*), bearded sakis (wisha) and titi monkeys (*hõkõhõkõmi*) are used for naming persons as well as dogs (Lizot 1973). Becher (1974), working among the Brazilian Yanomami at the village of *Ironasitéri* ("the place of the howler monkey"), noted that howler monkey (*iro*) was not hunted or eaten there, as it was the name of one of the shamans of this *shabono*. (Becher 1974). The people at Ironasitéri used to call themselves "howler monkey people": they stated that howler monkeys have souls, and that because of this their ancestors were able to turn into howler monkeys and vice versa (Becher 1974).

In more recent original fieldwork carried out by author JPB in the Maturacá region of Brazil, the preferred game species for the Yanomami were spider monkeys, currasows, and peccaries (Fig. 9.2a, b), although other animals such as tapirs, deer, spiny rats, agoutis, pacas, and smaller monkeys were readily taken if encountered during a hunting expedition. The main targets on these expeditions were larger monkeys such as spider and howler monkeys. However, they also hunted small primates such as titi monkeys and night owl monkeys if nothing bigger was found. In this region, no taboos around eating any monkeys were observed. As noted by other authors, the Yanomami find monkeys to be relatively easy to locate due to their conspicuous calls and noisy locomotion. Their arboreal habits also mean that it is easy for the Yanomami hunter to get within shooting range undetected. The spider monkey remains one of their preferred game species due to its alleged good taste compared to other species such as the howler monkey (see also Cormier and Urbani 2008). An avoidance or lower preference for howler monkey has been reported for other South American lowland indigenous societies (Shepard 2002; Urbani 2005; Cormier 2006, Urbani and Cormier 2015).

On one occasion, JPB observed a Yanomami hunting party returning with 23 spider monkeys from a single expedition to the foothills of Pico da Neblina, near Igarape Tukano (Figs. 9.2c and 9.5). The Yanomami of this region claim that the

Fig. 9.5 Pile of smoked game animals hunted on a large fortnight-long expedition to supply meat for the annual *pupunha* feast, in celebration for the peach palm harvest. There are 23 spider monkey carcasses in this pile together with carcasses of a tapir, caimans, peccaries, curassows among other game species. (Photograph: Jean P. Boubli)

uakari was extremely difficult to hunt with bow and arrow, so they only became a common prey item more recently, with the advent of shotguns as the main hunting tool. Now, uakaris are hunted as much as any other species (Fig. 9.2d). During a stay in Maturaca in 1994, JPB retrieved 45 recently eaten uakari skulls from fire hearths. Uakaris are hunted around Maturacá preferably from April to June when the Yanomami say the animals are fatter due to an abundance of forest fruits, in particular the fruit of the palm *Mauritia flexuosa*.

9.2.3 Monkeys and Material Culture

Monkey bones and skins are important elements in Yanomami material culture (Fig. 9.4a, b). However, monkey teeth are not generally used in body ornamentation, unlike the case for large feline and caiman teeth found frequently on necklaces and other adornments. The Yanomami manufacture characteristic arrows with harpoon-like tips, called *ẽthẽri* (Biocca 1966; Finkers 1986; Lizot 2004), carved from the long bones of monkeys (Fig. 9.3a). Finkers (1986) describes how the "harpoon" points, carved from spider monkey bones, are secured to the shaft made from a shrub known as *etheri there* (*Mouriri myrtofolia*) by wrapping with a cord made from fibers of *shiki* (*Cecropia* sp.) thread coated in beeswax (Fig. 9.3a). Hunters need to be careful that these types of arrows are not stolen, lest they lose their hunting ability (Finkers 1986). Lizot (2004) reports the use of bearded saki bones for this type of arrowhead (Lizot 2004). Among the Brazilian Yanomami, Milliken et al. (1999) report the use of arrows with tips made from the monkey's radial bone to hunt small vertebrates. Yanomami children use wood from the palm *Bactris gasipaes* to make practice arrow tips fashioned with the same design (Milliken et al. 1999). Also, quivers for arrow-tips made of bamboo tubes are sometimes sealed with covers made of deer or monkey skin (Fig. 9.3) (Biocca 1966; Chagnon 1968; Boubli this study).

Among the most characteristic objects of Yanomami body adornments are feather-ornamented armbands (Fig. 9.4b), headbands (Fig. 9.6), and belts made from the skins of various primate species. Fossi-Cedeño (1999) describes headbands made from the skins of howler monkey and capuchins tied with cotton twine. The tail of the bearded saki is also used as a head ornament (Fig. 9.6) (Anduze 1960).

Fig. 9.6 Yanomami man from Marari village in Brazil wearing a headband made from the tail of a male *Chiropotes israelita*. Inset shows a similar headband but from Venezuela. (Photograph: Yanomami man by Francisco Pontual. Inset headband (48 cm long, 4 cm wide) from Manuel Lizarralde coll)

According to Biocca (1965), the bearded saki tail is skinned, dried, perforated, tied around the head with cotton thread, and adorned with colorful bird feathers. Similar belts, called *wisha shina*, are made from the tails of male bearded sakis, which have fuller and longer fur than females (Lizot 2004). Photographic records of Yanomami men from different regions, especially in ritual settings, often reveal prominent armbands made of primate skins decorated with feathers (Biocca 1965, 1966; Steinvorth de Goetz 1968; Viguera 1968; Grossa 1975; Eguillor-García 1984; Cocco 1987; Chagnon 1992; Lizot 1992) (Fig. 9.4b). Likewise, among the Yanomami of Brazil, a headpiece made from bearded saki (*wisha*) tail is a predominant form of festive adornment for men (Fig. 9.6). Given this animal's restricted distribution within Yanomami territory, this item is widely traded across villages (JPB, personal observation).

9.2.4 Primate Ethnoecology

Despite the wealth of published material on Yanomami ethnography, especially shamanism and warfare, there are relatively few studies focusing on ethnobiological and ethnoecological knowledge (Taylor 1974; Fuentes 1980; Milliken et al. 1999; Shepard Jr 2006; Albert and Milliken 2009). Fuentes (1980), Milliken et al. (1999), and Albert and Milliken (2009) make specific reference to plants that are consumed or otherwise used by monkeys according to the Yanomami (Table 9.2).

Plant and animal names among the Yanomami sometimes reference primate species or aspects of primate behavior. One particular species of hawk (*Buteo* sp.), for example, is associated with the bearded saki: *wisha karakapɨ* (Lizot 2004), perhaps an indication of the bird's preference for hunting this species. Fuentes (1980) likewise reports several tree names associated with primates, such as "capuchin monkey tree" (*hoashi kë moka*) and "spider monkey owl tree" (*pasho efetami*). *Pasho ãhũ* (*Garcinia macrophylla*) is a tree with edible fruits that references the spider monkey, while the *hoashi mosi* palm tree is associated with the capuchin monkey (Lizot 2004). Moreover, the sound made by *wakata* trees during windstorms is said to emulate spider monkey vocalizations (Lizot 2004). Yanomami of the village of Toototobi in Brazil associate the spider monkey with the cultivation of the introduced peach palm (*raxa paxo kikɨ*), while the name of a variety of bitter manioc (*hutuwisasi koko*) references the tail of the capuchin monkey (Milliken et al. 1999).

A close reading of the classic ethnographies reveals a wealth of insightful if dispersed observations relevant to primate ethnoecology. Spider monkeys are considered by the Yanomami to be strong and agile, while uakaris are noisy. Lizot (2004) notes that certain shamans who identify with the spider monkey will imitate its behavior and vocalizations to gain healing power, described by the verb *pashomou*, "to do like the spider monkey." *Pashomou* is also the sharp cry made by people to announce the arrival of visitors or enemies (Lizot 2004). On the other hand, the expression *hoashimou*, "to do like the capuchin monkey," refers to mischievous

Table 9.2 Plants used by monkeys as reported by the Yanomami

Venezuela	Brazil
Fuentes (1980)	Milliken et al. (1999), Albert and Milliken (2009)
1. *Alouatta macconneli*:	1. *Alouatta macconelli*:
morã (Burseraceae)	Leguminosae
morokoi (Melastomataceae: *Mouritia grandiflora*)	(*Elizabetha leiogyne*)
paihirimi (n. r.)	Moraceae
watupara (Burseraceae: *Dacryodes* sp.)	(*Pourouma minor*)
2. *Ateles bezebuth*:	2. *Ateles belzebuth*:
arõwae (Rhamnaceae: *Zizyphus cinnamomum*)	Moraceae
kareshi (Palmae, *Maximiliana regia*)	(*Pourouma minor*)
õpõni (Anacardiaceae: *Spondias sp.*)	3. *Cebus olivaceus*:
wanari (u. f. g.)	Sterculiaceae
yei (Palmae, *Attalea speciosa*)	(*Theobroma subincanum*)
3. *Cebus olivaceus*:	4. Monkeys (general):
ama ãsi (Caesalpiniaceae: *Elizabetha princeps*)	Anacardiaceae (*Anacardium*
apia (Sapotaceae, *Pouteria* sp.)	*giganteum*)
haproa (Palmae, *Oenocarpus bataua*)	Anacardiaceae
hayahama thotho (Bignoniaceae: *Tynnanthus polyanthus*)	(*Spondias mombin*)
hayu (n. p.)	Chrysobalanaceae
hoko (Palmae, *Oenocarpus bacaba*)	(*Licania* aff. *heteromorpha*)
kokoa (u. f. g.)	Leguminosae
koyosi (u. f. g.)	(*Hymenaea parvifolia*)
kumato (Caryocaraceae: *Caryocar villosum*)	Leguminosae
misikiri (n. r.)	(*Inga alba*)
mokorama (Marantaceae: *Ischnosiphon* cf. *aruma*)	Leguminosae
momi (Sapotacea)	(*Inga paraensis*)
NAł, (Sapotaceae: *Manilkara bidentata*)	Moraceae
shawarakurimi (Sapotaceae)	(*Pourouma bicolor* ssp. *digitata*)
shoroshoro (Moraceae: *Cecropia* cf. *javitensis*)	Myristicaceae
sititi (Gentianaceae)	(*Iryanthera juruensis*)
weima (Palmae, *Euterpe precatoria*)	Sapotaceae
wẽkama / wakama (Mimosaceae: *Inga nobilis*)	(*Chrysophyllum argenteum*)
wito (Anacardiaceae, *Anacardium giganteum*)	Sapotaceae
4. *Chiropotes israelita*	(*Manilkara huberi*)
ashowa (Moraceae, *Pseudolmedia* sp.)	Sapotaceae
hayu (Moraceae, *Pseudolmedia* sp)	(*Micropholis melinoniana*)
moshima (Mimosaceae, *Inga sp.*)	
shiwaikirimi (n.r.)	

Abbreviations: n. r. (not reported, in Fuentes (1980): Appendix 1); u. f. g. (unknown family and genus, in Fuentes (1980): Appendix 1)

Abbreviations: Bra.: Brazil; Ven.: Venezuela

[a]Following the latest classification of primates from the Venezuelan Guayana (Urbani and Portillo-Quintero 2018)

[b]iro (Bra: Saffirio and Scaglion 1982, Ven: Migliazza 1972), ilo (Ven: Migliazza 1972)

[c]paxo (Bra: Saffirio and Scaglion 1982, Bra: Milliken et al. 1999), pasció (Ven: Biocca 1966), pasho (Ven: Fuentes 1980, Ven: Eguillor-García 1984), basho (Ven: Chagnon 1992), paso (Ven: Migliazza 1972), pašo (Ven: Migliazza 1972)

[d]yarim (Bra: Saffirio and Scaglion 1982; referred by the authors as *Cebus* [*Sapajus*] *apella*), hoashi (Ven: Fuentes 1980), hoaxi (Ven: Cocco (1987), howashi (Ven: Chagnon 1992)

[e]wixa (Bra: Saffirio and Scaglion 1982; misidentified as *Lagothrix* sp.), wisha (Ven: Fuentes 1980), wishia (Ven: Biocca 1965)

[f]Lizot (1975a, b) does not list a Yanomami name for *Cacajao*, the reason of this absence seems to be explained as Lizot (2004:110) later indicated that "[this monkey] is less frequent in the region inhabited by the central Yanomami," where Lizot conducted most of his field research, as did D. Grossa

behavior, while the related term *hoashiprou* means "lacking in discipline" (Lizot 2004). Both JPB (Maturacá) and HCA (upper Orinoco) observed naughty children and toddlers referred to as *hoashi*, the white-fronted capuchin (*Cebus albifrons*), considered mischievous and undisciplined. Other indigenous groups throughout Amazonia likewise consider capuchin monkeys to be mischievous and badly behaved, due to their curious, somewhat hyperactive nature when habituated with humans (Shepard Jr 2002; Cormier 2006). The Yanomami, like many indigenous groups in Amazonia, consider howler monkeys to be slow, lazy, and often infested with botfly larvae (Shepard Jr 2002, Cormier 2006). In fact, howler monkeys in Panama (*Alouatta palliata*) can spend more than 65% of their day resting or sleeping and moving on average only around 443 m (Milton 1980). In stark contrast, capuchins are much more active, with the white-fronted capuchins of Cosha Cashu in Peru on average spending only 12% of their day resting and moving 2 km (Terborgh 1983).

9.2.5 Monkeys in Yanomami Cosmology

Yanomami cosmology and shamanism is replete with supernatural beings or nature spirits known as *hekura*, some of which are associated with animal species (Eguillor-García 1984, Lizot 2004). The *hekura* have a type of arrow associated with the capuchin monkey, named *hoshiri shereka* (Lizot 2004). Fossi-Cedeño (1999) reports two primates *hekura*. *Sibná* is a large mythical monkey possessing great strength and the ability to fly, capable of facing any type of enemy, and announcing its presence with the sound of powerful wind as it passes through the forest. The second is *pasó* (generic term for monkey in this dialectical variation, related to the term for spider monkey, *pasho*) that consists of two monkey brothers, one larger than the other, who have the ability to cure illness by facing down illness-causing spirits with a club made from the white branch of a tree called *mahomoi*. This pair of *hekura*, associated with the setting sun, enters the Yanomami shaman through the left arm and lodges in his chest. Lizot (2004) notes other monkey *hekura*, such as *iroriwë*, the *hekura* of howler monkeys, which is ever vigilant and able to perceive the presence of enemies and unfriendly shamans. *Pashoriwë*, the spider monkey *hekura*, is of a distrustful nature, and like the howler monkey *hekura* and *hoashiwë*, the capuchin *hekura*, is alert at all times, announcing the presence of enemy shamans (Eguillor-García 1984; Lizot 2004). *Wishariwë* is the *hekura* of the bearded saki, which, aside from alerting people to the dangers of enemy shamans, also has a tendency to become enraged (Lizot 2004).

In his autobiography, Yanomami shaman Davi Kopenawa (Kopenawa and Albert 2014) provides a first-hand account of his encounters with various supernatural and mythical entities. He mentions *Paxori*, the brave and powerful spider monkey spirit, who helps shamans maintain the celestial order by holding up and repairing the fragile vault of the sky, which is in constant danger of collapsing and falling, as it once did in primordial times (Kopenawa and Albert 2014). Kopenawa describes how the animals they hunt in the forest are different from the spirits or "images" they appeal to in shamanic trance: "And so the *iro* howler monkey we arrow in the trees is other than its image, *Irori*, the howler monkey spirit which the shamans call" (Kopenawa and Albert 2014: 60). Just as animals possess spirit "images" or counterparts, the Yanomami themselves possess animal doubles, known as *rixi*, consisting of rare animal species, encountered only in distant lands, like forest dogs for women and harpy eagles for men (Kopenawa and Albert 2014). Sometimes, these animal doubles are hunted by distant peoples at the fringes of Yanomami territory, causing illness and potential death to the human possessed of that particular *rixi*. When this happens, shamans and their helper spirits (*xapiri*), as well as certain animal *hekura* including the *purupuru namo* monkey spirit, rush to their aid, pulling out the enemy's arrow, hiding the wounded *rixi* and fending off attacking spirits while the shaman completes the cure (Kopenawa and Albert 2014). In the upper Orinoco, the double of a human being embodied in an animal is called *noreshi*, and the spider monkey (*pasho*) is one of the *noreshi* that is embodied among men of the same family (original observation by HCA). Ramos (1990) likewise mentions that the Yanomami may possess individualized alter-animals in the forest that represent some of the physical and behavioral qualities of the person.

Yanomami mythology contains numerous references to various monkey species (Wilbert and Simoneau 1990). The capuchin monkey plays a significant and somewhat humorous supporting role in the Yanomami myth about the origin of copulation, part of the larger cycle of origin myths concerning the twin male ancestors or demiurges, *Omawë* and *Yoawë*, who created the social life of the Yanomami as well as all living beings. Though numerous variations have been recorded (Eguillor-García 1984; Cocco 1987, Lizot 1975b, 1989; Chagnon 1992; Lizot et al. 1993), the story focuses on how *Hoashi* (or *Howashi* in Chagnon 1992) the white-fronted capuchin monkey, who is cousin or son-in-law to the ancestor twins, becomes a victim of his own lust for *Raharaiyma*, the daughter of a giant river creature. The ancestor twins desire her but *Hoashi* rushes to take her first. However, she is possessed with a *vagina dentata* containing a piranha that, in most versions, bites off the tip of the unfortunate *Hoashi's* penis. *Hoashi* rushes off to the trees screaming "Ko, ko, ko!" and turns into a monkey. According to Chagnon (1992), the myth explains why the capuchin monkey has a stubby, nail-shaped penis. In some versions, the capuchin monkeys hide their face in shame because they are unable to reach the attractive girl before the twins arrive. In all versions, the ancestor twins are able to successfully remove the piranhas from her vagina before copulating.

A myth of origin describes how howler monkeys, spider monkeys, and bearded sakis were created when a *hekura* took the shape of a vine hanging from the sky (*hetu misi*), split in half and opened: the beings located at the level of the trees turned into these primates (Lizot 1975b). Another version of the myth describes how a certain kind of climbing vine, *rasirasi* fell from an *apia* (*Micropholis* sp.) tree, which howler monkeys, spider monkeys, capuchin monkeys, and bearded sakis then climbed. Bearded sakis and spider monkeys later dispersed from the tree, while the howlers sat there and began shouting. People also climbed the tree, but those who fell while climbing turned into peccaries, while those who walked on branches and fell due to their weight became tapirs (Lizot 1989). Lizot (2004) makes brief mention of a mythological place referred to as *irori*, "howler monkey village" or *hoashiri*, "capuchin monkey village."

Regarding the origin of curare, a Yanomami story says that a man decided to create a poison (Lizot et al. 1993). He first tried it on a monkey, which died, confirming its proper preparation and lethal toxicity. The myth of the ancestor of snakes, one of the mythological animals that are now called "sleeper snakes," placed the tail of a bearded saki on his head before going to a party (Lizot 1989). *Wataperariwe*, a Yanomami mythological being, also wears the tail of a bearded saki while walking. In the myth of the "petrified hunter," a man pursuing a group of spider monkeys reaches a rock that begins to turn red, and the man proceeds to turn to stone (Lizot 1989).

9.3 Observations on Primates in the Rio Branco—Rio Negro Interfluvium

9.3.1 Study Site and Study Period (JPB)

Some 2000 Yanomamis of the Kohoroxi-teri group live in and around the village of Maturacá within Pico da Neblina National Park on the upper Rio Negro of Brazil (Boubli 1997, Menezes 2010), the largest Yanomami population in the binational territory. The Yanomami of Maturacá are distributed in eight permanent villages: Maturacá, Ariabú, União and Auxiliadora along the Maturacá canal, Maiá on the Maiá river, Nazaré on the Iá river, Pohoro and Xamatá in small right bank tributaries of the Marauiá river. Ariabú and Maturacá are the two largest villages with about 500 people each (Boubli 1997, Menezes 2010, Boubli, personal observation). The other villages are somewhat smaller with 400 people or fewer. All observations come from a series of field trips taken by JPB to the Yanomami territory in Amazonas, Brazil, from 1986 to 2005, more specifically to the Rio Negro—Rio Branco interfluvial region of Amazonia (Boubli 1997).

In surveys of this region, JPB recorded ten primate species of which five occur throughout the entire territory, namely, *Ateles belzebuth, Alouatta macconnelli, Cheracebus lugens, Saimiri cassiquiarensis,* and *Aotus trivirgatus*. Three species are restricted to the eastern part of the territory: *Chiropotes israelita, Cacajao ayresi, and Cebus olivaceus*; while two are found only in the west: *Cacajao hosomi* and *Cebus albifrons*. A maximum of seven and a minimum of four primates are found in sympatry in different parts of this region (see Boubli 2006). Despite the apparently pristine state of the forest in this region, with a small human population overall, primate densities are naturally low. This is primarily due to the characteristic poor sandy soils in this region limiting forest primary productivity (Boubli 2005). Low primary productivity translates into low biomass at higher trophic levels. The region is covered by a forest mosaic consisting of white sand savanna forests (*campinarana*), *terra firme* uplands, swamp forests and seasonally flooded forests along black water rivers (*igapós*) (Boubli 2002). These forests are dominated by a limited number of Fabaceae and Euphorbiacieae species (Boubli 2002), most of which produce dry, barochorous fruits that are not favored by arboreal frugivores such as primates, except seed eaters such as uakaris and bearded sakis. However, forests on the slopes of Pico da Neblina mountain range appear to be more productive and species-rich. Although not systematically surveyed, the frugivore community of these slopes appears to be much richer and more abundant. Spider monkeys, in particular, are restricted to this habitat, being virtually absent or extremely rare throughout the flat lowlands below. This ecological fact is determinant to Yanomami choices of settlement and hunting patterns.

9.4 Conservation Outlook

In 2014, the IUCN Primate Specialist Group met to reevaluate the conservation status of all New World Primates. Of the 10 primate species found in the Yanomami territory, only two were classified as threatened, namely, white-bellied spider monkey, *Ateles belzebuth*, and Neblina uakari, *Cacajao hosomi*. *A. belzebuth* has a large distribution, spreading from the foothills of the Andes in Ecuador to the right bank of the Rio Branco in Roraima, Brazil. This species was classified as Endangered under IUCN Criteria A (A2acd + 3 cd + 4acd), which signifies a recent reduction in numbers due to habitat destruction and hunting. Yanomami hunters heavily target this primate as one of their preferred game species. Primates such as the spider monkey that have relatively extended life histories are especially sensitive to overhunting (see Urbani 2005).

Neblina uakaris are also heavily hunted and have been driven to near local extinction along the Maturacá channel, a place where the Yanomami once considered them the most abundant primate. A recent review of the taxonomy of black uakaris separated this taxon into three distinct species with Neblina uakaris restricted to a much smaller range than previously thought (Boubli et al. 2008). In a species assessment in 2012, Neblina uakaris were classified as vulnerable under criteria A2d as there is reason to believe the species has declined by at least 30% over the past 30 years, mainly due to hunting by the Yanomami.

The increased efficiency of firearm use, coupled with rapid population growth of Yanomami communities, has certainly contributed to this decline in primate populations. Guns are much more effective weapons than traditional technology (Yost and Kelley 1983), resulting in an almost tenfold increase in hunting success when compared with bow-and-arrow (Alvard and Kaplan 1991; Levi et al. 2009). Sustained hunting with firearms can reduce populations of sensitive species like large primates very quickly to the point of local extinction in the vicinity of human settlements (Peres 1990; Shepard et al. 2012). However, as long as reserves are large and the human population relatively low, even vulnerable animal species manage to persist by repopulating from distant, non-hunted zones in what is known as "source-sink dynamics" (Novaro et al. 2000; Sirén et al. 2004). Such natural mechanisms of species recovery can be leveraged in community-based management systems for subsistence hunting in areas where forest cover is still largely intact, such as indigenous and sustainable use reserves in Amazonia (Levi et al. 2009; Shepard et al. 2012; Antunes et al. 2016).

9.5 Final Considerations

The Yanomami live in one of the most remote areas of South America. They have rich cultural traditions, beliefs, and practices that express their intimate knowledge about and relationships with various animal species, including primates. Primates are important in Yanomami diet, material culture, and mythology. Their traditional way of life has undergone rapid change over the last 30 years due to contact with missionaries, researchers, and gold miners and the influx of Western goods, medicines, and technology. Although their forest remains largely intact, wildlife species, including primates appear to have declined in regions near roads and urban centers, as well as near larger Yanomami villages with access to firearms and ammunition. Large primate species can be seen as core, "flagship" species, both in the terms of their cultural importance to the Yanomami, and their role as indicators of hunting pressure and general ecological health. Elsewhere in Amazonia, computer-based modeling in conjunction with participatory monitoring and applied research has focused on large primates as keystone biocultural species for streamlining community-based conservation efforts and modeling alternative management scenarios (Levi et al. 2009, Shepard et al. 2012). Ongoing collaborative research among anthropologists, biologists, and the Yanomami themselves will be crucial to monitoring the health of primate populations and finding solutions for hunting sustainability and food security.

Acknowledgments We thank Bruce Albert (Institut de Recherche pour le Développement) and Henri Ramírez (Universidade Federal de Rondônia) for sharing their detailed information on Yanomami primate nomenclature from Brazil. We also thank Erika Wagner and Pamela Navarro at the Venezuelan Institute for Scientific Research (IVIC), Pedro Rivas and Ana María Resnik at the Fundación La Salle de Ciencias Naturales (FLSCN), and the library staff of FLSCN and IVIC in Caracas for their cooperation. We appreciate the ornithological identifications provided by John F. Kvarnbäck (Kvarnbäck Birding) for the birds associated with the Yanomami feather armbands (Fig. 9.4b). Thanks to Francisco Pontual (University of California, Berkeley) for the photograph of Fig. 9.5. For fieldwork in the Yanomami villages within the Brazilian Pantepui region, JPB is grateful to the Yanomami people and the riverine communities of the Rio Negro Basin for their help and support during surveys, Catholic and evangelical missionaries for assistance in their remote outposts, and local authorities including FUNAI, IBAMA, ICMBio, and the Brazilian Army. JPB is also thankful to colleagues Amanda Colombo, Francisco Pontual, Italo Mourthe, Leandro Salles, and field assistants Arnaldo, Negão, Branco, Dorismar, and Murilo for their invaluable help on the expeditions to the left bank tributaries of the Rio Negro. Funding for JPB was kindly provided by the National Science Foundation (USA), National Geographic Society (USA), ZSSD (San Diego Zoo, USA), Sustainable Development of the Brazilian Biodiversity Program (PROBIO/MMA/BIRD/GEF/CNPq), and the University of Auckland. BU and HC are thankful for the support of the Venezuelan Institute for Scientific Research. GHS acknowledges support from the Brazilian

National Research Council (CNPq) research productivity grant (Proc. 308991/2015-0, Proc. 312037/2018-0), thanks JPB and Francisco Pontual for the opportunity to conduct ethnoecological fieldwork with the Yanomami of the Marari River in 2004 with the PROBIO-Pantepui expedition, and recognizes the assistance of Lourenço Yanomami during that work (see Shepard 2006). The authors appreciate the comments of the external reviewer that served to improve this chapter.

References

Albert B (1989) Yanomami "violence": inclusive fitness or ethnographer's representation? Curr Anthropol 30:637–640

Albert B, Le Tourneau FM (2007) Ethnogeography and resource use among the Yanomami. Curr Anthropol 48:584–592

Albert B, Milliken W (2009) Urihi A. A terra-floresta Yanomami. Instituto Socioambiental – Institut de Recherche pour le Development, São Paulo

Alvard MS, Kaplan H (1991) Procurement technology and prey mortality among indigenous Neotropical hunters. In: Stiner MC (ed) Human predators and prey mortality. Westview Press, Boulder, pp 79–104

American Anthropological Association (2002) El Dorado task force papers, vol 1. American Anthropological Association, Washington, DC

Anduze PJ (1960) Shailliko-ko: Descubrimieto de las fuentes del Orinoco. Talleres Gráficos Ilustraciones, Caracas

Antunes AP, Fewster RM, Venticinque EM, Peres CA, Levi T, Rohe F, Shepard GH Jr (2016) Empty forest or empty rivers? A century of commercial hunting in Amazonia. Sci Adv 2(10):e1600936

de Barandiarán D (1965) Mundo espiritual y shamanismo Sanema. Antropológica 15:1–28

Barker JP (1953) Memoria sobre la cultura Guaica. Boletín Indigenista Venezolano 3-4:433–489

Becher H (1974) Poré/Perimbó. Einwirkungen der lunaren Mythologie auf den Lebensstil von drei Yanomámi-Stämmen: Surára, Pakidái und Ironasitéri. Kommissionsverlag Münstermann-Druck KG, Hannover

Biocca E (1965) Yanoáma: Dal racolto di una donna rapita dagli Indi. Leonardo da Vinci Editrice, Bari

Biocca E (1966) Viaggi tra gli indi: Ato Rio Negro – Alto Orinoco. Apunti di un biologo. Volume secondo. Yanoáma. Consiglio Nazionale delle Ricerchez, Rome

Boubli JP (1997) The ecology of the black Uakari monkey: Cacajao melanocephalus melanocephalus in Pico da Neblina National Park, Brazil. PhD dissertation. University of California, Berkeley

Boubli JP (2002) Floristic assessment of the lowlands of Pico da Neblina National Park, Brazil. Plant Ecol 160:149–167

Boubli JP (2005) Floristics, primary productivity and primate diversity in Amazonia: contrasting a Eutrophic Varzea Forest and an Oligotrophic Caatinga Forest in Brazil. In: Dew JL, Boubli JP (eds) Fruits and Frugivores: in search for strong interactors. Kluwer Academic/ Plenum Publishers, Netherlands, pp 59–73

Boubli JP (2006) Primatas (Capítulo 5). In: Boubli JP, Pontual FB (eds) Biodiversidade de Vertebrados do Pantepui--AM. PROBIO/Ministerio de Meio Ambiente, Brasilia, pp 87–95

Boubli JP, Silva MNF, Amado MV, Herbk T, Pontual FB, Farias IA (2008) Taxonomic reassessment of black uakari monkey, *Cacajao melanocephalus*, Humboldt (1811), with the description of two new species. Int J Primatol 29:723–741

Caballero-Arias H (2011) Los Yanomami. Ediciones El Perro y La Rana, Caracas

Caballero-Arias H (2014) Desencuentros y encuentros en el Alto Orinoco. Incursiones en territorio Yanomami, siglos XVIII-XIX. Ediciones IVIC, Caracas

Chagnon NA (1968) Yanomamö. The fierce people. Holt, Rinehart and Wilson, Inc, New York
Chagnon NA (1992) Yanomamö. The last days of Eden. Harcourt Brace & Company, Orlando
Cocco L (1987) Iyëwei-teri. Quince años entre los Yanomamos. Librería Editorial Salesiana SA, Caracas
Colchester M (1981) Myths and legends of the Sanema. Antropológica 56:25–126
Colchester M (1982) The Cosmovision of the Venezuelan Sanema. Antropológica 58:97–122
Cormier LA (2003) Kinship with monkeys: the Guajá Foragers of Eastern Amazonia. Columbia University Press, New York
Cormier LA (2006) A preliminary review of neotropical primates in the subsistence and symbolism of indigenous lowland South American peoples. Ecol Environ Anthropol 2(1):14–32
Cormier LA, Urbani B (2008) The ethnoprimatology of the spider monkey: from past to present. In: Campbell C (ed) Spider monkeys: the biology and behavior of the genus *Ateles*. Cambridge Studies in Biological and Evolutionary Anthropology. Cambridge University Press, Cambridge, UK, pp 377–403
Davis S (1977) Victims of the miracle: development and the Indians of Brazil. Cambridge University Press, Cambridge, UK
Eguillor Garcia MI (1984) Yopo, Shamanes y Hekura: Aspectos fenomenológicos del mundo sagrado Yanomami. Caracas, Librería Editorial Salesiana
Emmons LH (1990) Neotropical rainforest mammals. University of Chicago Press, Chicago
Ferguson BR (1995) Yanomami warfare: a political history. School of American Research Press, Santa Fe (New Mexico)
Finkers J (1986) Los Yanomami y su sistema alimenticio (Yanomami Nii Pë). Monografía n° 2. Vicariato Apostólico de Puerto Ayacucho, Puerto Ayacucho
Fleck DW (2013) Panoan languages and linguistics. Anthropol Pap Am Mus Nat Hist 99:1–112
Fossi-Cedeño LE (1999) Ceremonial y adornos sagrados: Acercamiento a la dinámica de la teatralidad en los principales rituales Yanomami del Alto Orinoco. Caracas, Fondo Editorial Fundarte-Alcaldía de Caracas
Fuentes E (1980) Los Yanomami y las plantas silvestres. Antropológica 54:3–138
Good KR (1989) Yanomami hunting patterns: trekking and garden relocation as an adaptation to game availability in Amazonia, Venezuela. PhD dissertation. University of Florida, Gainesville
Gregor T, Gross DR (2004) Guilt by association: the culture of accusation and the American Anthropological Association's investigation of Darkness in El Dorado. Am Anthropol 106(4):687–698
Grossa D (1975) Shori, caimë ya jama (Amigo, quiero visitarte). Ernesto Armitano Press, Caracas
Hames RB (1979) A comparison of the efficiencies of the shotgun and the bow in neotropical forest hunting. Hum Ecol 7:219–252
Herzog-Schröder G (1999a) Yanomami: Nacidos de la pantorrilla. In: Delgado L, Ortega A, Jiménez Z (eds) Orinoco-Parima – Comunidades indígenas del sur de Venezuela – La colección Cisneros. Caracas, Fundación Cisneros, pp 34–49
Herzog-Schröder G (1999b) Canjear y compartir: Sobre la importancia de la dádiva entre los Yanomami. In: Delgado L, Ortega A, Jiménez Z (eds) Orinoco-Parima – Comunidades indígenas del sur de Venezuela – La colección Cisneros. Caracas, Fundación Cisneros, pp 50–59
INE (2011) Resultados población indígena. Caracas, Instituto Nacional de Estadística
ISA (2017) Yanomami. Povos Indígenas no Brasil. Instituto Socioambiental. https://pib.socioambiental.org/pt/Povo:Yanomami (Consulted on 12-xii-2018)
ISA (2006) Povos Indígenas do Brasil: 2001–2005 / Editores gerais: Beto Ricardo e Fany Ricardo. São Paulo, Instituto Socioambiental
Koch-Grünberg T (1924) Von Roraima zum Orinoco. Strecker und Schrõder, Stuttgart
Kopenawa D, Albert B (2014) The falling sky: words of a Yanomami Shaman. Belknap Press of Harvard University, Cambridge, MA
Levi T, Shepard GH Jr, Ohl-Schacherer J, Peres CA, Yu DW (2009) Modeling the long-term sustainability of indigenous hunting in Manu National Park, Peru: landscape-scale management implications for Amazonia. J Appl Ecol 46:804–814

Lizarralde M (2002) Ethnoecology of monkeys among the Barí of Venezuela: perception, use and conservation. In: Fuentes A, Wolfe L (eds) Primates face to face: the conservation implications of human and nonhuman primate interactions. Cambridge University Press, Cambridge, pp 85–100

Lizarralde M (2019) Etnoprimatología Barí en la Sierra de Perijá del estado Zulia. In: Urbani B, Ceballos-Mago N (eds) La Primatología en Venezuela. Academia Nacional de Ciencias Exactas, Físicas y Naturales/Universidad Simón Bolívar (Colección Conjunta Acfiman/USB), Caracas

Lizot J (1973) Onomastique Yanõmami. L'Homme. Rev franç d'anthropol XIII:60–71

Lizot J (1975a) Diccionario Yanomami-español. Universidad Central de Venezuela, Caracas

Lizot J (1975b) El hombre de la pantorrilla preñada y otros mitos Yanomami. Caracas, Fundación La Salle de Ciencias Naturales

Lizot J (1988) Los Yanomami. En: Lizot J (ed), Los Aborígenes de Venezuela, Volumen III. Etnología Contemporánea. Fundación La Salle/Monte Avila Editores: Caracas pp 479-583

Lizot J (1989) No patapɨ tëhë: En tiempo de los antepasados. Texto de lectura –2. Vicariato Apostólico de Puerto Ayacucho: Puerto Ayacucho

Lizot J (1992) El círculo de los fuegos: Vida y costumbres de los indios Yanomami. Caracas, Monte Ávila Editores

Lizot J (2004) Diccionario enciclopédico de la lengua yãnomãmɨ. Vicariato Apostólico de Puerto Ayacucho, Puerto Ayacucho

Lizot J, Cocco L, Finkers J (1993) Los pueblos indios en sus mitos 4: Yanomami. Abya Yala, Quito

Mattei-Muller MC (2007) Lengua cultura Yanomami: Diccionario Yanomami-Español/Español-Yanomami. CONAC-UNESCO-Agencia Española de Cooperación Internacional, Caracas

Menezes R (2010) Yanomami na encruzilhada da conquista: Contato e Transformação na fronteira amazônica. PhD dissertation. Universidade de Brasilia, Brazil

Migliazza EC (1982) Linguistic prehistory and the refuge model in Amazonia. In: Prance GT (ed) Biological diversification in the Tropics. Columbia University Press, New York, pp 497–519

Migliazza EC (1972) Yanomama Grammar and Intelligibility. PhD. Dissertation. Indiana University, Bloomington

Milliken W, Albert B (1997) The construction of a new Yanomami round-house. J Ethnobiol 17(2):215–233

Milliken W, Albert B, Gomez GG (1999) Yanomami: a Forest people. Royal Botanic Gardens, Kew

Milton K (1980) The foraging strategy of the howler monkey: a study in primate economics. Columbia University Press, New York

Novaro AJ, Redford KH, Bodmer RE (2000) Effect of hunting in source-sink systems in the Neotropics. Conserv Biol 14:713–721

Peres CA (1990) Effects of hunting on western Amazonian primate communities. Biol Conserv 54:47–59

Ramos A (1987) Reflections on the Yanomami: ethnographic images and the pursuit of exotic. Cult Anthropol 2(3):284–304

Ramos AR (1990) Memórias Sanumá: Espaço e Tempo em uma sociedade Yanomami. Editoras Marco Zero e Universidade de Brasília, Brasília

Saffirio G, Scaglion R (1982) Hunting efficiency in acculturated Yanomama villages. J Anthropol Res 38:315–328

Shepard GH Jr (2006) Etnobiologia Yanomami. In: Boubli JP, Pontual FB (eds) Biodiversidade de Vertebrados Do Pantepui--AM. PROBIO/Ministerio de Meio Ambiente, Brasilia, pp 102–128

Shepard GH Jr (2002) Primates in Matsigenka subsistence and worldview. In: Fuentes A, Wolfe L (eds) Primates face to face: the conservation implications of human and nonhuman primate interconnections. Cambridge University Press, Cambridge, UK, pp 101–136

Shepard GH Jr, Levi T, Neves EG, Peres CA, Yu DW (2012) Hunting in ancient and modern Amazonia: rethinking sustainability. Am Anthropol 114(4):652–667

Sirén A, Hamback P, Machoa E (2004) Including spatial heterogeneity and animal dispersal when evaluating hunting: a model analysis and an empirical assessment in an Amazonian community. Conserv Biol 18:1315–1329

Signi-Sánchez AJ, Morales-Mago P (2008) Nuestras manos… savia de tradición. Museo Etnográfico de Amazonas "Monseñor Enzo Ceccarelli": Puerto Ayacucho

Smole WJ (1976) The Yanoama Indians: a cultural geography. Texas University Press, Austin

Sponsel L (1998) Yanomami: an arena of conflict and agression in the Amazon. Agress Behav 24:97–122

Steinvorth de Goetz I (1968) ¡Uriji jami! Impresiones de viajes orinoquenses por aire, agua y tierra. Volkswagen de Venezuela-Asociación Cultural Humboldt, Caracas

Taylor KI (1974) Sanuma Fauna: prohibitions and classifications. Caracas, Fundación La Salle de Ciencias Naturales

Terborgh J (1983) Five New World Primates: a study in comparative ecology. Princeton University Press, Princeton NJ

Urbani B (2005) The targeted monkey: a re-evaluation of predation on New World primates. J Anthropol Scien 83:89–109

Urbani B, Cormier L (2015) The ethnoprimatology of the howler monkeys (*Alouatta* spp.): from past to present. In: Kowalewski MM, Garber PA, Cortés-Ortiz L, Urbani B, Youlatos D (eds) Howler monkeys: behavior, ecology, and conservation. Springer Press, New York

Urbani B, Portillo-Quintero C (2018) Consideraciones sobre la distribución y estado de conservación de los primates de la Guayana venezolana. In: Urbani B, Kowalewski M, RGT C, de la Torre S, Cortés-Ortiz L (eds) La primatología en Latinoamérica 2 – A primatologia na America Latina 2. Tomo II: Costa Rica-Venezuela. Ediciones IVIC, Caracas, pp 677–689

Valero H (1984) Yo soy napëyoma. Relato de una mujer raptada por los indígenas Yanomami. Fundación La Salle de Ciencias Naturales, Caracas

Viguera JM (1968) Los guaicas. Logroño, Aldus, SA

Wilbert J, Simoneau K (1990) Folk literature of the Yanomami Indians. UCLA Latin American Studies, 73, UCLA. Latin American Center Publications, Los Angeles

Yost JA, Kelley PM (1983) Shotguns, blowguns, and spears: the analysis of technological efficiency. In: Hames RB, Vickers WT (eds) Adaptive responses of native Amazonians. Academic Press, New York, pp 189–224

Chapter 10
Kixiri and the Origin of Day and Night: Ethnoprimatology among the Waimiri Atroari Ameindians of the Central Amazonia, Brazil

Rosélis R. de Souza-Mazurek and Ana Carla Bruno

10.1 Introduction

Primate species are an integral part of the lives of many Amerindians in Lowland South America (Amazonia). The use of primates as food in Amerindian subsistence hunting practices has been the subject of several studies concerning interactions among human and primate species (Mittermeier 1991; Milton 1991; Prado et al. 2012). Particular emphasis has also been given in the literature to the negative effects of subsistence hunting by indigenous and nonindigenous people on wildlife, in which monkeys represent a special concern (Mittermeier 1987; Queiroz and Kipnis 1990; Peres 1990, 1991; Jerozolimski and Peres 2003; Endo et al. 2009; Mena et al. 2000; Souza-Mazurek et al. 2000). The relatively large size of some cebids makes them frequent hunting targets (Emidio-Silva 1998; Mena et al. 2000; Peres 2000). The low reproductivity rate in some species raises concerns about the long-term conservation of large monkeys under heavy hunting pressure, particularly of the genus *Ateles* (Milton 1981). Nevertheless, hunter choices not only follow optimum prey returns but also are influenced by cultural factors (Souza-Mazurek et al. 2000; Lizarralde 2002; Shepard 2002; Cormier 2003). Monkey species are not only an important component of the subsistence diets of many Amazonian indigenous groups as different patterns of hunting and subsistence occur among

R. R. de Souza-Mazurek (✉)
Coordenação de Biodiversidade, Instituto Nacional de Pesquisas da Amazônia, Manaus, Brazil

Local Heritage, Environment and Globalization Laboratory – IRD MNHN, Paris, France

Takahi Socioambiental, São Paulo, SP, France
e-mail: roselismazurek@yahoo.com

A. C. Bruno
NPA-Instituto Nacional de Pesquisas da Amazônia, Manaus, Brazil

Coordenação de Sociedade, Ambiente e Saúde – COSAS, Manaus, Brazil

B. Urbani, M. Lizarralde (eds.), *Neotropical Ethnoprimatology*, Ethnobiology,
https://doi.org/10.1007/978-3-030-27504-4_10

indigenous groups mediated and influenced by symbolic relationships that are culturally established (Da Silva 2005, summarized in Cormier 2006). They are frequently subject to taboos and cultural regulations that often figure in indigenous myth, sacred rites, and social symbologies. Understanding the cultural aspects influencing the relationships that different indigenous people have with nonhuman primates aids in assessing the disruptions of these patterns that occur over time as many of these communities and their territories are undergoing socioeconomic and environmental changes.

There are few ethnobiological studies available about the Waimiri Atroari Amerindians from central Amazonia. Ethnobotanical studies carried out by Miller et al. (1989) and Miliken et al. (1992) indicated that the Waimiri Atroari have extensive knowledge of forest trees, having named 95% of 135 plant species and reported indigenous use of 65% of a sample of 34 species present in the 1 hectare plot of upland forest (*terra firme*). A study carried out by Souza-Mazurek (2001) on subsistence hunting and fishing practices showed that hunting alone is responsible 76% the 143 T total annual offtake of terrestrial vertebrates and fish species combined. As regards the game composition, the Waimiri Atroari select some species including two large cebid species, spider monkey (*Ateles paniscus)* and red howler monkey (*Alouatta macconnelli)*, by taking more individuals than expected compared to the natural densities of nonhunted sites (Souza-Mazurek 2001).

In this chapter, we present some aspects of the role of primate species in subsistence practices, myths, cultural taboos, and avoidances as well as medicine in the Waimiri Atroari worldview. Data on primate hunting and consumption stem from the doctoral thesis of the first author, while data on the primate cultural use and symbolism were compiled from the available general literature on the Waimiri Atroari.

10.2 The Waimiri Atroari People

The Waimiri Atroari Amerindians refer to themselves as *kinja*, or "true people," and belong to the Carib linguistic group practicing a hunting/horticultural trekking mode of subsistence. Their homeland is located in the rainforest of central Amazonia, encompassing the valleys of Alalaú, Camanaú, Curiaú, and Santo Antonio do Abonari rivers, in the northern region of Brazil (between 0° and 2° S and 62° and 60° W). The Indigenous Reserve Waimiri Atroari was officially demarcated in 1987 with an area of 2,585,911 hectares (25,859 km^2). The area traditionally occupied by these Amerindians according to historical records (Bandeira 2009, Barbosa Rodrigues 1885, Monte 1992) also included the area of the Rivers Jauaperi, Urubu, Jatapu, Uatuma, Taruma Açú, Cuieiras, and Apuaú. Stable and peaceful contact with the nonindigenous population was established only since the seventies. Currently the upper portion of Jauaperi River limits the reserve in its northwest side, but it is not part of the Indigenous Land.

The Waimiri Atroari currently have a population of approximately 2013 individuals living in 30 villages scattered along the main river channels and Federal

Highway 174, which connects the cities of Manaus, capital of Amazonas state, and Boa Vista, the capital of in the state of Roraima (Fig. 10.1). The villages are organized around a single hut or communal round house, the *Mydy*, within which the families divide into sectors with a central common area. Around the *Mydy*, there is generally a communal school and a cassava (*Manihot esculenta*) processing house. Centralized power is absent with each village having political and economic autonomy. However, neighboring villages often work together in efforts that involve much labor such as building a *Mydy* or opening new swidden fields. Decisions involving broader issues are made in meetings that bring together all

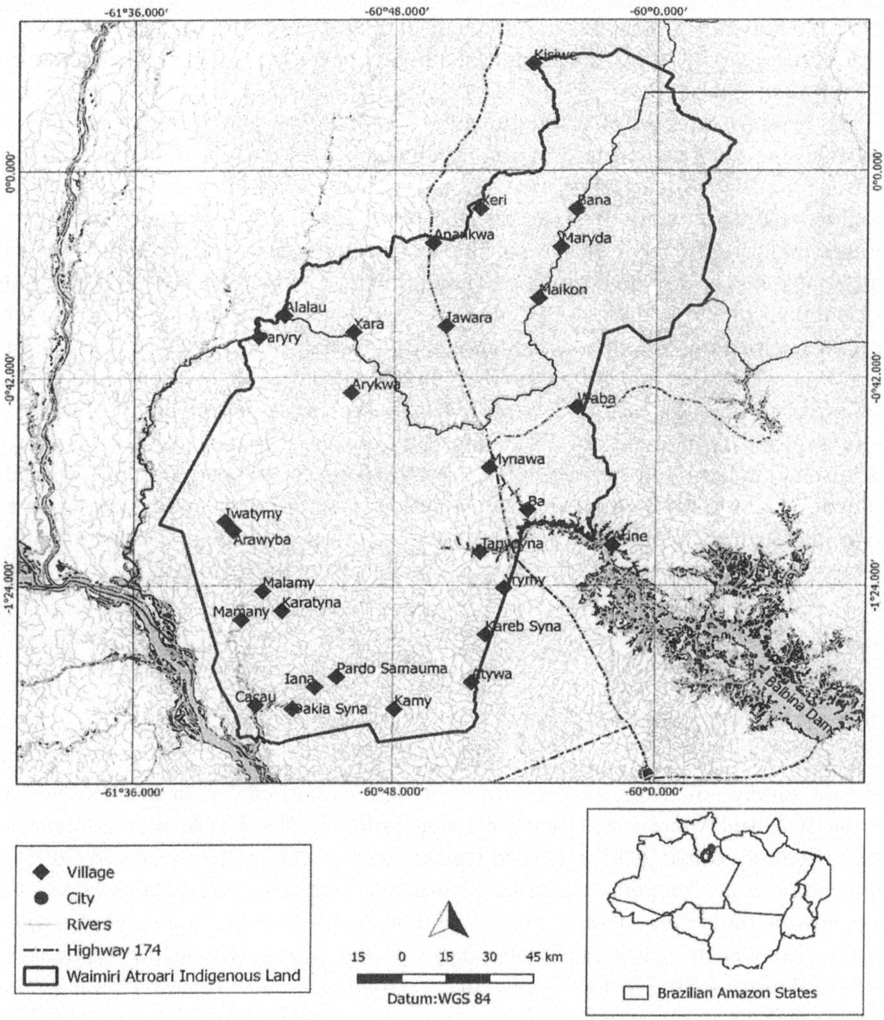

Fig. 10.1 Geographical distribution of the villages within the Waimiri Atroari Indigenous Land. (Source: Ecology Brasil 2014)

the village leaders. Subsistence is based on shifting cultivation, hunting, fishing, and gathering. The main staple of the Waimiri Atroari diet is cassava with both bitter and sweet varieties of this crop being cultivated in swidden fields. Sweet potatoes, yams, sugar cane, pineapples, and other crops are also cultivated. Work is communal, but men and women have distinctive roles: The felling and burning is done by men, while planting is undertaken by both men and women, and harvesting is only done by women (Miliken et al. 1992). Hunting is mostly carried out by men; although, women may hunt small mammals such as red humped agouti (*Dasyprocta agouti*) and red acouchi (*Myoprocta acouchy*), trapping them in hollowed logs (Souza-Mazurek pers. obs.). Hunting yields are split among the hunters in collective hunts, sharing with their relatives and keeping the choice parts of animals for a hunter's immediate family. Fishing is carried out during most of the year, but with more emphasis in the dry season (August to November) by both men and women.

The Waimiri Atroari would traditionally move their villages from time to time as a response to local game scarcity, soil productivity depletion, or political conflicts between families (Silva 1993). However, some changes in settlement patterns from mobile to more sedentary lifestyles have occurred due to a number of external interference in the 1970s and 1980s. A social assistance program for the Waimiri Atroari was established by the state-run power and utility company, ELETRONORTE, and the National Indian Foundation FUNAI (Brazilian agency for Indian affairs). This program helped mitigate the social and environmental impacts they suffered after part of their land was flooded by the Balbina hydroelectric dam's reservoir in 1987. The program provides health assistance, education, and technical support for agricultural production, territorial protection, environmental issues, and documenting the history and cultural heritage of the Waimiri Atroari people. Despite external influences, the Waimiri Atroari retain much of their original culture and obtain many forest resources needed for construction, crafts, and medicines, while still practicing intensively their rituals and ceremonies (Matarezio-Filho 2010; Do Vale 2002; Bruno 2003, 2010, 2014).

10.3 Primate Species

Primate surveys of naturally occurring populations carried out in the Rio Negro region of central Amazonia (Trolle 2003) and within the Waimiri Atroari Indigenous Land (Ecology Brasil 2014) recorded the presence of at least 10 species of callitrichids and cebids: *Saguinus midas* (golden-handed tamarin), *Alouatta macconnelli* (red howler monkey), *Ateles paniscus* (Guianan black spider monkey), *Sapajus apella* (Guianan brown capuchin monkey), *Cebus olivaceus* (weeper capuchin), *Cebus albifrons* (white fronted capuchin), *Chiropotes satanas* (brown bearded saki monkey), *Pithecia pithecia* (Guianan saki monkey), and species of the genus *Aotus* (night monkey).

10.4 Waimiri Atroari Hunting Practices and the Importance of Primate Species as Food

The Waimiri Atroari Amerindians are agriculturalists that hunt primarily for food consumption (Souza-Mazurek et al. 2000). They are active "central place" foragers and rarely hunt for more than one day per trip; the exception to this is during ritual hunts, when hunting groups travel to specific areas and hunt and smoke slain animals for days or weeks to ensure enough food for large amounts of people participating in the ritual gathering. Most hunters use bows and arrows, although a small number have shotguns, regularly, and at least one man carries a gun per hunting group (Souza-Mazurek, pers. obs.). The group size of hunters is variable but averages around five people, and each hunter carries a weapon. Hunting events are mostly diurnal and consist of using trails from the village on foot or taking boats or trucks to reach more distant known hunting spots. Hunting groups walk along the trail looking for animal tracks and often imitate animal sounds, particularly black spider monkey and tapir (*Tapirus terrestris*) (Souza-Mazurek pers. obs.).

In the case of primates, once a troop is encountered in trees, hunters jockey to encircle them. In turn, each hunter then chooses an individual target in the troop and shoots upward, nearly vertically, from below. Individual hunters disperse to chase monkeys that have escaped in different directions after the initial group shooting attempt (Souza-Mazurek pers. obs.).

Monkeys are an important part of the diet of the Waimiri Atroari Amerindians. Three cebid species represented 24.5% of the total vertebrates (N = 5537 individuals) and 35% of all individual mammals hunted (N = 3841). Red howler monkey (n = 638) ranked first followed by Guianan black spider monkey (n = 615) and Guianan brown capuchin *monkey* (n = 99). Red howler monkey and Guianan black spider monkey are the largest monkey species occurring in the area (Fig. 10.2).

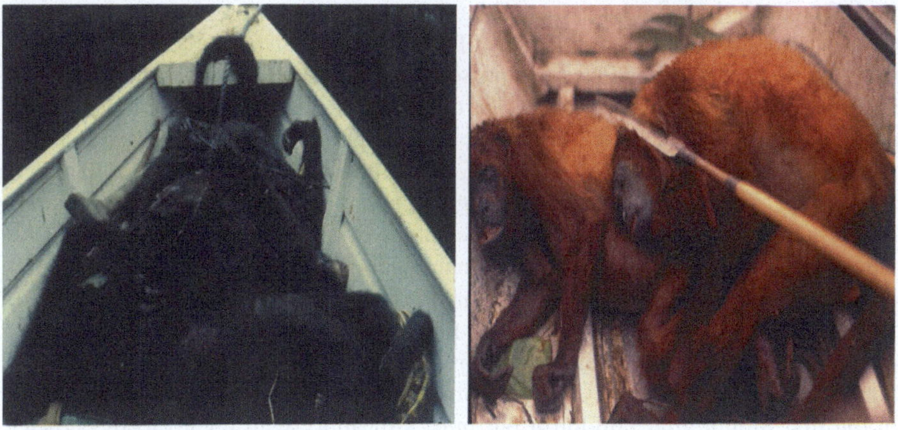

Fig. 10.2 Individuals of Guianan black spider monkey and red howler monkey killed during hunts by Waimiri Atroari

The three primate species together contributed with 8.9% of the total vertebrate biomass of 110 tons extracted over a 14-month period in the hunting catchment of 5000 km^2 (Souza-Mazurek 2001). More males than females of Guianan brown capuchin monkey and red howler monkey were hunted by the Waimiri Atroari. In the case of the spider monkey, the sex ratio of female biased in a proportion of 3:1.

10.4.1 Food Avoidance and Primate Medicinal Use

Taboos and temporal avoidance of certain primate species are related to the cosmology and cultural symbolism of the social order of a particular indigenous group. Certain species are not consumed because, culturally, they are not considered "human food," while others are not caught because they are believed to pose serious health and life threats to those who eat them because of their powerful spiritual strength. Additionally, other species are not eaten due to unwanted characteristics of the animals that can be transmitted to the person who consumes them or to his relatives. In the case of the Wamiri Atroari, these characteristics can include morphology and behavioral and/or feeding habits (Espinola 1995). Among the primate species not considered food by the Waimiri Atroari are golden-handed tamarin, common squirrel monkey, Guianan saki monkey, white fronted capuchin, and night monkey.

Some animal species including primates may be also avoided because they were *Kinja* (human beings) in mythical times before being transformed into animals by their own will or by the will of others. According to some myths recorded by Espinola (1995), in the past, it was prohibited to hunt Guianan black spider monkey because it was *Kinja* and considered a *yaska* or "relative" of the Waimiri Atroari. Those failing to follow the prohibition exposed themselves to the risk of early aging. Nevertheless, golden-handed tamarin which is a mythical figure and was also *Kinja* was often hunted for food together with other small-sized species.

In several cases, the consequences of eating forbidden animals affect the relatives of the person who violated this proscription more than the individual breaking the rule. Food interdictions are not permanent as people are more vulnerable in particular moments of their life cycles, such as pregnancy, primogeniture status, menarche, and during initiation rituals, when prescribed protocols are in order. Primate species are often avoided during pregnancy and the first years of the first-born children, particularly affecting the father of such children in the event he would break such rules. Likewise, Guianan black spider monkey cannot be hunted or eaten by men whose wives are pregnant, during the whole gestational period. This prohibition is extended over to the first month after childbirth, the same procedure is applied to the consumption of red howler monkey. In the case of second-born children, Guianan black spider monkey can already be eaten by the husband in the first month after childbirth but must still apply with regard to red howler monkeys and capuchins. If a howler monkey is consumed by the father, then the baby will grow a big belly or will have parasitic worms in the case of consuming capuchin monkeys.

In the case of first menses, food taboos for the young woman include Guianan brown capuchin monkeys along with other animals such as peacock bass, *pintado* (a cat fish species), redtail catfish, tapirs, tortoises, sting rays, red macaws, and also sugar cane (Espinola 1995). If the feeding rules are not respected, then the menstruation bleeding might never stop or the girl can become *Yaweri* (mentally disturbed and violent).

10.4.2 Primates in Myths and Rituals

The cultural perception of the world and its elements includes myths and relationships among human and nonhuman beings. According to Levi-Strauss and Eribon (1988: 193), "myths represent histories from a time when the differences between human and not human beings were not distinct." To the Waimiri Atroari, several animal species including the primates Guianan black spider monkey, golden-handed tamarin, and Guianan brown capuchin monkey were *kinja* (people) in the stories from the mythical times (*tahkome ikaa* – ancient stories), but were transformed voluntarily or as a punishment by a superior being into monkeys as a consequence of socially reprehensible attitudes. As in many Amerindian groups, the cultural universe of the Waimiri Atroari is also inhabited by human and nonhuman entities. Viveiros de Castro (2002a: 354) describes "an ontological intertwining of these beings that inhabit indigenous myths in that human and non-human aspects are entangled. Human beings are those who remained as humans, as such, animals are ex-humans and humans are ex-animals. It is possible to perceive that the indigenous cosmologies involve an ongoing engagement of alliances, relations, and disputes between human and non-humans."

The Waimiri Atroari myth *Wyie ika nenuwe many xiriki many* – history of the sun, the moon, and the stars – explains how the daily movement of the sun in the sky is perceived, and the origin of day and night, at a time when the golden-handed tamarin was still human, plays an essential part in showing these dynamics: "In ancient times there was no sunset and it was never dark. The Amerindians of that time would hunt all the time because they ate all the time because there was no night. The moon and the stars looked like the sun. When the sun was going to set the owner of the sun (*Mawa*, Waimiri Atroari mythical being) would send it back, to do the reverse path in the sky. One day the *Xiriki* (*Saguinus midas*), in a time when he was still a person, broke the sun. He wanted to find out what was inside the sun, because the sun was like a clock with a juice inside that made it work. Once the sun was broken it became dark and the hunters who were away hunting could not return to the village and stayed in the forest. *Xiriki* broke the Sun when the owner of the Sun was just arriving. 'You broke the sun! Why did you touch the sun?' asked the owner of the sun before throwing the little monkey away. *Xiriki* was a person before he broke the sun but the owner of the sun transformed him into a monkey with golden hands because he touched the juice of the sun" (*Wyie ikaa nenuwe many xiriki many* – history of the sun, the moon, and the stars told by Dauna Elzo in Fonseca and Fonseca, unpubl. Inf.)

After the sun had been broken by *Xiriki*, the narrative continues to describe the attempts by *Mawa*, the sun's owner, to fix the sun until reaching partial success. As only one side of the sun was fixed, *Mawa* made the moon and stars work on the other side, giving origin to the night, with alternate time periods. *Xiriki* was punished by *Mawa* to remind him of his inappropriate behavior.

10.4.3 Primates and the Maryba *Rituals*

One of the supports of the Waimiri Atroari culture is the group of *Maryba* rituals. According to Do Vale (2002), who studied the ritual, the word *Maryba* can be translated as feast, song, or dance. It is a moment when the communities suspend everyday existence and transport themselves to another time and space. They are important sociopolitical moments for the Waimiri Atroari as it is a time when people from all villages gather to establish and reaffirm alliances between themselves. There are three main *Maryba* rituals performed by the Waimiri Atroari: one associated with the deceased (*Iohy Maryba*), a second when a new communal hut is built (*Mydy Maryba*), and one for the male initiation rites (*Bahinja Maryba*) when boys are between ages of 3 and 5 years old. The *Bahinja Maryba* ritual lasts three days and two nights. During the three days of the *Bahinja Maryba*, the attributes, skills, power, behavior, and ecological aspects of animal species, cultivated and wild plants, and their byproducts are evoked, sang about, and performed through dances by the participants guided by the *Eremy* (a ritual singer). According to Do Vale (2002), the *Eremy* represents the links between the Waimiri Atroari with their spiritual world and mythical past. Through the songs and dances, the *Eremy* leads the Waimiri Atroari and establishes relationships involving the present and mythical times, uniting nature and culture. The dances and chanting are entwined forming a single whole implying that both song and choreography need to be known in order to be performed.

It is during the *Maryba* that the symbolic, cultural, and material worlds of the Waimiri Atroari come together. The dance sequences and instrument and music performances present all the symbolic food items (game species, fruits, and tubers) consumed during the ritual, connecting the parents of the initiated boys, the guests, and the nonhuman entities who are evoked throughout the three-day event. It is possible to perceive that indigenous cosmologies encompass the presence, alliances, and disputes between human and nonhuman beings. Viveiros de Castro (2002b) highlights that the perspectivism theory must be understood from a relational instead of relative perspective. In that sense, the real world of the different species depends on their own particular points of view since the world is composed by individual species. The *Maryba* rituals reflect the cultural symbolism of the everyday life of the Waimiri Atroari, during which time the boy being initiated acquires the knowledge about the wildlife, the forest plant species, the danger, the cultural roles, and obligations expressed by the chants. The human and social conditions for being a Waimiri Atroari person are linked to specific prescriptions such as eating banana porridge (considered human food or *Kinja ipa*), singing and dancing the *Maryba*, marrying only among themselves, accepting food taboos and avoidances, and speaking their own language (*Kinja Iara*) (Matarezio-Filho 2010).

Primates are among the several animal species evoked by the Waimiri Atroari by the *Eremy* during the *Bahinja Maryba* initiation ritual. The songs and behavior of the animal species are performed throughout the ritual, representing the way of being and the set of affection and abilities that characterized each of them as they are perceived as conceived by Viveiros de Castro (2002b). The jaguar (*Felis onca*) and the harpy eagle (*Harpia harpija*) are important figures in the ritual and are the first spirits evoked (Do Vale 2002). They are perceived only by the boys being initiated and the *Eremy*. The *Eremy* assumes and represents the behavior of the jaguar through singing and dynamic dancing performances. The Jaguar will protect the whole existence of the young hunter preventing him from feeling fear in the face of threats. The Guianan black spider monkey and Guianan brown capuchin are represented in the *Bahinja Maryba* songs among the fourteen mammals, seven birds, three reptiles, two fish, and two insect species. Some aspects of the ecology of black spider monkey are described in this chant: "... the song tells how the *kwata* (*Ateles paniscus*) eats, how it moves through the trees. The song tells how it eats in the thorny *tucumã* palm tree (*Astrocaryum aculeatum*), and eats *açaí* palm (*Euterpe precatoria*) leaves, which are monkeys' food. I learned that until the end of the *Maryba* monkey dance, when all people cross the hut eating from one side to the other imitating how the monkey eats" (Sekymy Pedrosa, pers. comm. to Do Vale 2002: 65).

The song of the *Meky* (*Sapajus apella*) describes the noisy and libidinous behavior of the primate species as perceived by Amerindians. It is performed prior and to counteract with the chant about the precepts of sexual morality of the Waimiri Atroari, having an important role in the ritual process of male initiation. During the *Meky Maryba*, the *Eremy* conducting the ritual requests that the boy's parents bring bracelets made of *karwa* fiber (*Ananas erectifolius*) and Guianan brown capuchin monkey teeth to be tied at the initiates' ankles. The anklets aim to protect the bodies of the boys during the *Behe* or "ritual whipping" that occurs at dawn of the third and final day of the male initiation ritual (Do Vale 2002).

10.5 Discussion

Throughout this chapter, we described some aspects of the ethnoprimatology of the Waimiri Atroari. Primate species, particularly cebids, represent an important source of food for the Waimiri Atroari, and they are hunted in larger numbers compared to other Carib speakers (Hames 1979; Milton 1991; Mittermeier 1991; Linke 2009) and other indigenous groups (Vickers 1991; Ouhoud-Renoux 1998; Cormier 2003; Prado et al. 2012).

The Waimiri Atroari focus most of their primate hunting on red howler monkey, Guianan black spider monkey, and Guianan brown capuchin monkey, and the first two are the largest species present in the study area. There is a preference for hunting the males of Guianan brown capuchin monkey and red howler monkey, and these returns do not correspond to the natural sex ratio found in unhunted population of these species (Rudran and Fernandez-Duque 2003; Carosi et al. 2005; Izawa 1980; Fragaszy et al. 2004; Queiroz 1995). The observed preference for red howler monkey and Guianan brown capuchin monkey males suggests that body size might play a role in hunters' choices among the primate species items that are considered "human food." Males are larger than females in both species (Crockett and Eisenberg 1987; Thorington et al. 1979; Souza-Mazurek unpubl. data). In the case of Guianan black spider monkey, hunting returns were biased toward females in a 3:1 ratio among hunted individuals, although the species lack body size sexual dimorphism (Souza-Mazurek 2001). Some of the observed patterns of use of primate species as food may have additional explanations besides hunting returns based on their body size or their natural availability. Females are targeted first when a group is found, and the Waimiri Atroari claim that females are considered "tastier" as they have more body fat than males. In spite of this slight preference, males are also frequently chased by the Waimiri Atroari (Souza-Mazurek pers. obs.), revealing that while females are preferred, prey males are not rejected. The Wayana and Aparai Amerindians from the same linguistic group as the Waimiri Atroari and the Matsigenka Amerindians of the Peruvian Amazon region also present female-biased sex ratio hunting returns for spider monkeys (Linke 2009; Da Silva et. al. 2005). In all these cases studied, the pattern is explained by cultural beliefs according to the social order of each group. The Matsigenka believe that certain monkeys (especially large adult males) and other game animals may pose serious health and life threats to those who eat them due to their powerful spiritual strength. Their vengeful spirits can "take revenge" on the hunter's family, causing illness among young children (Da Silva et al. 2005). The Wayana and Aparai avoid hunting male spider monkeys to prevent them from becoming bad hunters (Linke 2009). Nevertheless, female-biased sex ratios in Guianan black spider monkey are also documented in their natural population (van Roosmalen 1985; Symington 1987) which could also be reflected in hunting returns in the absence of strong, culturally established avoidances.

Primates fulfill other cultural roles and are not only seen as a source of food. Temporal avoidances of primate species are present for the three most consumed cebid species and are mostly related to periods of pregnancy and postpartum periods and affect the husband in the case of the Waimiri Atroari. The Tapirapé present similar avoidances concerning *A. macconnelli* (Wagley 1983) and among the Shipibo postpartum temporal restriction for *Cebus* for both parents (Behrens 1986). *A. paniscus* temporal avoidance is less common among other Amazonian indigenous groups, when compared to species of the genus *Alouatta*. The Tupi speaker Wapishana hunt eight species of monkeys, including howlers, but the prohibition applies only to spider monkeys (Henfrey 2002). Nevertheless, primates are also present in rituals and participate in important myths with multiple symbolic significances for the social and cultural order of the Waimiri Atroari.

The *Wye ika nenuwe many xiriki many* myth explains the origin of observed natural astronomical phenomena of the day and night among the Waimiri Atroari. It reminds them today that since ancient times, the appearance of stars in the celestial rain constellation guided them to prepare their agricultural fields, at a time when stone axes were used to fell trees and were sharpened with the teeth of spotted pacas (*Cuniculus paca*) and collared peccaries (*Tayassu tajacu*). In the same myth, the Amerindian is transformed into a monkey with golden hands (*Saguinus midas*) due to a culturally repugnant and reprehensible human attitude, with the ensuing punishment. The myth also provides elements that explain the presence of an evident phenotypical features of the golden-handed tamarin. Similarly, the Awá-Guajá share complex social relationships with primates including golden-handed tamarin (Cormier 2003).

Primate body parts are used in magical, and religious rituals make amulets to protect their users. Anklets made of teeth of *Sapajus* are tied around the ankle of infants Waimiri Atroari to protect their bodies during the whipping that is part of the male initiation rite. Hanson-Alp et al. (2003) reported similar uses of chimpanzee central incisors around the waist of infants in Sierra Leone as amulets to protect them and give them power over others in their cohort.

Species of the genus *Alouatta* are frequently present in the cosmology of several Amerindian groups of Amazonia (reviewed in Urbani and Cormier 2015). Among the Awá-Guajá, howler monkeys were once human beings who were transformed into monkeys so that other humans would be able to eat and survive. Howlers are said to be like humans because they "sing," which is intrinsically the way the Awá-Guajá travel into the spirit world (Cormier 2003). Among the Waimiri Atroari, red howler monkey symbolically appears as subject of temporal food avoidance. In the past, the Waimiri Atroari also avoided hunting the Guianan black spider monkey because it was once a human being and considered their relative, but this same symbolic role associated to the golden-handed tamarin did not seem to impose food restrictions. Conversely, in present days, Guianan spider monkeys and other medium- to large-sized primates are heavily hunted, while the golden tamarin is never found among hunted items (Souza-Mazurek et al. 2000, Souza-Mazurek 2001).

Some of the cultural patterns for the primate species described here are shared with other Amazonian groups (Shepard 2002, Da Silva et al. 2005). Nevertheless, there are more ethnozoological information available on subsistence hunting and its effects on primate populations than on other cultural aspects of the relationship among human–nonhuman primates for the Waimiri Atroari, as it is true for other indigenous people (see compilations in Urbani [2005] and Cormier [2006]). The impacts of the observed extraction levels and selective hunting pattern on the primate species population within the Waimiri Atroari territorial context is currently under analysis (Souza-Mazurek et al. unpubl. data). Additionally, another problem-directed research effort may reveal other relevant cultural aspects of the relationship between the Waimiri Atroari and the nonhuman primate species occurring in the shared region.

Acknowledgments We thank the Waimiri Atroari people for their support during Souza-Mazurek's field studies and data collection on subsistence hunting. The Waimiri Atroari Program provided logistical support in the indigenous area. Souza-Mazurek was supported by a PCI grant from the Brazilian National Research Council – CNPq. We thank Robert P. Miller and Louis Forline for their comments and for the English revision.

References

Bandeira A (2009) Jauapery. Edua, Manaus

Barbosa Rodrigues J (1885) Pacificação do Crichanás. Imprensa Nacional, Rio de Janeiro

Behrens CA (1986) Shipibo food categorization and preference: relationships between indigenous and Western dietary concepts. Am Anthropol 88:647–658

Bruno AC (2003) Waimiri-Atroari grammar: some phonological, morphological, and syntactic aspects. Ph.D. Dissertation, University of Arizona, Tucson

Bruno AC (2010) "How can I write my language?" Linguistic analysis and language revitalization: lessons from Waimiri Atroari syllable structure. Liames 10:85–99

Bruno AC (2014) (org) Vozes da Floresta: a arte de contar histórias – Histórias do Passado e do Cotidiano Indígena. Editora INPA, Manaus

Carosi M, Linn GS, Visalberghi E (2005) The sexual behavior and breeding system of tufted capuchin monkeys (*Cebus apella*). Adv Study Behav 35:105–149

Cormier LA (2003) Kinship with monkeys: the Guajá Foragers of Eastern Amazonia. Columbia University Press, New York

Cormier LA (2006) A preliminary review of Neotropical primates in subsistence and symbolism of indigenous lowland South American peoples. Ecol Environ Anthropol 2:14–32

Da Silva MNF, Shepard GH Jr, Yu DW (2005) Conservation implications of primates hunting practices among the Matsigenka of Manu National Park. Neotrop Primates 13(2):31–36

Do Vale MCR (2002). Waimiri-Atroari em festa é Maryba na floresta. Master thesis, Universidade do Amazonas, Amazonas

Crockett CM, Eisenberg JF (1987) Howlers: variations in group size and demography. In: Smuts BB, Cheney DL, Seyfarth RM, Wrangham RW, Strushsaker TT (eds) Primate societies. The University of Chicago Press, Chicago, pp 54–68

Endo W, Peres C, Salas E, Mori S, Sanchez-Vega JL, Shepard G et al (2009) Game vertebrate densities in hunted and non hunted forest sites in Manu National Park, Peru. Biotropica 42(2):251–261

Espinola CV (1995) O Sistema Médico Waimiri-Atroari: Concepções e Práticas, Master Thesis. PPGAS-UFSC, Florianópolis

Emídio-Silva C (1998) A caça de subsistência praticada pelos índios Parakanã (sudeste do Pará): características e sustentabilidade. Master Dissertation. Museu Paraense Emílio Goeldi & Empresa Brasileira de Pesquisa Agropecuária, Belém

Fragaszy D, Visalberghi E, Fedigan L (2004) The complete capuchin: the biology of the *genus Cebus*. Cambridge University Press, Cambridge

Jerozolimski A, Peres C (2003) Bringing home the biggest bacon: a cross-site analysis of the structure of hunter-kill profiles in Neotropical forests. Biol Conserv 111(3):415–425

Hames RB (1979) A comparison of the efficiencies of the shotgun and the bow in neotropical forest hunting. Hum Ecol 7:219–252

Hanson-Alp R, Bakarr MI, Lebbie A, Bangura KI (2003) Sierra Leone. In: Kormos R, Boesch C, Bakarr M, Butynski T (eds) West African chimpanzees: status survey and conservation action plan. IUCN/SSC Primate Specialist Group, Gland, pp 77–87

Henfrey TB (2002) Ethnoecology, resource use, conservation, and development in a Wapishana Community in South Rupununi, Guyana. Ph.D. dissertation. University of Kent, Canterbury

Izawa K (1980) Social behavior of the wild black capped capuchin *Cebus apella*. Primates 21(4):443–467

Lévi-Strauss C, Eribom D (1988) De perto e de longe: entrevista com Claude Lévi-Strauss. Nova Fronteira, Rio de Janeiro

Linke IHV (2009) Caracterização do uso da fauna cinegética em aldeias das etnias Wayana e Aparai na terra indígena Parque do Tumucumaque, Pará. Master Thesis. Museu Paraense Emílio Goeldi, Belém

Lizarralde M (2002) Ethnoecology of monkeys among the Barí of Venezuela: perception, use, and conservation. In: Fuentes A, Wolfe LD (eds) Primates face to face: the conservation implications of human-nonhuman primate interconnections. Cambridge University Press, Cambridge, pp 85–100

Matarezio-Filho ET (2010) Ritual e Pessoa entre os Waimiri Atroari. Master Thesis. University of São Paulo, São Paulo

Mena VP, Stallings JR, Regalado JB, Cueva RL (2000) The sustainability of current hunting practices by the Huaorani. In: Robinson J, Bennett E (eds) Hunting for sustainability in tropical forests. Columbia University Press, New York, pp 57–78

Miliken W, Miller RP, Pollard S, Wandeli EV (1992) Ethnobotany of the Waimiri Atroari Amerindians. Royal Botanic Gardens, Kew

Milton K (1981) Estimate of reproductive parameters for free ranging *Ateles geoffroyi*. Primates 22:574–579

Milton K (1991) Comparative aspects of diet in Amazonian forest-dwellers. Philos Trans R Soc Lond Ser B Biol Sci 334:253–263

Miller RP, Wandelli EV, Grenand P (1989) Conhecimento e utilização da floresta pelos índios Waimiri Atroari do Rio Camanaú, Amazonas (I). Acta Bot Bras 3(2) Supl 47–56

Mittermeier RA (1987) Effects of hunting on rain forest Primates. In: Marsh CW, Mittermeier RA (eds) Primate conservation in the tropical rain forest. Alan Liss Inc, New York, pp 109–146

Mittermeier RA (1991) Hunting and its effect on wild primates populations in Suriname. In: Robinson JG, Redford KH (eds) Neotropical wildlife use and conservation. University of Chicago Press, Chicago, pp 93–107

Monte PP (1992) Etno-história Waimiri Atroari (1883–1962). Master Dissertation, Pontifícia Universidade Católica de São Paulo, São Paulo

Ouhoud-Renoux F (1998) Se nourrir à trois sauts analyse diachronique de la prédation chez les wayâpi du haut-oyapock (guyane française). JATBA, Revue d'ethnobiologie 40(1–2):181–206

Peres C (1990) Effects of hunting on western Amazonian primate communities. Biol Conserv 54:47–59

Peres C (1991) Humbolt's woolly monkeys decimated by hunting in Amazonia. Oryx 25:89–95

Peres C (2000) Evaluating the impact and sustainability of sustainable hunting at multiple Amazonian forest sites. In: Robinson JG, Bennett EL (eds) Hunting for sustainability in tropical forests. Columbia University Press, New York, pp 31–56

Prado HM, Forline LC, Kipnis R (2012) Hunting practices among the Awá-Guajá: towards a long-term analysis of sustainability in an Amazonian indigenous community. Bol Mus Para Emílio Goeldi Cienc Hum 7(2):479–491

Queiroz HL (1995) Preguiças e guaribas: os mamíferos folívoros arborícolas do Mamirauá. Sociedade Civil Mamirauá/MCT/CNPq, Distrito Federal, Brasília

Queiroz HL, Kipnis K (1990) Os índios Awá-Guajá e os primatas da Amazônia maranhense: um caso de sustentabilidade de caça. In: Ferrari SF, Schneider H (eds) A Primatologia no Brasil. Universidade Federal do Pará, Belém, pp 81–94

Rudran R, Fernandez-Duque E (2003) Demographic changes over thirty years in a red howler population in Venezuela. Int J Primat 24:925–947

Shepard GHJ (2002) Primates in Matsigenka subsistence and worldview. In: Fuentes A, Wolfe LD (eds) Primates face to face: the conservation implications of human-nonhuman primate interconnections. Cambridge University Press, Cambridge, pp 101–133

Silva MF (1993) Romance de primos e primas: uma etnografia do parentesco Waimiri Atroari. Ph.D Thesis Social Anthropology. Universidade Federal do Rio de Janeiro/Museu Nacional, Rio de Janeiro

Souza-Mazurek RR, Pedrinho T, Feliciano X, Hilario W, Geroncio S, Marcelo E (2000) Subsistence hunting among the Waimiri Atroari Amerindians in Central Amazonia. Brazil Biodivers Conserv 5(5):579–596

Souza-Mazurek RR (2001) Kinja Txi Taka Nukwa Myrykwase: fishing and hunting among the Waimiri Atroari Amerindians from Central Amazonia. PhD Dissertation. University of Illinois at Chicago, Chicago

Symington MM (1987) Sex ratio and maternal rank in wild spider monkeys: when daughters disperse. Behav Ecol and Sociobiol 20:421–425

Thorington-Jr RW, Rundran R, Mack D (1979) Sexual dimorphism in *Alouatta macconnelli* and observations in capture techniques. In: Eisenberg JF (ed) Vertebrate ecology in the northern Neotropics. Smithsonian Institution Press, Washington, pp 97–106

Trolle M (2003) Mammal survey in the Jauaperi river region, rio Negro Basin, the Amazon, Brazil. Mamalia, t. 67. n° 1, 75–83

Urbani BL (2005) The targeted monkey: a re-evaluation of predation on New World primates. J Anthropol Sci 83:89–109

Urbani BL, Cormier LA (2015) The ethnoprimatology of the howler monkeys (*Alouatta* spp.): from past to present. In: Kowalewski MM et al (eds) Howler monkeys, developments in primatology: progress and prospects. Springer Science + Business Media, New York

van Roosmalen MGM (1985) Habitat preferences, diet, feeding strategy and social organization of the black spider monkey *(Ateles paniscus paniscus* Linnaeus 1758) in Surinam. Acta Amaz 15:1–238

Vickers W (1991) Hunting yields and game composition over ten years in an Amazonian Indian territory. In: Robinson JG, Redford KH (eds) Neotropical wildlife use and conservation. University of Chicago, Chicago, pp 53–81

Viveiros de Castro E (2002a) Imagens da Natureza e da Sociedade. In: Viveiros de Castro E (ed) A Inconstância da alma selvagem e outros ensaios de Antropologia. Cosac&Naify, São Paulo, pp 310–344

Viveiros de Castro E (2002b) Perspectivismo e Multinaturalismo na América Indígena. In: Viveiros de Castro E (ed) A Inconstância da alma selvagem e outros ensaios de Antropologia. Cosac&Naify, São Paulo, pp 225–254

Wagley C (1983) Welcome of tears, the Tapirapé Indians of Central Brazil. Prospect. Waveland Press, Heights

Chapter 11
Linguistic, Cultural, and Environmental Aspects of Ethnoprimatological Knowledge Among the Lokono, Kari'na, and Warao of the Moruca River (Guyana)

Konrad Rybka

11.1 Introduction

South American indigenous people possess extensive knowledge of the environment they inhabit (e.g., Voss and Fleck 2011, Posey 2002). This knowledge translates into a plethora of practices, involving subsistence, medicine, arts, crafts, and landscape management, which speak to the deep understanding of the local biotic and abiotic resources (Anderson et al. 2011). The ethnobiological vocabularies of indigenous languages, a reserve of the speakers' environmental knowledge, astound in turn with the number of terms and the diversity of principles according to which they are organized (e.g., Fleck and Harder 2000; Berlin 1992; Hunn 1982, 1976). Even more confounding is the mosaic of linguistic systems that crystalized in South America, fragments of which Amazonian scientists try to piece together (e.g., Eriksen 2011; Hornborg and Hill 2011).

The Moruca River in northwestern Guyana is one piece of this puzzle. The area, inhabited by linguistically unrelated peoples, the Lokono, Kari'na, and Warao, is an ideal setting for studying environmental adaptation and cultural contact among indigenous populations. It is from these two angles that ethnoprimatological knowledge and the vocabularies concomitant with it are analyzed here. The chapter offers the first ethnoprimatological account of northwestern Guyana, an understudied area (Lehman 2004: 90), contributing to research on human–primate interactions in South America (e.g., Mere Roncal et al. 2018; Urbani and Cormier 2015; Cormier 2006; Lizarralde 2002; Shepard 2002). At the same time, it is a domain-focused study of contact, a topic of import to Amazonian linguistics (e.g., Epps and Michael 2015; Aikhenvald 2010). Methodologically, the chapter documents terms for non-human primates in the three languages and practices associated with the species.

K. Rybka (✉)
Berkeley Linguistics Department, University of California at Berkeley, Berkeley, CA, USA
e-mail: konrad.rybka@berkeley.edu

B. Urbani, M. Lizarralde (eds.), *Neotropical Ethnoprimatology*, Ethnobiology,
https://doi.org/10.1007/978-3-030-27504-4_11

It then compares them with those known for the dialects of the same languages spoken in areas with different primate ecologies. Such contrastive distributions are determined in turn by the presence of keystone palm resources and wide rivers dissecting the Guianas that prevent species from spreading (Lehman 2004). Mapping the terms, practices, and species allows us to observe how language and culture adapt to the environmental and social niches.

The results show that the different dialects of the three languages retained, borrowed, and dropped terms for primates, or even changed their meanings, fine-tuning their lexical resources to the local environment. The results also unravel intricate systems of ethnoprimatological knowledge with noticeable cases of cultural convergence, some of which are, more strongly put, likely cases of cultural borrowing. One example of such cultural borrowing is the consumption of primates, a taboo among the Warao in Venezuela but a norm among the Warao in contact with the Lokono and Kari'na in Guyana who know no such restrictions. Particularly interesting is the shared knowledge surrounding the Guianan red howler monkey (*Alouatta macconnelli*), encompassing numerous intertwined domains, such as medicine, art, oral tradition, weather forecasting, and timekeeping. Here too cultural borrowing can be identified, for instance, in the spread of Venezuelan Warao tradition of making drums from the skin of the howler that spread to the nearby Lokono and Kari'na in Guyana, contrasting with the practices of the Lokono and Kari'na further east who prefer the skin of other animals. However, the linguistic and cultural borrowings do not align. The observed lexical borrowings are cases of classic borrowing motivated by the need to name unknown referents, independent of the borrowing of cultural practices related to the animals. The chapter ends with an evaluation of the results with the view to using the methods developed here to reconstruct ethnoprimatological proto-vocabularies and identify the areas where Amazonian protolanguages were spoken, thus advancing Amazonian historical linguistics.

11.1.1 Languages and Participants

The lower Moruca River in northwestern Guyana is dominated by seasonally flooded savannah and swamp forest, bordered by mangrove forests along the Atlantic coast and lowland forest to the south. The data for this chapter were collected during 2 months of fieldwork in 2017 in three communities, Santa Rosa, Manawarin, and Waramuri, spread across many "islands" on the seasonally inundated savannah. The communities are inhabited by speakers of unrelated languages: Lokono, Kari'na, and Warao, respectively. Lokono and Kari'na are spoken throughout the pericoastal Guyana, Suriname, French Guiana, and Venezuela (Rybka 2015; Courtz 2008) and belong to the Arawakan and Cariban language family, respectively. Warao is a *language isolate*, a language without known relatives, spoken predominantly in the Orinoco delta (Romero-Figueroa 1997). Figure 11.1 maps the location of the three Moruca communities in Guyana and the other dialects of the three languages compared in this chapter.

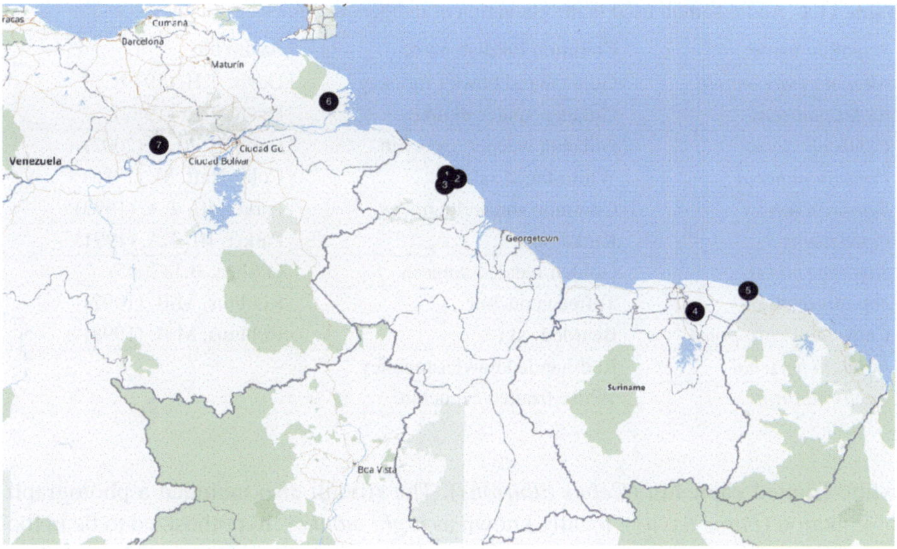

Fig. 11.1 Compared dialects of Lokono, Kari'na, and Warao: Moruca: Lokono, Santa Rosa (1); Warao, Waramuri (2); Kari'na, Manawarin (3); Suriname: Lokono, Cassipora (4); Kari'na, Galibi (5); Venezuela: Warao, Winikina (6); Kari'na, San José de Guanipa Municipality (7)

The indigenous languages of the Moruca are endangered. There are only a handful of Lokono speakers in Santa Rosa, a situation similar to Lokono settlements in Suriname (Rybka 2015). In Manawarin, only the eldest generation speaks Kari'na. The language is, however, still used by all generations in a few villages in Suriname and Guyana (Courtz 2008: 8). Warao is a vital language in the Orinoco delta (Romero-Figueroa 1997), but its Guyanese variety is on the brink of extinction, with a dozen speakers remaining in Waramuri. In addition to the indigenous languages, the inhabitants of Moruca speak Guyanese Creole English and, often less fluently, (Guyanese) English, the official language of Guyana. Language endangerment parallels the loss of traditional cultural practices, as reflected in the quantitative differences between the Moruca communities, where the Kari'na, speakers of the least endangered language and least affected by contact with outsiders, appear to have preserved more traditional practices involving primates.

11.2 Methods

To understand the relationships between primate terms, biogeography of primates, and the circulation of cultural practices associated with them, a three-step analysis was carried out. First, primates known to the speakers were identified using laminate photographs of all Guyanese primates and two (distractor) photographs of species not found in Guyana: red-handed howler monkey (*Alouatta belzebul*) and

Table 11.1 Audio stimuli used in the elicitation of indigenous names of primates

Scientific name	Common English name	Audio file
Alouatta macconnelli	Guianan red howler monkey	Davis, T.H. (1979)
Ateles paniscus	Guianan spider monkey	Davis, T.H. (1982)
Cebus olivaceus	Guianan weeper capuchin	Robbins, M.B. (2002)
Pithecia pithecia	White-faced saki	Cohn-Haft, M. (1988)
Saimiri sciureus	Common squirrel monkey	Parker, III, T.A. (1993)
Potos flavus	Kinkajou	Parker, III, T.A. (1991)
Saguinus midas	Golden-handed tamarin	O'Shea, B.J (2005)
Sapajus apella	Tufted capuchin	Robbins, M.B. (1997)
Chiropotes chiropotes	Bearded saki	Robbins, M.B. (1998)
Alouatta belzebul	Red-handed howler monkey	–
Cebus albifrons	White-fronted capuchin	–

white-fronted capuchin (*Cebus albifrons*). The stimuli also included a photograph of kinkajou (*Potos flavus*), locally known as *night monkey*, hypothesized to be in the same taxonomic category as primates in the indigenous classification. The speakers were also asked to identify the vocalizations of the animals from the Macaulay Library (Table.11.1). In each community, three men and three women participated in the interviews, except for the Lokono where only five consultants were found. The participants were selected by the community councils for their knowledge of the languages.

To identify dialectal differences, the terms collected on the Moruca were compared with those from the dialects of the same languages spoken in other parts of the Guianas: Surinamese Lokono from Cassipora (author's data), Surinamese Kari'na from Galibi (Courtz 2008), Venezuelan Kari'na from the San José de Guanipa Municipality (Mosonyi 2002; Linares 1998), and Venezuelan Warao from Winikina (Barral 1979; Fig. 11.1). The names were then compared with those used in related languages to identify words borrowed from neighboring languages (*lexical borrowings*), terms coined with the language's own resources (*lexical innovations*), and likely retentions from the protolanguages (*cognate candidates*). As opposed to *true cognates*, whose relatedness is demonstrated by regular sound changes, *cognate candidates* are formally and semantically similar terms that await such evidence. Given the paucity of diachronic linguistic research in Amazonia, it is at this stage best to speak of *cognate candidates*. The comparative sample included 30 Arawakan, 10 Cariban, and 10 Tupian languages (which though spoken further east may have been the source of several terms borrowed into the languages discussed here). For limitations of space, only one cognate candidate is given for each term. The data come from Apalaí (Camargo 2002), Bahuana (Ramirez 1992), Carijona (Robayo 1996), Manao (Goeje 1948), Mawayana (Coretta 2013), Pemón (Armellada and Salazar 1981), Piapoco, Yucuna (Kondo 1983), Tariana (Aikhenvald et al. 2001), Taruma (Farabee 1918), Trio (Amazon Conservation Team Suriname 2018), Wapishana (Henfrey 2002), Wayãpi (Grenand 1989), Yumana, and Maragua (Ramirez 2001).

Finally, open-ended interviews about primates were conducted on the Moruca. The results were compared with ethnographic records from Suriname, Venezuela, and Guyana (Olsen 1996, Barral 1979, Heinen 1973, Abbenhuis 1939, Roth 1924, 1915, Penard and Penard 1907). The observed variation in terms and practices was then mapped against the distribution of primates in the respective areas to determine whether the vocabularies are attuned to the biogeography of the species or mediated by the borrowing of cultural practices. The distribution maps are based on data from the International Union for Conservation of Nature, elaborated with more detailed Guianan sources (e.g., Lehman 2004; Linares 1998).

11.3 Results: Terms for Nonhuman Primates

The speakers easily recognized the familiar species from the laminates and vocalizations. The species known to the speakers were also identified in situ. The speakers did not recognize the kinkajou from the picture, as they were only familiar with its vocalization, but considered it an animal belonging to the same taxon in the indigenous classification, similarly to other indigenous people (e.g., Barí, Lizarralde 2002). The speakers were familiar with the Guianan spider monkey even though they reported it was not found in the area. For this reason, the species was excluded from the subsequent interviews. So were the red-handed howler, white-fronted capuchin, tufted capuchin, bearded saki, and golden-handed tamarin, which were unknown to the consultants. The vocabularies of Moruca Lokono, Kari'na, and Warao can be compared with those of the other dialects (Table 11.2).

Table 11.2 Primate terms in the Moruca (MO), Surinamese (SU), and Venezuelan (VN) dialects of Lokono, Kari'na, and Warao. Species: *Alouatta macconnelli* (1), *Cebus olivaceus* (2), *Potos flavus* (3), *Pithecia pithecia* (4), *Ateles paniscus* (5), *Saimiri sciureus* (6), *Saguinus midas* (7), *Sapajus apella* (8), *Chiropotes chiropotes* (9)

	Lokono (MO)	Kari'na (MO)	Warao (MO)	Lokono (SU)	Kari'na (SU)	Kari'na (VN)	Warao (VN)
1	*itorhi*	*arawata*	*wai*	*hitorhi*	*arawata*	*arawata*	*wai*
2	*howa*[a]	*yarakaru*[a]	*neku*	*howa*	*iwarakaru*	*iwarakaru*[a]	*neku*[a]
3	*wisowiso*	*kushinkushin*	?	*wisowiso*	*kupara*	?	*koraikorai*
4	*horhwe*	*ariki*	*horowe*	*horhwe*	*ariki*	–	*horobe*
5	*kwata*[b]	*kwata*[a, b]	*kuata*[b]	*adafe*	*kwata*	–	–
6	*kabwanama*	*karimia*	*kabuanama*[a]	*kabwanama*	*akarima*	–	–
7	–	–	–	*sûtu*	*kusiri*	–	–
8	–	–	–	*fodi*[a]	*meku*[a]	–	–
9	–	–	–	*bisa*	*kusiu*	*kusiu*	–

[a]Used also as a general category term
[b]Known by name but not found in the area
? Found in the area but not known by name or no term in the literature

The observed dialectal variation reflects the biogeography of the species; the names match the distributions expected for the different parts of the Guianas. First, three species, Guianan red howler monkey, Guianan weeper capuchin, and kinkajou, are found throughout the Guianas, and expectedly there is little variation in their names across the dialects. The Guianan red howler monkey (*A. macconnelli*, Fig. 11.2), known in Lokono dialects as *(h)itorhi*, has cognate candidates in Arawakan languages, fine-tuned to the locally available howler species (e.g., Maragua *ytury* "*A. seniculus*"). Similarly, Kari'na *arawata*, attested in the three dialects, is common in the Cariban family, typically referring to the Guianan red howler monkey (e.g., Apalaí *arrata*), but occasionally to other howler species (e.g., Carijona *arawata* "*A. seniculus*"). It is beyond the scope of this chapter to determine which species was the original referent of the Lokono and Kari'na terms. Finally, Warao *wai* is used in both Warao territories.

Similarly, Lokono *howa* "Guianan weeper capuchin" (*C. olivaceus*, Fig. 11.3), attested in both dialects, has cognate candidates in other Arawakan languages that sometimes refer to closely-related species (e.g., Tariana *halo* "*C. albifrons*"); which species was its original referent remains unclear. Kari'na *yarakaru*, stable across Kari'na dialects, is likely a retention as well that consistently refers to Guianan weeper capuchin (e.g., Pemón *iwarka*). Warao *neku* is most likely a native Warao term, less likely a borrowing from Kari'na *meku* "*S. apella*" that underwent a semantic shift to *C. olivaceus*. All in all, terms for the Guianan red howler monkey and Guianan weeper capuchin, species present in all seven locales, appear to be

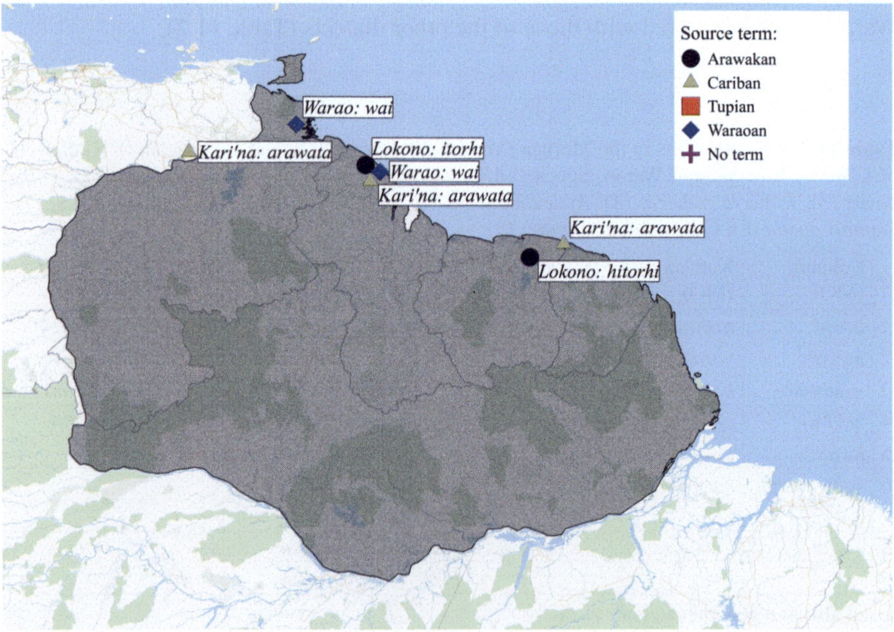

Fig. 11.2 Terms for *A. macconnelli* and the species distribution based on IUCN (2008a)

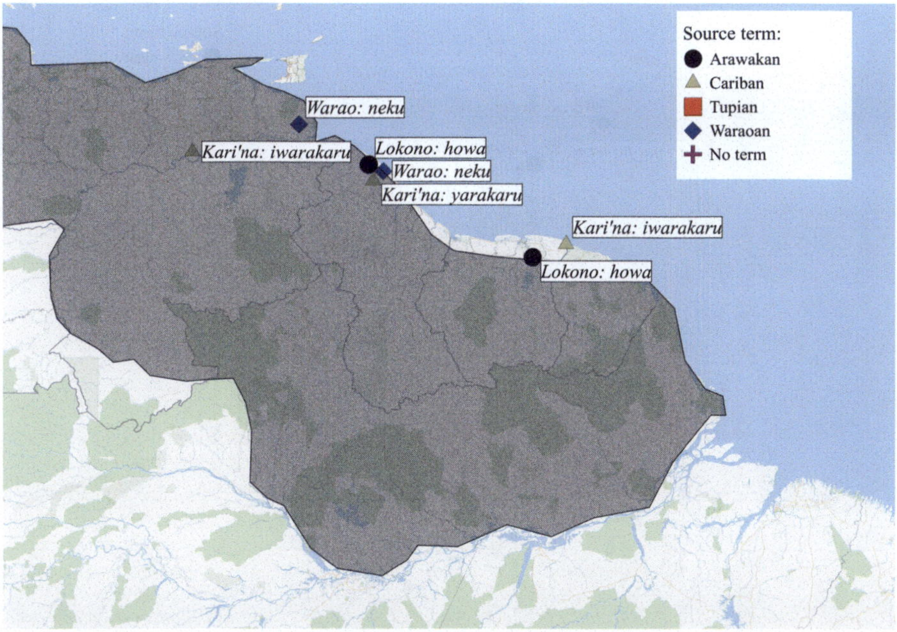

Fig. 11.3 Terms for *C. olivaceus* and the species distribution based on IUCN (Currently the definition of the species is being reassessed by IUCN, and neither the old nor the new entry are available at the IUCN portal; the distribution is based on the old entry, still reproduced on Wikipedia)

retentions that might have merely undergone semantic fine-tuning to the local population of the *Alouatta* genus and the Cebidae family within the larger Arawakan and Cariban language families.

The terms for kinkajou (*P. flavus*, Fig. 11.4), often reduplicated, are likely onomatopoeic; notably, the animal was easily recognized by its sound. Lokono *wisowiso* is found in both dialects but not in other Arawakan languages, which makes it a likely innovation. Warao *koraikorai* was documented only in Venezuela, while its name in Moruca Warao and Venezuelan Kari'na is unknown. However, two Kari'na terms were found in other dialects. Moruca Kari'na *kushinkushin* is a likely Cariban retention (e.g., Apalaí *kuxikuxi*), also used in Venezuelan Spanish (Linares 1998). Surinamese Kari'na *kupara*, on the other hand, is a possible borrowing from Wayãpi *yupala*, a language with which Kari'na was in contact at an earlier stage (Meira and Muysken 2017). The nocturnal nature of the animal and language attrition may have contributed to its low linguistic salience, resulting in the loss of the name in Moruca Warao and Venezuelan Kari'na and its reinvention in Lokono through coinage and in Surinamese Kari'na through borrowing.

The linguistic picture is different for the species with restricted distributions such as the white-faced saki (*P. pithecia,* Fig. 11.5). Lokono *horhwe* is likely Arawakan, but it is not common in the family, suggesting it is an innovation, perhaps exchanged with the Wapishana (Arawakan), who call it *oroa*. The term was borrowed as *horobe* into

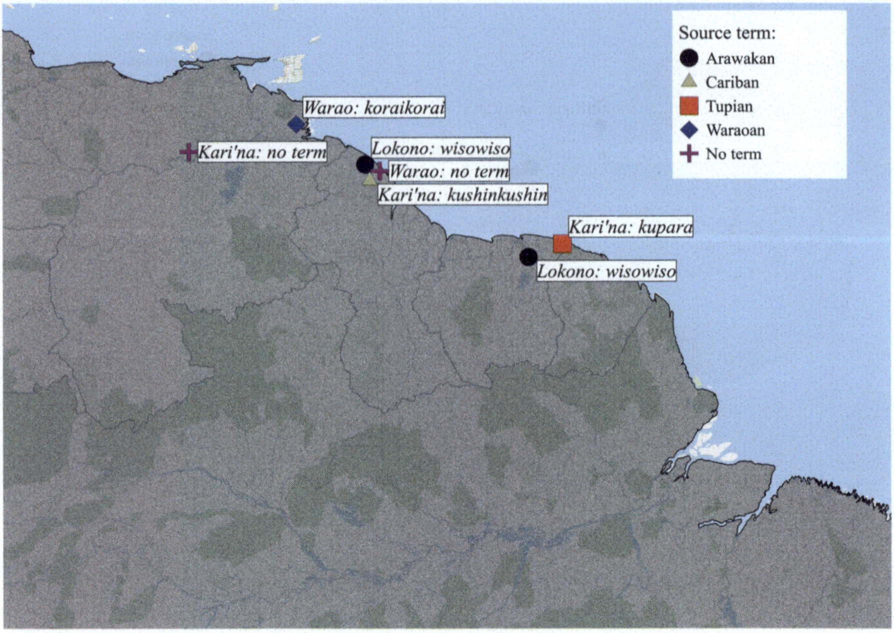

Fig. 11.4 Terms for *P. flavus* and the species distribution based on IUCN (2015)

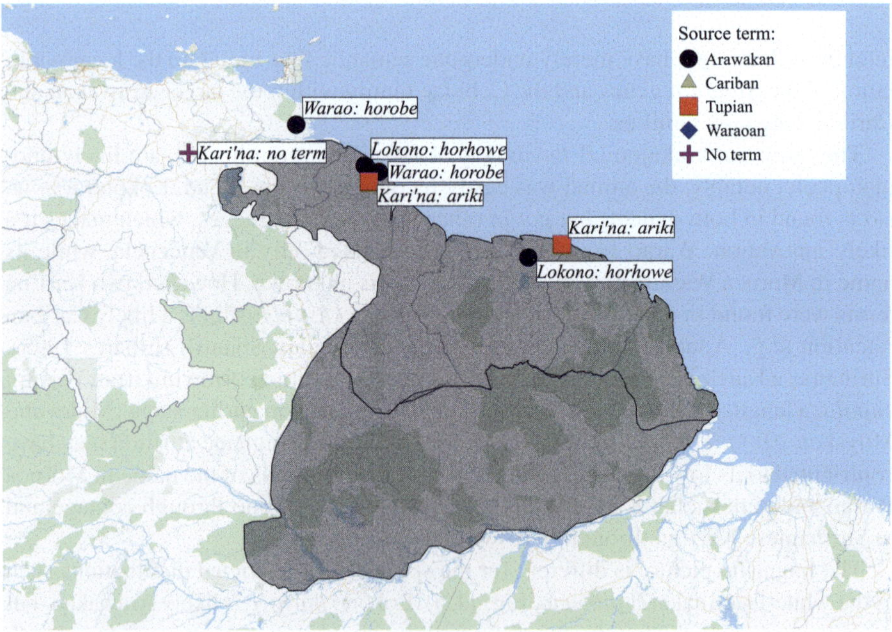

Fig. 11.5 Terms for *P. pithecia* and the species distribution based on IUCN (2008f)

Moruca Warao, spoken within the distribution of the species, and into Venezuelan Warao, spoken just outside of it. Venezuelan Kari'na, also outside its distribution, does not have a term for the animal, while Moruca and Surinamese Kari'na *ariki* are a likely borrowing from Wayãpi *yaliki*, which likely replaced *wanuku*, an older term documented for Kari'na and Island Carib (Courtz 2008; Breton and Besada Paisa 1999). It is noteworthy that several other Cariban languages likely borrowed the species' name from Wayãpi (e.g., Macushi, Trio, Wayana), suggesting their ancestors may have come from an area where it was unknown.

Surinamese Lokono *adafe* "Guianan spider monkey" (*A. paniscus*, Fig. 11.6) is transparent (*ada–fe* "tree–garbage"), suggesting it is an innovation. Surinamese Kari'na *kwata* is found in Arawakan (e.g., Yumana *kuwatá*), Cariban (e.g., Pemón *kwata*), and Tupian languages (e.g., Wayãpi *kwata*). The term may be of Tupian origin, but its spread is difficult to trace back as it was borrowed into local *lingua francas*, which may have dispersed it relatively recently (e.g., Guyanese Creole English, Brazilian Portuguese). *Kwata* is also used by the Moruca Lokono, Warao, and Kari'na even though the animal is not found in area. Moruca *kwata* may come from the creole, used within and outside the community, whose vocabulary reflects the distribution of primates in a larger area. An alternative interpretation is that the range of Guianan spider monkey was once larger and that the indigenous terms on the Moruca are the only markers of this past distribution. The distribution of the species by IUCN (2008b) does in fact include a second zone including the Moruca,

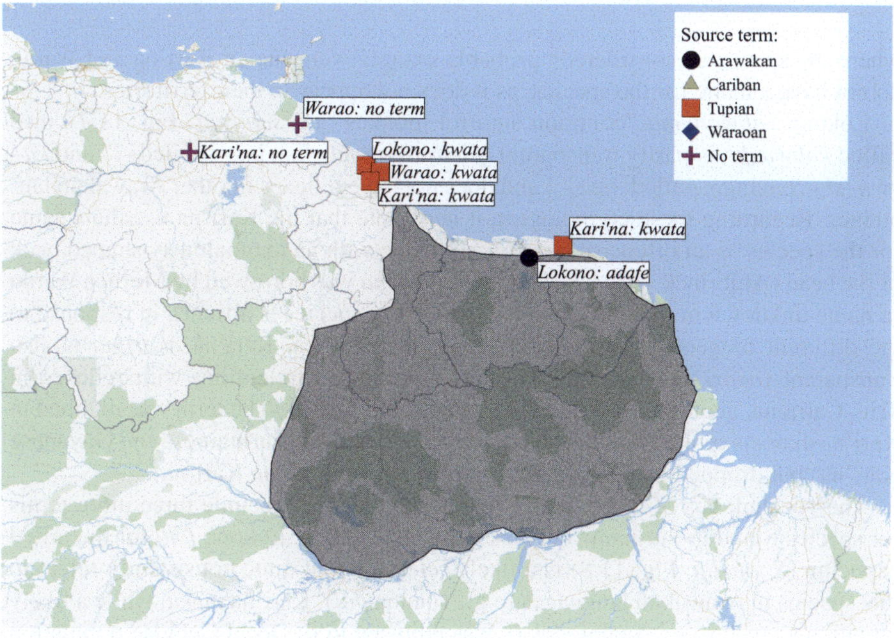

Fig. 11.6 Terms for *A. paniscus* and the species distribution based on IUCN (2008b), including only the area where the species is "extant" in keeping with Lehman et al. (2006)

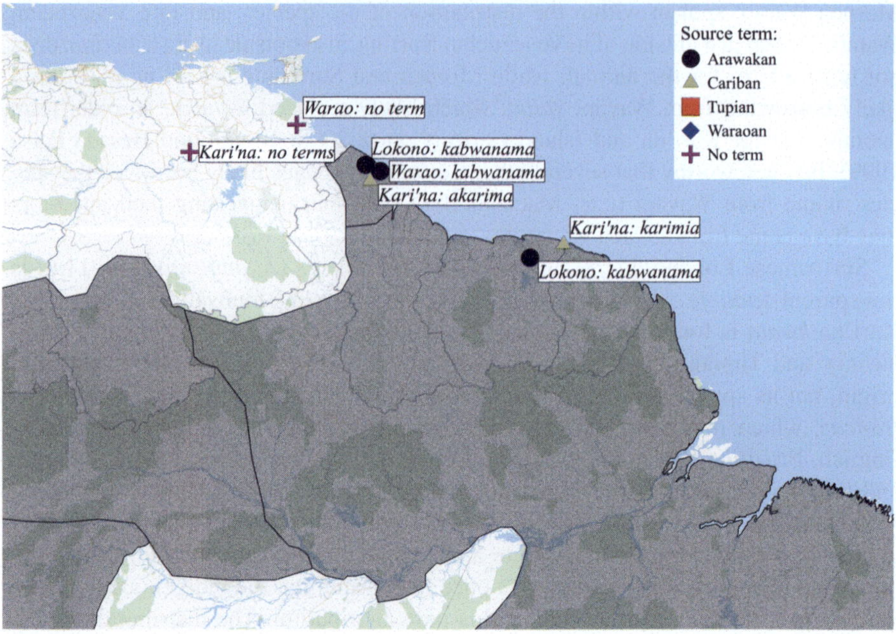

Fig. 11.7 Terms for *S. sciureus* and the species distribution based on IUCN (2008e)

where the species is considered "probably extant." Venezuelan Kari'na and Warao do not have a name for the species, as it does not appear in their territories.

Lokono *kabwanama* "common squirrel monkey" (*S. sciureus*, Fig. 11.7), also called *kabwashi*, is partly transparent (*ka–bwa–nama* "having–spoiled–?" and *ka–bwa–shi* "having–spoiled–head") and does not have reflexes in other Arawakan languages. Regarding these etymologies, it is of note that the Kari'na use their name for the species as an offensive term for someone with an anomalously shaped back of the head (Ahlbrinck 1931). Lokono *kabwanama* was borrowed by Moruca Warao (a name unknown in the Orinoco delta where the species is absent), in response to the different biogeography of primates on the Moruca. Kari'na *akarima* is also transparent (*akari–ma* "squirrel–big") and possibly a retention, with reflexes in Trio, Carijona, and Macushi, referring to the same species. The term was retained in Kari'na dialects spoken in areas where the species appear (Surinamese and Guyanese Kari'na) but dropped in areas where it is absent (Venezuelan Kari'na).

The vocabularies of Surinamese Lokono and Kari'na include three more terms for species found in Suriname but absent on the Moruca. Lokono *fodi* for the tufted capuchin (*S. apella*, Fig. 11.8) is a likely retention that quite consistently refers to this species in Arawakan languages (e.g., Yucuna *poí*). Kari'na *meku*, while a likely retention as well (e.g., Apalaí *meku*), was adjusted to the local Cebidae population (e.g., Carijona *meku* "*Cebus albifrons*"); it remains unclear which species was its original referent. Given that the terms are likely retentions, the Moruca Lokono and

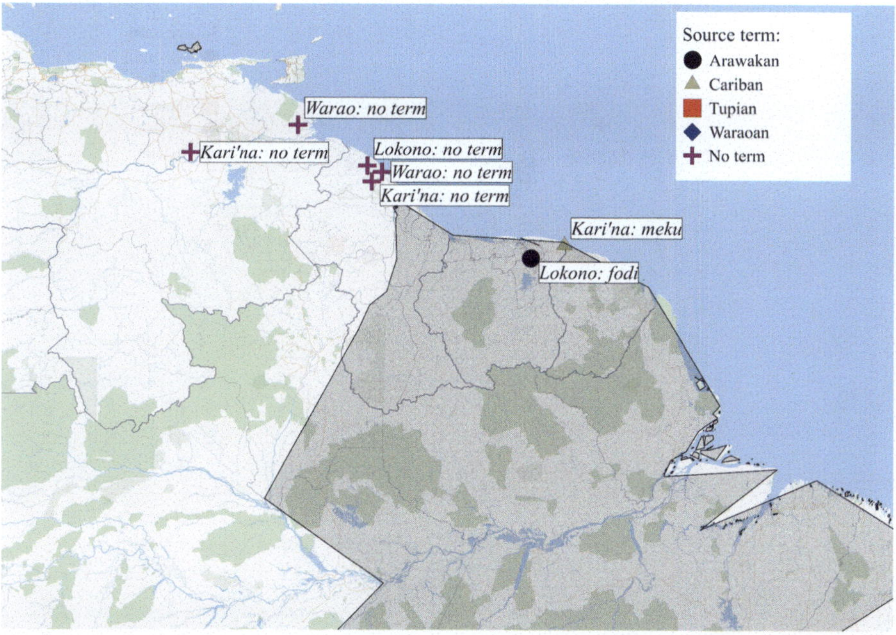

Fig. 11.8 Terms for *S. apella* and the species distribution based on IUCN (2008g) and Lehman et al. (2006)

Kari'na must have dropped them for lack of the referent. Logically, none of the dialects of Warao has a term for the species.

Surinamese Lokono *sûtu* (also *sururu*) for the golden-handed tamarin (*S. midas*, Fig. 11.9) is a possible retention (e.g., Piapoco *síiré*). On the one hand, the Arawakan term was, however, adjusted to the local conditions, as Piapoco *síiré* refers to the common squirrel monkey, which the Lokono call with partially transparent innovations. Kari'na *kusiri*, on the other hand, is a possible borrowing from Tupian languages, where the term is transparent (e.g., Wayãpi *kusili*, *kusi* "brown" and *–li* "small"). The Moruca and Venezuelan dialects of Lokono, Kari'na, and Warao, spoken outside the range of the species, do not have a term for the animal.

Finally, Surinamese Lokono *bisa* "bearded saki" (*Ch. chiropotes*, Fig. 11.10) is a possible Arawakan retention in closely related Wapishana (*wisa*), Bahuana (*wica*), and Manao (*huitcha*), which could, however, have been borrowed from Wapishana by the more distantly related Lokono, who used to be in contact with the Wapishana (Eriksen 2011). Taruma (isolate) *hisai* "*Ch. Chiropotes*" (Farabee 1918) could have been borrowed from the nearby Wapishana, Bahuana, or Manao or alternatively be the ultimate source of the borrowing into these languages. On the Moruca, for lack of the referent, the name was either dropped or never borrowed. The Moruca and Venezuelan Warao, living outside its range, do not have a term for it, but the term was borrowed by Surinamese Warao (Staffeleu 1975). Surinamese Kari'na *kusiu* is in turn a borrowing of Wayãpi *kusiu*, transparent in Wayãpi (*kusi* "brown" and *–u*

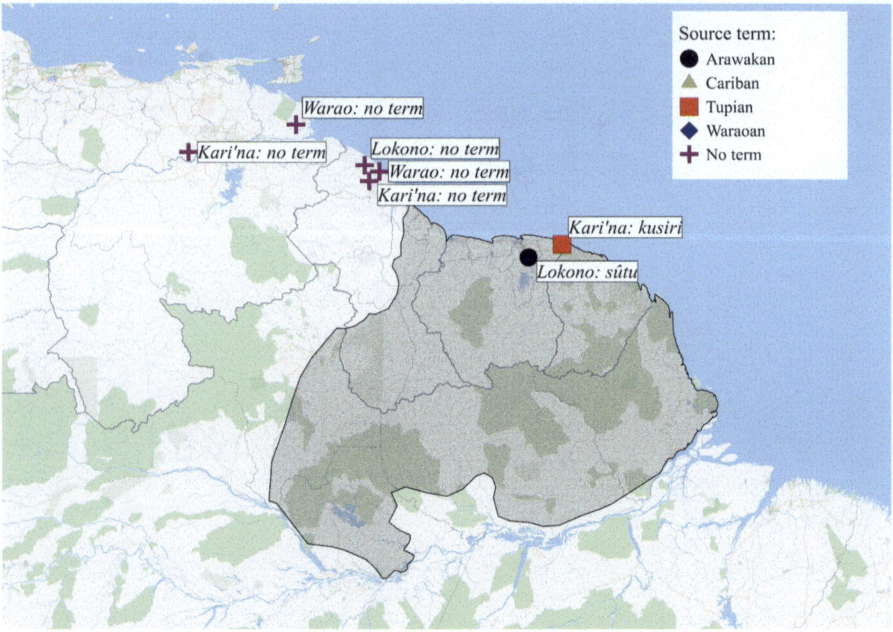

Fig. 11.9 Terms for *S. midas* and the species distribution based on IUCN (2008d)

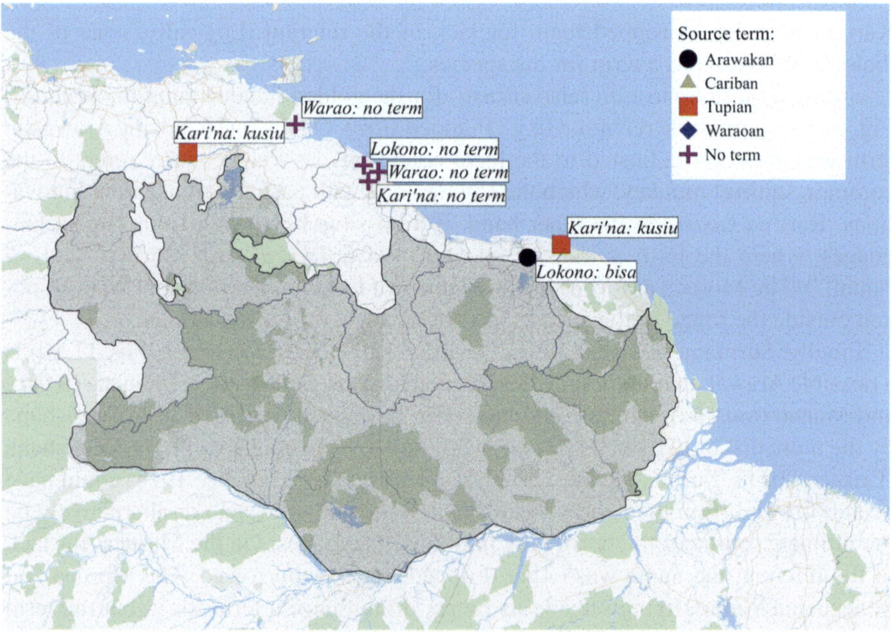

Fig. 11.10 Terms for *Ch. chiropotes* and the species distribution based on IUCN (2008c)

"big") and attested in other Tupian languages. Though absent on the Moruca, *kusiu* appears in Venezuelan Kari'na, spoken at the edge of the saki's area. *Nota bene*, from Tupian languages, the term was likely borrowed into Brazilian Portuguese as *cuxiú*, referring to different *Chiropotes* species (e.g., IUCN 2008h). Since Tupian languages are widespread south of the Amazon where the bearded saki is absent and from where the Wayãpi migrated north (Rose et al. 2012), Wayãpi *kusiu* must have first been extended to the bearded saki encountered north of the Amazon.

The vocabulary differences between the dialects are also conspicuous on the level of general terms. The three languages do not have an equivalent of the hypernyms *monkey* or *primate*, a distinct term that the species names are in a type-of relationship with. Instead, the languages employ for this purpose one of the species terms. On the Moruca, the term for the common squirrel monkey (Lokono and Warao) or the Guianan weeper capuchin (Kari'na) is used. In Moruca Lokono, one could say therefore that *howa* is a type of *kabwanama* since the latter term has a secondary hypernymic meaning. In Suriname, the term for the tufted capuchin (Lokono and Kari'na) and in Venezuela for Guianan weeper capuchin (Warao, and possibly Kari'na) serve as hypernyms. The differences reflect perceived species density. The consultants on the Moruca name the common squirrel monkey and Guianan weeper capuchin as the most common species, while in Suriname this place is given to the tufted capuchin by the Lokono. In the Orinoco delta, two species of primates are common: Guianan red howler monkey and Guianan weeper capuchin. As explained below, however, the former is culturally highly marked, making the latter a more natural choice for a general term.

11.4 Results: Cultural Practices

The practices associated with primates on the Moruca are summarized in Table 11.3, except those practices involving Guianan red howler monkey discussed separately (Sect. 11.4.1 – 11.4.12). The four species are important today mainly as pets, pests, source of food, medicine, leather, and a commodity on the wildlife market. Cultural practices mentioned in the literature but not by the consultants (the use of teeth in

Table 11.3 Cultural practices involving primates on the Moruca among the Lokono (L), Kari'na (K), and Warao (W). Numbers indicate the number of consultants that discussed a particular practice

	Cebus olivaceus	*Pithecia pithecia*	*Potos flavus*	*Saimiri sciureus*
Pet	L4, W4, K6	L4, W4, K6	W1, K2	L4, W4, K6
Pest	L4, W4, K5	–	–	L4, W4, K4
Food	L1, W3, K3	K4	K3	–
Accessories	K3	K1	–	L1, W1, K3
Commodity	L5, W6, K6	–	–	L5, W6, K6
Medicine	K1	K1	–	K1

necklaces; bones as spoons, containers, and ornaments; hair in armbands, belts, and brushes; primate patterns in basketry, earthenware, and string games, primate-derived clan names) are not discussed. The cultural practices show significant qualitative overlap across the three communities. Quantitative differences seem to reflect progressing acculturation, most advanced among the Lokono and least felt by the Kari'na.

All species except kinkajou (*P. flavus*) were consistently praised as good pets. Incidents of kinkajous being kept as pets were reported by the Warao and Kari'na but described as a curiosity. Some Warao called it a "spirit animal," because it moves quickly and noisily at night. Two species, Guianan weeper capuchin (*C. olivaceus*) and the common squirrel monkey (*S. sciureus*), were also often named as pests of corn fields and fruit trees, respectively, and as commodity on the wildlife market. One Kari'na consultant listed also medicinal uses of oils prepared with the burnt hair of the two species and the white-faced saki (*P. pithecia*), worth mentioning considering the medicinal and spiritual properties of the howler's hair (Sect. 11.4.6–11.4.7). Occasionally, the Lokono, Kari'na, and Warao use the skin of the common squirrel monkey to make watch bracelets or stuff the whole animal to make ornaments. The Kari'na also listed two more species for this purpose.

An interesting case of the development of new practices can be discerned in the food category. Today the Moruca Kari'na consider several species edible. Guianan weeper capuchin (*C. olivaceus*) was deemed palatable by the Moruca Warao and Lokono. The Lokono and Warao dietary patterns have a long history in the area. Already a century ago, mourning Warao women on the Moruca would cry out "Who will catch agouti, monkey, fish, and turtle for us now?" (Roth 1915: 74). The species must have been consumed by the Lokono with more frequency in the past, since it used to be referred to with an avoidance term, only attested for game species. *Mayeriki* "untrimmed one" was employed on the Moruca when traveling in a boat in order not to anger the water spirit (Roth 1915). For the same reason, among Surinamese Lokono, where primates are consumed as well, their blood cannot be dropped into the river (Abbenhuis 1939). The decrease in consumption of primates among the Moruca Lokono has been attributed by the consultants to acculturation. The fact that the Guianan weeper capuchin was considered palatable by the Moruca Warao is more surprising since primates are not consumed by Venezuelan Warao (Heinen 1973). Equally surprising are the Moruca Kari'na practices as Schomburgk (1847) and Roth (1915) assure that Guyanese Kari'na, in contrast to Surinamese Kari'na, do not eat primates. The change of dietary patterns of the Moruca Warao and Kari'na is therefore a possible Lokono influence. However, this cultural convergence in the food category on the Moruca does not align with linguistic borrowing: none of the terms for edible species was borrowed from Lokono. Crucially too, Kari'na data show that the such practices are localized. Generalization such as "Amerindians in Guyana prefer meat from spider monkeys and brown capuchins [as opposed to other primates]" should therefore be avoided (*pace* Lehman et al. 2006: 123). Similar patterns can be observed for the Guianan red howler monkey (*A. macconnelli*, henceforth, *Alouatta*, Table 11.4). Again, the

Table 11.4 Cultural importance of *Alouatta macconnelli* on the Moruca

Use	Lokono	Kari'na	Warao
Call used as alarm	2	3	3
Call used for weather forecasting	2	4	5
Call interpreted as praying	2	3	
Ludic dances and song inspired by behavior	2	6	4
Restrictions on ridiculing *Alouatta*	3		
Hair used to chase evil spirits away	2	1	3
Hair used as medicine for scorpion bites		1	
Skin used to make a drum	2	3	4
Skin used to make ornaments	2	1	2
Meat considered a delicacy	6	6	6
Throat used as medicine for whooping cough	3	6	6
Folklore tradition of *Alouatta's* cough	2	3	
Young animal kept as pet	1	2	2

quantitative differences between the communities appear to reflect the effects of acculturation.

As opposed to the above species, the knowledge about *Alouatta* encompasses a wider set of interconnected domains, forming templates at the intersection of subsistence, oral tradition, beliefs, medicine, and language. Though some aspects of such knowledge are based on general observations of the species that were likely made independently, others might have been exchanged. The following sections discuss each of the categories in Table 11.4 except for fact that *Alouatta* can be kept as a pet.

11.4.1 Call Used as Alarm

The Lokono use the verb *shimakun*, equivalent to Warao *koita* and Kari'na *eta*, to describe the sound made by *Alouatta.* The verbs mean "call" and do not have the doleful connotations of *howl.* The recognizable calls can be heard in the morning in the communities and form part of time-keeping practices, signaling it is time to wake up. Though mentioned in all three villages, it is a rare practice today as modern time-keeping devices are available. The Surinamese Kari'na produced also a charm from *Alouatta*'s larynx, rubbed into trumpets and flutes to imbue them with a stronger sound (Penard and Penard 1907). Thus enchanted, the instruments were used as a call to arms. Penard and Penard (1907) also mention *Alouatta* as an ingredient in charms increasing singing abilities (see Sect. 11.4.8 on drums). The practices speak to the saliency of the call, without the doleful overtones it has in English.

11.4.2 Call Used for Weather Forecasting

The Lokono, Kari'na, and Warao on the Moruca say that when *Alouatta* calls in morning, a hot day approaches, and when it calls in the evening, rain will come. The weather associations appear also in a Warao story, according to which *Alouatta* was too proud of its vocal abilities to influence the weather. Therefore, one day when it was calling the rain, a lightning struck it and burned its face black (see Sect. 11.4.5 for avoidance terms referring to *Alouatta*'s face). Its weather forecasting abilities are also mentioned by Roth (1915) as common in Guyana and may be a widespread observation, possibly arrived at independently.

11.4.3 Call Interpreted as Praying

The Lokono and Kari'na on the Moruca say that *Alouatta* come in a circle when they call and describe it as "praying," *khoyabwan* and *okunoma*, respectively. The Kari'na find an explanation for it in their folklore related to the whooping cough, speaking of *Alouatta* praying to God to save him from human predators (see Sect. 11.4.12). A similar religious interpretation of the call is encoded in Surinamese terms for the species. *Dominei* (Lokono) and *dominiɨ* (Kari'na), borrowings from Dutch *dominee* "a minister of Dutch Reformed Church," refer to the individual leading the calling. Surinamese Kari'na had two more terms for the individual leading the groups' calls, *masakari* and *wororoku* (Ahlbrinck 1931). These native terms suggest that the observation that the animals call in a group has a long history though its monotheist guise is likely a more recent Western influence.

11.4.4 Ludic Dances and Songs Inspired by Behavior

The Lokono, Kari'na, and Warao on the Moruca have a traditional dance called *itorhi ibinin, arawata kɨnuwanon,* and *wai ahoho*, respectively, meaning "*Alouatta*'s dance." The choreography mimics the behavior of the animal: the dancers walk clumsily in a circle, scratching themselves to make the audience laugh. The Kari'na also have a song that accompanies the dance, a different version of which is known from Suriname (Ahlbrinck 1931: 488). The peculiar behavior of *Alouatta* is also documented in Surinamese Kari'na simile: "you scratch yourself like *Alouatta*" (Penard and Penard 1907: 83). Both Ahlbrinck (1931) for Surinamese Kari'na and Mink (1992) for Surinamese Lokono discuss imitative dances, though *Alouatta*'s dance is not explicitly mentioned (see Sect. 11.4.9 for costumes made of *Alouatta* that may have been used during such dances). *Alouatta*'s song, but not a dance, is also documented for Venezuelan Warao but appears unrelated to the ludic dances described here (Olsen 1996). While likely a case of cultural convergence between Lokono and Kari'na, the trajectory of the exchange of the songs and dances cannot be demonstrated.

11.4.5 Ridiculing Alouatta

Ridiculing *Alouatta* is, however, not always allowed. The Moruca Lokono warn that pregnant women should not laugh at the animal because the child would be born hairy. Laughing at *Alouatta* can even be punished by death according to Lokono oral traditions no longer remembered on the Moruca (Roth 1915). Similar precautions are taken by pregnant Warao women in Venezuela, who call the animal *amuhoro hoko* "white face," so that their children should not be born with monkey fur (Barral 1979). The Moruca Warao in turn joke about *Alouatta*'s black face in relation to its abilities to call the rain (see Sect. 11.4.2). The Kari'na do not know such restrictions on ridiculing *Alouatta*. Moreover, in Surinamese Kari'na, *arawata* is an offensive term for someone with a particularly dark face (Ahlbrinck 1931).

11.4.6 Hair Used to Chase Away Evil Spirits

The Lokono, Warao, and Kari'na use *Alouatta*'s hair to repel evil spirits causing illnesses. Epileptic fits and "mystery illness," an unknown condition with symptomatic uncontrollable fits, were specifically mentioned. Such smoking practices are rarely mentioned in previous studies of the Kari'na and Lokono, although a smoking motif appears in Lokono oral tradition about the whooping cough believed to be caused by spirits (see Sect. 11.4.12). Moreover, in Suriname, a mad dog would be forced to inhale the smoke from monkey hair as a cure (Abbenhuis 1939: 38). On the Orinoco, among the Warao, Brown (1877) discusses smoking the patient as a remedy for an epileptic fit, likely a related treatment. Crucially, smoking practices should not be confused with blowing tobacco smoke on the patient to remove evil spirits practiced by medicine men, which are discussed in the literature at length.

11.4.7 Hair Used as Medicine for Scorpion Stings

The oil made with *Alouatta*'s hair has yet another medicinal application. One Kari'na participant mentions burning the hair of *Alouatta* and mixing it with oil to produce an anointment against scorpion stings. It is worth pointing out that the same consultant knew also the medicinal properties of the hair of three other primates (*C. olivaceus*, *P. pithecia*, and *S. sciureus*), all of which had to be prepared in the same way. Noticeable is the connection to the burning of the hair to repel evil spirits, the preparation of the remedy for whooping cough, one recipe for which also involves an anointment of burned hair, and the oral tradition linking these elements. Burned hair of the *Alouatta* is the active ingredient in all these medicaments (see Sects. 11.4.6, 11.4.11, and 11.4.12).

11.4.8 Skin Used to Make Drums

The Lokono, Kari'na, and Warao on the Moruca report that they manufactured a two-sided drum from the skin of *Alouatta* called *sambura* in Lokono and Kari'na (from Spanish *tambor*) and *eruru* in Warao. The Moruca and Venezuelan Warao consider *Alouatta*'s skin the best choice (Heinen 1973). While the oral traditions shared by the Warao and Lokono explain the origin of the drum (Roth 1915), the Lokono and Kari'na report that the skin of other animals is preferred, particularly that of the red-rumped agouti. This scenario is consistent with Surinamese sources which suggest deer, agouti, and peccary species, listing *Alouatta* as the last resort (Kambel and Jong 2006; Mink 1992; Ahlbrinck 1931). These differences do not correlate with the availability of resources; deer, agouti, peccary, and *Alouatta* are found throughout the Guianas (Husson 1978; Linares 1998). Given the preferences, lack of native terms for such drums in Lokono and Kari'na, and the oral tradition known to the Warao and Lokono, the use of *Alouatta*'s skin to make a drum likely spread from Venezuelan Warao to the Lokono and Kari'na on the Moruca.

11.4.9 Skin Used to Make Ornaments

The Lokono and Warao on the Moruca mention that the skin and tail can be made into a cap; the Kari'na reported a carnival mask instead. There are no special terms for such headpieces in the languages. Neither are they daily garments but costumes for a special occasion, such as New Year celebrations. The garment has, however, a long history as it has been mentioned by other authors who similarly saw it being worn on special occasions (Roth 1924). The hat is perhaps the last remnant of the many costumes that were once worn during the performances of the imitative animal songs and dances (§11.4.4). Penard and Penard (1907) give an account of the festivities during which they observed such animal costumes. Finally, the consultants also discussed stuffing the animal and using it as a decoration for the interior of their houses.

11.4.10 Meat Considered a Delicacy

Consultants from the three communities report that the meat of *Alouatta*, especially that of the hind legs, is a delicacy, though most participants reported not having eaten it in years. The meat is described as tastier than that of tapir and peccaries and only surpassed by that of paca and deer. The Moruca Lokono have restrictions on the consumption: pregnant women should not eat it since the baby will be as hairy as the animal, but the animal is considered a delicacy both in the Moruca and in Suriname. In a Surinamese Kari'na story, a comparison is made to "the teeth of a cooked howler" (Penard and Penard 1907: 26); charms for hunting monkeys made of *Alouatta*'s brain are also known among the Surinamese Kari'na (Penard and Penard 1907), suggesting that the species has been consumed in Suriname. Ethnohistorical sources, however, indicate that primates were not consumed by the Moruca Kari'na (e.g., Schomburgk 1847). The Moruca Warao joke about tricks to ascertain that *Alouatta* falls from the tree when hit, as it can wrap its tail around the branches. Venezuelan Warao, however, deem primates unpalatable (Heinen 1973). This suggests a change in diet of the Moruca Warao and Kari'na, possibly under the influence of the Lokono.

11.4.11 Larynx Used as Medicine for Whooping Cough

On the Moruca, the larynx of *Alouatta* is used as a medicine against whooping cough, known locally as *itorhi thonolia* (Lokono), *wai obo* (Warao), and *arawata atono* (Kari'na), meaning "*Alouatta*'s cough." The recipes vary; the most common one, however, consists in using the larynx as a cup. The medicinal properties of the larynx are also known among the Lokono and Kari'na in Suriname (Kambel and de Jong 2006), but not among the Warao in Venezuela. The relationship between the *Alouatta* and whooping cough is documented for other Arawakan (e.g., Baniwa *iit-shítta* "suffer from whooping cough," from *íitsi* "*Alouatta*," Ramirez 2001) and Cariban people (e.g., Makushi *arautaimî* "whooping cough," from *arauta* "Alouatta," Amódio and Pira 2007). Venezuelan Warao know it as *obo sabana* "bad cough" (Barral 1979), although Wilbert (2001) calls it also *wai obo*. The Moruca Lokono, Kari'na, and Warao also use a fern, whose root resembles *Alouatta's* tail to prepare a medicine for the disease (Reinders 1993). What is most likely the same fern is also used for this purpose by Venezuelan Warao, who also use several other plants to make a medicine against whooping cough (Wilbert 2001). This information and the fact that Venezuelan Warao do not consume primates suggest that the medicinal use of the larynx and the fern on the Moruca may be originally Lokono or Kari'na and was borrowed by the Warao.

11.4.12 **Folklore Tradition of *Alouatta*'s Cough**

Oral traditions connect many aspects of cultural knowledge. A good example is the story of the origin of the whooping cough. The Moruca Lokono know a story about an evil spirit that decimates children. The Lokono killed him and his family with smoke. As the spirits died, they coughed and fell from a tree in the shape of *Alouatta*. The story thus explains the Lokono name for whooping cough, a disease particularly dangerous to children, and the rationale behind the medicine. It also sheds light on the practice of burning the hair to scare off evil spirits: *Alouatta*, being an incarnation of those, is deterred by the smell of its own kind burning.

The Moruca Warao today do not know the story, but a Warao version of it was documented in the area a century ago (Roth 1915); a similar story was documented among the Venezuelan Warao (Wilbert 1970). Roth (1915) also gives a Kari'na story with a virtually identical plot. The modern Kari'na version has *Alouatta* over-killed by people, for which God punished them with whooping cough but also gave them the medicine. Since Venezuelan Warao do not eat *Alouatta*, nor use it as medicine, this medicinal knowledge was likely borrowed from their neighbors on the Moruca together with the folklore concomitant with it.

11.5 Conclusions

The observed picture of the linguistic and cultural aspects of knowledge about primates is one of environmental adaptation and cultural convergence. Primate terms reflect the local biogeography of species. The various dialects retained, borrowed, and dropped certain terms or even changed their meanings, fine-tuning their lexical resources to the niches in which they are spoken (Table 11.5). The findings are central to the discussion of Amazonian contact scenarios, showing that classic lexical borrowing motivated by the need to name new species is common in Amazonia, despite the known claims about restriction on lexical borrowing, typical of some parts of Amazonia (Haynie et al. 2014; Bowern et al. 2011).

With the ethnobiological terms finely attuned to the local environment, by identifying retentions, borrowings, and innovations, the animal lexicons of protolanguages can be reconstructed and plotted against the distribution of species to illuminate the homelands of the proto-speakers. This chapter shows that primates are a particularly appropriate taxon for such a study in Amazonia. First, primates form a natural semantic domain, populated with discrete terms in any language, yet small enough to render a large comparative study feasible. The species are easy to

Table 11.5 Lexical adaptation of the dialects of Lokono, Kari'na, and Warao. Etymological codes: likely retention (R), semantic shift (S), lexical borrowing (B), lexical innovation (I), deletion (D)

	Lokono (MO)	Kari'na MO	Warao MO	Lokono (SU)	Kari'na (SU)	Kari'na (VN)	Warao (VN)
A. macconnelli	R + S?	R + S?	R	R + S?	R + S?	R + S?	R
C. olivaceus	R + S?	R	R/B	R + S?	R	R	R/B
P. flavus	I	R	–	I	B	–	R
P. pithecia	R/B	B	B	R/B	B	–	B
A. paniscus	B	B	B	I	B	–	–
S. sciureus	I	R	B	I	R	D	–
S. midas	–	–	–	R + S	B	–	–
S. apella	D	–	–	R	R + S?	–	–
C. Chiropotes	–	–	–	R/B	B	B	–

recognize and culturally salient, hence a likely target for lexicographers and ethnographers, assuring the availability of data. They are found throughout the continent; however, their ranges are determined by large rivers creating areas with contrastive distributions, as opposed to more widespread animals, allowing to zoom in on the homelands of protolanguages. Based on the presented data, it can be concluded, for instance, that the homeland of Proto-Kari'na should be within the range of the common squirrel monkey (*S. sciureus*) and the Guianan weeper capuchin (*C. olivaceus*), the terms for which are likely retentions without semantic shifts, and outside the range of the white-faced saki (*P. pithecia*), Guianan spider monkey (*A. paniscus*), and the bearded saki (*Ch. chiropotes*), the names of which are borrowings in Kari'na dialects. When plotted, these distributions imply two potential homelands. For such analyses to be reliable, however, attention must be paid to the dialectal variation, reliable etymologies based on larger language samples, and definite species distribution. Close collaboration between linguists and primatologist is thus of mutual interest to advance such research.

The observed cultural practices reveal in turn a set of intertwined motifs at the intersection of language, medicine, beliefs, arts, crafts, oral tradition, subsistence, time-keeping, and weather forecasting. There are noticeable cases of areal cultural convergence, some of which are, more strongly put, cases of cultural borrowing (Table 11.6). Importantly, cultural borrowings do not map onto linguistic borrowings. Highest cultural convergence in fact appears for species whose names are never borrowed in the sample (i.e., *A. macconnelli*). It merits a mention too that the practices are clearly disappearing, most likely due to acculturation, a factor mentioned by the consultants themselves, and reflected in the quantitative differences between the consultants' responses in the three communities and the comparison with historical sources which list a number of other culturally important uses of nonhuman primates. Finally, tracing the linguistic and cultural aspects of ethnoprimatological knowledge through time may in the future turn out to be of interest to conservation efforts. Terms such as *kwata* for the Guianan spider monkey on the Moruca may be indicative of a once larger range of the species, informing environmental research and policies aimed at the preservation of primate diversity in Amazonia.

Table 11.6 Shared cultural practices involving primates on the Moruca and their possible origin

Shared practices	Possible origin
Call used as alarm	–
Skin used to make ornaments	–
Call used for weather forecasting	–
Young animal kept as pet	–
Ludic dances and songs	–
Hair used to chase evil spirits away	–
Skin used to make a drum	Warao
Meat considered a delicacy	Lokono
Medicine for whooping cough	Lokono or Kari'na
Folklore tradition of *Alouatta's* cough	Lokono or Kari'na

Acknowledgments This chapter has been produced in close cooperation with the speakers of Lokono, Kari'na, and Warao from Santa Rosa, Manawarin, and Waramuri, respectively, whom the author wants to command on their knowledge and thank for sharing it with him. The author is also grateful to Lev Michael for discussions of the topic and comments on the first draft, as well as to the editors of the volume and the external reviewer for their feedback. The research leading to this publication was funded by the Netherlands Organization for Scientific Research (project number 446-15-012).

References

Abbenhuis MF (1939) Arawakken in Suriname: enquête-materiaal voor een volkenkundige studie. Leo Victor, Paramaribo

Ahlbrinck W (1931) Encyclopaedie der Karaïben. Koninklijke Akademie van Wetenschappen, Amsterdam

Aikhenvald AY (2010) Language contact in Amazonia. Oxford University Press, Oxford

Aikhenvald AY, Brito C, Brito L et al (2001) Diconário Tariana - Português, Português - Tariana. Boletim Do Museu Paraense Emilio Goeldi Serie Antropologia 17:1

Amazon Conservation Team Suriname (2018) Biodiversity database Suriname. http://www.ethnobiobase.act-suriname.org. Accessed 24 May 2018

Amódio E, Pira V (2007) Makusi Maimu. Língua Makuxi: guia para a aprendizagem e dicionário Makuxi. Valer Editora, Manaus

Anderson EN, Pearsall D, Hunn ES, Turner N (2011) Ethnobiology. Wiley-Blackwell, Hoboken

Armellada C, Salazar MG (1981) Diccionario Pemón. Universidad Católica Andrés Bello, Caracas

Barral BM (1979) Diccionario Warao-Castellano, Castellano-Warao. Universidad Católica Andrés Bello, Caracas

Berlin B (1992) Ethnobiological classification: principles of categorization of plants and animals in traditional societies. Princeton University Press, Princeton

Bowern C, Epps P, Gray R et al (2011) Does lateral transmission obscure inheritance in hunter-gatherer languages? PLoS One 6:e25195. https://doi.org/10.1371/journal.pone.0025195

Breton R, Paisa M (1999) Dictionnaire caraïbe-français: 1665. Karthala, Paris

Brown BC (1877) Canoe and camp life in British Guiana. E. Stanford, London

Camargo E (2002) Léxico bilingüe aparai-português/português-aparai. Lincom Europa, Muenchen

Coretta S (2013) The phonology of Mawayana. MA thesis. Università degli Studi di Pavia, Pavia

Cormier L (2006) A preliminary review of neotropical primates in the subsistence and symbolism of indigenous lowland South American peoples. Ecol Environ Anthropol 2(1):14–32

Courtz H (2008) A Carib grammar and dictionary. Magoria Books, Toronto

Epps P, Michael L (2015) The areal linguistics of Amazonia. In: Hickey R (ed) Cambridge handbook of areal linguistics. Cambridge University Press, Cambridge, pp 934–963

Eriksen L (2011) Nature and culture in prehistoric Amazonia: using G.I.S. to reconstruct ancient ethnogenetic processes from archeology, linguistics, geography, and ethnohistory. Ph.D. dissertation. Lund University, Lund

Farabee WC (1918) The Central Arawaks. The University Museum Anthropological Publications Vol. IX. Philadelphia: University of Pennsylvania.

Fleck DW, Harder JD (2000) Matses indian rainforest habitat classification and mammalian diversity in Amazonian Peru. J Ethnobiol 20:1):1–1)36

Goeje GH (1948) La langue manao (famille Arawak-Maipure). In: Actes Du XXVIII Congrès International de Américanistes, pp 157–171

Grenand F (1989) Dictionnaire wayãpi-français: lexique français-wayãpi: Guyane française. SELAF, Paris, Peeters

Haynie H, Bowern D, Epps P et al (2014) Wanderwörter in languages of the Americas and Australia. Ampersand 1:1. https://doi.org/10.1016/j.amper.2014.10.001

Heinen HD (1973) Adaptive changes in a tribal economy: a case study of the winikina-Warao. Ph.D dissertation. University of California, Los Angeles

Henfrey TB (2002) Ethnoecology, resource use, conservation and development in a Wapishana community in the south Rupununi, Guyana. Ph.D. dissertation. University of Kent, Canterbury

Hornborg A, Hill JD (2011) Ethnicity in ancient Amazonia: reconstructing past identities from archaeology, linguistics, and ethnohistory. University Press of Colorado, Boulder, CO

Hunn ES (1976) Toward a perceptual model of folk biological classification. Am Ethnol 3:508. https://doi.org/10.1525/ae.1976.3.3.02a00080

Hunn ES (1982) The utilitarian factor in folk biological classification. Am Anthropol 84:830. https://doi.org/10.1525/aa.1982.84.4.02a00070

Husson AM (1978) The mammals of Suriname. Zoölogische monographieën van het Rijksmuseum van natuurlijke historie. Brill, Leiden

IUCN (2008a) *Alouatta macconnelli*: Boubli, J.-P., Di Fiore, A. & Mittermeier, R.A.: The IUCN Red List of Threatened Species 2008: E.T40642A10347360. https://doi.org/10.2305/IUCN.UK.2008.RLTS.T40642A10347360.en

IUCN (2008b) *Ateles paniscus*: Mittermeier, R.A., Rylands, A.B. & Boubli, J.-P: The IUCN red list of threatened species 2008: E.T2283A9392691. https://doi.org/10.2305/IUCN.UK.2008.RLTS.T2283A9392691.en

IUCN (2008c) *Chiropotes chiropotes*: Veiga, L.M., Silva Jr., J.S., Mittermeier, R.A. & Boubli, J.-P.: The IUCN Red List of Threatened Species 2008: E.T43891A10829879. https://doi.org/10.2305/IUCN.UK.2008.RLTS.T43891A10829879.en

IUCN (2008d) *Saguinus midas*: Mittermeier, R.A., Rylands, A.B. & Boubli, J.-P.: The IUCN red list of threatened species 2008: E.T41525A10489882. https://doi.org/10.2305/IUCN.UK.2008.RLTS.T41525A10489882.en

IUCN (2008e) *Saimiri sciureus*: Boubli, J.-P., Rylands, A.B., de La Torre, S. & Stevenson, P.: The IUCN Red List of Threatened Species 2008: E.T41537A10494364. https://doi.org/10.2305/IUCN.UK.2008.RLTS.T41537A10494364.en

IUCN (2008f) *Pithecia pithecia*: Marsh, L.K., Mittermeier, R.A. & Veiga, L.M.: The IUCN Red List of Threatened Species 2015: E.T43942A115201263. https://doi.org/10.2305/IUCN.UK.2015-1.RLTS.T43942A70609046.en

IUCN (2008g) *Sapajus apella*: Rylands, A.B., Boubli, J.-P., Mittermeier, R.A., Wallace, R.B. & Ceballos-Mago, N.: The IUCN red list of threatened species 2015: E.T39949A70610943. https://doi.org/10.2305/IUCN.UK.2015-1.RLTS.T39949A70610943.en

IUCN (2015) *Potos flavus*: Helgen, K., Kays, R. & Schipper, J.: The IUCN red list of threatened species 2016: E.T41679A45215631. https://doi.org/10.2305/IUCN.UK.2016-1.RLTS.T41679A45215631.en

IUCN (2008h) *Chiropotes albinasus*: Veiga, L.M., Pinto, L.P., Ferrari, S.F., Rylands, A.B., Mittermeier, R.A. & Boubli, J.-P.: The IUCN Red List of Threatened Species 2008: E.T4685A11085894. https://doi.org/10.2305/IUCN.UK.2008.RLTS.T4685A11085894.en

Kambel E-R, de Jong C (2006) Marauny na'na emandobo, Lokono shikwabana. I. Rosgal S.A, Uruguay

Kondo R (1983) La clasificacion de mamiferos y reptiles por los indigenas Guahibo, Cuiba, Piapoco y Yekuana. Articulos En Lingüistica y Campos Afines 12:93–131

Lehman SM (2004) Distribution and diversity of primates in Guyana: species-area relationships and riverine barriers. Int J Primatol 25(1):73–95

Lehman SM, Sussman RW, Phillips-Conroy J et al (2006) Ecological biogeography of primates in Guyana. In: Lehman SM, Fleagle JG (eds) . Primate biogeography. Springer, New York, pp 105–130

Linares OJ (1998) Mamíferos de Venezuela. Sociedad Conservacionista Audubon de Venezuela, Caracas

Lizarralde M (2002) Ethnoecology of monkeys among the Barí of Venezuela: perception, use and conservation. In: Fuentes A, Wolfe LD (eds) Primates face to face. Cambridge University Press, Cambridge, pp 85–100

Meira S, Muysken PC (2017) Cariban in contact: new perspectives on trio-Ndyuka pidgin. In: Yakpo K, Muysken PC (eds) Boundaries and bridges. De Gruyter, Berlin, pp 197–227

Mere Roncal C, Bowler M, Gilmore MP (2018) The ethnoprimatology of the Maijuna of the Peruvian Amazon and implications for primate conservation. J Ethnobiol Ethnomed 14:19. https://doi.org/10.1186/s13002-018-0207-x

Mink E (1992) De muziek-, zang-, en danstraditie van de Arowak-Indianen; van Suriname naar Nederland. Ph.D. Dissertation. University of Amsterdam, Amsterdam

Mosonyi J (2002) Diccionario básico del idioma kariña. Fondo Editorial del Caribe, Barcelona (Venezuela)

Olsen DA (1996) Music of the Warao of Venezuela: song people of the rain forest. University Press of Florida, Gainesville

Penard FP, Penard AP (1907) De menschetende aanbidders der zonneslang. H.B. Heyde, Paramaribo

Posey DA (2002) Kayapó ethnoecology and culture. Studies in environmental anthropology 6. Routledge, London

Ramirez H (1992) Le bahuana une nouvelle langue de la famille arawak. Amerindia 17 (supplement 1)

Ramirez H (2001) Línguas arawak da Amazônia setentrional: comparação e descrição. Editora da Universidade, Manaus

Reinders M (1993) Medicinal plants and their uses and the ideas about illness and healing among the Warao of Guyana MA thesis. Utrecht University, Utrecht

Robayo C (1996) Datos actuales de la lengua carijona, equivalentes al documento de de Wavrin donde utiliza el cuestionario del instituto etnológico de París. In: Landaburu J (ed) Documentos sobre lenguas aborígenes de Colombia del archivo de Paul Rivet. Uniandes, CCELA, COLCIENCIAS, Santafé de Bogotá, pp 521–538

Romero-Figueroa A (1997) A reference grammar of Warao. Lincom Europa, Muenchen

Rose F, Vanhove M, Stolz T et al (2012) Borrowing of a Cariban number marker into three Tupi-Guarani languages. In: Vanhove M, Stolz T, Urdze A, Otsuka H (eds) Morphologies in contact. Akademie Verlag, Berlin, pp 37–69

Roth WE (1915) An inquiry into the animism and folk-lore of the Guiana Indians. Annual Report. Smithsonian Institution Bureau of American Ethnology 30. Government Printing Office, Washington

Roth WE (1924) An introductory study of the arts, crafts, and customs of the Guiana Indians. Government Printing Office, Washington

Rybka K (2015) State-of-the-art in the development of the Lokono language. Language Docum Conserv 9:110–113

Schomburgk R (1847) Reisen in Britisch-Guiana in den Jahren 1840–1844 : nebst einer Fauna und Flora Guiana's nach Vorlagen von Johannes Müller, Ehrenberg, Erichson, Klotzsch, Troschel, Cabanis und Andern. J. J. Weber, Leipzig

Shepard GH (2002) Primates in Matsigenka subsistence and world view. In: Fuentes A, Wolfe LD (eds) Primates face to face. Cambridge University Press, Cambridge, pp 101–136

Staffeleu P (1975) Surinaamse zoogdiernamen. Zoologische Bijdragen 18:1–74

Urbani B, Cormier LA (2015) The ethnoprimatology of the howler monkeys (*Alouatta* spp.): from past to present. In: Kowalewski MM, Garber PA, Cortés-Ortiz L et al (eds) Howler monkeys. Springer New York, New York, pp 259–280

Voss RS, Fleck DW (2011) Mammalian diversity and Matses ethnomammalogy in Amazonian Peru. Part 1: Primates. Bull Am Mus Nat Hist 351:1

Wilbert J (1970) Folk literature of the Warao indians. Narrative material and motif content. University of California, Los Angeles

Wilbert W (2001) Dau yarokota. Plantas medicinales Warao. Caracas: Fundación La Salle de Ciencias Naturales. ICAS, Caracas

Chapter 12
Relationships Between Scientific Ecology and Knowledge of Primate Ecology of Wapishana Subsistence Hunters in Guyana

Thomas Henfrey

12.1 Introduction

While difficulties over definition and theoretical orientation have ensured that traditional, local or indigenous knowledge (hereafter referred to as local ecological knowledge, or LEK) has become a problematic concept (Purcell 1998; Ellen and Harris 2000), its role in the debate on development has advanced from radical critique to established orthodoxy (Dove 2002). LEK, which I here define as ecological knowledge acquired by resource users independently of any formal scientific training (in contrast with scientific ecological knowledge, or SEK), exhibits substantive and epistemological continuity with scientific approaches (Agrawal 1995; Ellen 2004). Research on applications in agriculture especially has provided strong demonstrations of its scientific validity and practical value (e.g. Richards 1985, Warren et al. 1995). However, LEK also differs in significant, if not uniform, ways from SEK (Sillitoe 2002a). In addition, a strictly scientific treatment neglects some important features of LEK (Escobar 1995, Stirrat 1998; Agrawal 2002).

Because of this partial overlap, LEK and SEK are complementary along several dimensions key to practical problems in resource management (DeWalt 1994; Moller et al. 2004). From the perspective of traditional resource users, SEK offers a body of knowledge and range of techniques with which they might extend their management capacities as they seek to adapt to changing circumstances (Sillitoe 1998; Puri 2001, for practical cases see Hanna 1998; Pinkerton 1998; Becker 2003; Becker and Ghimire 2003). Some scientists now seek equitable partnerships with traditional resource users in creating joint research programmes based upon common interests in ecological conservation (Bodmer and Puertas 2000; Ticktin and Johns 2002; Ticktin et al. 2004; Moller et al. 2004).

T. Henfrey (✉)
Centre for Ecology, Evolution and Environmental Change (cE3c), Lisbon University, Schumacher Institute for Sustainable Systems, Bristol, UK
e-mail: tom@schumacherinstitute.org.uk

B. Urbani, M. Lizarralde (eds.), *Neotropical Ethnoprimatology*, Ethnobiology,
https://doi.org/10.1007/978-3-030-27504-4_12

From the perspective of scientific research in ecology, LEK can be employed as a source of hypotheses via which to focus research more efficiently on important ecological issues (Posey 1986, 1990; Townsend 1995). For example, collaborations between scientists and Inuit hunters have extended scientific data sets on the ecology of beluga whales (Myrmin et al. 1999; Huntington et al. 1999) and long-term changes in abundance and distribution of caribou (Ferguson et al. 1998; also see Huntington 2000 and references therein). Indigenous ecologists have also contributed information on entomology (Posey 1986), ichthyology (Ponte Johansons 1995), food habits of game animals (Balée 1994; Cormier 2004), herpetology (Nabhan 2003), primate behavioural ecology (Townsend 1995) and interspecies mutualisms (Vasquez-Davila 1995; Donovan and Puri 2004).

Comparisons between such data and information in the scientific literature have generally shown close matches. Ethnoprimatological data provided by a single Murui informant corresponded closely in detail with information published in the scientific primatological literature (Townsend 1995). Balée, although he does not give details, reports that a biologist's field tests of information provided by Ka'apor hunters on the food plants of game animals proved it was highly accurate (Balée 1994). However, at the time of this study, such examples were few in number, and empirical backing for the lofty claims often made on behalf of local ecological knowledge surprisingly scarce (Donovan and Puri 2004). One project to have conducted studies in LEK alongside simultaneous studies in scientific ecology endorses the value of LEK in providing scientific information relevant to resource management, but cautions that the information supplied requires verification via scientific methods (Gilchrist et al. 2005). Accordingly, this chapter compares the ethnoprimatological knowledge Wapishana hunters in Guyana, South America, contextualised in relation to the wider cultural significance of primates, with findings from corresponding areas of scientific ecology.

12.2 Background: Wapishana Settlement and Cultural Ecology

Research was part of a pre-doctoral study on the applications in subsistence of the local ecological knowledge of Wapishana people in southwestern Guyana (Henfrey 2002, 2017, 2018). Fieldwork took place over a total of 20 months during 1998, 1999 and 2000 (see Fig. 12.1). The 5000 or so Guyanese Wapishana (part of a wider population also resident in adjacent areas of Roraima State in northern Brazil) reside in nine main villages and numerous smaller settlements. These settlements are mostly located along an arc at the boundary between Guyana's South Rupununi savannah and adjacent forested areas in the Kanuku Mountains and the basins of the Kwitaro and Kujuwini Rivers. The main research site was one of the more remote of these villages, in the Kwitaro River basin and approximately 100 miles from the district capital at Lethem.

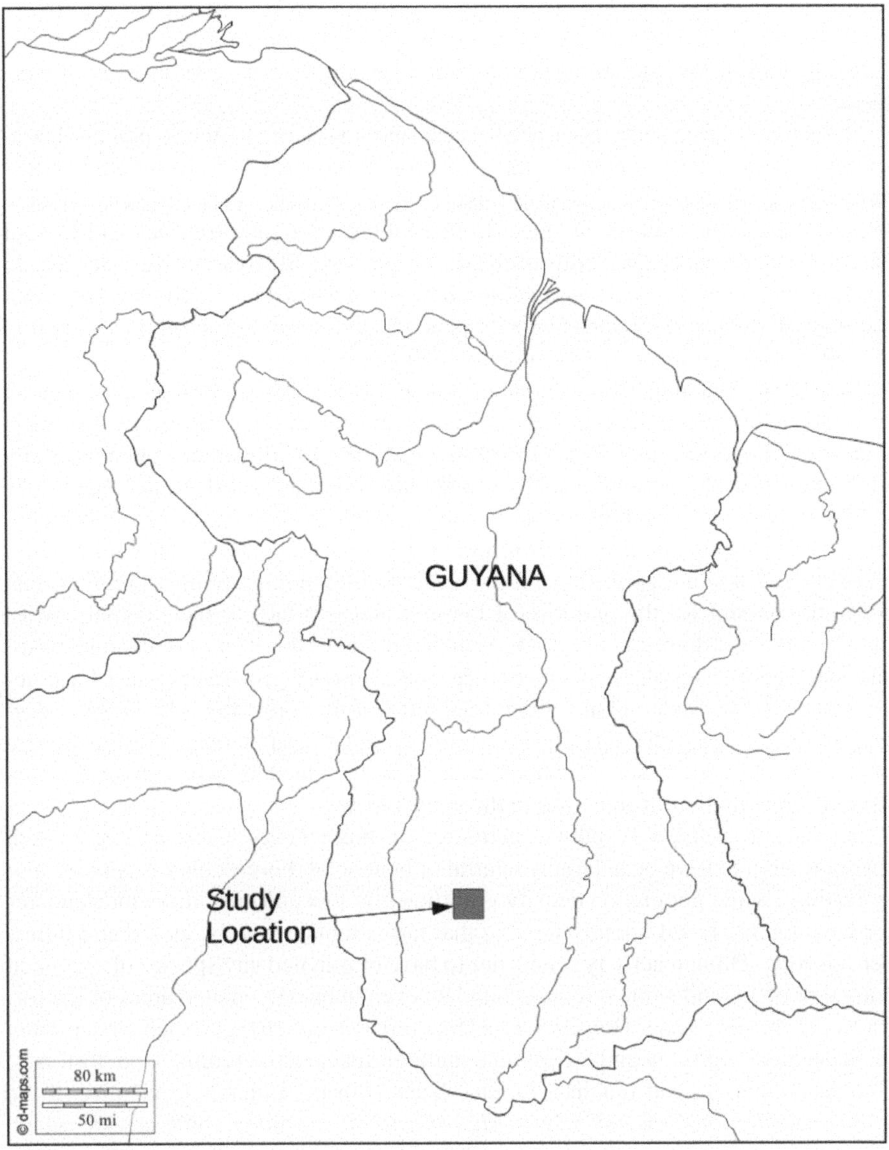

Fig. 12.1 Location of the study site

Sustained contact between Wapishana people and non-Amerindians in the South Rupununi began with the arrival of cattle ranchers in the final decade of the nineteenth century (Baldwin 1946: 36–39) and of Roman Catholic missionaries in the 1920s (Bridges 1985). Most villages, including the study village, have been incorporated into the national administrative system. Its very limited state-built infrastructure includes a primary school at which most current residents obtained some formal education, including learning spoken and written English. However, the dominant local language is Wapishana, spoken as a first language by the vast majority of local residents.

Remoteness, extremely poor road access and a lack of economic potential conspired to limit outside interest in the area over the course of the twentieth century, although this scenario is changing rapidly with Guyana's increasing engagement with global politics and economics. Limited alternative opportunities and lack of external interference have both obliged and allowed lifestyles to remain largely unchanged in many respects, particularly in terms of subsistence activities. However, the pace of change in Guyana since it opened its economy to foreign investment in the 1990s makes it unlikely this situation will persist.

Settlement straddles the savannah–forest boundary. Most families have a main dwelling in the central village on the savannah and one or more subsidiary dwellings at farms or hunting camps in the forest, the main site of subsistence pursuits. Daily life and livelihoods are dominated by subsistence activities, based around long fallow swidden agriculture combined with, in varying degrees, hunting (primarily of ungulates, large rodents and ground-dwelling birds), fishing, collection of wild animals and plants, and home gardening. At the research site, hunting remains an important economic pursuit for the majority of families and is a largely male activity. Most families also rear domestic livestock, a small number of these on a commercial basis, and some people supplement subsistence agriculture by growing peanuts for commercial sale. The trade in balata, the dried latex of the forest tree *Manilkara bidentata*, in the past provided an opportunity for a regular cash income. Most men were involved in the trade in some capacity, usually as the 'bleeders' who extract the latex from the tree, up until its collapse in the early 1980s.

To date, the effects of human activities on biodiversity levels and ecosystem function seem to have been largely neutral or beneficial. Forest cover is persistent or expanding, while population density is sufficiently low and customary mechanisms for land tenure are sufficiently flexible that farms remain widely spaced and fallow periods long. Human activity seems not to have eliminated any species of exploited animal or plant, and game remains abundant even in heavily visited areas of the forest in the farming zone. However, residents report local changes such as depletion of sedentary and/or heavily exploited animal species (including land tortoises, *Geocheleone* spp., and iguanas, *Iguana iguana)* in the immediate vicinity of the village and removal of etai palms *(Mauritia flexuosa)* along some creeks, which some people associate with a deterioration of hydrological cycles.

Awareness of dependence on biological resources, and their vulnerability to overexploitation, is behind an emerging conservation ethic expressed in public meetings, interviews and informal decisions by many, in the village, particularly leaders, teachers and other progressive persons. This phenomenon is invariably cast in utilitarian terms, as a means to secure and improve local livelihoods, retain political and economic independence and continue to provide the option of living along traditional lines. While generally aware that the very fact of making sustainable (more or less) use of the forest demonstrates a local management capacity, people also recognise that this capacity has limitations, and that expanding the skill base and management capacity could become increasingly important in the future. In 2012, the indigenous people of the South Rupununi produced a collective management plan, based on existing patterns of resource use and local ecological knowledge and management capacity (Gomes and Wilson 2012).

The potential for scientific ecology to contribute to such an expansion of local capacity depends on a complex intersection of factors. One is the compatibility between local ecological knowledge and scientific approaches to the study of ecology, including the extent to which the two overlap in terms of substantive content, epistemology and practical skills. This area of potential intersection of knowledge systems, ethnoecological knowledge in its most limited sense, represents a latent potential for traditional resource users to engage scientific approaches on their own terms. Whether this potential can be realised additionally depends on political conditions, specifically the capacity of traditional resource users to exert economic and hence sociocultural, self-determination (e.g. Toledo 2001), and the coherence of a scientifically informed ethnoecology with the wider sociocultural context of understandings of and interactions with the natural world (Posey et al. 1984).

The political circumstances for Guyanese Wapishana are relatively favourable. State recognition of indigenous land tenure, though imperfect and incomplete, does to a large degree allow autonomy in local governance and decisions over land use (Henfrey 1999). Active petitioning of the national government for an extension of titled lands, ongoing since the Amerindian Lands Commission report at the time of Guyana's independence and more recently supported by extensive self-documentation of land use in the affected areas, reflects a strong local and regional capacity for political self-assertion.

The study reported here addresses the epistemological dimension of Wapishana ethnoprimatology. It first describes the cultural ecology of Wapishana relationships with the local primate fauna. It then provides a detailed account of a study of the overlap between the substantive ethnoprimatological knowledge of Wapishana hunters and corresponding scientific information.

12.3 Cultural Significance and Understanding of Primates

The area inhabited and used by the Wapishana is home to eight species of wild primate: black spider monkey, *roomi* (in Wapishana) (*Ateles paniscus*); red howler monkey, *soboru* (*Alouatta seniculus*); common capuchin, *powato* (*Sapajus apella*); wedge-capped capuchin, *oao* (*Cebus olivaceus*); brown-bearded saki, *wishi* (*Chiropotes satanas*); Guianan white-faced saki, *oroa* (*Pithecia pithecia*); squirrel monkey, *chaumaa* (*Saimiri sciureus*); and midas tamarin, *witaro* (*Saguinus midas*). All were seen and heard, except *P. pithecia*, in the study area (the areas of forest used exclusively by Wapishana residents of Maruranau) during the course of research, including close to swiddens and in other areas heavily frequented by people.

Wapishana ethnoclassification recognises these as distinct natural types, each with a unique name, whose referent in the course of normal usage corresponds with the biological species. Both males and females of the highly sexually dimorphic *P. pithecia* form part of the segregate referred to as *oroa*; informants questioned about this said it is because they are the male and female of the same kind. However, terminologically distinguished subcategories exist within both *soboru* (a smaller kind referred to as *sooman siki*) and *powato* (a larger kind termed *wainsari*). I interpreted this as possibly reflecting the marked size differences between males and females of *Alouatta seniculus* and *Sapajus apella*, although no Wapishana informant identified them as markers of sex-specific differences.

Depending on circumstances, these primary segregates are grouped into several cross-cutting larger categories. The term *powato* is polysemous. Its plural form *powatonnao* also refers to a larger group comprising the eight primate species plus various other arboreal mammals; this grouping appears to be based on perceptual similarity. *Dukornainao*, translated by informants as meaning tree-dwelling creatures, is one of several higher-order categories based on habitat type. Its usage is inconsistent – sometimes apparently interchangeable with *powatonnao*, while at other times referring to a wider category also including various tree-dwelling birds. The term *aimaakannao* is also inconsistently used as a broad, but inexhaustive, collective category within the zoological domain, sometimes apparently restricted to mammals or quadrupeds, or sometimes a wider range of biological taxa. The zoological domain has no unique beginner or discrete collective term as such. Its cultural salience is demonstrated by its exhaustive partitioning into two binary categories based on perceived edibility: *wunii* (or edible animals) and *mawuniki* (animals which are not eaten). Assignment of less inclusive segregates to these groups is flexible according to changes in both cultural preferences and personal dietary choice. Wapishana classification of the zoological domain thus corresponds less to the rigid taxonomies described by some researchers (e.g. Berlin 1992; Atran and Medin 2008), which I consider to be artefacts of the elicitation context, and more to the flexible and dynamic frameworks described in more ethnographically situated studies (Ellen 1993; Sillitoe 1980, 2002b).

Most Wapishana hunters consider the larger six primate species as potential game – hence within the category *wunii*. The small size of squirrel monkeys and tamarins means hunters tend to disregard them as not worth pursuing, though both may be opportunistically captured as pets; I also observed one specimen of *C. olivaceous* being kept as a household pet. Hunting pressure on primates seems to be declining for both technological and cultural reasons. The cost of guns and ammunition in this highly cash-poor society, and to a lesser extent a government reluctance to issue firearm licences following an armed uprising in the Rupununi region in 1969, means that most hunters rely on bows and arrows made from natural materials, with which arboreal animals are harder to kill. Some people also reported changing dietary preferences, especially amongst some young people who express aversion to consumption of primate meat (hence, to them, becoming *mawuniki*, i.e. inedible meat). Ad hoc observations on my part suggested hunting pressure is not severe, with most species, including *Ateles paniscus*, evident even in areas subject to regular human use.

Hunting pressures on large primates, and other popular game species, are mitigated by a series of dietary prohibitions applied to the entire household (usually a tri-generational extended family group) following birth of a child or certain illnesses. These are particularly extensive in the case of spider monkeys, which are particularly sensitive to hunting pressure and, according to both ecological and ethnoecological reports, of great ecological importance as a key disperser of many tree species. The nature of this mechanism is thus suggestive of regulatory functions akin to a traditional form of conservation, part of a wider range of symbolically mediated restrictions on resource use, flexibly applied via customary mechanisms and the intervention of specialist spiritual practitioners (Henfrey 2002, 2018). Spider monkeys in particular are also important in various forms of traditional medicinal practice and various folk tales, some of them alluding to times of greater affinity with humans including shared language and ancestry. While of less direct economic importance than the main game animals, primates are thus recognised by Wapishana people as having both cultural significance and ecological value.

12.4 Ethnoecological Methods

Data reported here derive from a wider study of the ethnoecological knowledge of Wapishana hunters, mainly focussing on the six largest primate species found locally plus six other animal species of dietary, economic, ecological and/or cultural significance (Henfrey 2002, 2017). Of a total of 130 interviews with 18 individual hunters on the ecology of these twelve species, 45 covered primates and comprise the data set addressed in this chapter: *Ateles paniscus* (interviews with 12 different individuals), *Alouatta seniculus* (9), *Cebus apella* (7), *C. olivaceus* (5), *Pithecia* (7) and *Chiropotes satanas* (5).

I identified and recruited interviewees via peer recommendation (cf. Davis and Wagner 2003), targeting those locally regarded as most knowledgeable about the ecology of forest animals. This reflected the aims of the study: To obtain the best possible ethnoecological data set for comparison with scientific findings, not to examine patterns of variation in ethnoecological knowledge or document it comprehensively or systemically. Interviewees therefore came from a small subset of the population: mature men, regular hunters with a local reputation for skill in this regard. The majority had been involved in the balata (latex of the tree *Manilkara bidentata*) trade, and many claimed to have acquired much of their ecological knowledge during extended stays in the forest connected with this work.

Following earlier ethnoecological studies (Townsend 1995; Huntington 1998), the main data collection method was semi-structured interview. I also collected significant quantities of ethnoecological data by other methods: ad hoc recording of comments made during trips to the forest, observations of people's behaviour while hunting, their interpretations of animal signs and their explanations of how they track and hunt animals. However, for the sake of analytical uniformity, this chapter reports only data collected in interviews.

Interviews were conducted in English, as the strongest common language and the more effective in which to frame the categories of information in which I was interested. All animal and plant segregates mentioned in the course of interviews were named in Wapishana; in addition, both interviewer and interviewees commonly employed Creolese names.

Each interview focused on one species of mammal, named in Wapishana in the opening sentence of the interview, usually by the interviewer in a statement along the lines of 'What do you know about [segregate X]'. The interviewee first talked freely for as long as he wished. I subsequently asked specific questions: First clarifying any ambiguous or otherwise unclear points and expanding on points of particular interest and then following a predetermined question schedule reflecting the basic data collection goals I would have set for a preliminary synecological study. The questions covered diet, dispersal behaviour of frugivores, sociality, predation, reproduction, classification and human use. Finally, a series of leading questions invited the interviewee to add any further information on any of the points raised. I subsequently pooled interview data using a basic form of consensus analysis (including points common to two or more interviewees, rejecting those mentioned only once) to obtain overall ethnoecological profiles for each species.

I systematically compared the ethnoecological profiles for each species with data published in ecological studies. For each plant food for which I was able to assign a scientific gloss to the Wapishana name, I scanned the ecological literature for records of consumption of plants of the same genus and family. For other subject areas, I compared ecological and ethnoecological data to determine whether the two were compatible. For each observation in the ethnoecological data set, I also scanned the ecological literature for information on the same subject. When the latter was available, I noted whether or not the corresponding information in the two data sets was consistent.

12.5 Ethnoecological Findings

12.5.1 Summary of Ethnoecological Findings by Species

The following are brief accounts of ethnoecological findings on the six primate species in question, reported in full elsewhere (Henfrey 2017: 187–194):

12.5.1.1 Black Spider Monkey

All interviewees identified fruit as the major food. Most referred to seasonal variation in diet, animals being fatter (and more attractive targets for hunting) during the rainy season glut in fruit production, which is also when the single young is born. Some mentioned that spider monkeys call more often during the dry season when fruit is scarce and that they increase their consumption of leaves to compensate.

Most interviewees reported endozoochorous dispersal of seeds. A few also mentioned a habit of drinking water from pools that form in hollows in trees.

Reported group sizes ranged from 1 to 15, with a modal value of 4–6. One interviewee reported that groups separate to forage during the day and aggregate at sleeping trees at night. Those who considered habitat use concurred that they are found largely in high forest, where they occupy fixed home ranges and sleep on emergent trees. Most interviewees identified harpy eagles as the main predator. Several also mentioned their habit of pelting people with rotten wood.

12.5.1.2 Red Howler Monkey

Fruits and young leaves were identified as the main types of food in almost all cases, with opinion evenly split as to whether fruits or leaves are the most important food. Most identified a seasonal food shortage during the dry season, at which time the diet consists mostly of leaves. Most interviewees said that seeds are dispersed endozoochorously, although a few others contradicted this, saying that seeds are not swallowed.

Suggestions as to group size were quite consistent: most respondents reported groups of 4–6 animals, with wider answers ranging between 2 and 12. Several interviewees said that groups include both larger and smaller individuals (the latter being terminologically distinguished in the Wapishana language as *sooman sik).*

Views varied as to territoriality – several interviewees claimed that howler monkeys occupy fixed home ranges, several others claimed that they do not – and daily activity patterns. All interviewees reported that treetops are used to rest, and the majority further specified a preference for sites with substantial epiphyte cover. Several interviewees suggested falls to be quite common, for a variety of reasons, and carcasses of hunted animals often to show evidence of broken bones. All identified harpy eagles as a major predator, some also mentioned big cats, while several mentioned chronic external worm infestations as a continuous condition.

12.5.1.3 Brown Capuchin

All interviewees identified fruits as the major food source; many noted certain palm fruits as being of particular importance. Most interviewees also mentioned arthropods as an important food: particularly, according to some, in the dry season when fruit is scarce. Several observed that they break open hard-shelled fruits or nuts by banging them against branches. Many also mentioned farm raiding, especially for maize and sugar cane.

Suggested group sizes varied, from approximately 4–10 to 20–30, which I interpreted as meaning that group size itself is variable. Several interviewees distinguished larger individuals via the term *wainsari*, identified by one as the leader of the group, by another as a large male, and observed by another as sometimes being found alone. Many interviewees referred to the formation of mixed-species groups with other species, in particular squirrel monkeys.

Several interviewees observed that breeding coincides with fruiting of the *koram* tree (*Inga alba* and perhaps other species of *Inga*) during the rainy season. All identified kokerite palms (*Attalea regia*) as the preferred resting place, some specifying the use of the large woody spathes to shelter from rain. Eagles, usually harpy eagles, were consistently noted as the major predator. Capuchins are reputed to outwit the attempts of jaguar to capture them, including in Wapishana folklore, which often alludes to the intelligence of this species.

12.5.1.4 Wedge-Capped or Weeping Capuchin

Interviewees all reported fruits and arthropods as the major foods, with several specifying the fruits of various species of *Inga* as of particular importance. Most also noted a seasonal food shortage during the dry season; a couple identified the fruit of *Parinari excelsa* as a key food source during this time. Some interviewees suggested seeds may be dispersed exozoochorously, at least of some species, when fruits are carried some distance from the parent tree before they are eaten and the fruit discarded.

Reported group sizes range from 4 to 12. Some interviewees mentioned (but did not terminologically distinguish) a larger kind that occasionally form groups of one or two individuals. Several interviewees mentioned that groups rest in the spathes of kokerite palms (*Attalea regia*). Most interviewees identified eagles as the main predator, in most cases the harpy eagle.

12.5.1.5 Guianan Saki

All interviewees agreed that fruits are eaten, in some cases adding either leaves and flowers, or insects. Several identified as the major foods the fruits of the kokerite palm (*Attalea regia*) and/or those of various *Inga* species. While some interviewees claimed selective endozoochorous dispersal of seeds, opinions varied as to whether or not seeds or dispersed.

Observations on group size ranged from two to six individuals, some specifying that groups include both sexes. All interviewees recognised the difference between males and females of this highly sexually dimorphic species, which are not terminologically distinguished. While some interviewees reported aggressive intergroup interactions, one interviewee said groups might temporarily aggregate to feed at the same kokerite tree.

Some interviewees noted that this species is common in the farm area (much of which comprises thick secondary growth in fallows). Many drew attention to their distinctive locomotory pattern based more on jumping between tree trunks than climbing through branches. A number of eagle species were named as predators, with some interviewees noting that sakis will hide from predators rather than attempt to flee.

12.5.1.6 Brown-Bearded Saki

All interviewees agreed that the diet is composed of fruit alone, with most specifying that unripe fruits are eaten. Several labelled this behaviour destructive, on the grounds that the seeds are often masticated and, in any case, picked before they are mature and so have no chance to grow. This information was contrasted with the behaviour of other monkeys known to disperse the seeds of their food plants and so – like people – contribute to their propagation.

Observations on sociality suggested groups of anything from 8 to 40 individuals, most commonly between 15 and 20. Two interviewees described how groups disperse into smaller subgroups to forage and aggregate at particularly ample food sources. Both suggested this information might reflect the uneven distribution of fruit and the fact that few single trees supply enough fruit to feed the entire group at once. Observations on habitat use suggested a preference for large areas of continuous high forest.

12.5.2 Comparison of Ethnoecological and Ecological Data

Space does not permit the inclusion here of a full description of the ethnoecological data set or its comparison with scientific data. For details of this and full references to scientific studies consulted, see Henfrey (2002: 179–212, 2017: 185–226). Table 12.1 provides a summary overview.

Table 12.1 Summary of comparison between ethnoecological and ecological data sets

Species	*Ateles paniscus*	*Alouatta seniculus*	*Sapajus apella*	*Cebus olivaceous*	*Chiropotes satanas*	*Pithecia pithecia*
Number of interviews	12	9	7	5	5	7
Proportion of food plants corroborated in ecological studies to family level	95% (20/21)	73% (8/11)	83% (10/12)	82% (9/11)	91% (10/11)	83% (5/6)
Proportion of food plants corroborated in ecological studies to genus level	81% (17/21)	36% (4/11)	42% (5/12)	64% (7/11)	45% (5/11)	50% (3/6)
Number of other observations for which comparable ecological data available	15	18	11	8	8	7
Proportion of other observations for which ethnoecological and ecological data are compatible	0.87	1.00	1.00	1.00	0.88	1.00

The ethnoecological and ecological data sets show high levels of substantive overlap. The majority of food species identified via ethnoecological methods were corroborated in ecological studies. Percentage corroboration at family level ranged from 73% (red howler monkey, *Alouatta seniculus*) to 95% (black spider monkey, *Ateles paniscus*), at genus level from 36% (red howler monkey, *Alouatta seniculus*) to 81% (black spider monkey, *Ateles paniscus*).

These figures are impressive considering the incompleteness and, in some cases, scarcity of ecological data on these species at the time. Ecological data sets are far from complete, and for many species, comparison relied on data from locations geographically distant and ecologically very different from the setting of this study. It is perhaps noteworthy that the highest correspondence at both genus and family level came from the species with the best quality data set, a multi-year field study of the ecology of *Ateles paniscus* (black spider monkey) in Surinam (Roosmalen 1985). This correspondence suggests that food lists derived from ethnoecological research are largely reliable, in which case food items occurring only in ethnoecological data sets will in most cases correspond to those eaten locally but not recorded in conventional ecological studies. However, this conjecture is impossible to prove without conducting formal ecological research in the same area over periods sufficiently long to assemble comprehensive food lists.

Considering qualitative observations, correspondence was very strong. I found directly comparable information in the ecological literature for 67 distinct observations in the ethnoecological data set. In 64 cases, the observations are at least consistent; in most of these, they are identical. For many ethnoecological observations, the ecological literature provided no comparable information: ethnoecology may thus extend the range of existing ecological data sets (cf. Ferguson et al. 1998; Huntington et al. 1999; Myrmin et al. 1999) or at least point towards new lines of investigation (Posey 1986), treating novel information from ethnoecological studies as hypotheses for testing by formal methods in scientific ecology.

Substantive correspondence between ethnoecological and ecological data sets is unsurprising, given the common ground between hunters and scientists in both aims and available techniques. Scientific researchers seek to reveal the behavioural ecology of the animal species of interest. Hunters apply detailed knowledge of their behaviour to increase the effective availability of animals that are often scarce and usually furtive in their behaviour.

Most participants in ethnoecological interviews qualified statements of knowledge with some reference to its acquisition. Almost all information reported was based on direct experience. Their reported methods were a subset of those employed by field biologists: direct observation of behaviour, interpretation of tracks and other spoor such as feeding signs and droppings, and examination of stomach contents of hunted animals. Correspondences between ecological and ethnoecological data sets therefore reflect overlap in aims and methods. Differences in opportunities for observation mean this correspondence is only partial.

12.5.3 Limitations of the Ethnoecological Data Set

While ethnoecological data appear to be accurate, they are also limited in various respects. Dietary information is incomplete, and in some other topics, ethnoecological methods yielded minimal or no data of direct scientific value. Some of these limitations I believe to be, at least in part, the result of specific weaknesses in the methodology employed here, upon which future ethnoecological studies may improve. Others may be inherent to ethnoecological studies.

Lists of corroborated food species in the ethnoecological data set were far shorter than those in the most complete of the ecological studies. Published studies list 171 food plants for *Ateles paniscus* (Roosmalen 1985) compared with 26 in the current work, 97 versus 15 for *Alouatta seniculus* (Julliot and Sabatier 1993) and 66 versus 18 for *Sapajus apella* (Guillotin et al. 1994). Food lists were comparable in length for *Cebus olivaceus* (Wright 2002) and *Chiropotes satanas* (Norcock and Kinzey 1994), but only the *C. olivaceus* study covers at least a full year.

The case of *Ateles paniscus* strongly suggests gaps in the ethnoecological data set. The fruit of *Bagassa guianensis* is an important component of spider monkey diets in Surinam (Roosmalen 1985; Norcock and Kinzey 1994) and French Guiana (Simmen and Sabatier 1996). Its referent in Wapishana, for which I made a reliable field identification, was only mentioned by one ethnoecological informant and hence discarded from the data set, although the tree itself was familiar to all interviewees. This may reflect a biogeographical difference: *B. guianensis* is relatively uncommon in the study site and locally absent from many people's primary hunting areas, limiting their opportunities for observation. The discrepancy may also result from the relatively small number of interviewees and the weakness of the method used for consensus analysis.

For food items other than fruits, the discrepancy between the two data sets is even greater. Ethnoecological interviews identified of only one species whose leaf is consumed by each of *Ateles paniscus* and *Alouatta seniculus* (although most interviews on these species identified leaves as a category of food), whereas ecological studies report 28 for *Ateles paniscus* (Roosmalen 1985: 74) and 98 for *Alouatta seniculus* (Julliot and Sabatier 1993: 536). Ethnoecological data on other food categories such as flowers and invertebrate foods are similarly impoverished. For arthropods and other invertebrate foods, the difficulty of field identification means the same is often true in ecological studies (e.g. Freese and Oppenheimer 1981).

In some respects, the Wapishana biological lexicon constrains the potential of the ethnoecological data set. For example, very few named terminal categories in the Wapishana language refer to plants of liana habit: most are subsumed under a single residual category. In one ecological study of the feeding behaviour of *Ateles paniscus* in Surinam, 25.6% of food species reported were lianas (Roosmalen 1985). In ethnoecological interviews on this and other species, interviewees often reported that the fruits of several kinds of lianas were eaten, but that the plants concerned either did not have Wapishana names or that if they did, they did not know them. The same may well apply to invertebrate foods, as Wapishana classification clumps many categories of invertebrates into groups corresponding to relatively high taxonomic ranks, often order (Henfrey 2002). I found the Wapishana terminology for ecological zones not to be very detailed, so lexical factors might also limit ethnoecological observations on habitat use.

Further weaknesses in the ethnoecological data set were apparent in subject areas not included among the data reported here. My earliest phase of interviews included questions on group dynamics and breeding rates. In the former case, answers given were invariably to the effect that juvenile animals, on maturity, remain in the natal group to breed with either parents or siblings. Such answers clearly contradict basic biological theory on inbreeding avoidance and are of little biological value, though may be part of significant cultural narratives.

It is hardly surprising that ethnoecological enquiry does not produce accurate information on these topics, as they are not accessible to the opportunistic observation that is its main method. The collection of such information by biologists depends on regular observations, sustained over extended periods, of particular animals recognised as either individuals or groups. Wapishana and other indigenous naturalists whose main immediate concerns are much more practical do not interact with animals on such a basis and so cannot reproduce data sets collected under the conditions in which most biological field research takes place.

12.6 Discussion: Suggestions for Improvement in Ethnoecological Research Methods

In the light of both the findings of this research and methodological prescriptions of other ethnoecological studies since published, in this section I suggest various methodological improvements. I believe their implementation would substantially improve data on diets in particular. Some weaknesses of the ethnoecological data set in this study, however, I believe to be inherent to this type of research, however good the methodology.

Sample size is one factor: clearly, the greater the number of people providing information, the more the information they can collectively provide. Simple linear regression using the full data set from this study (i.e. also including animals other than primates) indicated a strong positive correlation between the number of interviews on any particular animal species and the number of corroborated food plants elicited ($r^2 = 0.691$, $p < 0.001$). The same analysis did not indicate any relationship between numbers of corroborated food plants in the ethnoecological data set and the numbers of food species reported in the most complete ecological study available for each species ($r^2 = 0.043$, $p = 0.6$). This strongly suggests that further interviews on any animal species would have extended the list of food plants. For no species did we conduct sufficient interviews to reach a point of diminishing returns where further interviews introduced few or no new food plants. Acting upon this finding may not always be possible, as any study is constrained by limits of time and resources and can only incorporate as many interviews as these allow. However, additional interviews are in most circumstances a more efficient way to accumulate ecological data than conventional field research methods. Given sufficient field time, I would suggest following the recommendation of Davis and Wagner (2003) that peer recommendations be followed comprehensively in order to identify all those regarded as experts within a particular local setting.

Increasing the number of interviewees would both necessitate and facilitate a more sophisticated method for determining which responses to include. With the rough method employed, increased sample size raises the possibility of including erroneous observations, if two or more interviewees provide identical, factually inaccurate, information. It is also important to identify rare, accurate information provided by collaborators with particularly extensive or specialised knowledge. Consensus analysis (Romney et al. 1986) gave unsatisfactory results with this data set, which deals with dispersed knowledge in which most information was provided by only a small number of informants, but could perhaps be modified to deal with a larger data set.

Another useful method for corroborating doubtful information and dealing with contradictory responses is group interviews (Huntington 1998). Individual interviews remain crucial, as they compel the interviewee to respond. It is likely that many people provide information that they would not in a group context, in which people are effectively competing for speaking time and may be inhibited by the prospect of censure for inaccurate responses. Discussion in a group context may also encourage people to modify their opinions in the light of what other people are saying; the dynamics of this process are complex and not necessarily based upon deference to superior knowledge (Ellen 1993). I therefore propose that group interviews would be most effective as a follow-up to a programme of individual interviews. Pooled information from individual interviews could form the starting point for discussion in group sessions concerned with establishing a consensus on controversial or infrequently mentioned points.

Dietary lists might be more complete if food plants, rather than the animal species themselves, are the starting point of interviews. While simply naming the plants may be acceptable, more effective would be to observe them in situ or provide either specimens or high-quality pictorial representations. In the present study, interviewees often volunteered information, within and outside of the interview context, in response to the sight of a fruit or a plant known to be food for a particular animal. I also conducted a small number of interviews on tree ecology, which yielded much information on animal consumers and dispersers absent from those based upon the animal species as their starting point, even in cases where the same interviewees had spoken about the same animals. This suggests that interviewee recall is an important factor: focusing on the plant eaten rather than the animal stimulated the recall of different information.

Conducting ethnoecological interviews on every single plant that might possibly be an animal food could be rather laborious, particularly if only a few species of consumers are of interest. More efficient is the use of botanical voucher specimens to serve as concrete stimuli for responses. Interviewees could be asked to pick out the plants eaten by a particular animal, asked whether a particular animal eats the species represented by each voucher specimen in turn or asked to list the animals who feed on each species in turn, depending on the aims and setting of the study and the time available. It is clear that for ethnoecological information to be scientifically useful requires translation of the local botanical and zoological lexicons. For this purpose, as well as the potential methodological application of voucher specimens, a study such as this one would be most effectively conducted in conjunction with a thorough study of ethnonomenclature, especially of plants.

One study of Guajá hunters' knowledge on the diets of monkeys used botanical voucher specimens as the basis for elicitation, obtaining far longer lists of food plants for several primate species than those obtained in the present study (Cormier 2004). In the Guajá study, 90 food plants were listed for *Alouatta belzebub*, 88 for *Sapajus apella* and 74 for *Chiropotes satanas.* This data may partially reflect differences in knowledge and sample size: monkeys are far more important to Guajá than Wapishana hunters, and in the Guajá study, data came from 25 informants. However, I strongly suspect that using plants rather than animals as the starting point of interviews is also a significant factor.

12.7 Conclusion

This chapter has addressed the relationship to scientific ecology of a limited aspect of Wapishana ethnoprimatology: ethnoecological information on various primate species, organised into categories derived from the concepts of scientific ecology and compared with information collected in ecological studies. Ethnoecological data is largely consistent in both form and substance with that of scientific ecology, is accurate in detail when assessed in scientific terms and for many species extends the existing ecological data set. It is also limited in various respects, in which scientific ecology provides a possibility for extending its epistemological and analytical range.

The area of overlap also encompasses scientific epistemology: The generation of hypotheses and the collection of field data according to which these hypotheses can be tested. Applying ethnoecological data within its cultural context, by combining it with information on resource use, generated hypotheses concerning the ecological consequences of resource use amenable to testing via scientific methods (Henfrey 2002, 2017). The demands of collecting data associated with such testing can be partly fulfilled by applying practical and technical skills associated with performative aspects of local ecological knowledge to ecological research, an endeavour to which the skills of indigenous hunters are well suited and readily transferable.

Many of the subject areas inaccessible to ethnoecological enquiry include precisely those of most importance in conventional management programmes based upon scientific ecology. Traditional resource management systems include a variety of mechanisms for monitoring populations and regulating harvest based on different premises from scientific management, requiring far less data input (Johannes 1998). These represent a set of methodologies and practical measures available in LEK but beyond the range of SEK. LEK and SEK are thus complementary not only in terms of producing partially overlapping data sets, but also in that each can extend the epistemological and practical scope of the other.

There thus exists a strong basis for the complementary integration of LEK and SEK, able to extend the range of each without eliminating any of their essential features. The greatest likelihood of such a synergistic combination arises from employing LEK in its original context: interactions of resource users with their biotic environment in the conduct of routine domestic and subsistence tasks. This complementarity means that scientific ecologists concerned with resource management can apply their skills most effectively by placing them at the disposal of traditional resource users, as tools to extend their management capacity and provide a stronger base for local decision-making, within the context of equitable relationships based on mutually respectful dialogue.

Acknowledgements Thanks to the people of Maruranau village for allowing me to live and work there, particularly those who collaborated as interviewees and many others who provided direct assistance during my stay there, and Guyanese government officials at the Environmental Protection Agency and Ministry of Amerindian Affairs for expediting permission to stay and conduct research in Guyana. Roy Ellen, Simon Platten and Victoria Reyes-Garcia provided helpful comments on earlier versions of this chapter. Fieldwork in Guyana was funded by an APFT research studentship originating with D.G.VIII of the European Commission. Writing was supported by an individual research grant from the Wenner-Gren Foundation.

References

Agrawal A (1995) Dismantling the divide between indigenous and scientific knowledge. Dev Chang 26(3):413–439

Agrawal A (2002) Indigenous knowledge and the politics of classification. Int Soc Sci J 173:287–297

Atran S, Medin D (2008) The native mind and the cultural construction of nature. MIT Press, Cambridge

Baldwin R (1946) The Rupununi record. The Barbados Advocate Company, Barbados

Balée WL (1994) Footprints of the forest: Ka'apor ethnobotany; the historical ecology of plant utilization by an Amazonian people. Columbia University Press, New York

Becker CD (2003) Grassroots to grassroots: why forest preservation was rapid at Loma Alta, Ecuador. World Dev 31(1):163–176

Becker CD, Ghimire K (2003) Synergy between traditional ecological knowledge and conservation science supports forest preservation in Ecuador. Cons Ecol 8(1):1. [online] URL: http://www.consecol.org/vol8/iss1/art1

Berlin B (1992) Ethnobiological classification. Princeton University Press, Oxford

Bodmer RE, Puertas PE (2000) Community-based comanagement of wildlife in the Peruvian Amazon. In: Robinson JG, Bennett EL (eds) Hunting for sustainability in tropical forests. Columbia University Press, New York, pp 395–409
Bridges J (1985) Rupununi mission. The story of Cuthbert Cary-Elwes SJ among the Indians of Guiana 1909–1923. Jesuit Missions, London
Cormier LA (2004) Monkey ethnobotany. In: Stepp JR, Wyndham FS, Zarger RK (eds.), Ethnobiology and cultural diversity. Proceedings of the seventh international congress of ethnobiology, International Society of Ethnobiology, pp 313–325
Davis A, Wagner J (2003) Who knows? On the importance of identifying "experts" when researching local ecological knowledge. Hum Ecol 31(3):463–489
DeWalt BR (1994) Using indigenous knowledge to improve agriculture and natural resource management. Hum Organ 53(2):123–131
Donovan D, Puri RK (2004) Learning from traditional knowledge of non-timber forest products: Penan Benalui and the autecology of Aquilaria in Indonesian Borneo. Ecol Soc 9(3):3. [online] URL: http://www.ecologyandsociety.org/vol9/iss3/art3
Dove M (2002) Hybrid histories and indigenous knowledge among Asian rubber smallholders. Int Soc Sci J 54(173):349–359
Ellen RF (1993) The cultural relations of classification. Cambridge University Press, Cambridge
Ellen RF (2004) From ethno-science to science, or 'what the indigenous knowledge debate tells us about how scientists define their project'. J Cogn Cult 4(3–4):409–450
Ellen RF, Harris HJ (2000) Introduction. In: Ellen RF, Bicker A, Parkes P (eds) Indigenous environmental knowledge and its transformations. Harwood, Amsterdam, pp 1–33
Escobar A (1995) Encountering development: the making and unmaking of the third world. Princeton University Press, Chichester
Ferguson MAD, Williamson RG, Messier F (1998) Inuit knowledge of long-term changes in a population of Arctic tundra caribou. Arctic 31(3):201–219
Freese C, Oppenheimer JR (1981) The capuchin monkeys, genus Cebus. In: Coimbra-Filho AF, Mittermeier RA (eds) Ecology and behaviour of Neotropical primates, vol 1. Academia Brasiliera de Ciências, Rio de Janiero
Gilchrist G, Mallory M, Merkel F (2005) Can local ecological knowledge contribute to wildlife management? Case studies of migratory birds. Ecol Soc 10(1):20. [online] URL: http://www.ecologyandsociety.org/vol10/iss1/art20/
Gomes P, Wilson H (eds) (2012) Thinking together for those coming behind us: an outline plan for the care of Wapichan territory in Guyana. South Central and South Rupununi Districts Toshaos Councils, South Rupununi, Region 9, Guyana
Guillotin M, Dubost G, Sabatier D (1994) Food choice and food competition among the three major primate species of French Guiana. J Zool 233:551–579
Hanna SS (1998) Managing for human and ecological context in the Maine soft shell clam fishery. In: Berkes F, Folke C (eds) Linking social and ecological systems. Management practices and social mechanisms for building resistance. Cambridge University Press, Cambridge, pp 190–211
Henfrey T (1999) Land conflicts and cultural change in Southern Guyana. In: Grenand P, Grenand, F (eds) Les peuples des forêts tropicales aujord'hui. APFT Final Report Volume IV: Volume Regional Caraibes, pp 328–333
Henfrey T (2002) Ethnoecology, resource use, conservation and development in a Wapishana community in the South Rupununi, Guyana. Ph.D. dissertation. University of Kent, Canterbury
Henfrey T (2017) Wapishana ethnoecology: a case study from the South Rupununi, Guyana. Wapichan Wadauniinao Ati 'o, Maruranau
Henfrey T (2018) Edges, fringes, frontiers. Integral ecology, indigenous knowledge and sustainability in Guyana. Berghahn, New York, Oxford
Huntington HP (1998) Observations on the utility of the semi-directive interview for documenting traditional ecological knowledge. Arctic 51(3):237–242
Huntington HP (2000) Using traditional ecological knowledge in science: methods and applications. Ecol Appl 10(5):1270–1274

Huntington HP, The Communities of Buckland, Elim, Koyuk, Shaktoolik and Point Lay (1999) Traditional knowledge of the ecology of beluga whales (Delphinapterus leucas) in the eastern Chukchi and northern Bering Seas, Alaska. Arctic 52(1):49–61

Johannes RE (1998) The case for data-less marine resource management: examples from tropical nearshore fisheries. Trends Ecol Evol 13:243–246

Julliot C, Sabatier D (1993) Diet of the red howler monkey (Alouatta seniculus) in French Guiana. Int J Primatol 14(4):527–550

Moller H, Berkes F, Lyver PO, Kislalioglu M (2004) Combining science and traditional ecological knowledge: monitoring populations for co-management. Ecol Soc 9(3):2. [online] URL: http://www.ecologyandsociety.org/vol9/iss3/art2

Myrmin NI, The Communities of Novoe Chapling, Sireniki, Uelen, Yanrakinnot, Huntington HP (1999) Traditional knowledge of the ecology of beluga whales (Delphinapterus leucas) in the northern Bering Sea, Chukotka, Russia. Arctic 52(1):62–70

Nabhan GP (2003) Singing the turtles out to sea. The Comcáac (Seri) art and science of reptiles. University of California Press, Berkeley, Los Angeles and London

Norcock MA, Kinzey WG (1994) Challenges of neotropical frugivory: travel patterns of spider monkeys and bearded sakis. Am J Primatol 34:171–183

Pinkerton E 1998. Integrated Management of a Temperate Montane Forest Ecosystem through Wholistic Forestry: A British Colombia Example, in F. Berkes and C. Folke (eds), Linking Social and Ecological Systems: Management Practices and Social Mechanisms for Building Resilience. Cambridge: Cambridge University Press, pp. 363–89.

Ponte Johansons V (1995) Contributions of the Warao Indians to the Ichthyology of the Orinoco Delta, Venezuela. In: Heinen H, Dieter UA (eds) Naturaleza y Ecología Humana en el Neotrópico/nature and human ecology in the neotropics, Scientia Guaianæ No 5. IVIC, Caracas, pp 371–392

Posey DA, J Frechione J. Eddins, L. Francelino da Silva (1984) Ethnoecology as Applied Anthropology in Amazonian Development. Human Organization 43(2): 95–107

Posey DA (1986) Topics and issues in ethnoentomology with some suggestions for the development of hypothesis-generation and testing in ethnobiology. J Ethnobiol 6(1):99–120

Posey DA (1990) The application of ethnobiology in the conservation of dwindling natural resources: lost knowledge or options for the survival of the planet. In: Posey DA, Overal WL (eds) Ethnobiology: implications and applications, Proceedings of the first international congress of ethnobiology, Vol. 1

Purcell TW (1998) Indigenous knowledge and applied anthropology: questions of definition and direction. Hum Organ 57(3):258–272

Puri RK (2001) Local knowledge and manipulation of the fruit Mata Kucing (Dimocarpus longan) in East Kalimantan. In: Victor M, Barash A (eds) Overview of an international seminar on cultivating forests: alternative forest management practices and techniques for community forestry, 23–25 September 1998. RECOFTC Report no. 17, Bangkok

Richards P (1985) Indigenous agricultural revolution: ecology and food production in West Africa. Hutchinson, London

Romney AK, Weller SC, Batchelder WH (1986) Culture as consensus: a theory of culture and informant accuracy. Am Anthropol 88(2):313–338

van Roosmalen MGM (1985) Habitat preferences, diet, feeding strategy and social organisation of the black spider monkey (Ateles paniscus paniscus Linnaeus 1758) in Surinam. Acta Amazon:15, 7–238. número 3/4 (suplemento)

Simmen B, Sabatier D (1996) Diets of some French Guianan primates: food composition and food choices. Int J Primatol 17(5): 661–693.

Sillitoe P (1980) Confusions in the classifications: how the Wola name their plants. Ethnos 45(3–4):133–156

Sillitoe P (1998) The development of indigenous knowledge. Curr Anthropol 19(2):223–235

Sillitoe P (2002a) Participant observation to participatory development: making anthropology work. In: Sillitoe P, Bicker A, Pottier J (eds) Participating in development: approaches to indigenous knowledge (Association of Social Anthropologists Monograph 39). Routledge, London, New York, pp 1–23

Sillitoe P (2002b) Contested knowledge, contingent classification: animals in the highlands of Papua New Guinea. Am Anthropol 104(4):1162–1171

Stirrat RL (1998) Reply to sillitoe. Curr Anthropol 39(2):242–243

Ticktin T, Johns T (2002) Chinanteco management of Aechmea magdalenae: implications for the use of TEK and TRM in management plans. Econ Bot 56(2):177–191

Ticktin T, de la Peña Valencia G, Illsey Granich C, Dalle S, Ramirez F (2004) Participatory ethnoecological research for conservation. In: Stepp JR, Wyndham FS, Zarger RK (eds.), Ethnobiology and cultural diversity. Proceedings of the seventh international congress of ethnobiology, International Society of Ethnobiology, pp 575–584

Toledo V (2001) Biocultural diversity and local power in Mexico: challenging globalization. In: Maffi L (ed) On biocultural diversity. Smithsonian Institute Press, Washington, DC, pp 472–488

Townsend W (1995) Cultural teachings as an ecological data base: Murui (Witoto) knowledge about primates. Latinamericanist 31(1):1–7

Vasquez-Davila MA (1995) El amash y el pistoloque: un ejemplo de la etnoecólogia de los Chontales de Tabasco, Mexico. Ethnoecólogica 3:59–69

Warren DM, Slikkerveer LJ, Brokensha D (eds) (1995) The cultural dimension of development. Intermediate Technology Publications, London

Wright BW (2002) The relationship between jaw morphology and the physical properties of fruits in "robust" (Cebus apella) and “gracile” (Cebus olivaceus) capuchins in central Guyana. Ph.D. thesis. University of Illinois, Chicago

Chapter 13
Past, Present, and Future of Secoya Ethnoprimatology in the Ecuadorian Amazonia

Stella de la Torre, Pablo Yépez, and Alfredo Payaguaje

13.1 The Past

Nonhuman primates have been the focus of our attention since our origins as a species. Their notable similarities with us, humans, in their behaviors and morphology have been widely represented and interpreted in many different ways. As a result, nonhuman primates have been an important component of the myths and religions of many human cultures around the world in past and present times (Estrada et al. 2017). The importance of nonhuman primates in human cultures is evidenced by a large variety of archeological work that includes several types of artistic representations of nonhuman primates. The Nazca figures in Perú (200 BC–900 AD) and the ceramics of the Machalilla culture in Ecuador (1800–1500 BC) are two examples of this old and strong relationship between our species and the nonhuman primates (Zardini 1991; Meggers and Evans 1962).

Before the European colonization, the myths and traditions of the native cultures in the Neotropics represented nonhuman primates as the result of failed attempts to create humans or as humans that were transformed after confronting their gods (Gutiérrez 2007). Monkey representations in ceramics of this period have been also related to reproduction and fertility since primates were considered

S. de la Torre (✉)
Universidad San Francisco de Quito, Colegio de Ciencias Biológicas y Ambientales. Av. Interoceánica y Jardines del Este, Cumbayá, Quito, Ecuador
e-mail: sdelatorre@usfq.edu.ec

P. Yépez
Fundación Raíz Ecuador. Vía Intervalles, Pasaje Benjamín Carrión, Quito, Ecuador

A. Payaguaje
Nacionalidad Secoya del Ecuador. Comunidad Secoya San Pablo de Catesiaya, San Pablo, Ecuador

B. Urbani, M. Lizarralde (eds.), *Neotropical Ethnoprimatology*, Ethnobiology,
https://doi.org/10.1007/978-3-030-27504-4_13

a link between the material and the spiritual worlds (Uribe 2016). In the Neotropics, Amazonian cultures, such as the Secoya, may well have been the ones that had the most direct and strong relationship with nonhuman primates. The diverse primate community of Amazonian ecosystems (de la Torre 2000; Peres 1997) and the diurnal and conspicuous behavior of these animals very likely attracted the attention of these first people since they occupied these habitats. We present a temporal analysis of the knowledge and perceptions of the Secoya about nonhuman primates. The results of this analysis point to the need of preserving their deep traditional knowledge as a means to conserve nonhuman primates, Amazonian forests, and the Secoya culture.

13.1.1 The Secoya

The Secoya people are members of the western Tukano group that occupied large areas in the upper Amazon basin, from the Putumayo to the Amazonas rivers in Colombia, Ecuador, and Peru (Vickers 1989). In present times, the Secoya people are an ethnic minority of about 600 people living in some areas of the Aguarico River in Ecuador (Fig. 13.1). A similar number of Secoya people lives in Peru, in the upper Napo Basin (Yépez et al. 2010). These Amazonian ecosystems are the environment where the Secoya culture originated and evolved. Their survival, success, and persistence as a culture depended on the profound knowledge they had about the different elements of these complex ecosystems (Cerón et al. 2011; Vickers 1989).

13.1.2 Secoya Ethnoprimatology

For the Secoya, nonhuman primates were allies, partly responsible of their success as a culture. They were considered as forest guides that taught humans what could and could not be eaten. They were also seen as forest guardians, alerting humans about the presence of predators. Monkeys were also kept as pets and, very importantly, were a valuable source of protein (Cipolletti and Payaguaje 2008). Some species of nonhuman primates, especially the larger ones, such as the woolly monkeys (*Lagothrix lagotricha*), have been a highly valued hunting prey in past and present times. Evidently, this close relationship was based on a considerable knowledge of all nonhuman primate species occurring in their territory, identifying each by morphological and behavioral characteristics (Vickers 1989) (Table 13.1).

This intimate relationship between the Secoya people and the Amazonian primates, that began several hundred years ago, is expressed in the Secoya beliefs about the ability of monkeys to move between the material world, the forest, and the spiritual world. This theory was stated by one of their elders: “Since the beginning of life, monkeys have helped our god, *Ñañe-Paina*, to create people and other monkeys.

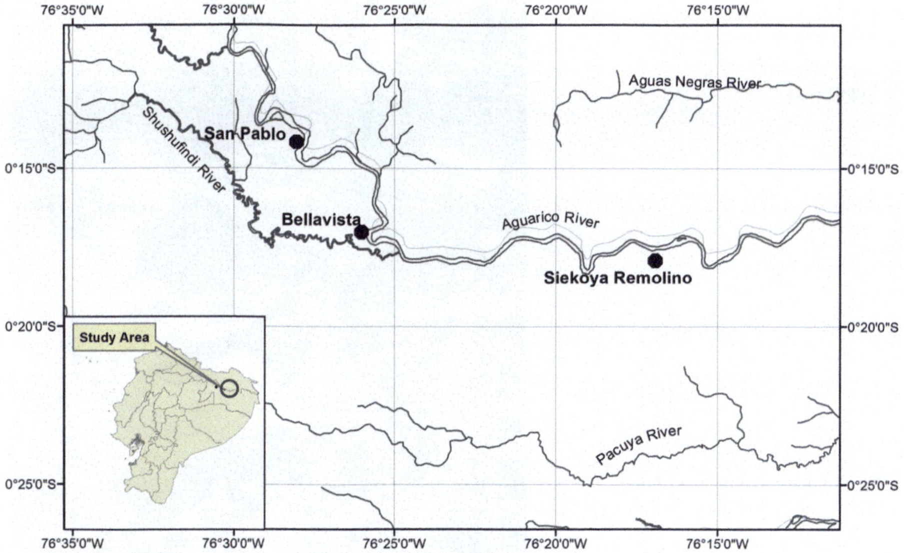

Fig. 13.1 Secoya communities in Ecuador, San Pablo de Catesiaya, Bellavista, and Siekoya Remolino

Table 13.1 Secoya taxonomy of nonhuman primates

Secoya name	Scientific name	Common name (English)	Meaning of the Secoya name
Ñunkwa sisi	*Cebuella pygmaea*	Pygmy marmoset	Little monkey of the *Astrocaryum* palm
Nea sisi	*Leontocebus nigricollis*	Black-mantled tamarin	Black little monkey
Bo sisi	*Saimiri cassiquiarensis*	Squirrel monkey	White little monkey
Bo take	*Cebus yuracus*	White capuchin monkey	White capuchin monkey
Ma wa'o	*Plecturocebus discolor*	Dusky titi monkey	Red hairy monkey
Nea wa'o	*Cheracebus lucifer*	Yellow-handed titi monkey	Black hairy monkey
Wa'o su'tu	*Pithecia milleri*	Saki monkey	Grey hairy monkey
Yami naso	*Aotus vociferans*	Owl monkey	Night woolly monkey
Emu	*Alouatta seniculus*	Red howler monkey	Howler monkey
Yuwi naso	*Lagothrix lagothricha lagothricha*	Common woolly monkey	Ashy woolly monkey
Ma naso	*L. l. poeppigii*	Poeppigii's woolly monkey	Red woolly monkey
Pai take	*Ateles belzebuth*	Spider monkey	Humanlike monkey

Modified from Vickers (1989)
All the listed taxa occur in the Secoya territory in Ecuador, with the exception of *Ateles belzebuth*.
Primate taxonomy follows Mittermeier et al. (2013)

Fig. 13.2 *Inga acuminata* Benth., called *sisi pene* in Secoya language (credits: P. Yépez)

They have helped our god several times, when other animals of the forests were trying to kill him…" (D. Payaguaje, Secoya shaman, pers. comm.).

By carefully observing and recording the plant species used by monkeys as food sources, the Secoya people were able to identify some of their potential foods. They also used monkeys as references to name culturally important plant species:

- *Inga acuminata* Benth. is called *sisi pene*, meaning "squirrel monkey's fruit" (*sisi* is squirrel monkey and *pene* refers to a legume fruit) (Fig. 13.2).
- *Matisia obliquifolia* Standl. is called *take apasi*, meaning "capuchin monkey's fruit" (*take* is capuchin monkey (Fig. 13.3) and *apasi* refers to a drupe fruit).
- *Plukenetia polyadenia* Müll. Arg., is called *take tsima*, meaning "capuchin monkey's curare" (*tsima* means "venom/curare").
- *Theobroma subincanum* Mart. is called *take pona*, meaning "capuchin monkey's cocoa (*pona* is cocoa).
- *Pouteria glomerata* (Miq.) Radlk. is called *naso toa*, meaning "woolly monkey's fruit" (*naso* is woolly monkey and "*toa*" refers to a drupe fruit) (Cerón et al. 2011).

Other plant species have been named because they share some features with monkeys. *Inga velutina* Willd. is called *emu pene*, meaning "red howler monkey's fruit" ("*emu*" is howler monkey and *pene* is a legume fruit). This fruit has a red and hairy cover that looks similar to the fur of this monkey. Other fruits have some similarity to Secoya cooking tools so they were named as tools for the monkeys. *Eschweilera* spp. is called *take cua'co* (Fig. 13.4), meaning "capuchin monkey's

Fig. 13.3 *Cebus yuracus*, called *Bo take* in Secoya language (white capuchin monkey) (credits: P. Yépez)

Fig. 13.4 *Eschweilera* spp. *take cua'co* in Secoya language (credits: P. Yépez)

pot" ("*take*" is capuchin monkey and *cua'co* is pot) conveying that they used them (Cerón et al. 2011). These examples are not only evidence of the Secoya knowledge on the feeding behavior of monkeys, as in the case of the fruits that are eaten by the monkeys, but also of their belief that monkeys are so similar to humans that they even use venoms and pots.

Based on careful observations of monkeys and of the flowering and fruiting times in the Amazonian forests, the Secoya associated the period of high fruit production to an increase in weight of monkeys. The Secoya called the month of high fruit productivity as *naso huiyape ñañe* which means "fat woolly monkeys' month." They focused their hunting efforts on woolly monkeys in this month, that corresponds to April, a rainy season month in northern Ecuadorian Amazonia where the Secoya people live.

The traditional knowledge that the Secoya had about nonhuman primates and their acknowledgement of the complexity and similarity of their behaviors with that of humans was the basis of myths about the power of nonhuman primates and their importance to gods and people. The myth of the birth of *Ñañe*, the god of the Secoya, evidences this, stated in the following: "It is said that *Ñañe* was born from a stone that was kept by the daughters of the *potoo*, a nocturnal bird. When *Ñañe* was a child, he cried like a small bird, so the girls put him in a bowl to protect him. When *Ñañe* was young, he visited the people of these old places and transformed them into peccaries and other animals. This is how *Ñañe* created all the diversity of animals of the Amazon. When *Ñañe* was adult, there were people that lived in the underground. They cooked and ate red clay that they called peach palm. One day, a wise man came from the underground to look for firewood to cook the clay. *Ñañe* was hidden in the forest watching the man. He stepped on the wood and asked "What are you doing?" The man answered "I am going to cook this peach palm." *Ñañe* laughed at him and said "'I am going to give you the real peach palm!" He gave the man a pack of maize leaves with a fermented mass of real peach palm inside and told the man how to do *cono*, a fermented drink. The wise man went back to the underground and made *cono*. After drinking the *cono*, people began to emerge to the surface through a tunnel in a creek. These people had tails. *Ñañe* cut the tail of all of them as they emerged. With the tail of white people, he created the white capuchin monkeys. There were different people coming out from the underground, black people, yellow people, red people, and so on. With the tails of these people, *Ñañe* created all the monkey species. There was one group of underground people with colorful dresses, *Ñañe* called them *Siecopai*. He cut the tails of these people and with these tails he created the woolly monkeys. *Ñañe* named the creek where all this took place *Siecoya*; this is the place where we, the *Siecopai*, originated" (H. Payaguaje, pers. comm. – the myth was narrated in Spanish with some Secoya words that were maintained in this translation).

13.2 The Present

The close relationship that the Secoya had with nonhuman primates allowed them to even recognize the spiritual leader of a monkey troop and to reduce or stop their hunting when this troop leader asked them to do so. In current days, this close relationship has been lost. The use of firearms, instead of blowguns or traps, increased their effectiveness as hunters and, apparently, ended this relation of respect for nonhuman primates and other animal preys. "Monkeys are like people. One day, we found a troop of woolly monkeys feeding on *airo toa*, a forest fruit, and we stayed under the tree. One monkey cried from far away and all the others ran to see him. On a branch of a large tree, this monkey became a person. He was huge. He was the leader of the woolly monkeys, a *iowáiëjaë*. He had a white tunic. One of us wanted to shoot the monkeys with the blowgun; but when he saw the leader, he was scared and could not blow. In all monkey troops there was a leader *iowáiëjaë*. Now the Secoya don't use blowguns, we shoot the monkeys with firearms and scared them with the noise. This is why the *iowáiëjaë* are no longer in the troops" (Cipolletti and Payaguaje 2008).

In most of the Secoya territory, intense and uncontrolled hunting in addition of high deforestation rates in the past 40 years (Josse 2001) has caused the local extinctions of some monkey species, such as the woolly monkey (de la Torre pers. obs.) (Fig. 13.5). Most Secoya people today are no longer living in direct contact with nonhuman primates. "Most Secoya children have never seen a woolly monkey. The morning choruses of howler and titi monkeys in the gallery forest of the Aguarico

Fig. 13.5 *Lagothrix lagothricha*, called *naso* in Secoya language (woolly monkey), juvenile kept as pet in San Pablo de Catesiaya, 2005 (credits: P. Yépez)

River are only in the memory of older people. Their calls have been replaced by the noise of motor boats, chainsaws, TVs and loudspeakers. Now, the Secoya children imitate the sounds of cocks, pigs, horses, and dogs but are not able to imitate the sounds of the monkeys" (H. Payaguaje, pers. comm.).

Most Secoya still have a preference for monkey meat, but in current times, very few of them do hunt monkeys. In a study about hunting patterns that we carried out from April through December 2006 in the Secoya communities of San Pablo, Bellavista, and Siekoya Remolino, few monkeys of only three species were hunted, red howler monkeys (four individuals hunted in San Pablo and one in Siekoya Remolino), white fronted capuchins (one individual hunted in Bellavista and six in Siekoya Remolino), and saki monkeys (one individual hunted in San Pablo) (Table 13.2). No woolly monkeys were hunted. Woolly monkeys were the most hunted primate species and one of the most hunted mammals by the Secoya in the early 1970s. Woolly monkey meat is still considered a delicacy, the best among all monkeys (Vickers 1989, D. Payaguaje, pers. comm.). The absence of woolly monkeys as hunting prey and the overall low hunting rate of other primate species in 2006 seems to be related to the fact that primates are now rare or absent in the forests close to the Secoya settlements. In this same year (2006), we carried out biweekly censuses to estimate mammal diversity in different forest types in the Secoya territory. We complemented the data from the censuses with records from camera traps. During that year, we recorded 9 of the 11 primate taxa that could be found in the Secoya territory (Table 13.3), but the frequency of recordings was low for all species (mode: 0–1 record per month). We did not record any of the two subspecies of woolly monkeys. Given the low densities or absence of primates in the areas close to Secoya,

Table 13.2 Species and number of individuals for the mammals hunted between April and December 2006 in the Secoya communities of San Pablo de Catesiaya (SP), Bellavista (BE), and Siekoya Remolino (SR)

Family	Species	Common name (English)	SP	BE	SR
Cervidae	*Mazama* spp.	Brocket deer	2	0	1
Tayassuidae	*Pecari tajacu*	Collared peccary	4	7	5
	Tayassu pecari	White-lipped peccary	1	4	0
Felidae	*Panthera onca*	Jaguar	0	1	0
Procyonidae	*Nasua nasua*	Coati	0	9	0
Dasypodidae	*Dasypus novemcinctus*	Nine-banded armadillo	1	14	0
Myrmecophagidae	*Tamandua tetradactyla*	Ant eater	0	3	1
Tapiridae	*Tapirus terrestris*	Tapir	0	0	1
Atelidae	*Alouatta seniculus*	Red howler monkey	4	0	1
Cebidae	*Cebus yuracus*	White capuchin monkey	0	1	6
Pithecidae	*Pithecia milleri*	Saki monkey	1	0	0
Agoutidae	*Agouti paca*	Paca	12	7	5
Dasyproctidae	*Dasyprocta fuliginosa*	Agouti	16	31	4
Dasyproctidae	*Myoprocta acouchi*		1	0	2

Modified from de la Torre et al. (2007)

Table 13.3 Mammal species recorded in biweekly censuses and camera traps in *terra firme* and *várzea* forests, between April and December 2006 in the Secoya community of Siekoya Remolino

Family	Terra firme forest species	Várzea forest species
Didelphidae	*Caluromys lanatus*	
	Didelphis marsupialis	*Didelphis marsupialis*[a]
Cervidae	*Mazama goauzoupira*	*Mazama goauzoupira*
	Mazama americana	*Mazama americana*
Tayassuidae	*Pecari tajacu*	*Pecari tajacu*[a]
	Tayassu pecari[a]	
Canidae	*Atelocynus microtis*	*Atelocynus microtis*[a]
	Speothos venaticus	
Felidae	*Herpailurus yagouaroundi*	
	Leopardus pardalis	*Leopardus pardalis*
	Panthera onca[a]	*Panthera onca*[a]
		Puma concolor[a]
Mustelidae	*Eyra barbara*	*Eyra barbara*[a]
	Lontra longicaudis[a]	
Procyonidae	*Nasua nasua*	*Nasua nasua*
	Procyon cancrivorus	
Dasypodidae	*Cabassous unicinctus*	
	Dasypus novemcinctus	*Dasypus novemcinctus*
	Priodontes maximus[a]	*Priodontes maximus*[a]
Megalonychidae		*Choloepus didactylus*[a]
Myrmecophagidae	*Myrmecophaga tridactyla*[a]	*Myrmecophaga tridactyla*
		Tamandua tetradactyla[a]
Tapiridae		*Tapirus terrestris*[a]
Aotidae[b]	*Aotus vociferans*[a]	*Aotus vociferans*[a]
Atelidae[b]	*Alouatta seniculus*[a]	*Alouatta seniculus*[a]
Callitrichidae[b]		*Cebuella pygmaea*[a]
	Leontocebus nigricollis graellsi[a]	*Leontocebus nigricollis graellsi*[a]
Cebidae[b]	*Cebus yuracus*	*Cebus yuracus*
	Saimiri cassiquiarensis[a]	*Saimiri cassiquiarensis*[a]
Pithecidae[b]	*Plecturocebus discolor*[a]	
		Cheracebus lucifer[a]
	Pithecia milleri[a]	*Pithecia milleri*[a]
Agoutidae	*Agouti paca*	*Agouti paca*
Dasyproctidae	*Dasyprocta fuliginosa*	*Dasyprocta fuliginosa*
	Myoprocta acouchi	*Myoprocta acouchi*
Echimyidae		Echimyidae sp.
Hydrochaeridae		*Hydrochaeris hydrochaeris*[a]
Sciuridae	*Microsciurus flaviventer*[a]	*Microsciurus flaviventer*[a]
	Sciurus igniventris	*Sciurus igniventris*

Modified from de la Torre et al. (2007)

[a]Species only recorded in censuses; families marked with the [b]sign are primates)

hunters have to travel several kilometers to find some monkey prey. Few people are willing to travel that far and, if they do so, it is only sporadically. Additionally, the price of bullets has increased in past years; hence, not everyone can afford to hunt with firearms. Last, but not least, many young Secoya have jobs in the nearby towns and cities; therefore, they do not have the time or the interest to carry out hunting expeditions.

13.3 The Future

The profound knowledge about nonhuman primates that the Secoya culture gathered over centuries is in the brink of disappearing. This knowledge only remains in the mind of the old Secoya people, the *ñenk'e*, and needs to be preserved for future generations. Several studies have compiled important pieces of traditional knowledge from the Secoya eldest, contributing to its preservation (Vickers 1989; Cipolletti and Payaguaje 2008, Cabodevilla 1990; Yépez et al. 2010; Cerón et al. 2011). In this analysis of the temporal changes of the Secoya ethnoprimatology, we are presenting information not only from these studies but also from interviews of Secoya leaders (*ñenk'e*) and from our previous work (e.g., de la Torre & Yépez 2007, de la Torre et al. 2009).

If the current trend of reduced hunting is maintained within the Secoya territory in the next decades, it may be possible that primate populations would recover. However, this would only occur if the forest is preserved and the resources that monkeys need are maintained. Since the deforestation rate in the area is considerably high (de la Torre et al. 2009), the education system of the Secoya needs to be strengthened to help reduce this rate. An improved education system should include the rescue and valorization of the traditional Secoya knowledge to increase environmental and cultural awareness, especially in the young. It should also allow the Secoya to design and implement alternative productive and economic strategies that are not based on logging. There have been some attempts in that direction (de la Torre et al. 2007, Yépez et al. 2005, 2010); however, more efforts are certainly needed. We hope our study will motivate other researchers to collaborate with the Secoya to work with them in preserving their deep traditional knowledge about nonhuman primates and ecosystems in Ecuadorian Amazon. Their persistence as a culture and the conservation of nonhuman primates and of the Amazonian forests depend on it. Similar efforts should be carried out for other Amazonian cultures, such as the Wao in Ecuador, who appear to suffer from a similar loss of knowledge across generations (Papworth et al. 2013).

> Before foreigners came, there were many monkeys. We knew a lot about them. There were so many monkeys that we used monkey names to name several areas of the vast Secoya territory. Monkeys have provided us with food. I liked to see them jumping from tree to tree; sometimes they went to the ground in the salt leaks and fed on clay. They were fast and agile. Some of them still come to my *tsio,* my Secoya garden, but only the small ones. I wish it were now like in the past, but I am old now and I don't think I will see again the Secoya singing happily with the monkeys (C. Piaguaje, Secoya shaman, pers. comm).

Acknowledgments We thank all the Secoya people, especially the *ñenk'e* and Hernán Payaguaje, for their willingness to share with us their knowledge, memories, and worries about monkeys and their forests. We also thank the editors of this book and one anonymous reviewer for their thoughtful comments and suggestions to improve the manuscript. Our research on Secoya ethnobiology has been supported by Fundación VIHOMA, Proyecto CAIMAN-USAID, WCS, Ecofondo, Fundación Raíz, and USFQ.

References

Cabodevilla MA (1990) El bebedor de yajé, Fernando Payaguaje. Ediciones CICAME, Shushufindi

Cerón C, Reyes C, Payaguaje D, Payaguaje A, Payaguaje H, Piaguaje E, Piaguaje R, Yépez P (2011) Mil y más plantas de la Amazonía ecuatoriana utilizadas por los Secoyas. Cinchona 11:13–205

Cipolletti S, Payaguaje F (2008) La Fascinación del Mal. Historia de un shaman Secoya de la Amazonía ecuatoriana. Ediciones Abya-Yala, Quito

de la Torre S (2000) Primates de la Amazonía Ecuatoriana. Proyecto PETRAMAZ/ SIMBIOE, Quito

de la Torre S, Yépez P (eds) (2007) Caminando en el sendero, hacia la conservación del ambiente y la cultura Secoya. Fundación VIHOMA, Imprenta Noción, Quito

de la Torre S, Payaguaje H, Payaguaje P, Payaguaje A (2007) Investigación participativa y conservación de la fauna en tierras Secoya. In: de la Torre S, Yépez P (eds) Caminando en el sendero, hacia la conservación del ambiente y la cultura Secoya. Fundación VIHOMA, Imprenta Noción, Quito, pp 77–90

de la Torre S, Yépez P, Snowdon CT (2009) Ecology and conservation of pygmy marmosets in Amazonian Ecuador. In: Davis LC, Ford SM, Porter L (eds) The smallest anthropoids: the Marmoset/Callimico radiation. Springer, New York, pp 451–463

Estrada A, Garber PA, Rylands AB, Roos C, Fernandez-Duque E, Di Fiore A et al (2017) Impending extinction crisis of the world's primates: why primates matter. Sci Adv 3(1):e1600946. https://doi.org/10.1126/sciadv.1600946

Gutiérrez J (2007) Conservación y medicina de primates. Bellaterra. http://ddd.uab.cat/record/158785

Josse C (2001) La biodiversidad del Ecuador: Informe 2000. Ministerio del Ambiente; EcoCiencia; UICN, Quito

Meggers BJ, Evans C (1962) The Machalilla culture: an early formative complex on the Ecuadorian coast. Am Antiq 28:186–192

Mittermeier RA, Rylands AB, Wilson DE (eds) (2013) Handbook of the mammals of the world. Volume 3: Primates. Lynx Editions, Barcelona

Papworth S, Milner-Gulland EJ, Slocombe K (2013) The natural place to begin: the ethnoprimatology of the Waorani. Am J Primatol 75:1117–1128

Peres CA (1997) Primate community structure at twenty western Amazonian flooded and unflooded forests. J Trop Ecol 13:381–405

Uribe S (2016) *La representación* zoomorfa en la cultura Guangala: Un análisis pre-iconográfico en el Período de Desarrollo Regional de la costa central ecuatoriana. Ediciones AbyaYala / Universidad Politécnica Salesiana, Quito

Vickers W (1989) Los Sionas y Secoyas, su adaptación al ambiente amazónico. Ediciones AbyaYala, Quito

Yépez P, de la Torre S, Payaguaje H, Payaguaje A (eds) (2010) ¿Al final del sendero? Aportes a la conservación del ambiente y la cultura Secoya. Imprenta Noción Fundación VIHOMA, Quito

Yépez P, de la Torre S, Cerón C, Palacios W (eds) (2005). Al inicio del sendero: Estudios Etnobotánicos Secoya. Ed. Arboleda. Quito

Zardini E (1991) Ethnobotanical notes on "Yacon", *Polymnia sonchifolia* (Asteraceae). Econ Bot 45:72–85

Chapter 14
The Importance of Nonhuman Primates in Waorani Communities of the Ecuadorian Amazon

Margaret Franzen Levin

14.1 Introduction

Ethnoprimatology is an interdisciplinary approach to the study of the interactions between humans and nonhuman primates (see Sponsel 1997; Fuentes and Wolfe 2002; Fuentes 2006). Given the biological and behavioral similarities shared by humans and nonhuman primates, the consequences of their interactions could be of special significance (Fuentes and Wolfe 2002; Fuentes 2006). Ethnoprimatology has been used as a framework for exploring the effects of human predation on nonhuman primate populations (Urbani 2005) and the conservation of nonhuman primate species (Fuentes and Wolfe 2002; Fuentes and Hockings 2010). The framework also allows for the consideration that nonhuman primates influence "human ecologies" (Fuentes 2006). Other researchers have studied the importance of nonhuman primates in human cultures by looking at how humans perceive their primate neighbors (Papworth et al. 2013) and how nonhuman primates are represented symbolically through mythology and ritual (Cormier 2006).

In this chapter, I use data and observations from two field seasons spent in three Waorani communities in the Ecuadorian Amazon. I consider the importance of nonhuman primates, relative to other species, to the subsistence, and meat-sharing practices of the Waorani and question whether primates are the most important species for maintaining a culture of food sharing in Waorani communities. I briefly discuss the importance of primates as pets in Waorani households. I also consider the conservation implications of primate hunting among the Waorani and propose some future areas of inquiry regarding the dynamics between Waorani hunters and their primate prey. This chapter does not address the symbolic importance of primates in Waorani culture. I draw from research that had a quantitative focus and was not conducted with ethnoprimatology questions in mind. This chapter is therefore lacking in insights that can be gained through a more

M. F. Levin (✉)
Graduate Group in Ecology, University of California, Davis, CA, USA
e-mail: m.f.levin@gmail.com

B. Urbani, M. Lizarralde (eds.), *Neotropical Ethnoprimatology*, Ethnobiology,
https://doi.org/10.1007/978-3-030-27504-4_14

culturally oriented approach, including relevant details concerning the role of primates in Waorani mythology and cosmology. For a detailed discussion of the symbolic differences in the hunting of primates versus ground-dwelling peccary among the Waorani, see Rival (1996), and for a review of the symbolic importance of primates among a sample of Neotropical subsistence hunters, see Cormier (2006).

14.2 The Waorani

The Waorani are an indigenous hunter–gatherer–horticulturalist group living in the Amazon region of eastern Ecuador. Since missionary contact in the 1950s, their population has increased, their traditional homeland has been reduced in size, and there have been significant changes to their settlement and subsistence patterns. The Waorani population numbered around 500 individuals at the time of contact (Yost 1991; Lu 2001) and has since grown to an estimated 1700–2000 individuals (Beckerman et al. 2009; Lu 2010). Originally estimated at 20,000 km^2 (Yost 1991), the ancestral homeland of the Waorani has been incorporated into the Waorani Ethnic Territory and the Yasuni National Park and Biosphere Reserve, covering a total area of 16,820 km^2 (Finer et al. 2009).

The Yasuni National Park and Biosphere Reserve is bordered by the Napo River to the north and the Curaray River to the south. It is moist tropical forest consisting of 80% *terra firme* forest (Pitman et al. 2001) with floodplains and swamps near the rivers (Valencia et al. 2004). Where the forest meets the eastern base of the Andes, it reaches elevations of between 200 and 500 meters (Pitman et al. 2001; Valencia et al. 2004). The climate is relatively aseasonal, and the area receives high rainfall. Monthly average precipitation from 1997 to 1999 was 248 mm (range 102–639 mm), and monthly average temperature was 28.2 °C (range 21.5 °C–34.7 °C) (Nabe-Nielson, 2001). The park is considered an area of "mega-diversity," and some speculate that it could be the most biodiverse place on earth (Finer et al. 2009). It is estimated to contain over 1576 species of plants, 173 species of mammals, and 610 species of birds (Koester 2001). Of the at least 10 species of primates, 3 are atelines: *Ateles belzebuth* (spider monkey), *Lagothrix lagothricha* (woolly monkey), and *Alouatta seniculus* (howler monkey), the largest of the primate species found in Amazonian forests (Di Fiore 2001).

The Waorani were traditionally seminomadic, moving locations frequently between several existing home sites that were built in areas of relatively high elevation and away from rivers (Yost 1991). Their past mobility has largely been attributed to the high intra-ethnic homicide rate within the group and the risk of having the location of a home site discovered by other hostile Waorani or by outsiders (Yost 1991). Today the majority of the Waorani are living in permanent settlements, many of which are concentrated in the western part of the territory, and most of which are near to landing strips, rivers, and/or roads. As of 2006, an estimated 35 Waorani settlements had been established; although this number fluctuates due to the tendency of the Waorani to leave current settlements and form new ones (Beckerman et al. 2009).

The Waorani communities that are the focus of this chapter are located along an oil road that was built in 1993 into the northern part of the Yasuni Biosphere Reserve, beginning at the Napo River. The road is known as the "Maxus Road" as it was constructed for oil development by Maxus Ecuador, Inc., the oil company operating there at the time. Access to the park via the Maxus Road is controlled at a checkpoint on the southern bank of the Napo River where the road begins. With the exception of the Waorani and Quichua residents of the park, permission is required to enter the park along this road.

In this chapter, I present data and observations based on fieldwork conducted in the year 2000 in the Waorani communities of Guiyero and Dicaro, located along the Maxus Road, and again from August to December 2002 in Guiyero, Dicaro, and a third community, Tiimpuca (Fig. 14.1). Tiimpuca was formed in 2001 when five families left the community of Guiyero and established a new settlement approximately 20 km away. A total of 20 households participated in the study (Guiyero = 3, Tiimpuca = 5, Dicaro = 12). Households included all located along the road with the exception of two households that were not associated with any of the three communities. Information on diet was collected during household meal interviews (N = 62) in Guiyero and Dicaro in the year 2000. Data on hunting and sharing were collected during hunting interviews (N = 413) in the three communities and corresponding interviews on sharing (N = 394) during the 2002 study period. In 2002 I also recorded all the pets that were present in each household from August to December.

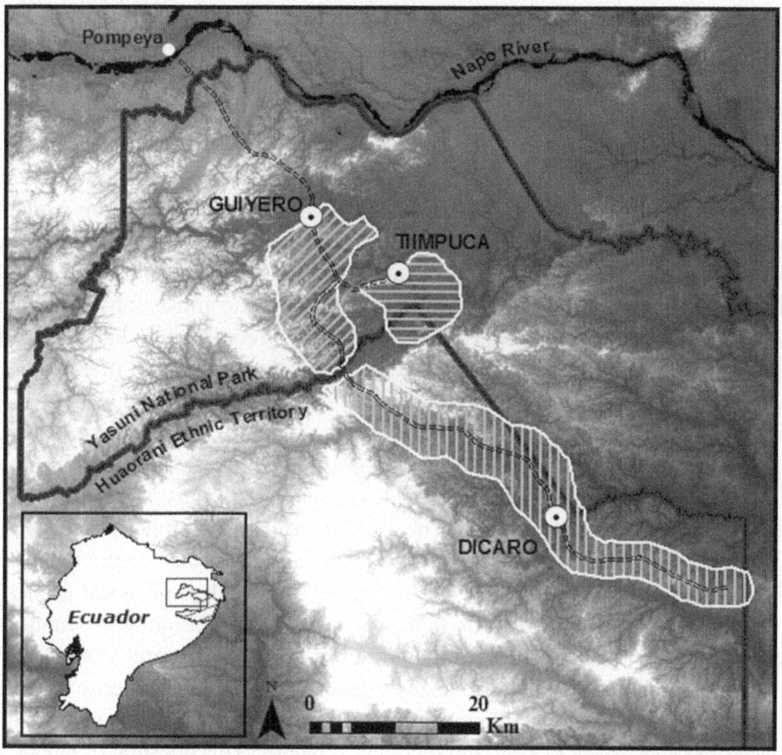

Fig. 14.1 The location of the Waorani communities of Guiyero, Tiimpuca, and Dicaro, and their hunting areas, within the Yasuni Biosphere Reserve, Ecuador. (Reprinted from Franzen 2006)

14.3 Primates and Waorani Subsistence

Studies of Neotropical subsistence hunters have demonstrated that large-bodied animals, including large-bodied primates, are typically the most preferred prey species (see Mittermeier 1991; Vickers 1991; Peres and Lake 2003). Primates are an important prey species among the Waroani (Yost and Kelley 1983; Rival 1996; Lu 1999; Mena et al. 2000; Franzen 2006). In household diet interviews conducted during fieldwork in 2000 (Guiyero and Dicaro, N = 62), primates were the most frequently reported meat consumed in households. Primate meat was reportedly eaten in 37% of the interviews, followed by fish (29%), tapir and birds (26%), peccary (21%), deer and rodents (6%), and turtle (2%) (Franzen, unpubl. data).

Papworth et al. (2013) found that the woolly monkey and spider monkey were both named by Waorani informants as preferred species for consumption, although the white-lipped peccary (*Tayassu pecari*) received the highest preference score. Among the large-bodied primate species, the woolly monkey was the most preferred, followed by the spider monkey, and then the howler monkey, which was a favorite among some and avoided by others (Papworth et al. 2013). Lu (1999) also reports that peccary, woolly monkey, and howler monkey were named by Waorani informants as favorites for consumption. According to Rival (1996), the Waorani especially value woolly monkeys because they believe they are more prolific than the other large-bodied primates and because they share certain social similarities with humans.

In a preliminary review of 70 Neotropical subsistence-hunting studies, Cormier (2006) found that large-bodied primates tended to be hunted more than smaller-bodied primates. Large-bodied primates are hunted by the Waorani with greater frequency than smaller-bodied primates, supporting this general finding (Lu 1999; Mena et al. 2000; Franzen 2006). During fieldwork conducted between August and December 2002, I recorded a total of 413 hunting interviews in the Waorani communities of Guiyero, Tiimpuca, and Dicaro. Among Waorani hunters in the newly established community of Tiimpuca, primates were the most commonly hunted prey type, with the large-bodied primates (weighing greater than 6 kg and including *Ateles belzebuth*, *Lagothrix lagothricha*, and *Alouatta seniculus*), making up 86% of the primates harvested (Franzen 2006).

Tiimpuca hunters hunted a total of 92 primates, 40 ungulates, and 88 birds over the study period (Franzen 2006). Due to the recent establishment of this community in a previously unhunted area of forest, it provides a baseline for harvest profiles where no depletion of wildlife has yet occurred. The harvest profile of Tiimpuca hunters suggests that the large-bodied primates are the most important prey species. In the longer established communities of Guiyero and Dicaro, where hunting had occurred for 10 years prior, ungulates (Guiyero) and birds (Dicaro) were the most commonly hunted prey species, suggesting local depletion of the preferred primates (Franzen 2006).

Although large-bodied primates are likely to be avoided in certain cultures, or instances, or to have taboos associated with them (Cormier 2006), I did not observe any instances of avoidance of primate meat or hear any mention of taboos. This question of taboos, however, was not a component of my study, and other researchers have found evidence for occasional avoidance. Among the Waorani informants in Papworth et al.'s (2013) study, spider monkeys and saki monkeys (*Pithecia pithecia*) were reportedly avoided during pregnancy because they will cause the child to be thin. These two species as well as the capuchin (*Cebus albifrons*) were also mentioned as making people feel unwell, and some people avoided howler monkey because of worms and bad taste (Papworth et al. 2013). However, Rival (1993) reports that breastfeeding Waorani women are believed to require nourishing food such as monkey meat so that their breast milk is both nutritious and abundant. Primate meat may also be preferred during certain times of the year. According to Cormier (2006), there appears to be a seasonal pattern to the hunting of primates in the Amazon region that corresponds to a period of time when the monkeys are fatter. The "season of fat monkeys" is recognized by the Waorani as a period of time between May and August when monkeys become fatter and their meat tastes better (Rival 1993).

14.4 Primates and Waorani Food Sharing

There has been a long-standing discussion among researchers pertaining to the reasons for the extensive sharing of meat documented among hunter–gatherer societies. Many explanations have been proposed, and it is likely that multiple factors are important in motivating an individual to share meat (Hill and Kaplan 1993; Winterhalder 1996a; Gurven 2004b; Franzen and Eaves 2007). Individuals may be benefitting their kin (Hamilton 1963; Hamilton 1964; Morgan 1979; Palmer 1991), forming alliances (Patton 2000; Patton 2005) or ensuring future assistance from others in times of illness (Gurven et al. 2000). Through widespread sharing of meat, men may be demonstrating their skill as hunters in an effort to increase mating opportunities (Hawkes 1991; Smith and Bliege Bird 2000; Bliege Bird et al. 2001). Men may hunt cooperatively so as to increase hunting success and subsequently share any meat that is harvested (Hames 1990). Meat may be shared when others in the community witness the hunter returning home, and, having none, they ask the hunter to share (Blurton Jones 1984; Winterhalder 1996b; Tucker 2001). Furthermore, the reciprocal sharing of meat can reduce the risk of not having any meat on a given day (Kaplan and Hill 1985; Hames 1990; Winterhalder 1990; Gurven et al. 2000; Hames 2000; Gurven 2004a, 2004b).

The idea that sharing may act as a signal (Gurven et al. 2000) is supported by evidence that the Waorani are aware of how often others share. Significant correlations were found between individuals' perceptions of how often others shared and actual sharing behavior (Franzen 2005; Franzen and Eaves 2007). In addition, people made comments during interviews such as someone "knows how" or "does not know how" to share (Franzen and Eaves 2007). Whatever the primary motivation is for sharing, it is a widespread and important component of Waorani society. Generosity is valued, and food and drink are readily shared with visitors (High 2015). The ability to go into the forest and bring back food to "give away" is a measure of independence for growing Waorani children, and chants aimed at adolescents during ear piercing ceremonies speak of bringing home extra food from hunting and gathering trips in order to share it (Rival 1993).

In diet interviews conducted in 2000, the Waorani reported that primate meat came from another household 39% of the time. Other types of meat were also reported to have come from another household, including tapir (88% of the time), deer (75%), peccary (62%), birds (44%), rodents (25%), and fish (11%) (Franzen, unpubl. data). This extensive sharing is expected for the very large species such as tapir, deer, and peccary for which a single household might have difficulty consuming the entire animal.

Detailed sharing information including the species shared, amount shared, and the recipients of sharing was available for 394 hunts recorded during fieldwork in 2002. The majority of sharing events involved two peccary species (*Tayassu pecari* and *Tayassu tajacu*), large-bodied primates weighing greater than 6 kg (*Ateles belzebuth*, *Lagothrix lagothricha*, and *Alouatta seniculus*), and large birds (*Mitu salvini*, *Aburria pipile*, *Psophia crepitans*, *Penelope jacquacu*, *Ara macao*, *Tinamus major*, *Ara ararauna*, *Amazona farinosa*, and *Ramphastos cuvieri*). The frequencies with which these animals were hunted, and the frequency and magnitude of sharing are summarized in Table 14.1.

In the combined sample, the large birds were hunted the most frequently; however, they were not shared with the same frequency as the large-bodied primates and the peccary species. Whereas birds were shared on 46% of occasions, peccary species were shared 78% of the time when they were hunted, and large-bodied primates were shared 70% of the time when they were hunted. On average, approximately two households, other than the hunter's, received meat when peccary or large-bodied primates were shared.

The amount of sharing that occurred is largely explained by the frequency with which certain species were captured (Fig. 14.2). However, the peccary species (squares) and the large-bodied primate species (the three diamonds above the trendline) were shared more, while the majority of bird species (circles) were shared less.

Very large animals such as the tapir (*Tapirus terrestris*), which can weigh as much as 125.8 kg (Mena et al. 2000), tend to be shared most extensively. During the 2002 field season, hunters reported catching a tapir on three occasions, and they shared the meat on each occasion. After each tapir hunt, an average of 7.6 other

Table 14.1 Sharing of meat from peccary, large-bodied primates, and large birds in the communities of Guiyero, Tiimpuca, and Dicaro during August–December 2002 (N = 394 hunts for which detailed sharing information was available)

	Peccary species	Large-bodied primates (>6 kg)	Large birds (>0.73 kg)
Total # times hunted[a]	74	94	268
Total kg hunted	3194 kg	1077 kg	513 kg
Total # times shared	58	66	123
Total kg shared	816 kg	354 kg	92 kg
Frequency hunted (# times hunted/total # hunts)∗100	19%	24%	68%
Frequency of sharing (# times shared/# times hunted)∗100	78%	70%	46%
Avg. # of additional families receiving meat when it was shared	2	1.9	1.4

[a]This refers to the total number of times a hunter returned home with one or more individuals of a given species

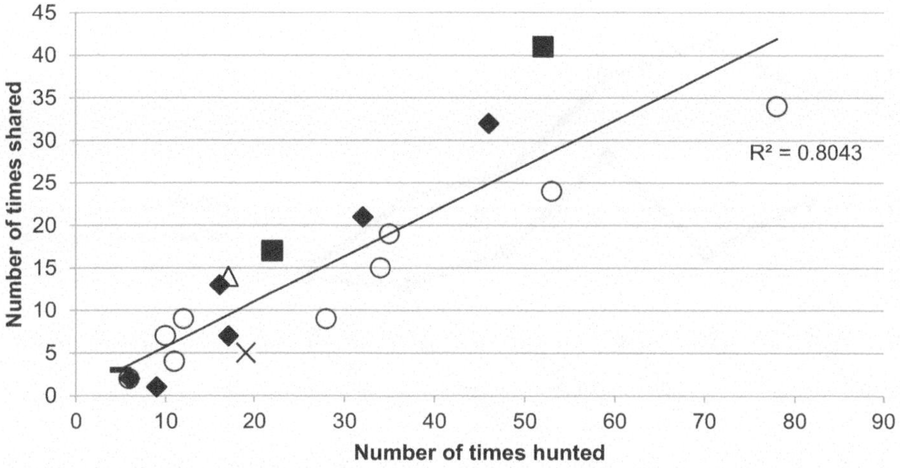

Fig. 14.2 Relationship between the number of times hunted and the number of times shared for species with five or more recorded hunts over the August–December 2002 field season. In this figure, ■ = peccary, ◆ = primates, ○ = birds, Δ = deer, X = squirrel, and — = agouti. The three ◆ above the trendline represent the large-bodied primates

households received a portion of meat. The red brocket deer (*Mazama americana*), which averages 20.23 kg (Mena et al. 2000), was hunted 17 times and was shared 82% of the time with an average of three households receiving meat when it was shared. However, these very large sharing events were less frequent than the smaller sharing events involving peccary, primates, and birds.

Among the large-bodied primate species, the frequency of sharing decreased with decreasing numbers of captures, with the exception of the howler monkey. It was shared with the greatest frequency among all the primates and was captured the fewest number of times among the large-bodied primates (Fig. 14.3).

14.5 Primates as Companions

Pets are common in Waorani households and are treated like "dependent members" of the house, fed well, and reportedly buried when they die (Rival 1993). During the period from August to December 2002, I recorded all the pets observed in Waorani households in the communities of Guiyero, Dicaro, and Tiimpuca. Primates were the most frequently observed pets, followed by birds (Table 14.2). All primate pets were juveniles with the exception of one adult spider monkey. Spider monkeys were the most common primate pets in households, followed by the golden-mantled tamarin (*Saguinus tripartitus*). All of the primate species hunted by the Waorani were observed as pets in at least one household with the exception of the tropical night monkey (*Aotus vociferans*). Papworth et al. (2013) also observed primate pets in Waorani households, although the woolly monkey was the most common pet at that time.

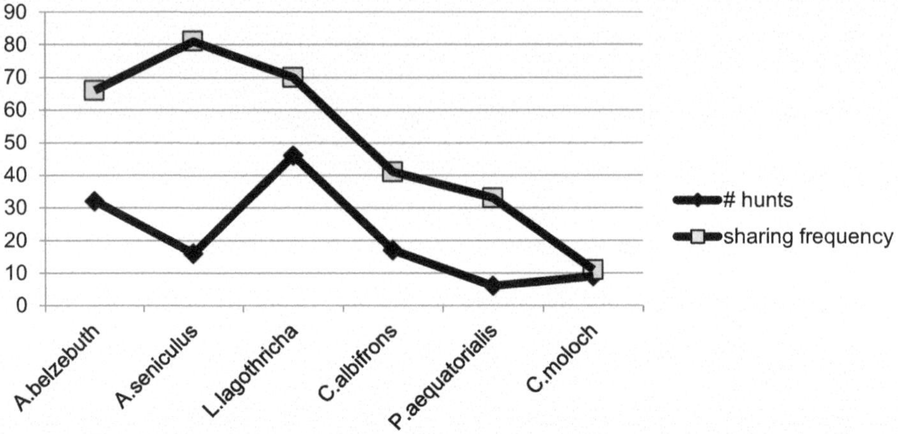

Fig. 14.3 The sharing frequency for the primate species calculated as the percentage of times a species was shared after it was captured compared with the number of times the species was hunted. This computation includes only the species with more than five hunts recorded over the study period (August–December 2002)

Table 14.2 List of species observed as pets in the communities of Guiyero, Dicaro, and Tiimpuca during the field season from August–December 2002

Species	Total
Primates	*17*
Spider monkey	7
Woolly monkey	2
Capuchin monkey	1
Saki monkey	1
Dusky titi monkey	1
Squirrel monkey	2
Golden-mantled tamarin	3
Birds	*14*
Scarlet macaw	2
Blue and yellow macaw	1
Chestnut-fronted macaw	1
Red-bellied macaw	1
Mealy amazon parrot	2
Spix's guan	1
Cobalt-winged parakeet	5
Unidentified bird	1
Other	*5*
Squirrel	1
White-lipped peccary	3
Giant anteater	1

Typically, pet monkeys are tethered with a rope to a post inside or outside the house during the day. They may then be brought along when family members leave the house. I often saw young girls carrying pet monkeys in slings, the same way Waorani mothers carry their babies. On one occasion, I observed an elderly man whose pet woolly monkey rode around gripping the top of the man's head by the hair. Another elderly man visiting the community of Dicaro arrived with his adult pet spider monkey, which he held by the hand. This monkey acted aggressively toward everyone but the man. On two occasions, I observed pet monkeys that were free to roam around. One was a woolly monkey, and the other a golden-mantled tamarin. The tamarin frequently visited the house where I stayed, several houses away from its owners, and it would explore the room and sometimes curl up and nap on the mattress.

The tamarin that was free to roam captured and ate a large grasshopper on one occasion, and I observed an untethered woolly monkey eating a jumping spider. The insect-eating behavior of pet monkeys is known and appreciated by the Waorani. In Papworth et al.'s study (2013), tamarins were reported to be the preferred pet because they consume cockroaches and are considered clean. Pet monkeys were often fed *chicha* (a mashed and masticated drink made from yucca) and plantains. Rival (1993) reported that monkey babies were breast-fed; however, this was not observed during my study or that of Papworth et al. (2013) and thus may have been a practice more common in the past.

Birds, the second most common pet in Waorani households, were treated similarly to monkeys; however, they were held less, and they rarely accompanied their owners on excursions outside of the house. Among the white-lipped peccary juveniles that were captured, I observed one being chased around a house by several young boys with spears. In this case, it appeared that the animal's purpose was to provide an opportunity for hunting practice, as opposed to being a true pet. This information supports Rival's (1996) assertion that baby peccaries are not adopted as pets but, rather, if captured, will be eaten.

14.6 Primate Conservation

Primates, with their relatively low rates of population increase and long generation times are especially susceptible to overexploitation when heavily hunted (Robinson and Redford 1991; Bodmer et al. 1997). The fact that the Waorani have gone from seminomadic to living in permanent settlements could have important consequences for the conservation of primates. In the past, the pattern of changing residence locations every few months resulted in reduced hunting pressure at a given location (Yost 1991) and allowed for recovery of heavily hunted species, whereas with permanent settlements, hunting takes place in the same areas over long periods of time.

New technologies, such as shotguns and rifles, have largely replaced the traditionally used blowguns and spears in many Waorani communities. Adoption of firearms improves the efficiency of Waorani hunters, allowing for an increase of 15% in kilograms and number of animals hunted per hour (Yost and Kelley 1983; Mena et al. 2000). Even in the distant community of Bameno, hunters own such technology; however, they are limited in its use due to the difficulty of acquiring ammunition regularly (Franzen, pers. obs.).

As predicted by optimal foraging theory (see Hames and Vickers 1982; Alvard 1993), prey appears to be chosen by Waorani hunters based on encounter rates and the size of the prey encountered (Lu 1999; Mena et al. 2000). Thus, it is predicted that when large-bodied primates are encountered, they will be pursued with high frequency. Interestingly, Rival (1993) found that the Waorani believe monkeys should be permitted to feed on peach palm (*Bactris gasipaes*) fruit, because it enables them to get fatter and to reproduce. When the trees are fruiting, the Waorani ensure that enough fruit is left on the trees for the monkeys to eat (Rival 1993, 1996). Whether this also means a hunter may pass by a monkey foraging for peach palm fruit is unclear, and to my knowledge, this behavior has not been reported by other researchers. Rival (1996) also reports that the Waorani will spare an individual monkey if it makes eye contact with the hunter and in this way communicates to the hunter that it wants to live. The possibility that Waorani hunters may choose not to pursue certain primates at certain times is intriguing from a conservation perspective. Evidence suggests, however, that the large-bodied primates are heavily hunted by the Waorani and may be at risk of local depletion (Mena et al. 2000; Franzen 2006).

In a comparison of hunting studies in the Waorani communities of Quehueiri-ono and Huentaro, one in 1996–1997 and the other in 2001, Lu (2010) found that howler monkeys actually made up a greater percentage of prey killed in 2001, contrary to what would be expected if these primates were being overhunted. In addition, the hunting of woolly monkeys, capuchin monkeys, and saki monkeys stayed fairly consistent over the two time periods. She suggests that the persistent presence of such highly preferred (and susceptible to overhunting) prey in the harvests in 2001 is indicative of a possible source area that repopulates the hunting area surrounding these communities. In this case, however, it is not clear where the hunting was taking place, and the possibility that the hunters were travelling farther to encounter the more preferred prey species is not addressed. In the same community of Quehueiri-ono, Mena et al. (2000) conducted an earlier study between July 1994 and May 1995 during which they compared prey densities in two hunting zones, one adjacent to the community (49 km^2) and one 16 km away (81 km^2). They discovered the complete absence of spider monkeys, woolly monkeys, and squirrel monkeys (*Saimiri sciureus*) in the zone closest to the community, while all the primate species were present in the more distant zone. If hunters are willing to travel farther for their preferred prey, then it is possible for them to continue harvesting prey that is depleted closer to the community.

The spider monkey was completely absent from the later studies reported by Lu (2010), while Mena et al. (2000) reported that ten individuals were hunted over the 1994–1995 study period. Hunting data from the Waorani community of Tiimpuca, collected in the year following its establishment in a previously unhunted area of forest, showed that the spider monkey was the most commonly hunted primate, with a total of 42 individuals hunted during the 5-month period between August and December 2002 (Franzen 2006). The fact that it was completely absent in the harvests of Guiyero and was the least commonly hunted large-bodied primate in Dicaro, where only 16 individuals were hunted, suggests that the population of spider monkeys may be facing local depletion near these two older communities (Franzen 2006).

In contrast to the spider monkey, more woolly monkeys were harvested in Dicaro (46 individuals), as well as more howler monkeys (20 individuals), than in the newly established community of Tiimpuca, where 30 woolly monkeys and only seven howler monkeys were harvested over the study period (Franzen 2006). In Guiyero, seven woolly monkeys and one howler monkey were harvested (Franzen 2006), making the woolly monkey the most commonly hunted large-bodied primate species in the two communities with the longest history of hunting.

The possibility that the spider monkey is not typically found in the same areas as the woolly monkey is not supported by the hunting data because the Tiimpuca hunters also hunted 30 woolly monkeys during the study period. Furthermore, their reported hunting area was a 57.2 km^2 zone adjacent to the community, the smallest hunting area of the three communities (Franzen 2006). In addition, Mena et al. (2000) report that the spider monkey and the woolly monkey were encountered in the same area and even along the same transect while they were measuring prey densities outside of Quehueiri-ono.

If the conservation of primate populations at risk for local depletion near communities is to become a priority among the people, there must first be a sense of scarcity (Lu 2001). The Waorani have traditionally viewed their environment as providing an abundance of food that must simply be acquired (High 2015), and Holt (2005) explains why we would not expect conservation practices to emerge among indigenous groups when human population density remains low and hunting is primarily for subsistence. She states: "Conservation awareness arises when people exert use pressure on resources and recognize the potential for overexploitation, conditions concurrent with population growth, adoption of Western technologies, and market production" (Holt 2005; 201).

The Waorani are now beginning to experience these exact changes, and Holt (2005) reports that some individuals are starting to see that hunting pressure combined with the external pressures of oil development and logging could result in a future of less abundance. It is at this time she says, when the idea of "natural plenty" is changing, that conservation biologists have a great opportunity to begin collaborating with local communities (Holt 2005). However, High (2015) cautions that the popular image of indigenous people as natural conservationists can lead to disappointment if the choices made by the people, as they navigate their changing world, do not fit with our expectations of conservationist behavior.

14.7 Discussion

Large-bodied primates and peccaries are preferred prey species that are heavily hunted by the Waorani. They are also the most frequently shared species in the Waorani communities I studied. When there is reciprocal sharing of meat between households, it allows individuals to smooth-out the consumption of meat over time and to reduce the chance of not having any meat available (Kaplan and Hill 1985). This process allows for increased food security in addition to other likely benefits of sharing, such as benefitting kin and promoting cooperative relationships and alliances (Hill and Kaplan 1993; Winterhalder 1996a; Gurven et al. 2000; Patton 2000, 2005; Franzen and Eaves 2007).

If the abundant harvest of these particular species allows for the maintenance of sharing networks, what happens when large-bodied primates and/or peccary are captured with decreasing frequency? It appears that in the community of Guiyero, where large-bodied primates were hunted on only five occasions over the study period (eight individuals total), individuals may be compensating by sharing a greater percentage of the birds that are captured (Fig. 14.4). This suggests that sharing networks may be resilient to decreasing harvests of the most preferred species. The sample of hunts for the community of Guiyero was small however (N = 39), and this finding is not conclusive, only suggestive. The resilience of sharing networks to changes in harvest profiles with local depletion of preferred prey could be an area of future research.

Peccary was shared with the greatest frequency when looking at the combined data (78%) and for the communities of Tiimpuca (72%) and Dicaro (90%) separately (Fig. 14.4). This could be explained by the finding that the two peccary species are the most preferred for consumption (Papworth et al. 2013), and/or they are shared due to their large size and diminishing marginal value to the hunter (Winterhalder 1996a; Winterhalder 1996b). Large-bodied primates were shared with similarly high-frequency relative to the peccary species, even in Guiyero where they were only captured on five hunts.

The large-bodied primates do not appear to have greater importance as food, or sharing resource, than the peccary species. Other researchers have come to a similar conclusion (Papworth et al. 2013). However, Tiimpuca hunters took 79 large-bodied primates over the 2002 study period versus 37 peccaries (Franzen 2006). It is possible that there were simply fewer opportunities to hunt peccary over this time

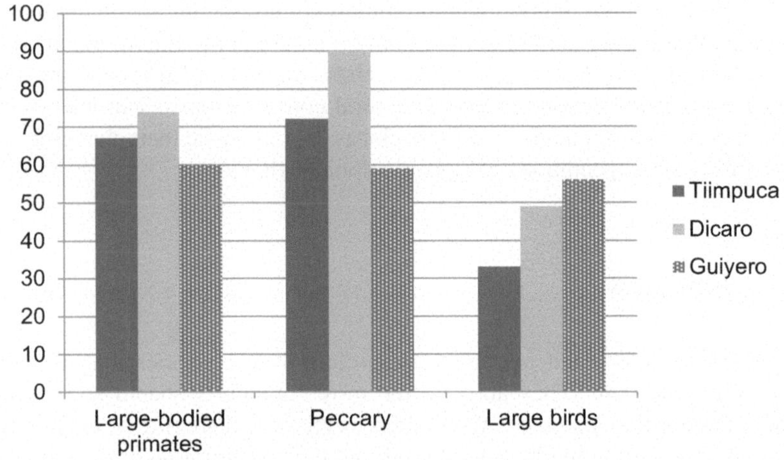

Fig. 14.4 Comparison of the three Waorani communities in terms of the percentage of large-bodied primates, peccaries, and large birds that were shared during the study period (August–December 2002)

period, or this could be an accurate portrayal of what would be hunted in an intact forest. If the harvest profile of Tiimpuca hunters is truly a reflection of the species that hunters would pursue given an unhunted environment, then the large-bodied primates may indeed be the most important prey species when their populations are intact (see Franzen 2006). This information would support Rival's (1996: 150) assertion that "monkeys are by far the most favoured game."

Among the large-bodied primates, the howler monkey was shared with greatest frequency even though it was not as commonly hunted. This might be explained by the strong preferences for or against the taste of the meat as noted by Papworth et al. (2013). A hunter may give more meat away if certain family members do not look favorably upon this type of meat, or conversely, if it is truly a favorite and rarely acquired, then individuals may come around hoping for some when it is captured.

In addition to being an important resource, the Waorani value primates as pets. Waorani individuals embrace the opportunity to have primate pets in their homes, and these pets provide companionship to many individuals, from young girls to old men.

In the context of conservation, it is important to remember that the Waorani are hunters and that this is fundamental to their subsistence and an important part of their identity (Lu 2010). If the primates that they rely on for hunting and sharing were to disappear, then it would be an enormous loss. Conservation awareness does appear to be growing among the Waorani. During the course of fieldwork, many individuals expressed to me an interest in tourism. These individuals spoke about the importance of conserving wildlife, especially primates, so that visiting tourists would be able to see them. They hoped for future work as tour guides, leading visitors to see the wildlife in their forests. This vision of the future also depends on healthy primate populations.

There are important questions that remain regarding the conservation of vulnerable prey species, such as primates, within the Yasuni region. For example, are source–sink dynamics enabling Waorani hunters to continue harvesting certain preferred prey in hunting areas that would otherwise have been depleted of these vulnerable prey species, as suggested by Lu (2010)? If so, then it would be important to understand where these source areas are located, how large they must be to continue populating sink areas, and how far they must be from communities so that they are left intact with little to no hunting pressure.

In addition, the complete absence of spider monkeys from the harvests of some of the longer-established communities mentioned in this chapter warrants further consideration. Are spider monkeys more susceptible to overhunting than the other large-bodied primates? Or is there a behavioral or cultural explanation for their more rapid disappearance from harvests? Are spider monkeys more likely to leave areas with hunting pressure or less likely to repopulate areas where they have been depleted? The hunters in Tiimpuca demonstrated a clear preference for spider monkeys in terms of harvest profile, suggesting that, when available, spider monkeys will be heavily hunted. Did the Tiimpuca families settle in an area with a higher density of spider monkeys than is typical, or does their large harvest reflect normal densities in unhunted areas?

14.8 Conclusion

The magnitude of sharing that occurs in Waorani communities demonstrates that it is an important social phenomenon. Due to the frequent nature of sharing of the large-bodied primates and peccaries, it could be argued that these species play an important role in allowing for consistent sharing between households in a community. These frequent sharing episodes likely signal some form of cooperative intent and, whether intentional or not, have the outcome of reducing the variance in meat consumption for the participating households. Primates are clearly an important food source and resource for sharing, although they do not seem to have greater importance than the peccary species in this regard. However, the harvest profile of hunters in an unhunted environment suggests that primates may be the most important prey species when their populations are still intact. Primates have a special role as household pets and are perhaps uniquely important in symbolic ways not addressed in this chapter. The spider monkey, which was heavily hunted in a newly settled area, is conspicuously absent from reported harvests of some longer-established communities. The relative vulnerability of this species warrants further consideration, as does the possibility that source–sink dynamics are enabling the continued hunting of primate species that are potentially at risk of overexploitation.

Acknowledgments This research was funded at various stages by a Fulbright Fellowship to Ecuador, a National Science Foundation Cultural Anthropology Dissertation Improvement Grant (0129826), a Wenner-Gren Dissertation Fieldwork Grant (6887), and a Gifford Center Grant (Program for Research on Population, Food and the Environment). I also received support from the Ecology Graduate Group at the University of California, Davis, as well as the Wildlife Conservation Society, Ecuador. I thank my dissertation advisor, Monique Borgerhoff Mulder, and Loukas Barton for his assistance with the map. I also wish to thank Bernardo Urbani, Manuel Lizarralde, and an anonymous reviewer for their thoughtful comments. I am extremely grateful to the Waorani residents of Guiyero, Tiimpuca, and Dicaro who generously shared their time and knowledge with me.

References

Alvard MS (1993) Testing the ecologically noble savage hypothesis – interspecific prey choice by Piro hunters of Amazonian Peru. Hum Ecol 21(4):355–387

Beckerman S, Erickson PI, Yost J, Regalado J, Jaramillo L, Sparks C, Iromenga M, Long K (2009) Life histories, blood revenge, and reproductive success among the Waorani of Ecuador. Proc Natl Acad Sci (PNAS) 106(20):8134–8139

Bliege Bird RL, Smith EA, Bird DW (2001) The hunting handicap: costly signaling in human foraging strategies. Behav Ecol Sociobiol 50:9–19

Blurton Jones NG (1984) A selfish origin for human food sharing: tolerated theft. Ethol Sociobiol 5:1–3

Bodmer RE, Eisenberg JF, Redford KH (1997) Hunting and the likelihood of extinction of Amazonian mammals. Conserv Biol 11:460–466

Cormier L (2006) A preliminary review of neotropical primates in the subsistence and symbolism of indigenous lowland South American peoples. Ecological and Environmental Anthropology (University of Georgia). Paper 21. http://digitalcommons.unl.edu/icwdmeea/21

Di Fiore A (2001) Investigación ecológica y de comportamiento de primates en el Parque Nacional Yasuní. In: Jorgenson JP, Coello Rodríguez M (eds) Conservación y desarollo sostenible del Parque Nacional Yasuní y su área de influencia: memorias del seminario-taller Yasuní. Editorial SIMBIOE, Quito, Ecuador, pp 165–173

Finer M, Vijay V, Ponce F, Jenkins CN, Kahn TR (2009) Ecuador's Yasuni biosphere reserve: a brief modern history and conservation challenges. Environ Res Lett 4:1–15

Franzen M (2005) Huaorani resource use in the Ecuadorian Amazon: hunting, food sharing, and market participation. Ph.D. dissertation. University of California, Davis

Franzen M (2006) Evaluating the sustainability of hunting: a comparison of harvest profiles across three Huaorani communities. Environ Conserv 33(1):36–45

Franzen M, Eaves J (2007) Effect of market access on sharing practices within two Huaorani communities. Ecol Econ 63:776–785

Fuentes A (2006) Human-nonhuman primate interconnections and their relevance to anthropology. Ecological and Environmental Anthropology (University of Georgia) Paper 1. http://digitalcommons.unl.edu/icwdmeea/1

Fuentes A, Hockings KJ (2010) The ethnoprimatological approach in primatology. Am J Primatol 72:841–847

Fuentes A, Wolfe LD (eds) (2002) Primates face to face: the conservation implications of human-nonhuman primate interconnections. Cambridge University Press, Cambridge

Gurven M (2004a) Reciprocal altruism and food sharing decisions among Hiwi and Ache hunter-gatherers. Behav Ecol Sociobiol 56:366–380

Gurven M (2004b) To give or not to give: an evolutionary ecology of human food transfers. Behav Brain Sci 27(4):543–559

Gurven M, Allen-Arave W, Hill K, Hurtado M (2000) "It's a wonderful life": signaling generosity among the Ache Paraguay. Evol Hum Behav 21(4):263–282

Hames RB (1990) Sharing among the Yanomamo: part I, the effects of risk. In: Cashdan E (ed) Risk and uncertainty in tribal and peasant economies. Westview Press, Boulder, pp 89–106

Hames RB (2000) Reciprocal altruism in Yanomamo food exchange. In: Cronk L, Chagnon N, Irons W (eds) Adaptation and human behavior. Aldine de Gruyter, New York, pp 397–416

Hames RB, Vickers WT (1982) Optimal diet breadth theory as a model to explain variability in Amazonian hunting. Am Ethnol 9(2):358–378

Hamilton WD (1963) The evolution of altruistic behavior. Am Nat 97(896):354–356

Hamilton WD (1964) The genetical evolution of social behavior. J Theor Biol 7:1–52

Hawkes K (1991) Showing off: tests of an hypothesis about men's foraging goals. Ethol Sociobiol 12(1):29–54

High C (2015) Victims and warriors: violence, history, and memory in Amazonia. University of Illinois Press, Urbana, Chicago, and Springfield

Hill K, Kaplan H (1993) On why male foragers hunt and share food. Curr Anthropol 34(5):701–706

Holt FL (2005) The catch-22 of conservation: indigenous peoples, biologists, and cultural change. Hum Ecol 33(2):199–215

Kaplan H, Hill K (1985) Food sharing among Ache foragers: tests of explanatory hypotheses. Curr Anthropol 26(2):223–246

Koester F (2001) El hombre y la biosfera en Yasuni. Nuestra Ciencia 3:23–29

Lu F (1999) Changes in subsistence patterns and resource use of the Huaorani Indians in the Ecuadorian Amazon. Ph.D. dissertation. University of North Carolina, Chapel Hill

Lu F (2001) The common property regime of the Huaorani Indians of Ecuador: implications and challenges to conservation. Hum Ecol 29(4):425–447

Lu F (2010) Patterns of indigenous resilience in the Amazon: a case study of Huaorani hunting in Ecuador. J Ecol Anthropol 14(1):5–21

Mena VP, Stallings JR, Regalado BJ, Cueva LR (2000) The sustainability of current hunting practices by the Huaorani. In: Robinson JG, Bennett EL (eds) Hunting for sustainability in tropical forests. Columbia University Press, New York, pp 57–78

Mittermeier RA (1991) Hunting and its effect on wild primate populations in Suriname. In: Robinson JG, Redford KH (eds) Neotropical wildlife use and conservation. The University of Chicago Press, Chicago, pp 93–107

Morgan CJ (1979) Eskimo hunting groups, social kinship, and the possibility of kin selection in humans. Ethol Sociobiol 1:83–86

Nabe-Nielson J (2001) Diversity and distribution of lianas in a neotropical rain forest, Yasuni National Park, Ecuador. J Trop Ecol 17:1–19

Palmer CT (1991) Kin selection, reciprocal altruism, and information sharing among Maine lobstermen. Ethol Sociobiol 12:221–235

Papworth S, Milner-Gulland EJ, Slocombe K (2013) The natural place to begin: the ethnoprimatology of the Waorani. Am J Primatol 75:1117–1128

Patton JQ (2000) Reciprocal altruism and warfare: a case from the Ecuadorian Amazon. In: Cronk L, Chagnon N, Irons W (eds) Adaptation and human behavior. Aldine de Gruyter, New York, pp 417–436

Patton JQ (2005) Meat sharing for coalitional support. Evol Hum Behav 26:137–157

Peres CA, Lake IR (2003) Extent of nontimber resource extraction in tropical forests: accessibility to game vertebrates by hunters in the Amazon basin. Conserv Biol 17:521–535

Pitman NCA, Terborgh JW, Silman MR, Nunez VP, Neill DA, Ceron CE, Palacios WA, Aulestia M (2001) Dominance and distribution of tree species in upper Amazonian *terra firme* forests. Ecology 82(8):2101–2117

Rival L (1993) The growth of family trees: understanding Huaorani perceptions of the forest. Man New Ser 28:635–652

Rival L (1996) Blowpipes and spears: the social significance of Huaorani technological choices. In: Descola P, Pálsson G (eds) Nature and society: anthropological perspectives. Routledge, London, pp 145–164

Robinson JG, Redford KH (1991) Sustainable harvest of neotropical forest mammals. In: Robinson JG, Redford KH (eds) Neotropical wildlife use and conservation. The University of Chicago Press, Chicago, pp 415–429

Smith EA, Bliege Bird RL (2000) Turtle hunting and tombstone opening: public generosity as costly signaling. Evol Hum Behav 21:245–261

Sponsel LE (1997) The human niche in Amazonia: explorations in ethnoprimatology. In: Kinzey WG (ed) New world primates: ecology, evolution and behavior. Aldine de Gruyter, New York, pp 143–165

Tucker B (2001) The behavioral ecology and economics of variation, risk, and diversification among Mikea Forager Farmers of Madagascar. Ph.D. dissertation. University of North Carolina, Chapel Hill

Urbani B (2005) The targeted monkey: a re-evaluation of predation on New World primates. J Anthropol Sci 83:89–109

Valencia R, Foster RB, Villa G, Condit R, Svenning JC, Hernandez C, Romoleroux K, Losos E, Magard E, Balslev H (2004) Tree species distributions and local habitat variation in the Amazon: large forest plot in eastern Ecuador. J Ecol 92:214–229

Vickers WT (1991) Hunting yields and game composition over ten years in an Amazon Indian territory. In: Robinson JG, Redford KH (eds) Neotropical wildlife use and conservation. The University of Chicago Press, Chicago, pp 53–81

Winterhalder B (1990) Open field, common pot: harvest variability and risk avoidance in agricultural and foraging societies. In: Cashdan E (ed) Risk and uncertainty in tribal and peasant economies. Westview Press, Boulder, pp 67–87

Winterhalder B (1996a) Social foraging and the behavioral ecology of intragroup resource transfers. Evol Anthropol 5:46–57

Winterhalder B (1996b) A marginal model of tolerated theft. Ethol Sociobiol 17(1):37–53

Yost J (1991) People of the Forest: the Waorani. In: Ecuador in the shadows of the volcano. Libri Mundi, Quito, pp 97–115

Yost JA, Kelley PM (1983) Shotguns, blowguns, and spears: the analysis of technological efficiency. In: Vickers WT (ed) Adaptive responses of native Amazonians. Academic Press, New York, pp 189–224

Chapter 15
Monkeys in the Wampis (Huambisa) Life and Cosmology in the Peruvian Amazonian Rainforest

Kacper Świerk

15.1 Introduction

The Wampis (also known as Huambisa and Peruvian Shuar) are a Jivaroan people inhabiting the northern part of the Peruvian Amazon near the border with Ecuador. They live mainly in the upper and middle courses of the Santiago (Kanus) and the Morona Rivers (on the Peruvian side of Ecuadorian–Peruvian border), in the administrative regions of Amazonas (in the west) and Loreto (in the east). Between those two rivers, there is a forested mountain range named Kampankis. A few Wampis live there presently; nonetheless, they still consider the mountains as a part of their traditional territory and use the natural resources of the Kampankis (such as game, fishes, leaves for thatching and lianas for baskets and bindings). According to different estimates, the Wampis number 6000–10,000 persons (Fabre 2016). Wampis are horticulturalists–hunters. Some of them also earn incomes as schoolteachers and regional officials. However, even individuals performing such functions usually do not give up entirely the traditional modes of subsistence which is also very common among other indigenous peoples (e.g., Lizarralde and Lizarralde 2018).

The Wampis is part of the Jivaroan linguistic family. The Jivaroans inhabit parts of lowland Ecuador and Peru. Besides the Wampis, the members of that group are the Awajún (Aguaruna) in Peru, the Shuar in Ecuador, and the Achuar (including Shiwiar – see Seymour Smith 1988) in Ecuador and Peru, all sharing quite similar culture. The status of the Wampis as a separate people/ethnic group is, nonetheless, a bit controversial. Descola (2006: 276) already noted that "with exception of the state border nothing, practically, distinguishes the Huambisa from the Shuar." Indeed, the Wampis with whom I talked often expressed opinions that "they [the Ecuadorian Shuar], are just Shuar like us." One of the indigenous

K. Świerk (✉)
Centro de Estudios Andinos de la Universidad de Varsovia en el Cusco (CEAC), Warsaw-Cusco, Poland
e-mail: kacpersw@yahoo.com

B. Urbani, M. Lizarralde (eds.), *Neotropical Ethnoprimatology*, Ethnobiology,
https://doi.org/10.1007/978-3-030-27504-4_15

organizations/federations of the Wampis bears the name *Organización Shuar de Morona* (Shuar Organization of the Morona River). Wampis assured that they speak the same language as the Ecuadorian Shuar. Nonetheless, they do not reject the name Wampis, which, according to some informants, comes from the *Wampis* fish (scientifically known as *Salminus* sp.).

The Peruvian Shuar state that they are divided into two groups based on dialect differences. Informants of the author and anthropologist Filip Rogalski in the Santiago River basin stated that there are two groups of Shuar speaking slightly different dialects. Those from middle Kanus, downriver from the Candungos native community are named Tsumu Shuar or Tsumu-nmaya Shuar (which is translatable as the Downriver Shuar). Meanwhile, the people living upriver from the Candungos are named Yakiin Shuar or Nujinmaya Shuar (translatable as upriver Shuar). This second group (the upriver Shuar) includes both the people from the upper course of the Peruvian Santiago River and the "Shuar proper" in Ecuador. An analogical division is evident on the side of the Morona River.

During the author's presence in Wampis territory in 2004, while collecting ethnozoological data, F. Rogalski and myself have noted that some animal names are different in upper Shuar and lower Shuar dialects. For example, the many-banded aracari (*Pteroglossus pluricinctus*) is named *pirisat* or *pristian* by the upper Shuar and *pininchi* in the lower Shuar dialect. Similarly, the brown-mantled tamarin (*Saguinus fuscicollis*) is named *tseepai* in upper dialect and *pinchich* in the lower Shuar.

In conclusion, it can be stated that, at least according to the linguistic criteria, the ethnolinguistic division inside the Shuar people (sensu *lato*) is not between the Ecuadorian Shuar versus the Peruvian Wampis but between northern (upper) Shuar in Ecuador and Peru and southern (lower) Shuar in Peru. The identity of the Peruvian Shuar versus Ecuadorian Shuar is a complex and somewhat complicated question, and there is no space here to discuss it fully.

The great majority of the Wampis live today in riverine villages, which since 1974 have had the legal status of nationally recognized native communities (*comunidades nativas*). Until the 1940s, and in many cases later, however, the Wampis population lived, according to the traditional settlement pattern, in dispersed hamlets and households, many of them in the Kampankis mountains and adjacent areas. The Wampis hunt and gather in various habitats. Those from the Santiago River exploit the resources of the nearby Kampankis Mountain Range, while those from the Morona (which is separated from the mountains by a broad plain area) mainly use resources they encounter in the Amazonian lowlands.

Other Jivaroan peoples related culturally and linguistically to the Wampis have been studied by many anthropologists. A few of those investigators dedicated a significant part of their work to Jivaroan–nonhuman primate relations. Guallart (1962) working with the Awajún (Aguaruna) neighbors of the Wampis people created a list of Awajún names of the mammalian species of the upper Marañón River. Many monkey names in that text are similar to Wampis names – many in effect are the same words, differing only in minor phonological features. For example, Wampis *tsere* (white-fronted capuchin) is *tseje* in Awajún.

Brown (1984, 1986) who also studied the Awajún in the Alto Mayo River basin, in the title of his main book *Tsewa's Gift: Magic and Meaning in an Amazonian Society* (Brown 1986), already mentions the mythical spider monkey which taught an Awajún man how to hunt monkeys with blowguns. According to the myth, before that the only manner to hunt monkeys was to climb the trees during night and club the sleeping monkeys with sticks (Brown 1984, 1986). The end of the story will not be recounted here because it is quite similar to the tale about Shuar *Mukucham* and old spider monkey which is presented later in this chapter. Brown (1984, 1986) also describes and quotes magical songs for monkey hunting given to the mythical hunter. Descola (1988) mentions and analyzes many cultural aspects associated with monkeys among the Achuar. Among other subjects, he writes about relationships the hunter establishes with the *amana* (chief) of the woolly monkeys by means of singing magical songs (*anent*).

This text is a contribution to the growing corpus of Neotropical ethnoprimatology (e.g., Lizarralde 2002; Shepard 2002; Cormier 2002, 2003a, 2003b, 2006; Urbani and Cormier 2015). Anthropologists studying the relationships between Amerindians of tropical lands and New World primates produced, especially during the last 10–15 years, many ethnoprimatological texts that confirm the significance that monkeys have for subsistence and cosmological beliefs of lowland South American indigenous peoples. This study represents the first study exclusively focused on Wampis ethnoprimatology.

15.2 Research Methods

The author visited Wampis territory three times – first in 2004, participating in a study ordered by the AIDESEP (*Asociación Interétnica de Desarrollo de la Selva Peruana* – a Peruvian indigenous organization), and again in 2009 and 2011, taking part of investigations organized by the Field Museum of Natural History of Chicago. The aim of the study made by AIDESEP was to document the links the Wampis and Awajún peoples have with the Kampankis mountains (Fig. 15.1). The report written on the basis of this study (Rogalski 2005) was intended to support the Jivaroans' effort to reclaim the Kampankis as a part of their traditional territory, while the Peruvian state expects to establish in the mountains a national park. The Wampis and Awajún disagree with this idea, fearing that it would limit their access to the natural resources of the Kampankis. The activities for the Field Museum were part of a rapid inventory, engaging anthropologists and naturalists. The aim of the anthropological component of the research was mainly to recollect information about indigenous knowledge of the Kampankis mountains as well as its flora, fauna, and ecology.

The data were collected in various native communities on the Santiago and Morona Rivers and in two hamlets still existing in the Kampankis mountains (Fig. 15.1). F. Rogalski and I used several methods of collecting the information

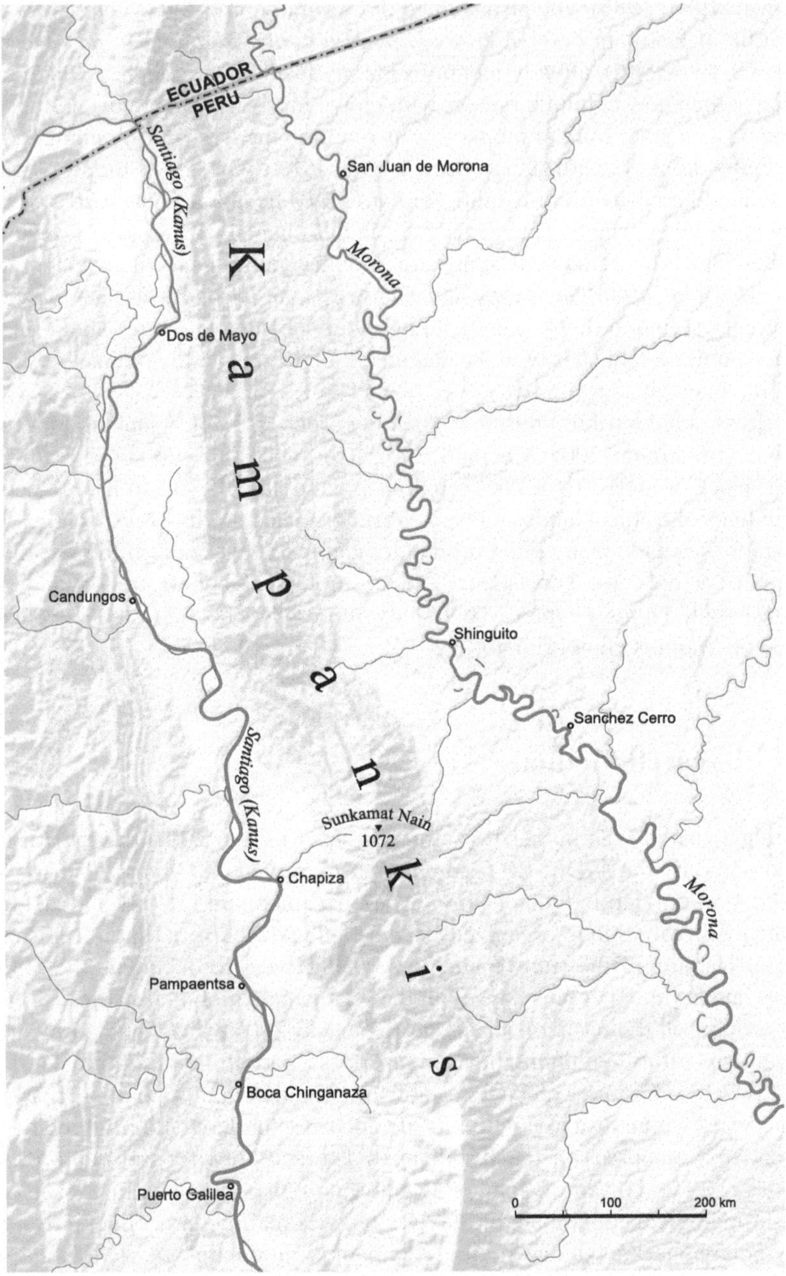

Fig. 15.1 Map of the Wampis territory showing the native communities mentioned in the text and/or in the photograph descriptions. Only Puerto Galilea is not listed in the text but on the map because it is the capital of the district of the Santiago River and one of the southernmost Wampis communities. The location of the mountain *Sunkamat Nain* ("the peak of white-tailed titi monkey") is also shown (Map by Radosław Przebitkowski)

including formal and informal interviews as well as unstructured conversations. In 2004, we crossed the Kampankis mountains from the Santiago River to the Morona River accompanied by several indigenous men. While in the mountains, Wampis companions often spontaneously commented on the animals we met or heard and on other subjects of interest for the field study.

The mountains were walked with indigenous people again in 2011. In 2004, Rogalski and I conducted mapping workshops in the majority of visited communities. Members of each *comunidad nativa* were asked to draw a map of what they considered to be their traditional territory, naming the rivers, streams, and places and indicating the parts where relevant plants and animals could be found. In this way, information was obtained about where various species were abundant including the "supernatural" ones. We also obtained information about the locations of mountains and watercourses named after animal species, among other ethnographic information. The result of that research was an ethnocartographical output that accompanied our report for AIDESEP.

15.3 The Wampis Nomenclature for their Their Monkeys

Apparently, in Wampis language, there is not a single or composite term or word for "monkey." A Wampis asked how would he call this group of animals that are lumped under the Spanish word *mono* (monkey) would usually respond with the term *ikeiñu*. However, the *ikeiñu* term includes not only all their monkeys but also other arboreal mammals such as squirrels (*kunam*, *Sciurus* spp.), dwarf squirrels (*wichim*, *Microsciurus* sp.), porcupines (*kuru*, *Coendou* spp.), kinkajous (*kuji*, *Potos flavus*), and olingos (*Bassaricyon alleni*). Therefore, the term *ikeiñu* is better translated as "arboreal mammal," which includes monkeys. Sometimes the term *ikamau ainna* ("those that climb the trees") is also used. This is quite common on other South American indigenous peoples such as the Barí in Colombia and Venezuela who do not have a term for monkeys in their language, but the Spanish word for monkeys does include kinkajous and olingos too (Lizarralde 2002).

The Wampis territory might have what could likely be the greatest diversity of primates recorded for a single region of the world. A total of 14 species of monkeys are present in this region (see Table 15.1), with the previous highest record of 13 species monkeys reported by Shepard (2002) for the Matsigenka in Manu, Peru. Therefore, the Wampis has what is the highest number of reported monkey species associated with an indigenous culture in the tropical Americas (M. Lizarralde 2018: pers. comm.). In addition, regarding Wampis names for monkeys, this section provides regional Spanish names, additional information about toponymy (derived from monkey names), and the approximate range for some species within Wampis territory. The species are ordered by family.

Table 15.1 List of monkeys present in the Wampis territory (ordered by Wampis folk-generic taxa to illustrate their ethno-taxonomic system)

English common name	Wampis name	Peruvian name	Scientific name
1. Pygmy marmoset	?	Leoncito	*Cebuella pygmaea*
2. Common woolly monkey	Chuu	Mono choro	*Lagothrix lagotricha*
3. Poeppig's woolly monkey	Chuu	Mono choro	*Lagothrix poeppigii*
4. Ecuadorian saki monkey	Sepur/ pentsepents[a]	Huapo	*Pithecia aequatorialis*
5. Monk saki monkey	Sepur/ pentsepents[a]	Huapo negro	*Pithecia monachus*
6. White-tailed titi monkey	Sunkamat	Tocón	*Callicebus discolor*
7. Common squirrel monkey	Tseem	Fraile/mono fraile	*Saimiri sciureus*
8. Brown-mantled tamarin	Tseepai/ pinchich[a]	Mono de bolsillo/ pichico	*Saguinus fuscicollis*
9. White-fronted capuchin monkey	Tsere	Mono Blanco/machín Blanco	*Cebus albifrons*
10. Nancy-Ma's night monkey	Ujukam	Musmuqui/mono nocturno	*Aotus nancymaae*
11. Spix's night monkey	Ujukam	Musmuqui/mono nocturno	*Aotus vociferans*
12. White-bellied spider monkey	Wáshi	Maquisapa	*Ateles belzebuth*
13. Juruá red howler monkey	Yakum	Coto/mono coto	*Alouatta seniculus juara*
14. Tufted capuchin monkey	Yukapkia	Mono negro/machín negro	*Sapajus apella*

[a]Dialectical variations

15.3.1 *Callitrichidae*

There are at least two Callitrichidae species in Wampis territory. Among them is the world's smallest monkey. The pygmy marmoset (*Cebuella pygmaea*) is named *leoncito* ("small lion") in regional Spanish. Unfortunately, the Wampis name was neither recorded in the field nor it appears in the Field Museum report from the Kampankis (Castro Vergara 2012b). This species was mentioned to the author by one Wampis informant in the context of the pet trade. He mentioned a "white Creole Peruvian" who came to the Wampis territory and asked the Amerindians for the possibility of obtaining a pygmy marmoset. At the end, a Wampis man caught an individual in the forest and sold it to the Creole man. Some informants told that this monkey inhabits only the Loretan part of the Wampis territory.

The brown-mantled tamarin or saddleback tamarin (*Saguinus fuscicollis*) is another small monkey. This species is named *tseepai* in the upriver dialect and *pinchich* (*pinchichi*) in the downriver dialect. In regional Spanish, it is known as *mono de bolsillo* ("pocket monkey") or *pichico* (that second name is also sometimes applied to the squirrel monkey in the northwestern Peruvian Amazon). I was told that brown-mantled tamarin is uncommon on the banks of the Santiago River and in the western slopes of the Kampankis, while it is rather common on the eastern slopes and in lowland forest in the side of the Morona River. According to the informants of Castro Vergara (2012a: 286) "*S. fuscicollis* is widely distributed in a variety of habitats but common in lowland secondary forests" and is "common along the banks of the Morona and Santiago Rivers."

15.3.2 Aotidae

Night monkeys (*Aotus* spp.) are also known as owl monkeys. According to Castro Vergara (2012b), there are two species of this genus in the area: Spix's night monkey (*Aotus vociferans*) and Nancy Ma's night monkey (*Aotus nancymaae*). However, these species differ in coloration; it seems that the Wampis recognized only one generic term for both Linnaean species. That name is *ujukam*, both in upriver and downriver dialect. The regional Spanish name for night monkeys is *musmuqui*.

15.3.3 Cebidae

This group is represented by three species within Wampis territory. The common squirrel monkey (*Saimiri sciureus*) is a relatively small monkey named *tseem* by the Wampis (in both dialects). Its regional Spanish name is *fraile* or *mono fraile*. In northern Peru, it is also sometimes named *pichico* that could lead to confusion because it is also applied to *Saguinus fuscicollis*. The white-fronted capuchin (*Cebus albifrons*) is a monkey of importance in the Wampis mythology. Its Wampis name is *tsere*, and its regional Spanish names are *mono blanco* and *machín blanco*.

The tufted capuchin (*Sapajus apella*) is named *yukapkia* by the Wampis. Castro Vergara (2012b) and Pitman et al. (2012) use also the transcription *yukapik*. Its regional Spanish names are *mono negro* and *machín negro*. It seems that this monkey is uncommon in the western part of the Wampis territory (the Santiago River basin). One of our Wampis cooperators from Chapiza accompanied our team to the Morona River (in 2004). When we arrived at the community Sánchez Cerro, he went hunting and shot a tufted capuchin. When asked about this monkey, the informant was unable to give me its Wampis name. Only when asked to a local Wampis the same question, the answer was *yukapkia* monkey (Fig. 15.2).

15.3.4 Pitheciidae

The white-tailed titi monkey (*Callicebus discolor*) is a species of small monkey. Its Wampis name is *sunkamat*, while in regional Spanish, it is known as *tocón*. In author's earlier publications (Świerk 2005; Świerk et al. 2012), the scientific

Fig. 15.2 Tufted capuchin (*yukapkia*, *Sapajus apella*) killed by a Wampis hunter in the lowland forest of the Morona River, Sánchez Cerro native community 2004 (Photograph by Kacper Świerk)

name recorded for this monkey was *Callicebus cupreus* because *Callicebus discolor* was formerly considered a subspecies of *C. cupreus*. In the central part of the Kampankis range, there is a mountain named *Sunkamat Nain* ("mountain of the white-tailed titi") because according to the Wampis there is abundance of titi monkeys there as it is in its vicinities.

According to Castro Vergara (2012a, b), there are two species of sakis in the Wampis territory. These are monk saki monkey (*Pithecia monachus*) and Ecuadorian saki monkey (*Pithecia aequatorialis*). These species differ in coloration. *P. aequatorialis* "has orange fur from the neck to the belly" (Castro Vergara 2012a: 287), while *P. monachus* lacks this coloration. However, it seems that the Wampis do not have distinct names for these two species and recognize one folk-generic taxa. The saki monkeys are named *sepur* (in the upriver dialect) and *pentsepents* (in the downriver dialect, used interchangeably with *sepur*). The regional Spanish name for these monkeys is *huapo*. It is interesting that the neighboring and related Awajún people apparently distinguish in their nomenclature at least two species of *Pithecia* monkeys. Guallart (1962) registered the name *puentse-muents* (similar to Wampis *pentsepents*) for *Pithecia monachus*. This author also provides the name *watac* for another *Pithecia* species not identified in his study. Guallart (1962) also supposes that another Awajún name – *wawan* – also may refer to an unidentified saki species. The sakis observed by me in the Kampankis mountains were of *P. monachus* species (without orange coloration). In one of the downriver Wampis communities (on the Santiago river), a pelt of *P. monachus* was also observed.

15.3.5 Atelidae

Juruá red-howler (*Alouatta juara*) is a large and heavy monkey. This species was formerly considered a subspecies of the *A. seniculus*. Castro Vergara (2012a, b) identified red-howler monkey as *A. juara* in the Wampis territory. Its Wampis name is *yakum*, and it is known in regional Spanish as *coto* or *mono coto*. The name of this monkey traditionally functioned as a personal name among the Wampis, and now it is possible to contact Amerindians with the "*yakum*" surname (Figs. 15.3, 15.4, and 15.5). The white-bellied spider monkey, *Ateles belzebuth*, is a large monkey named *washi* by Wampis and *maquisapa* in regional Spanish.

The woolly monkeys (*Lagothrix* spp.) also occurred in the area. According to Castro Vergara (2012b), there are two species of these monkeys in the Kampankis mountains. These are the brown woolly monkey (*Lagothrix lagothricha*) and the Poeppig's woolly monkey or silvery woolly monkey (*Lagothrix poeppigii*). It seems that the Wampis do not distinguish these two species. Their term for woolly monkey is *chuu*, while its regional Spanish name is *choro* or *mono choro* (Fig. 15.6).

Fig. 15.3 Wampis family from the Morona River with their pet young Juruá red howler (*yakum*, *Alouatta juara*), 2004 (Photograph by Kacper Świerk)

Fig. 15.4 Wampis children of the Morona River putting a killed Juruá red howler (*yakum*, *Alouatta juara*) into fire in order to remove its hair, San Juan de Morona native community, 2004 (Photograph by Kacper Świerk)

15.4 Hunting for Monkeys and Monkeys as Pets

During my stays among the Wampis, I never witnessed them hunting for monkey; however, I witnessed them hunting some other animals, mainly birds. This section on monkey hunting is mainly based on interviews and conversations with my Wampis informants. According to them, all but one of the monkeys listed above (see Table 15.1) are hunted by the Wampis (only the pygmy marmoset was not mentioned in that context). The preferred species is spider monkeys (*Ateles belzebuth*),

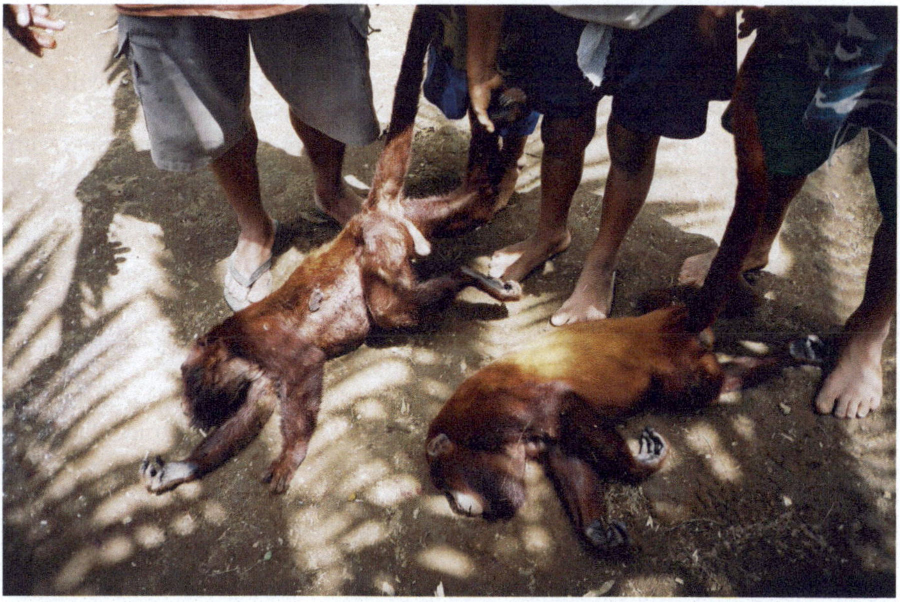

Fig. 15.5 Juruá red howlers (*yakum*, *Alouatta juara*) killed by Wampis hunters of the Morona River, San Juan de Morona native community, 2004, (Photograph by Kacper Świerk)

Fig. 15.6 Woolly monkey (*chuu*, *Lagothrix* sp.), a pet of a Wampis family from the Santiago River, Dos de Mayo native community, 2004 (Photograph by Kacper Świerk)

second woolly monkeys, and then howlers. Among preys brought to visited villages, capuchins and squirrel monkeys were also observed and consumed.

It seems that the majority of Wampis, especially young ones, currently hunt with rifles; however, I was told that there still remain individuals hunting monkeys with blowgun (named *uum* in Wampis and *pucuna* in the regional Spanish). Many of the Wampis own blowguns in their houses. It was told that the most prized ones are those made by the Achuar people which live further east from the Wampis territory. The Wampis of the Morona River obtain the blowguns (in barter or paying with money) from their eastern neighbors, the Achuar. The Santiago River Wampis, in turn, buy them from the Morona River Wampis.

The hunters seek their preys walking through the forest or await it in hiding, close to the mineral licks or drinking locations. Some Wampis construct platforms of palm wood and install them on a tree in the vicinity of a mineral lick. The hunter hides in such platforms and shoots the animals coming to eat the mineral-rich clay or to drink water. Some such places (licks, parts of the streams where the animals come to drink water) are commented to be visited mainly by monkeys, while others are associated with different animals such as peccaries and parrots; however, monkeys sometimes also come and eat or drink from them. During a conversation with an elder inhabitant of Candungos, I asked him if the mineral lick described by him as "a lick of the peccaries" was also frequented by other animal species. The man answered: "It's a mineral lick of the peccaries but the spider monkey is such a villain. How would it not come to steal [the mineral clay] from the lick of the peccaries?"

A Wampis hunter tends to usually go alone looking for games, sometimes with one companion, and rarely in larger hunting parties. Sometimes they hunt close to their villages, but at other times they go to the remote areas of the Kampankis Mountains or of the lowland forest where game is more abundant. In such cases, they sleep in small huts they keep in the forest or in provisional sheds thatched with palm leaves they construct on the spot. Hunting undertakings in remote areas take time, it is necessary to arrive there, usually walking many hours or even a day or two, but they are usually more profitable. The areas close to the villages, especially in the lower and middle Santiago River where the communities are larger, tend to be overhunted, while in more distant areas, monkeys and other game animals are still numerous.

There is also another remarkable aspect regarding primate hunting. As my informant pointed out, in the vicinity of villages animals are not only scarcer but also more wary than their "cousins" living in more remote parts. The Wampis indicate that animals living closer to human settlements are "timid" (in Wampis: *kuntin ashamain*) or "deceitful," "sneaky" (regional Spanish: *animal mañoso*), while those from distant areas are described as "without fear" (Wampis: *kuntin ashamainchau*) or "tamed" (regional Spanish: *animal manso*). As expected, monkeys from those far away regions are less wary and less afraid of people not because they are literally "tamed" but because they live in areas where encounters with hunters are uncommon, so the animals are not aware what a danger the humans pose to them. It can be confirmed that the monkeys in more remote areas are less wary. While walking with indigenous people in remote parts of the Kampankis mountain range, I witnessed many times spider monkeys, woolly monkeys, and white-fronted capuchins that were not shy. Some of them even approached us closer and intended to drive us away by shaking branches.

The hunted monkeys are usually cooked, smoked, or roasted. If smoked, then they can be stored for a longer time, weeks, or even months. To remove the hair of dead monkeys, the Wampis briefly put them above the fire. In other cases, the monkeys are skinned, and the skins of larger species (mainly of howlers) are used for making the membrane of the *tampug*, a drum used by the Wampis.

As mentioned above, there are overhunted areas in the Wampis territory, but generally the monkeys are doing well in the traditional territory of these Amerindians. This state of things is related not only to relative inaccessibility of its lands but also to the fact that the Wampis usually do not practice large-scale logging, using trees mainly for their own needs (e.g., for the construction of a house). Moreover, the majority of the Wampis native communities do not allow the *mestizo* loggers to operate in their territories. Here I mean not only the communal territories but also other areas such as major parts of the Kampankis mountains which are not part of the legally recognized communal lands but are still considered to be parts of their traditional territory. Thanks to such indigenous policies, large-scale habitat loss is avoided.

Like many other Neotropical Amerindians, the Wampis keep many kinds of monkey pets in their households. When they encounter young monkeys in the forest, they usually bring them to their communities and feed them and treat them like pets. Many of these animals are orphans whose mothers were killed during hunting. During my stays with the Wampis, I saw brown-mantled tamarins, squirrel monkeys, white-fronted capuchins, woolly monkeys, and howlers kept as pets in households. They were usually treated well. In some cases, the animals were tethered to one of the posts of the house, and in others they were allowed to roam freely in the house and its vicinities. They were given human food (plantains, bananas, cooked manioc) and also manioc beer (*nijamanch*). The brown-mantled tamarins are appreciated by Wampis because they eat spiders and insects, especially those that live in their thatched roofs. A Wampis informant from the Chapiza community (Santiago River) who accompanied us to the Morona River had a pet monkey belonging to this species and took it with him. In the community of Shinguito, I witnessed the efficiency of his pet as an exterminator of various invertebrates, especially large spiders. The tamarin ceaselessly patrolled the interior part of thatched roof of the house in which we were being accommodated and preyed on many kinds of invertebrates. The smaller preys were eaten in the roof. In case of the large spiders, however, the monkey bit them and then dropped them. The spiders fell into the wooden platform of the house, and the monkey went down to finish its kill and eat it.

15.5 Monkeys and Monkey-Like Creatures in Mythology and Cosmology

Monkeys occupy an important place in the Jivaroan mythical corpus, and Wampis myths are no exception to this fact. In many Wampis myths concerning hunting trips, spider monkeys – their preferred hunting game – are mentioned (see García Rendueles 1994). García Rendueles (1993) collected a myth regarding the human–spider monkey relationship. The protagonist of this narrative is a Wampis mythological figure named *Mukucham*, who was a very poor hunter. Once in a time while walking in the forest, he meets a group of *washi* (spider monkeys). One of them, an old male monkey, gave him a tobacco for smoking and taught him one of their magical songs (*anent*) which made him a good hunter. However, the monkey elder warned him that he must not hunt his children and should tell his people to not do it. After that encounter, *Mukucham* became a very good hunter, killing a lot of animals. One day, oblivious of the promise he gave to the monkey elder, he came to the place of the monkeys and killed the children of the old *washi*. When he was leaning over, in order to tie up his prey, the old spider monkey grabbed the *Mukucham*'s blowgun and inserted it into his anus, magically transforming the hunter into a *washi* (with his own blowgun becoming his prehensile tail). However, *Mukucham* did not become a common monkey but a *washi amana* (chief of the spider monkey). He lives in a rocky cave where he hides the spider monkeys from human hunters, above all in the season when the *washi* are fat and especially sought by the Wampis.

There is also a myth recorded by García Rendueles (1993) in the community of Pampaentsa in western bank of the Santiago River. It explains why some species of monkeys are absent from this particular side of the river. The myth tells about a Master of Animals who made a kind of bridge out of a liana in order to allow animals to cross from the eastern to western bank of the Kanus. The animals started to cross the river, but unfortunately someone cut the liana bridge, and a number of monkeys could no longer get to the other side. As the storyteller from Pampaentsa indicated, there are only medium-sized spider monkeys on his side of the river, while larger individuals live on the eastern side of the Santiago River. These larger spider monkeys are named *washi awatat* in the narrative. Besides the medium-sized spider monkeys, the animals which made their way to the western bank were the white-fronted capuchins, howlers, titis, and squirrels (Wampis: *kunam*, *Sciurus* sp.) (even squirrels are not considered monkeys proper; however, in Wampis view, they are related because of their similarity due to their arboreal lifestyle). The creatures that could not cross the river, because of the premature cutting of the liana that served as a bridge, were the larger spider monkeys, tufted capuchins, and sakis. Hence, on this side of the Kanus, there are only some kinds of monkeys (García Rendueles 1993).

One of the most important characters in Wampis (and other Jivaroan) myths is *tsere* – the white-fronted capuchin, characterized as an intelligent and astute person. According to the Wampis myths, early Jivaroan humanity lived in constant threat of being eaten by cannibals named *uya* (also known as *iwa*). Fortunately, the people of that time had several animal helpers who defended them and made them possible to survive those dangerous times. Among these helpers were the hummingbird (*jempe*, family Trochilidae), a freshwater crab (probably genus *Hypolobocera*), various insects, and, most of all, the *tsere* (the white-fronted capuchin monkey). According to a tale told by an informant from the community of Boca Chinganaza, once *uya* pursued a group of people who found their shelter on a high rock, they managed to climb. *Uya* had a powerful axe with which he intended to fell the rock. Fortunately, the *tsere* intervened, stealing the axe and concealing it under a stone in a stream. *Uya* started to look for the axe, but the capuchin monkey rapidly moved it from one stone under another until the cannibal became totally disoriented and confused. Additionally, *tsere* gave to the *uya*, a soap made of a kind of a liana which caused stinging in the cannibal's eyes. The *uya* got nervous and started to pursue the *tsere* with the intention of killing him. Finally, the monkey (which addressed the cannibal as *apachur* – grandfather) convinced the *uya* that instead of killing him, he can beat his penis with a stone. *Uya* agreed, and since that time the white-fronted capuchins were known to have flat penises. The informant finished that story with the words: "historically the white-fronted capuchin defended us a lot."

Animals not only defended the mythical Jivaroans from *uya* but also stole various things including arts and technological skills not known to the early humans from those cannibals and other beings. For example, the hummingbird stole fire from *uya*, setting in it his own tail. I collected a tale in which the *tsere* "steals" the art of roof thatching from *uya*. García Rendueles (1993) collected another version of this myth in which the thatchers are the *kunam* (squirrels,) that were humans in mythical time.

The *uya* knew how to thatch, while the human beings did not know that art. The white-fronted capuchin decided to go to the house of the *uya* in order to observe the work of thatching and to learn it in order to teach it to the people. He went to the house of the *uya* and greeted him by addressing him as grandfather. The *uya* responded: "Go away *tsere* [white-fronted capuchin]. We will thatch the roof and you surely will steal our art and teach it to the humans." The monkey responded: "I made such long way in order to visit you. Perhaps I could stay?" The *uya* responded: "Well, so you can stay but I will cover you with something." The monkey agreed. Therefore, the *uya* covered the *tsere* with a big *pinig* – a clay bowl for serving the manioc beer. The *uya* then asked: "Do you see something?" "Yes, I can see everything" – responded the *tsere*. Hence, the *uya* removed the *pinig* and put over the monkey a tightly woven basket named *suku*. "Can you see anything?" asked again. "Yes, I see everything" responded the monkey. So, the *uya* removed the *suku* and replaced it with loosely woven basket named *chankin* which has big meshes. "Now, can you see anything?" – asked. "No, now I see nothing" – responded the *tsere* which now could see everything. The *uya*, calmed down now, started the thatching with the leaves of the *Phytelephas macrocarpa* palm (Wampis: *chapi*, regional

Spanish: *yarina*) and *tsere* observed his work from under the basket learning how to do it. After returning to the people he taught them the thatching skills."

The Wampis also have the belief of a large and dangerous animal known as the *tsunkutsunk* (or *tsugkutsugku*) that lives in the Kampankis mountains. Their name is translated into Spanish as *mono-tigre* or monkey jaguar (the translation is not literal). According to my informants, these beings have patterns on their skin similar to those of jaguars. They move swiftly through the forest jumping from one tree trunk to another, catching on a tree with their claws, so their presence can be confirmed by traces of their claws on the trees (stripped bark). According to the informants, these animals were once widespread in the Kampankis Mountains but now are quite rare, and their main stronghold is the area of the mountains relatively close to the Chapiza community. I was told that ancient Wampis fought against these animals. Both Lizarralde (2002) and Shepard (2002) also indicated the presence of "imagined" animals in Barí and Matsigenka cosmology, respectively, suggesting that these indigenous peoples' believes may represent a folk memory of extinct giant ground sloths. Perhaps the same could be suggested for the *tsunkutsunk*; however, the behavior of this arboreal quick jumper probably does not resemble that of a ground sloth.

Finally, it can be added that the *iwanch* (spirits of the dead) is also compared to monkeys, specifically spider monkeys. According to one of informant, *iwanch* has a small head, similar to that of spider monkey, and shaggy arms and hands which also make him similar to this species. *Iwanch* are considered dangerous creatures which can kidnap women and children. According to informants, Wampis historically used to fight against them using spears and *tantar* (a kind of wooden shield).

During my stays among the Wampis, I did not collect any *anent* or magical songs concerning monkeys. In this kind or genre of songs, metaphors from animal world are frequently used. It is very probable that monkeys are mentioned in Wampis *anent*. Taylor and Chau (1983) collected many *anent* among related Achuar, and in two of them a reference to monkeys is made. In one case, it is a squirrel monkey and in another, a pygmy marmoset. In both cases, the singer is a woman who intends to appease the anger of her husband presenting herself metaphorically as a jolly squirrel monkey and a woeful pygmy marmoset. Descola (1988) quotes and analyses Achuar hunting *anents* directed to the *amana* (chiefs) of the woolly monkeys. Brown (1984) also quotes and describes magical hunting songs about spider monkeys sung by Awajún men. These *anen* (Awajún version of the word *anent*), according the natives, attract their favorite preys: spider monkeys.

The information presented above shows that monkeys and the narratives about them are of importance in Wampis mythology and cosmology. The same could be stated for other Jívaroan peoples and, in general, for upper Amazonian Amerindians. Versions of the majority of Wampis myths and tales presented in this text have their counterparts in the mythology of other Jivaroans. Thus, for example, in the field I heard a story about *tsere* stealing the art of roof thatching told by an Awajún man. This tale and its Wampis version were almost identical. It seems that the *tsere* stories are pan-Jivaroan. The same can be told about many other myths.

For example, the Awajún tale of *Tsewa* mentioned in the introductory section (Brown 1984, 1986) is very similar to the story of the Wampis *Mukucham*. Karsten (2000 [1935]) writing about the Ecuadorian Shuar mentions dead's spirits (*iwanch*) in form of a large, terrestrial spider monkey (*washi iwanch*). This belief clearly corresponds to the Wampis idea of similarity between dead spirits and spider monkeys. In Jivaroan cosmology, people become monkeys, and monkeys have human traits such as speech and altruism. They were the helpers and rescuers of endangered humans by the *uya* predation.

15.6 Conclusion

The information presented confirms that monkeys are animals of considerable importance in the culture of the Wampis people. They are hunted for food and kept as pets. They also have a relevant place in Wampis mythology and cosmology. In spite of the fact that in some areas of the Wampis territory, mainly in the vicinity of their larger communities, primates are overhunted and scarce, generally monkeys are quite common and widespread in the region, both in the mountains and in the lowlands. Still there are numerous areas of difficult access such as remote mountainous forests and lowland swamps where primates are able to refuge and hide from human hunters. The Wampis do not practice large-scale woodcutting for profit and rarely and reluctantly let the *mestizo* loggers enter their territory, so the monkeys' habitat is not currently endangered. All these facts are positive for having the hope that the relationship between the Wampis and their monkeys will last through future generations.

Acknowledgments I wish to thank Stephen Beckerman, Mariusz Kairski, Gerónimo Petsain Yacum, Filip Rogalski, Lino Max Santa Tsamaren, Bernardo Urbani, and Alaka Wali and the institutions AIDESEP (Asociación Interétnica de Desarrollo de la Selva Peruana) and the Field Museum of Natural History. Especially I am grateful to two persons: first, Manuel Lizarralde for editing my texts, correcting my English, preparing the table with monkey names, and writing my abstract; and second the cartographer Radosław Przebitkowski for making the map of the Wampis territory free of charge.

References

Brown MF (1984) The role of words in Aguaruna hunting magic. Am Ethnol 11(3):545–558

Brown MF (1986) Tsewa's gift: magic and meaning in an Amazonian society. Smithsonian Institution Press, Washington & London

Castro Vergara L (2012a) Mammals. In: Pitman N, Ruelas Inzunza E, Alvira D, Vriesendorp C, Moskovits DK, del Campo Á, Wachter T, Stotz DF, Noningo Sesén S, Tuesta Cerrón E, Chase Smith R (eds) Peru: Cerros de Kampankis. Rapid biological and social inventories. Report no 24. The Field Museum, Chicago, pp 282–288

Castro Vergara L (2012b) Apéndice/appendix 7: Mamíferos medianos y grandes/large and medium-sized mammals. In: Pitman N, Ruelas Inzunza E, Alvira D, Vriesendorp C, Moskovits DK, del Campo Á, Wachter T, Stotz DF, Noningo Sesén S, Tuesta Cerrón E, Chase Smith R (eds) Peru: Cerros de Kampankis. Rapid biological and social inventories. Report no 24. The Field Museum, Chicago, pp 388–397

Cormier LA (2002) Monkey as food, monkey as child: Guajá symbolic cannibalism. In: Fuentes A, Wolfe LD (eds) Primates face to face: conservation implications of human-nonhuman primate interconnections. Cambridge University Press, Cambridge, New York, Melbourne, Madrid, Cape Town, Singapore, São Paulo, pp 63–84

Cormier LA (2003a) Kinship with monkeys: the Guajá foragers of eastern Amazonia. Columbia University Press, New York

Cormier LA (2003b) Animism, cannibalism, and pet-keeping among the Guajá of eastern Amazonia. Tipití 1(1):81–98

Cormier LA (2006) A preliminary review of Neotropical primates in the subsistence and symbolism of indigenous lowland south American peoples. Ecol Environ Anthropol 2(1):14–32

Descola P (2006) Powinowactwo z wyboru. Sojusze małżeńskie, wojna i drapieżność wśród Indian Jívaro. In: Buliński T, Kairski M (eds) Sny, trofea, geny i zmarli. Wojna w społecznościach przedpaństwowych na przykładzie Amazonii – przegląd koncepcji antropologicznych. Wydawnictwo Naukowe UAM, Poznań, pp 273–302

Descola P (1988) La selva culta: Simbolismo y praxis en la ecología de los Achuar. Instituto Francés de Estudios Andinos & Ediciones Abya Yala, Lima, Quito

Fabre A (2016) Jívaro, http://www.ling.fi/Entradas%20diccionario/Dic=Jivaro.pdf

García Rendueles M (1993) Yaunchuk: wampisá aújmattairi. Vol. I. Centro Amazónico de Antropología y Aplicación Práctica, Lima

García Rendueles M (1994) Yaunchuk: wampisá aújmattairi. Vol. II. Centro Amazónico de Antropología y Aplicación Práctica, Lima

Guallart JM (1962) Nomenclatura jíbara-aguaruna de especies de mamíferos en el alto río Marañón. Biota 32(4):155–164

Karsten R (2000) [1935] La vida y la cultura de los shuar. Cazadores de cabezas del Amazonas occidental: La vida y la cultura de los jíbaros del este del Ecuador. Ediciones Abya Yala, Quito

Lizarralde M (2002) Ethnoecology of monkeys among the Barí of Venezuela: perception, use and conservation. In: Fuentes A, Wolfe LD (eds) Primates face to face: conservation implications of human-nonhuman primate interconnections. Cambridge University Press, Cambridge, New York, Melbourne, Madrid, Cape Town, Singapore, São Paulo, pp 85–100

Lizarralde M (2018) Etnoprimatología Barí en la Sierra de Perijá del estado Zulia. In: Urbani B, Ceballos-Mago N (eds) La Primatología en Venezuela. Academia Nacional de Ciencias Exactas, Físicas y Naturales/Universidad Simón Bolívar (Colección Conjunta Acfiman/USB, Caracas

Lizarralde M, Lizarralde R (2018) Los Barí. In: Pereira M, Rivas P (eds) Los Aborígenes de Venezuela. Volumen V (Monografía 52). Fundación La Salle de Ciencias Naturales, Caracas, pp 725–886

Pitman N et al (2012) Apéndice/appendix 10: Nombres comunes de plantas y animales/common names of plants and animals. In: Pitman N, Ruelas Inzunza E, Alvira D, Vriesendorp C, Moskovits DK, del Campo Á, Wachter T, Stotz DF, Noningo Sesén S, Tuesta Cerrón E, Chase Smith R (eds) Peru: Cerros de Kampankis. Rapid biological and social inventories. Report no 24. Chicago, The Field Museum, pp 410–431

Rogalski F (ed) (2005) Territorio indígena Wampis-Awajun: Cerro de Kampankis. CIPTA-AIDESEP, Iquitos

Seymour Smith C (1988) Shiwiar: Identidad étnica y cambio en el Río Corrientes. Ediciones Abya Yala, Quito, Lima

Shepard GH (2002) Primates in Matsigenka subsistence and worldview. In: Fuentes A, Wolfe LD (eds) Primates face to face: conservation implications of human-nonhuman primate interconnections. Cambridge University Press, Cambridge, New York, Melbourne, Madrid, Cape Town, Singapore, São Paulo, pp 101–136

Świerk K (2005) Uso de recursos naturales de la Cordillera de Kampankis por los indígenas Huambisa. In: Rogalski F (ed) Territorio indígena Wampis-Awajun: Cerro de Kampankis. CIPTA-AIDESEP, Iquitos, pp 66–84

Świerk K, Rogalski F, Wali A, Alvira D, Pariona M, Tuesta E, Treneman A (2012) Resource use and traditional ecological knowledge. In: Pitman N, Ruelas Inzunza E, Alvira D, Vriesendorp C, Moskovits DK, del Campo Á, Wachter T, Stotz DF, Noningo Sesén S, Tuesta Cerrón E, Chase Smith R (eds) Peru: Cerros de Kampankis. Rapid biological and social inventories. Report no 24. The Field Museum, Chicago, pp 310–317

Taylor AC, Chau E (1983) Jivaroan magical songs: Achuar *anent* of connubial love. Amerindia 8:87–127

Urbani B, Cormier L (2015) The Ethnoprimatology of the howler monkeys (*Alouatta* spp.): from past to present. In: Kowalewski MM, Garber PA, Cortes-Ortiz L, Urbani B, Youlatos D (eds) Howler monkeys: behavior, ecology, and conservation. Springer, New York, pp 259–280

Chapter 16
The White Monkey and the *Pelejo* Monkey: Primates in the Social and Cultural Configurations of the Shawi People of Northwestern Peru

Luisa González-Saavedra

16.1 Introduction

The Shawi of the upper Peruvian Amazon, also known as Chayahuitas, have an ample relation with their environment. However, the presence of monkeys in Shawi domestic spaces is practically null, possibly because of the fact of low densities of primates in that region and acculturation. The Shawi tend to avoid animals inside their dwellings, while sometimes dogs, which are used as hunting companions, are allowed staying in home premises. This is because they consider that entities of the forest should remain there. Thus, most monkeys, like many other animals at home, are then placed as food sources.

16.2 The Shawi of the Upper Amazon

Chayahuita is perhaps the most common name used in the literature as an exonym applied by the local *mestizo* population to them. In their own language, they prefer to use Shawi as their autonym or self-denomination. They belong to the Kawapana linguistic family (Rojas-Bercia 2013). The Shawi occupy an area that extends between the eastern Andean mountains and the upper western Amazonian basin (Renard-Casevitz et al. 1988). Today, their territory extends for more than 10,000 km^2 that combines plains with hills and valleys and a complex hydrographic system dominated by the presence of three major rivers: the Paranapura, the Cahuapanas, and the Sillay (Fig. 16.1). Part of this vast territory, traveled by the Spaniard Alonso de Mercadillo in 1538 (Golob 1982), is entitled under the Peruvian

L. González-Saavedra (✉)
Miembro del Grupo Etnografía de América, Universidad Complutense de Madrid, Madrid, Spain
e-mail: gs.luisa@gmail.com

B. Urbani, M. Lizarralde (eds.), *Neotropical Ethnoprimatology*, Ethnobiology,
https://doi.org/10.1007/978-3-030-27504-4_16

Fig. 16.1 Location of the Shawi territory in the Upper Amazon of Peru. (Base map: Wiki Commons)

law to native and peasant/creole communities, having been demarcated and recognized by the Ministry of Agriculture at the request of the communities themselves. At present, their population is approximately around 21,000 Shawi people (Rojas-Bercia 2013) spread in more than 90 communities throughout their territory. It is interesting to point out that the archaeological complex of Balsapuerto (900 BP–1200 AD) (Rivas-Panduro 2014) is located within today's Shawi territory. At this moment, this territory is also a region of touristic interest and where several archaeological research projects are currently carried out. Within this cultural complex stands out in importance the Stone of Cunpanama, a landmark of ample cultural relevance among Shawi.

The Shawi are horticulturalists and gatherers, hunters and fishers, and inhabitants of an ecologically poor area, marked by the scarcity of animals due to the extreme hunting of people linked to logging and oil companies. In this regard, their diet is basically concentrated in the intake of small fishes, medium rodent species such as the *majs* (lowland paca, *Cuniculus paca*), some monkeys, corn, bananas, and manioc. In fact, this last tuber crop is the most essential food in their diet. With manioc or cassava, they make *masato* (manioc fermented drink), which is popularly consumed among many native peoples of the forest and even among the nonindigenous population. In the case of the Shawi, it also has a special symbolic value in addition to the purely nutritional one. The Shawi *masato*, undoubtedly, is one of the most important elements of the social life of this society.

The Shawi are today considered Christians (Roman Catholics), as they were named as "old Christians" in the Marañón Missions (Figueroa 1904; Chantre Herrera 1901), and their festive calendar is actually the calendar of Catholic holidays (Fuentes 1988). There is no community without a church, and they like to differentiate themselves from the *Awajun* (another ethnic group also known as Aguarunas or Jivaros that are their neighbors and likely alleged enemies), among other indigenous peoples of the area like the *Shiwilu* or *Jebero*, the first Amerindians to be evangelized in the region, and today assimilate by the local *mestizo* society (Joulu 2006), having a language of similar Shawi stock (Grohs 1974). This has not led the Shawi, however, to leave aside their indigenous traditional cultural heritage, as expressed in rituals such as the first menstruation of women or the initiation into the animal world by men (González-Saavedra 2013). The Shawi value showing their particular universe, their special way of conceiving the world, of conceiving themselves within a universe of beings (some of them humans, some not) with which they share their essentiality, their ontological being. It is this particular way of understanding the Shawi conception of themselves together with the beings that inhabit their surroundings, from which this chapter is going to show both, the place and the task, that they had in the beginning of the times of this society: two "primate characters" such as the *white* and the *pelejo* monkeys.

16.3 Primates and the Shawi

There is no doubt that forest animals are found everywhere within the Shawi universe. The world of the *canpo piyapi* ("us", "us the Shawi," or "our people" in Shawi language [Ochoa 2008]) is a human space that cannot be conceived if it is not in continuity with that of the *tanan huayan* or "people" of the forest, where also animals are included. With this qualification, the Shawi refer to a range of beings living outside the space of *canpo piyapi*. It includes animals but also other beings (some human and some not) with whom Shawi have different forms of interaction such as "spirits of the forest."

There are eight species of monkeys under Linnaean taxonomy recognized by the Shawi and four other species of arboreal mammals that are also recognized under their category of "monkey" (Table 16.1). As can be seen in Table 16.1, some of the animals classified as monkeys that do not correspond to primates in Linnaean classification are included as such because of their arboreal life and their similar feeding habits of "Linnaean primates." For example, the brown-throated three-toed sloth (*Bradypus tridactylus*) is one of those cases. All these arboreal species, like the rest of the forest animals, are also considered "people" by the Shawi, with particular customs and habits. The word "people" is commonly used along the Amazon, as giving the forest life a sense of broad humanity, where it is nature (not culture) that separates species. They celebrate their holidays as Shawi celebrate the *canpo piyapi*, they have their own meals and drinks and also their traditional clothes, and sometimes they appear before us with the same appearance as our relatives and with

Table 16.1 Classification of "monkeys" according to the Shawi of the Upper Amazon (after Emmons 1990, García-Paredes 1994, González-Saavedra pers. obs.)

Name	Size	Color	Edible	Characteristics
Achuni (shu'shu)[a] *Nasua nasua* Coati	High: 15 cm Long: 50 cm Width: 10 cm	Black	Yes	–
Mono Blanco (ahui') *Cebus albifrons* White-headed capuchin monkey	H:17 cm L: 40 cm W: 9 cm	White and brown	Yes	It is very naughty; it breaks everything it sees
Coto (nu'nu) *Alouatta seniculus* Red howler monkey	H: 20 cm L: 70 cm W: 13 cm	Scarlet	Yes	When it is going to rain it screams
Chosna (cuhuasha') *Potos flavus* Kinkajou	H: 13 cm L: 45 cm W: 9 cm	Ash	Yes	–
Choro (suni') *Lagothrix flavicauda* Yellow-tailed woolly monkey	H: 20 cm L: 70 cm W: 13 cm	Black	Yes	Men do not eat their tail because when they go out to hunt it, this monkey gets entangled in the branches of the trees
Huapo (tequerenan) *Pithecia monachus* Monk saki monkey	H: 12 cm L: 50 cm W: 9 cm	Black and ash	Yes	–
Leoncito (ishi') *Cebuella pygmaea* Pygmy marmoset	H: 7 cm L: 12 cm W: 6 cm	Black	Yes	–
Maquisapo *Ateles paniscus* Black spider monkey		Black		Only some are left in the mountains
Musmuqui (cuhui`) *Aotus nancymaae* Nancy Mae's night monkey	H: 11 cm L: 30 cm W: 7 cm	Ash	Yes	The musmuqui makes a living at night and sleeps for the day. The pregnant woman cannot eat musmuqui. In the morning, this monkey is sleeping, and at night it wakes up crying
Pichico (isën) *Saguinus fuscicollis* Saddleback tamarin	H: 10 cm L: 30 cm W: 7 cm	Brown	Yes	–
Pelejo (tihuin) *Bradypus variegatus* Brown-throated three-toed sloth	H: 22 cm L: 60 cm W:18 cm	White and black	Yes	With the claws it is fastened to the ropes of the trees and to sleep does it with the head down; when going up is clinging to the branches of the trees or to the lianas. When a person eats a *pelejo*, he does not die, because the *pelejo* is hard to die
Shihui (suruntë) *Tamandua mexicana* Collared anteater or southern tamandua	H: 16 cm L: 45 cm W: 13 cm	Black and ash	Yes	It walks very slowly. When the woman eats her flesh during pregnancy, the baby becomes sick while the newborn's ribs penetrates his/her body a lot; it breathes quickly and dies

[a]Denomination in Shawi language

intentions to guide us and help us in our walks or in our outings to the mountains and forests. On other occasions; however, their presence is a sign of illness or death. Among those animals, there are boas, jaguars, peccaries, and birds such as parrots or toucans. There is also the *shapsico*, half animal-half man, and the *tunchi*, which presents to humans like skeletons, or the *shansho* (*Opisthocomus hoazin*; hoatzin or stinkbird), which is a pestilent and solitary bird that announces death. *Shapshico* is a fairly popular character in the forest, not only in indigenous areas but also among the coastal people (also named *mestiza*) living in the region where this study was carried out, a rather small being with hat, gnawed shirt and pants, and with his feet pointing backward in the opposite direction to its body. This last feature is the most relevant characteristic of this being which is particularly feared because it is a kidnapper. *Tunchi* is described as the "forest spirits."

16.4 The White Monkey and the Sloth or *Pelejo* Monkey

The white-headed capuchin monkeys or *ahui* (*Cebus albifrons*) are also the forest "people" among the Shawi. They have such a noisy and restless character that provoke the god *Cunpanama'*, the most relevant Shawi deity, to divide the land between the Shawi and the Jíbaro Awajun, giving to the latter a larger territory. For that reason, white-headed capuchin monkeys happen to be associated to the *gringo* (Americans) and Europeans, or whitish-skinned and pink-faced foreigner, an archetype that includes Westerner persons today.

On the other hand, the popular sloth or *pelejo* (brown-throated three-toed sloth) is classified as a monkey. Sloths are named *Tihuin*. The *pelejo* monkey, for example, has its particular characteristics because of *Apu* (a Quechua word that means person with authority in the community) as the ability "to see" (which is also called the power of shamans to achieve a connection with what is beyond natural, equivalent to having visions), which made it possible for the Shawi people to be saved from their own disappearance back in the early days. Undoubtedly, the *pelejo* was the first great shaman or *pënoton* among Shawi. *Pelejo* made it possible for the people to be saved and for the wealth of this people to be preserved to our time. For this, as seen below, it had to make the decision to turn many of these people into animals to feed today's Shawi men and women. Those women and men have to respect these animals as people. Also, they must appear before their "mothers" prior to hunting them, without petting them or killing them indiscriminately. These are nowadays the guidelines of a so-called good Shawi hunter.

The Shawi view and qualify the *pelejo* by the concept that they use, as a monkey, but also a being that is solitary, taciturn, and feeds on leaves and herbs, with claws that serve as a defense element. Thus, the Shawi perceive the sloth monkey as a predator over its own predators, which are the harpy eagle and jaguar. It is not to appear as an animal of agile and strong defense like the other monkeys. The *pelejo* is a being that uses silence as a mechanism of defense. Just like Shawi shamans, the *pelejo* are silent, taciturn, and needy of their medicinal plants which include tobacco. Also, among those relevant plants, the hallucinogenic *ayahuasca* (*Banisteriopsis caapi*) is widely used by various western Amazonian societies. It is the mother plant that allows them to have a particular and powerful vision of the world. This plant is also one that monkeys used in early times of the Shawi and thanks to which they were able to perpetuate as a group to the present. In this chapter I would like to develop a narrative on the role of these two relevant beings, the white monkey and the *pelejo* monkeys, according to the Shawi social and cultural configuration.

16.5 The *Huito* tree and the Hunguyacu Hill: The Actions of the White Monkey

The Shawi tell that in the days of their ancient times, there was within their territory a tree called *huito* (*Genipa americana*) so large and leafy that it shaded all their lands without allowing the sun to grow the fruits of their gardens. In need of food, they decided to ask *Cunpanama'* to cut this tree. *Cunpanama'* gathered around him carpenter birds, monkeys, parrots, and other animals, and next to them he prepared to turn it down. But so hard and tall was that *huito* that the ax of *Cunpanama'* ended up breaking. To fix it, it sent the white monkey to his house for tar. There was his wife who, seeing the white monkey arriving, asked him to accompany her so that he could see for himself at the source of his tar. She showed the monkey her vagina. Thus, the white monkey took so long to return that *Cunpanama'* went back to his house to see what could be happening. When he arrived, he found the white monkey over his wife "bothering" her, that is, having sex with her. *Cunpanama'* gets so angry when he discovers the scene that his fury causes the white monkey to run away so fast that he almost disappeared. The wrath of *Cunpanama'* cursed him thus making him a monkey so that the men would get used to eating them, and, besides, with so much anger it was that delayed *Cunpanama*'s return to the *huito*. Once *Cunpanama* returned, he discovers that the tree had already been cut, but not as he had ordered, but in the opposite direction. The branches that should go up were looking down, and the crown of the tree had given rise to the creation of a remote and abrupt hill. The erroneous fall of the *huito* also gave rise to the creation of narrow gorges, for which, however, the Awajun and the *mestizos* would soon enter their territory, not only abducting their members but also stealing Shawi lands.

Seeing such a disaster, *Cunpanama'* decided to retire into the hill created by the fall of the *huito* tree, which was named Hunguyacu hill. Once there, saddened by the action of the white monkey, it climbed the hill with tears in his eyes and some encapsulated fruits stored in his *shicra* (bags made from chambira palm, *Astrocaryum chambira*) as his only food. From those fruits, *Cunpanama'* would actually create the Shawi. And from his tears, they say, created the Awajun (Aguarunas or Jivaros) who appeared by their rivers kidnapping their wives and occupying Shawi lands. This is how the Shawi were created: the result of *Cunpanama*'s tremendous anger caused by the white monkey.

This narrative serves to the Shawi as an account for both the disposition of their territory and their own creation as human beings. The time in which the narrative is placed, the time of the ancients, is nevertheless a time still determined by a basic principle which is repeated with insistence in the references to previous generations of the Shawi as "when the animals were people... and the people animals." This served to configure Shawi's space and universe as it is known by them today.

The myth of the fallen *huito* tree gives entrance to the first great transformation of the Shawi past: the conversion of animals into people and also people into animals. This means that up to those times, animals and people shared the same ontological status. This aspect, precisely, is what was broken after the action of the white monkey. At that period when there was no difference between the beings that inhabited the earth, the white monkey was a protagonist of a series of stories that combine the teachings with the determination of several principles that today are considered main references of the social norms of Shawi coexistence. For example, the white monkey taught to build houses; in addition, with his transvestite and mischievous attitude (like the one he showed to *Cunpanama*'s wife), he perpetuated a basic rule of Shawi society: respect for their women, especially if they belong to another man. The actions of the white monkey, on the other hand, earned him the designation of a concrete nature, the nature of today's white-headed capuchin monkey. And at the same time, they formed a nation that, since that time, all beings were people that stand territorially and culturally between these two human groups: Awajun and *mestizo*.

But among the Shawi, who is the white monkey? It is an animal that annoyed *Cunpanama*'s wife, provoking his anger, and turns into a white person with a slightly pink face, someone who used to "bother" Shawi women and at the same time enjoyed learning how their ancestors built their houses. For this, also in times when the animals still were people, the white monkey went into a house pretending to be nice. There they gave this monkey shelter, until its mischief again forced the women to cover the monkey with a basket. Even so, the white monkey continued to enjoy looking at the roofs of Shawi houses, the weaving of its branches and its crosshairs. The women put baskets on baskets seeing that the white monkey was not but enjoyed and felt curiosity by their houses. They almost kill him by suffocation. Then they let him out. And so, it was, as the Shawis tell nowadays, how the monkey learned to build these houses that they build now thanks to the white monkey's teachings.

It is not especially difficult – though not lacking in audacity – to combine Shawi's past times with the history that appears in the written sources of the first years of contact between white and indigenous people. In this respect, the combination of the two previous stories (the one of the *huito* and the other that shows how the white monkey learned to build houses) could well sustain a reading on the colonization that arrives and breaks the basic fundamental norms of coexistence (like "getting in" with their women) of the indigenous peoples while at the same time contributing with new materials and ideas that are consolidated over the years in such a way that they are rooted in the primary history of the whole, the one that is situated in the ancient times. Thus, "the white monkey is just like you, missed, like a *gringo*." This response was repeated to the author in different occasions and asking about those people who became a white monkey after *Cunpanama*'s anger when meeting his wife. That was the white monkey: that noisy, transgressive being who lived in the trees, walked along its branches with its white and brown hair, and ate the leaves without any other need, never touching the ground.

The importance of this animal, and only this one, is that it is identified with the white men. If the Shawi are asked why the white monkey became a *gringo*, they will surely answer that because he is white as foreigners It upset *Cupanama'* and pushed him up to Hunguyacu Hill. Actually *Cupanama'* lives there since that moment, according to the Shawi. He entered that land with his wife and later facilitated the inclusion of foreign peoples. To this day, the Awajun are the main alleged enemies of the Shawi, and even today both nations continue to dispute over those territories that were outside the natural disposition originally occupied by *Cunpanama'* because of the white monkey. But there is something else, because it was in the wake of this ill-concealed *huito*, when *Cunpanama'* announces another transcendental fact for the life of the ancient Shawi: the arrival of men, in some versions, the arrival of the Christians, a reason for which many Shawi would be transformed into animals that served as food for those Christians.

In fact, the myth about the tree of the *huito* does not end with the definitive retreat of the creator god to Hunguyacu hill, from where *Cunpanama'* created the Shawi. He, after wiping his tears and giving shape to them, decided to return again to the place where the tree was placed originally. That was when the following happened: "Returning *Cunpanama'* from the place where the *huito* tree was, his canoe capsized. This is why he left again and went to the Cahuapanas River and kept marching up to the last hill. So he went to weep, and looking down from the very top of the hill, he called all the beasts of the mountain and said to them: You, from this moment will be animals so that the men can eat you. They are about to come." This end, as we also said above, can easily lead us to make a translation of what was collected in written sources about the first contacts between indigenous peoples and the white man: the appearance of the Spaniards by their rivers, the territorial displacements, the abuse of their women, and even the appearance of the Awajun that were gaining ground throughout the centuries placing themselves in the same whereabouts as the Shawi. Until today, there is a territorial struggle between them, a struggle that clearly started, as observed, with the arrival of Christianity. At this point emerges the first major transformation of the Shawi past, a transformation from animals to people and people to animals. From here begins the entrance of the *pelejo* monkey and its transcendental role in the Shawi culture.

16.6 When Animals Became People: The *Pelejo* Monkey and Its Power of Vision

There are several stories in the Shawi mythological repertoire that reflects the transformation of animals into people. In this section the protagonist is the *pelejo* (sloth), which is also accompanied by other monkeys such as the *maquisapo* (black spider monkey) and the *chosna* (kinkajou). It must be said that the leading role of the *pelejo* is granted by the creator god *Cunpanama'*. From here, the role traditionally occupied by *Cunpanama'* is then attained by the *pelejo* monkey's hands, now referred to as *Apu* (or *Hu'an* in Shawi language). A new era emerged on this change.

As the *Apu* became the highest authority of a Shawi community, a community which is usually made up of a large group of relatives where women will only get together with men who are not in the same line of consanguinity, the Shawi is thus a matrilineal society. The person who looks after the welfare of a group of families, which is a community, is the *Apu*. And this is how the *pelejo* is recognized in the myths that tell about the historical transformation of animals. It means, at the end, that this transformation actually explains the sociability of the *canpo piyapi*.

But how is this transformation? A Shawi informant explained in this way: "Formerly the *Apu*, was the *pelejo*, the sloth; he had in his house a hammock where he would lie down all day, and from there he would command his people. He was the one who knows everything. By that time men were to be transformed into animals, the *pelejo* was at home; there were also other authorities like the *maquisapo* (black spider monkey); but the major authority was the *pelejo*, the sloth. He knew very well that the god *Cunpanama'* was going to bring in new people. Then they, according to story, say that they will be transformed into animals so that new Christians who are about to come will eat them. Thus, they would serve as food to feed them. Being the whole group together, the *pelejo* monkey communicates to them: That certain day a transformation will occur. At this moment, each of them, the *chorito* (yellow-tailed whooly monkey), the black monkey (spider monkey), the white-headed capuchin monkey, their women, and others, started to gather cotton to make their *pampanilla* (traditional Shawi women skirt) so that when they become animals, then they can carry their babies. Women were the first ones to start collecting the cotton. All the animals worried about gathering enough cotton, all of them except the marsupial and the *chosna* [kinkajou]. They say that the *chosna* was lazy and spent the day doing nothing, while the other monkeys and other animals collected cotton and wove *pampanillas*. When finished, it was the time announced by the chief: the sloth. Meanwhile, the spider monkey was in charge of requesting the gathering. Together they took *ayahuasca* (the hallucinogenic *Banisteriopsis caapi*). By taking that plant, they transformed. Everyone took it, got "drunk," and from there the transformation occurred, and they became animals."

Everyone by then had their own *pampanilla*, except the *chosna* (kinkajou). What to do? How are you going to carry your babies? *Chosna* (kinkajou) has not made his *pampanilla* because of its laziness, so it will have to carry the babies using its mouth and teeth; and until today the *chosna* carries them out just like that. The same occurred with the marsupials that were forced to carry their offspring inside their belly.

In the meanwhile, at the communal house they were also ready to transform. Some became white, tall, white monkeys, just like the *gringos*. The black monkeys would transform into *choros* (yellow-tailed wooly monkey). "You into squirrel monkey, you in squirrel… you are going to eat pure *chambira* (*Astrocaryum chambira*) and nothing else. You, -they say, into the *isula* or bullet ant (*Paraponera clavata*) you are going to sting people without killing anyone. You a viper (*Bothrops asper)*, you are going to bite them and suddenly they are going to die. Just like the boa: you're going to hunt Christians and swallow them. So did with everyone. At twelve o'clock they were all transformed."

On another occasion and under the effect of *ayahuasca*, when they were dancing, the *pelejo*, who was actually the boss, was accidentally pushed by a lady. At that point the boss grabbed the woman by her arm and told her: "You're going to be old too, and as a woman, you're going to be older than your husband. You're going to grow up faster; you will have lots of kids and will get old quickly. And you (to the man standing beside her), as a man, will slowly grow up, even if you are eating too much. You're not going to grow old so fast." He blew a spell on them and spread his will all over the group. So, at present days, Shawi women have one child, and they already feel like they want to die, while watching their breasts getting saggy. As to the male, when they are 35 or 40, they still feel young and healthy.

This is how things happened in ancient or mythological times. By midnight they were already changing; they take that plant and also had coffee: they say they take it abundantly so that with its help, they could be transformed (W. Chanchari [Río Sillay], 2006, personal communication). The first societal codes appear with this myth: a system of authorities, a set of rules and behaviors, and tasks differentiated by gender. The ability to contact with *Cunpanama*' and the power of vision, so characteristic of a shaman, also appeared. The Shawi society as we know it today had its inception at this point. A group of beings became a society. The distinction between animals and persons is already a reality, and the characters that define men and women are already set (on the notion of "person" in Amazonian indigenous societies, see, for example, Carneiro da Cunha 1978; Hugh-Jones 1979; Taylor 1996, 2000; Erikson 1996; Goulard 1998; Gow 1991; Surrallés 2002; Surrallés and García-Hierro 2004). The idea of the white monkey that tempted the wife of *Cunpanama'* was a white man is confirmed. Here it is observed how he becomes a *gringo*.

With this transformation, the Shawi actually step away from the world of nature. Something that, in a parallel reading, can also be interpreted as distancing from the "gentility" and a more or less receptive approach to the Christian doctrines. So, it is relevant not to forget the announcement of *Cunpanama*' on the Hunguyacu hill because the men (Christians) are to come. This is, in fact, what historical sources tell us that actually happened. The Shawi, converted into Christians, "into people," moved away from what is "gentile," where they automatically place the alleged Awajun enemies.

In this way, after the conversion directed by sloth "people" became part of the world of human beings, the rest happens to be part of a society that remains under the expression "when they were people." That is, they became part of the world of the ancients. Now, these people do not disappear. They only change the way they appear today.

16.7 Conclusion

The fact that the Shawi rely in both, the white monkey and the *pelejo* (sloth), to give their own notion of society leads to suppose that it is on them, and not in other animals of their own environment, where they placed a greater degree of humanization and actions. In other words, with the white-headed capuchin and the brown-throated three-toothed-sloth, the Shawi considered themselves to have a greater degree of similarity with. But this is not simple or even obvious. All animals of the forest, practically without exception, have a certain similarity with the people, in this case with the Shawi, a similarity based in the most vital aspect: their humanity. They are all human beings but also particular bodies that make them appear to be different. This, likely, seems to be another example of Amazonian perspectivism (e.g., Viveiros de Castro 1998). Certainly, the physiognomy of some species of monkeys can best match a person's body, but this does not appear to concern the Shawi. Shawi defend that one's body is what makes them appear specifically different and not their humanity itself.

The white-headed capuchin monkey and the *pelejo* monkey, however, seem to contribute to the Shawi cosmogony with a degree of sophistication in their humanity, something that leads them from "animality to society." In one case, it is the white man, with its colonization process, and in the other, it is that of their own concept of self.

Acknowledgments I thank Bernardo Urbani for his efforts in counting on me for this edited volume. To Alejandro García-Villanueva for the translation into English of an early draft of this chapter and to the anonymous external reviewer and editors for working in the English text and making comments that enrich this contribution. Specially, to the Shawi, always, for their knowledge and their great humanity.

References

Carneiro da Cunha M (1978) Os mortos e os outros: uma análisis do sistema funerário e da noção de pessoa entre os indios Krahó. Hucitec, São Paulo

Chantre Herrera J (1901) Historia de las Misiones de la Compañía de Jesús, en el Marañón español. Avrial, Madrid

Emmons LH (1990) Neotropical rainforest mammals: a field guide. University of Chicago Press, Chicago

Erikson P (1996) La griffe des aïeux. Marquage du corps et démarquages ethniques chez les Matis d'Amazoni. Peeters, Lovaina/París

Figueroa F (1904) Relación de las Misiones de la Compañía de Jesús. Colección de Libros y Documentos referentes a la Historia de América, vol I. Librería Victoriano Suárez, Madrid

Fuentes A (1988) Porque las piedras no mueren. Historia, sociedad y ritos de los Chayahuitas del Alto Amazonas. Centro Amazónico de Antropologia y Aplicación Práctica, Lima

García-Paredes MD (1994) Buscando nuestras raíces. Centro Amazónico de Antropologia y Aplicación Práctica, Lima

Golob A (1982) The Upper Amazon in historical perspective. Ph.D. dissertation. City University of New York

González-Saavedra L (2013) Entre la predación y la docilidad. Padecimiento Shawi en la Alta Amazonía. Ph.D. dissertation. Complutense University, Madrid

Goulard JP (1998) Le genre du corps. Ph.D. Dissertation. Ecole des Hautes Etudes en Sciences Sociales, Paris

Gow P (1991) Of mixed blood: kinship and history in Peruvian Amazonia. Clarendon Press, Oxford

Grohs WH (1974). Los Indios del Alto Amazonas del siglo XVI al siglo XVII. Poblaciones y Migraciones en la antigua provincia de Maynas. *Amerikanistisch Stud* Bonner 2

Hugh-Jones S (1979) The palm and the Pleiades: initation and cosmology in northwest Amazonia. Cambridge University Press, Cambridge

Joulu R (2006) L'autre en soi. Interiorisation et de l'altérité metamorphose identitaire Jebero chez les d'amazonie peruvienne. Ph. D. dissertation. École des Hautes Études en Sciencies Sociales, Paris

Ochoa N (2008). Entre plusieurs mondes. Les Chayahuita de l'Amazonie. Ph.D. dissertation. École des Hautes Études en Sciencies Sociales, Paris

Renard-Casevitz FM, Saignes T, Taylor AC (1988) Al este de los Andes: relaciones entre las sociedades amazónicas y andinas entre los siglos S.XV y XVII. Instituto Francés de Estudios Andinos, Lima

Rivas-Panduro S (2014) Arqueología de la cuenca del Paranapura. In: Pitman N, Vriesendorp C, Alvira D, Markel JA, Johnston M, Inzunza ER, Pizango AL, Valencuela GS, Alvarez-Loayza P, Homan J, Wachter T, del Campo Á, Stotz DF, Heilpern S (eds) Rapid biological and social inventories report 26. Cordillera Escalera-Loreto, Perú, pp 204–210. The Field Museum Publication, Chicago

Rojas-Bercia, LM (2013) La sintaxis y semántica de las construcciones causativas en el chayahuita de Balsapuerto. Honor thesis. Pontificia Universidad Católica del Perú, Lima

Surrallés A (2002) De la percepción en antropología. Algunas reflexiones sobre la noción de persona desde los estudios amazónicos. Indiana 19/20 (59–72)

Surrallés A, García-Hierro P (2004) Tierra adentro. Territorio indígena y percepción del entorno. IWGIA- Grupo Internacional de Trabajo sobre Asuntos Indígenas, Copenhague

Taylor AC (1996) The Soul's body and its states: an Amazonian perspective on the nature of being uman. J R Anthropol Inst (N.S.) 2/2 (201–215). Londres

Taylor AC (2000) Le sexe de la proie: représentations jivaro du lien de parenté. L'Homme 154/155 (309–334)

Viveiros de Castro E (1998) Cosmological Deixis and Amerindian Perspectivism. J Royal Anthrop Inst 4(3):469–488

Chapter 17
Importance of Primates to Tacana Indigenous Subsistence Hunting in the Bolivian Amazon

Wendy R. Townsend, Robert B. Wallace, Kantuta Lara-Delgado, and Guido Miranda-Chumacero

17.1 Introduction

Subsistence hunting has been an integral part of indigenous livelihood strategies for millennium. The earliest evidence of the presence of people in the Amazon includes hunting tools. Filling nutritional needs is the obvious reason for this practice, but in the 11,000+ years since people are projected to have populated most of the Amazonian region (Roosevelt et al. 1991), wildlife has been inextricably integrated into the multiple cosmovisions of the region. Wildlife contributes to cultural histories (myths), controls (taboo and hunting zones), rituals (couvade and dances), and social cohesion (ancestry and reciprocity) (Cormier 2003; Lara-Delgado 2003; Urbani and Gil 2001; Karadimas 1999; Van Akkeren 1998; Braakhuis 1987; Townsend and Macuritofe-Ramírez 1995; Baker 1992; García del Cuero 1989; Reichel-Dolmatoff 1971, 1976, 1989; Baleé 1985). Mammals account for between 60% and 80% of the biomass harvested by most Amazonian indigenous people (Ojasti 1993), with the exception of a few indigenous groups such as the Ayoreo, who focus on land tortoises (Ayala 1997), or the Kalapalo of the Upper Xingu, who consider all terrestrial mammals disgusting except for monkeys (Basso 1973). Ungulates, primates, and rodents usually fill the top three positions of biomass and individuals harvested by Amazonian hunters (Townsend 2009), but this may be influenced by habitat type and the land use history of an area, as well as cultural dimensions.

W. R. Townsend (deceased) (✉)
Noel Kempff Mercado Natural History Museum, Santa Cruz de la Sierra, Bolivia

R. B. Wallace · K. Lara-Delgado · G. Miranda-Chumacero
Wildlife Conservation Society, Bolivia Program, La Paz, Bolivia

B. Urbani, M. Lizarralde (eds.), *Neotropical Ethnoprimatology*, Ethnobiology,
https://doi.org/10.1007/978-3-030-27504-4_17

Of these three taxa, primate hunting is the most controversial. Primate hunting is questioned due to sustainability concerns, but also because they are emotionally appealing to humans, and are not normally eaten by the majority of people in Bolivia. In addition, even though in some tropical areas humans may be the only remaining predators (Cowlishaw and Dunbar 2000), primates are surprisingly poorly adapted to human hunting practices (Urbani 2005). A combination of their high visibility and low reproductive rates, also makes them particularly vulnerable to overhunting (Estrada et al. 2017). Peres (1990, 1991) suggested that local extinctions of woolly monkeys (*Lagothrix* sp.) were a result of hunting, and conservation publications stress the vulnerability of the taxa, and cite, low reproductive rates, declining populations, and local extinctions in some Amazonian forests (Estrada et al. 2017; Chapman and Peres 2001). Primate populations, along with the rest of the biodiversity, also face the threat of habitat destruction through logging, mineral extraction, and large-scale energy projects (petroleum and hydroelectric dams), as well as an expanding agricultural frontier. Another threat is the increasing pressure on local people to legally and illegally commercialize resources and become participants in the global economy as access to markets becomes easier. Exacerbating these changes is the demand for wild meat from large urbanized populations in the Amazon (van Vliet et al. 2015), although these markets may be more interested in large-bodied ungulates or rodents (Castro et al. 1975). Increased access to previously distant and isolated parts of the Amazon basin may extend the impacts of primate hunting into areas that previously functioned as population replenishment sources (Navarro et al. 2000).

Primate hunting by indigenous people is a traditional activity protected by law in all Latin American countries that ratified the ILO Art 169, as indigenous people have the rights to use and self-manage their renewable natural resources, including wildlife. The suggested incompatibility between conservation and hunting rights (Peres 1991) urges a dialog between what may seem like divergent visions. The application of key partnerships for community participatory management efforts may hold the key to uniting these visions, as some projects in the Amazon have resulted in self-imposed hunting rules and regulations (CIPTA 2008; Townsend et al. 2005). Other efforts have accumulated data sets of self-registered game harvests providing a baseline to achieve an understanding of the importance of hunting and wildlife harvest (Constantino 2016; Iwamura et al. 2014). One such initiative is the WCS-CIPTA partnership, where the Tacana indigenous people of northern Bolivia, via their organization CIPTA (Consejo Indigena del Pueblo Tacana), have joined forces with the Wildlife Conservation Society (WCS) to work toward a sustainable future for their people, as well as for the Madidi National Park. The self-registry of hunting from five communities during 88 months enabled the compilation of a database, which could evaluate the importance of primates in the subsistence of those Tacana communities.

This dialog between visions is complicated by the intricate and systemic relationships of primates in Amazonian cosmovisions (Cormier 2003; Lara-Delgado 2003; Urbani and Gil 2001; Karadimas 1999; Van Akkeren 1998; Braakhuis 1987; Townsend and Macuritofe-Ramírez 1995; Baker 1992; Baleé 1985; García del Cuero 1989; Reichel-Dolmatoff 1971, 1976, 1989). The cultural importance of primate species varies between indigenous groups, localities, and history, including changes accrued through acculturation. In lowland Bolivia there are prehistoric artworks (petroglyphs) depicting humans hunting monkeys (Strecker 1987, cited in Salinas 2010) near Santiago de Chiquitos and figures of monkeys on rocks along the Beni River, at the confluence of the Quiquibey River in the ancestral lands of the Tacana (Salinas 2010). Oral histories that depict nonhuman primates include heroes and villains, teach moral lessons, and convey natural history details to the next generation (Townsend and Macuritofe-Ramírez 1995). Some of these stories include transformations of humans to nonhuman primates and back (Townsend and Macuritofe-Ramírez 1995). Contact with primates, in person or in dreams, can bring problems or even luck depending on a variety of factors and local situations. The Murui Amerindians of the Colombian Amazon believe that consumption of some primate meat, especially by children, can cause them to become lazy, or a jokester, or even infected with parasites, depending on the species consumed (Townsend and Macuritofe-Ramírez 1995).

Urbani (2005) reviewed various cultural beliefs about primates as human prey. While not all indigenous groups consume monkeys, he found reference to 56 indigenous groups in Latin America that reported primates within the top 10 harvested species. Nonhuman primates are preferably targeted for hunting, as people consider the meat tastier and more tender (Urbani 2005). Some indigenous groups may actually organize hunting forays for the sole purpose of returning with as many primates as possible. Others specifically target females that are carrying young that can be sold at the pet trade (Urbani 2005). Understanding subsistence strategies and the culture of primate use will improve dialog between indigenous hunters and conservationists and promote successful partnerships beneficial to both humans and nonhuman primates.

It is also important for community, landscape, or regional planning and conservation efforts to reflect the holistic values of hunting and wildlife. In a rapidly changing world, the Tacana of lowland Bolivia are pondering the potential for sustainable primate harvest into the future in their ongoing planning strategy (CIPTA-CIMTA 2014). In this chapter, we aim to (a) analyze self-reported hunting harvest of primates by the Tacana between 2001 and 2008, (b) explore the nutritional and cultural dimensions of their hunting, (c) evaluate the sustainability of that hunting activity, and (d) facilitate the design of mechanisms to conserve this culturally relevant resource.

17.2 The Tacana

The Tacana indigenous people have inhabited the Andean foothills of northern Bolivia for centuries, since at least 1200 AD, prior to the Incan empire (Diaz Astete and Murillo 1998). At one time their territory included three major freshwater systems, the Beni, Madre de Dios, and Tuichi Rivers, and possibly extended into Peru and Brazil. As part of the *Antisuyo* of the Incas (Paredes 1997), the Tacana commercialized skins, feathers, fruits, resins, vanilla, chocolate, and other natural resources from the Amazonian lowlands (Chiovoloni 1996). The first recorded contact with the Spanish explorer Pedro Anzures in 1539 was followed by about 20 other expeditions searching for the famed golden city *El Dorado*. Subsequently, the military and religious missions began to spread their influence in the Tacana territory, obliging them to use their chocolate and other natural resources to pay tribute to the colonial government (Wentzel 1991). The Tacana opted for a peaceful strategy to guarantee their liberty and the continued ownership of their most important possession, the land (Chiovoloni 1996).

Today, the Tacana own an indigenous land that totals 389,304 ha (Fig. 17.1), and CIPTA represents 20 Tacana communities in the titled Tacana Indigenous Territory or Tierra Comunitaria de Origen (TCO) Tacana I. The Tacana population in communities that form part of the TCO is stable with no discernable change between 2001 and 2014 (CIPTA-CIMTA 2014), although younger people are increasingly migrating to nearby towns. In 2000 a partnership between CIPTA and WCS was born out of interest in natural resource management and the urgent need to support the land-titling process in the region to ensure recognition of the Tacana TCO.

The Tacana Life Plan (CIPTA-CIMTA 2014) resulted from multiple community workshops in each Tacana community run by their own organization, CIPTA, and accompanied by WCS. The goal of CIPTA is to implement projects that strengthen community management capacity and improve and/or safeguard livelihoods. During this process the Tacana communities stressed the importance of wild game toward subsistence, and from this concern came a participatory game harvest registry with hunters from five of the most remote communities of the Tacana TCO I (CIPTA 2001). From 2001 to 2008, 117 volunteer hunters from 5 communities participated in recording hunting results.

Although agriculture is the main livelihood activity, the Tacana also hunt, fish, collect resources from their territory, and engage in day labor (Lehm et al. 2017; Chiovoloni 1996). The Tacana livelihood strategy is diversified and respects ecosystem carrying capacities and the local limitations of the soils. The Tacana practice slash-and-burn agriculture to open forest for planting manioc, rice, corn, yams, squash, and other products, and after 3–5 years these are mostly abandoned allowing secondary forest growth. The Tacana also raise chickens, ducks, pigs, and cattle, but the principle animal protein consumed comes from hunting and fishing. Hunters mostly use 0.22 rifles, but there are still a few shotguns (Lara-Delgado 2003). Hunting trips are made to community-designated hunting zones, accessible within a 24-hour period. Successful day-hunting trips are *a volver*, and those with an all-night stay are called *el mecheo* (Lara-Delgado 2003).

The five Tacana communities, Cachichira, Carmen de Emero, San Antonio de Tequeque, Villa Fatima, and Esperanza de Enapuera, are located in the vicinity of Madidi National Park (Fig. 17.1), one of the most biodiverse protected areas in the world. The area extends from the inundation-prone savanna of the Beni River alluvial plain (150 m a.s.l.) to the northeast and southwesterly into the forested Andean foothills (400 m a.s.l.). Annual precipitation amounts to 2000 mm and is concentrated in the wet season (November to April), with average temperatures of

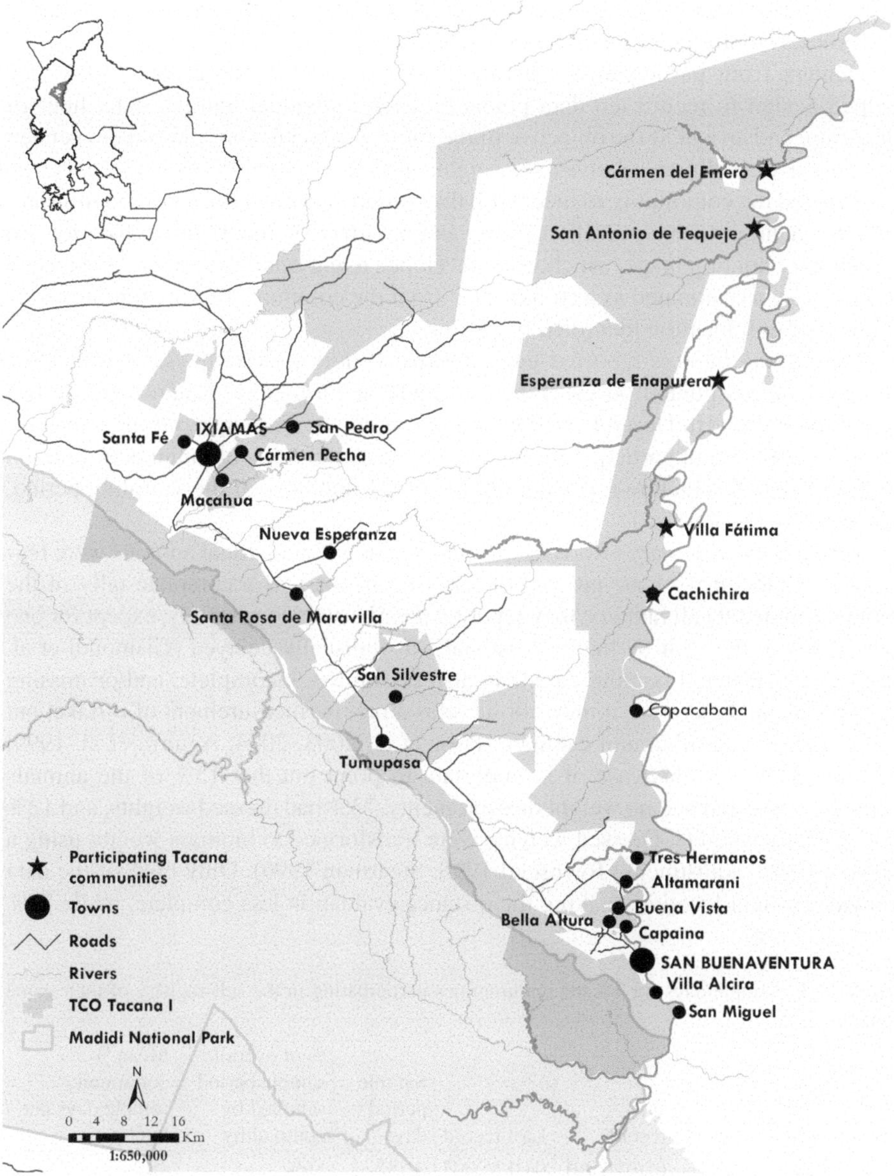

Fig. 17.1 Location of five participating Tacana communities in the Tacana Indigenous Territory or TCO

26 °C. Southerly cold winds can occasionally lower temperatures to 10 °C from July to August. Three forest types are found in the Tacana territory: humid foothill forests, seasonally humid Amazon forest, and Amazonian riverine forest, as well as seasonally flooded savanna and savanna woodland (Ribera 1992). The communities

along the river mostly inhabit riverine forest, but also use the inundated savannas and Andean foothills for their resource allocation.

Hunters from participating communities filled out the structured forms they helped design to record ten data points on each individual animal: date, hour of departure and arrival, if the objective of the foray was to hunt, species taken, gender, reproductive condition, relative age, weight, and kill location. Periodically, a technician visited the community to meet with the hunters and copy the recorded information by hand into an overall registry and to identify the hunting location on community hunting maps, which were developed using satellite images. Distance to hunt areas was estimated as a straight line from the community to the middle of the 1 km^2 grid cell location identified by the hunter.

The Tacana hunters reported game harvested during 2664 days, with the first reported animal hunted on the 17th July 2001 in Carmen de Emero and the last reported on the 11th December 2008 in San Antonio de Tequeque. Of the five communities, two began participating early in this time period, with hunters in Carmen de Emero and Cachichira reporting during 99.4% and 86% of the sampling period, respectively (Table 17.1).

Through the voluntary efforts of the hunters, 16,976 individual animals were registered. Although we have no way of knowing if this was a complete tally of the animals hunted by all hunters, they reported their harvest consistently, except for one period where the visit by the professional was unusually delayed (Gismondi et al. n.d.). Self-registered hunting data may also be limited, incomplete, and/or missing data points, and therefore it may not serve as a precise measurement of offtake, but rather indicates a minimum estimate (Noss et al. 2003, 2004; Aguirre et al. 1999; Morales 1999). In this case, it is important to point out that 15% of the animals reported harvested had no weight measurements, 73% had dressed weights, and 12% had biomass weights. Dressed weights were transformed to biomass weight using a standard 30% adjustment (Townsend 1995; Stearman 1990). Only 66% of the data provided a hunt location, making the distance evaluation less complete, while 76%

Table 17.1 Sample days for Tacana communities participating in the self-registry of their game harvest

Community	Hunter	First record	Last record	Sample period days	% of overall sample period sampled by community	Mean % community sample days per hunter
Cachichira	14	02/08/2001	04/12/2007	2,282	86%	48%
Carmen de Emero	65	17/07/2001	24/11/2008	2,647	99%	46%
Esperanza del Enapurera	5	17/06/2004	05/07/2006	738	28%	65%
San Antonio del Tequeque	11	27/06/2004	11/12/2008	1,604	60%	63%
Villa Fatima	24	26/03/2002	31/08/2008	2,315	87%	41%
Total	119	17/07/2001	11/12/2008	2,664		

of the data lines included temporal information. We considered hunter days as the denominator for the catch per unit effort estimates (kg/hunter days). A hunter day averaged 6.8 hours. A total of 8244 successful hunter days were reported by participating Tacana hunters. The game harvest provided by these hunters served the population of each community, which represents 855,277 consumer days, calculated using community sample periods multiplied by the number of consumers derived as the number of people over 3 years old in official 2000 and 2012 censuses.

17.3 Importance of Primates to Tacana Culture

When the Tacana consume primates, they are obtaining more than protein. Primates are an integral part of the Tacana cosmology, and eating any animal, especially a primate, provides all of its significance, essence, and life. Primates can have positive or negative meaning in Tacana dreams, for example, the spider monkey (*Ateles*) usually has negative connotations and may be a presage of a death in the family. However, if a hunter dreams of *Ateles*, he will have good hunting (Hissink and Hann 2000). Clearly dreaming of *Ateles* may have different meanings depending on the situation and circumstance of the person. Some hunters practice a ritual with primate and other favorite game species' skulls, first cleaning them and then placing them outside their houses. The Tacana believe that this will keep the spirit of the animal close to the house, allowing hunters easier access to game in the future.

Tacana cosmology is expressed in their oral history. These stories often tell of when humans were transformed to animals and vice versa. One story is called the *Wichi* story and is about a man who always fished alone and away from others, but regularly came back with plentiful fish for his family. His brother-in-law became very curious as to how he had such good fortune fishing, and so he followed him to discover the secret of his success: the man used his intestines as bait. Discovering that his secret was out, the fisherman asked his brother-in-law to tell his wife to meet him at the riverbank to say goodbye. After they said their farewells, he transformed into a howler monkey (*Alouatta*) and disappeared into the forest (Cárdenas 2003).

Meanwhile spider monkeys are said to steal human women from villages to take them to their forest hideouts. Sometimes the women escape with the help of other animals (Cárdenas 2003). Many of the stories feature humans and animals speaking naturally, and for the most part, the oral histories involving primates are about spider monkeys and howler monkeys, and most seem to be vengeance driven. Spider monkeys are particularly related to negativity and evil, and it is said that the spider monkey and the devil chase each other around but avoid water, so that as a way to avoid them when they are very angry, the Tacana will go into the water (Cárdenas and Jemio 2001). These stories consolidate the Tacana worldview about the forest as a space of social relationships.

As with all foods, different game species are preferred and considered more valuable than other species by different people. In an interview study conducted as part of a major Tacana planning process (CIPTA-CIMTA 2014), 302 men and women were asked which game meat they preferred. In total, 17% of people stated a preference for primate meat, especially *Saimiri* (squirrel monkey) and *Alouatta* (howler monkey) and most especially *Ateles* (86% of positive responses). Interviewees stated the following preference categories in order of frequency: meat flavor, tenderness, game size, and abundance. In an earlier study, four primate species were listed as favorites due to their flavor, the same three, and the capuchin monkey (*Sapajus*) (CIPTA 2001).

As one of the most preferred game species, hunters reserve *Ateles* to share with their closest family members, dividing it into small pieces. If a hunter is courting a woman, he might make an extra effort to hunt *Ateles* to invite the woman's family (Lara-Delgado 2003). On special occasions like a birthday or community celebrations, especially between February and June, when spider monkeys are fattest during and following fruit abundance peaks in the forest, hunters attempt to bring primates to the feast, particularly *Ateles*. The Tacana occasionally make extended multi-night hunting trips in search of *Ateles*. In short, the importance of primates for the Tacana is not only for food, but also contributes to social roles and reciprocity, since providing primate meat is perceived as prestigious.

17.4 Primates in the Tacana Wildlife Harvest

Tacana hunters registered 54 wildlife species, including reptiles, birds, and mammals; 7 of these species are primates. Fifteen species were consistently hunted, including ungulates, large rodents, land and river turtles, and four primate species. Mammals made up about 87% of the biomass harvested (Table 17.2). The largest proportion (62%) of the Tacana harvested biomass was composed of ungulate species, and primates followed with 13%, but it is critical to note that larger average sizes for ungulates meant the greater biomass proportion was derived from less individuals (62%, $n = 3502$) than the primate biomass proportion (13%, $n = 4693$). Thus, the Bolivian red howler monkey (*Alouatta sara*) is only fifth in harvest biomass ranking, with four large-bodied ungulates – white-lipped peccary (*Tayassu*

Table 17.2 Description of Tacana self-registered wildlife harvest from 2001 to 2008

Taxa	Individuals	Biomass (kg)	% of total
Mammals	11,239	132,165.66	86.6%
Ungulates	3,502	95,571.64	62.6%
Primates	4,693	19,938.16	13.1%
Carnivores	1,469	7,034.01	4.6%
Rodents	1,214	6,859.98	4.5%
Edentates	361	2,761.86	1.8%
Reptiles	2,116	12,471.35	8.2%
Birds	3,624	8,052.97	5.3%
Total	16,979	152,689.98	100%

pecari), lowland tapir (*Tapirus terrestris*), collared peccary (*Pecari tajacu*), and red brocket deer (*Mazama americana*) – all providing more harvest biomass. The black spider monkey (*Ateles chamek*), brown capuchin monkey (*Sapajus apella*), and Bolivian squirrel monkey (*Saimiri boliviensis*) were ranked 9th, 12th, and 14th in biomass, respectively. These four most hunted primates accounted for 98% of primate biomass reported by Tacana hunters.

More individual primates were reported harvested (n = 4693) than any other order, with 3 species exceeding 1000 individuals: *A. sara*, *S. boliviensis*, and *S. apella* representing 32%, 29%, and 22% of primate individuals hunted, respectively (Table 17.3). Spider monkeys represented 16% of primate individuals harvested. The three remaining species of primates hunted were small bodied (<1.5 kg) and rarely captured, making up only 2% of total individual primates hunted.

More males than females were harvested for all but one harvested primate species (Table 17.4). Twice as many females were reported harvested then males for *Ateles chamek*. Assuming sex ratios in primate populations are close to 1:1, a rate of 0.49 for *Ateles* might lead to the conclusion that hunters are targeting females to capture young for the pet trade (Urbani 2005). However, spider monkeys show female-biased sex ratios in most studied populations to date (Shimooka et al. 2008), and as such female harvest bias would also be expected in hunting results.

According to the Tacana data, 21–28% of females of the top four harvested primates were in a reproductively active state (Table 17.4). This includes females with young, lactating females and pregnant females. Harvested *Aotus azarae* were observed with a considerably lower reproductive rate, but this is probably due to a

Table 17.3 Primate species reported as harvest by Tacana hunters between 2001 and 2008

Tacana name	Local name	Scientific name	Individuals harvested	% of primates
D'hu	Manechi	*Alouatta sara*	1,483	31.6%
Shahui	Chichillo	*Saimiri boliviensis*	1,341	28.6%
Cushu	Silbador	*Sapajus apella*	1,033	22.0%
Bihua	Marimono	*Ateles chamek*	742	15.8%
Didia	Nocturno	*Aotus azarae*	81	1.7%
Ducaduca	Lucachi	*Plecturocebus aureipalatii*	12	0.3%
Chichulubasume	Leoncito	*Leontocebus weddelli*	1	0.02%

Table 17.4 Sex ratios and reproductive state of nonhuman primates reported harvested by Tacana hunters from 2001 to 2008

	Sex Ratio		Reproduction		
Species	M/F	Rate M/F	% F with young	% gravid	% F active
Alouatta sara	806:554	1.45	19.1%	9. %	28.2%
Ateles chamek	218:443	0.49	16.7%	4.1%	20.8%
Saimiri boliviensis	917:329	2.79	17.0%	4.6%	21.6%
Sapajus apella	658:285	2.31	16.1%	5.3%	21.4%
Aotus azarae	40:38	1.05	13. %	2.6%	15.8%

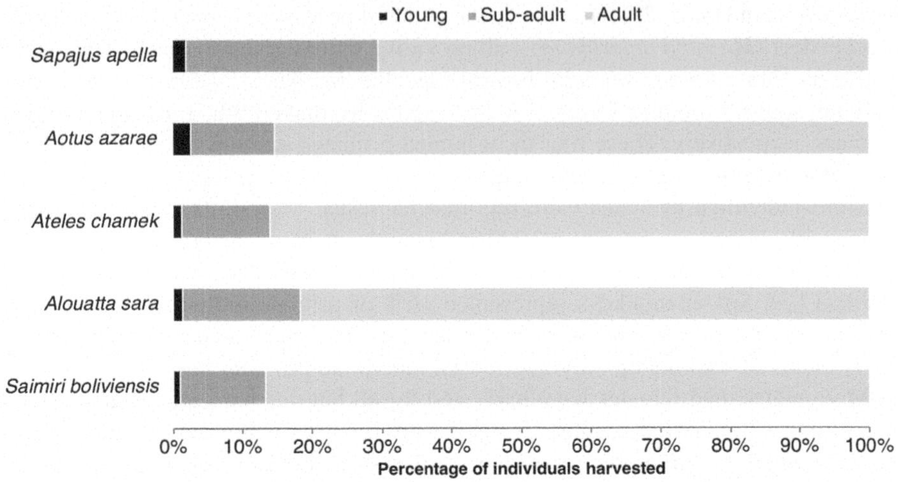

Fig. 17.2 Relative age structure of Tacana reported primate harvest from 2001 to 2008

limited sample size for the species. On the whole the sex ratio and female productivity observed by Tacana hunters were not markedly high or low. Of the females reported, *Alouatta* had the most pregnant females in the harvest, just 9%, and also the most females with young (19%). The sex ratio of the *Alouatta* harvest does not suggest that Tacana hunters were targeting females. Less than 20% of female *Ateles* captured were carrying young. Since females with young are slower to move away than the other monkeys, they may be more susceptible to hunters. However, the relatively low percentage of females with young in the offtake may indicate that Tacana hunters avoid taking female-carrying young.

The age structure of harvested primates (Fig. 17.2) reveals that young animals do not appear to be important, making up less than 20% for all species except *Sapajus*. Most younger primates hunted were subadults, with young or babies being less than 5% of any of the species. *Sapajus* subadults were the largest group of young animals in the harvest, but these were only about 25% of the total harvest of this species.

The important contribution of primates to the total biomass harvested underlines the need to evaluate how primates contribute to a hunter's family subsistence. We can estimate that contribution if we consider that wildlife produces about 70% edible or dressed weight (Gobierno de Santa Cruz 2009; Townsend 1995; Stearman 1990). Applying that conversion rate across the entire study period, on average participating Tacana hunters brought 1283 kg of meat into their communities, although not all hunters participated for the entire sampling period (Table 17.1). To estimate consumer days, we multiplied the number of people in each community from a census (CIPTA-WCS 2012) by the number of days of each community's sampling period. Across the 5 communities, this totaled 855,277 consumer days. By dividing the total dressed weight by the number of consumer days, we obtained an overall extraction rate per consumer/day. The same calculation was made for the registered primate harvest. The result is 23 g of total primate biomass and 16 g of edible primate

Table 17.5 Extrapolation of animal protein harvested by Tacana hunters between 2001 and 2008

	kg harvested biomass	kg biomass/855,277 consumer days	kg meat/ consumer day	g protein/ consumer day
Total harvest	152,689.98	0.179	0.125	25
Primate harvest	19,938.16	0.023	0.016	3

Formula = ((kg of Biomass ∗ 70%) ∗ 20%) ∗ 1000 = g of protein, Biomass estimation method = Average weight ∗ Number individuals

Table 17.6 Opportunity cost to replace wild meat obtained through Tacana hunting

	\$ opportunity cost at *\$5/kg*	\$ opportunity cost/ family	\$ family/day
Primate harvest	99,690.81	1261.91	0.47
Total harvest	763,449.9	9663.92	3.63

biomass per day per consumer (Table 17.5). Considering this value in protein (ca. 20% of edible parts: Wu Leung and Flores 1961), the Tacana people gained about 3 g of protein per day from the primates they extract or about 10% of the recommended daily allowance (FAO) of animal protein for a relatively small and medium active adult, although the FAO recommended daily allowance is probably less than the active Tacana lifestyle requires.

The reported hunting results also demonstrate the importance of wildlife to the family economy (Table 17.6). The opportunity cost of reported wild game meat for 79 Tacana families represents \$3.63 per day if they replaced wild meat with purchased meat (\$5 per kg of biomass). Annual opportunity costs of wild game meat provided a benefit of at least \$1250 per family. For primates the opportunity cost was \$0.47 per day per family, or \$171.5 a year.

17.5 Tacana Primate Hunting Effort

Of the 117 participants, 112 (94%) reported primates in their game harvest. Of 8244 hunting trips reported, 2625 (32%) registered primates as part of their return. No primate species was harvested on more than 10% of the hunting trips (Table 17.7). Hunters bagged one individual per hunting trip 75–88% of the time depending on the primate species, with a slight tendency for more individuals per trip for *Ateles*, with two individuals reported on less than 20% of hunting trips, three individuals on less than 5% of hunting trips, and more than three individuals on less than 4% of hunting trips.

More primates were reported in the austral autumn, between March and May. This is especially true for *Saimiri* and *Alouatta* (Fig. 17.3). The Tacana report that this is at the end of the fruiting period when animals will be fatter and in their best condition (CIPTA-CIMTA 2014). The harvest rate for *Ateles* decreases at the height of the dry season, August to September, perhaps because the animals are thin due to a lack of

Table 17.7 Number of hunting trips with number of individuals harvested by Tacana hunters between 2001 and 2008

	Hunt trips		Total trips with each quantity of individuals								
Species	Trips with individual	% Hunts	1	2	3	4	5	6	7	8	9 +
Alouatta	862	9.8%	651	151	30	22	5	1		1	
Saimiri	849	9.6%	637	150	42	14	2	3	1		
Sapajus	739	8.4%	654	68	14	3					
Ateles	436	4.9%	347	53	19	10	3	1	1		1 (12)
Aotus	60	0.7%	49	10	1						

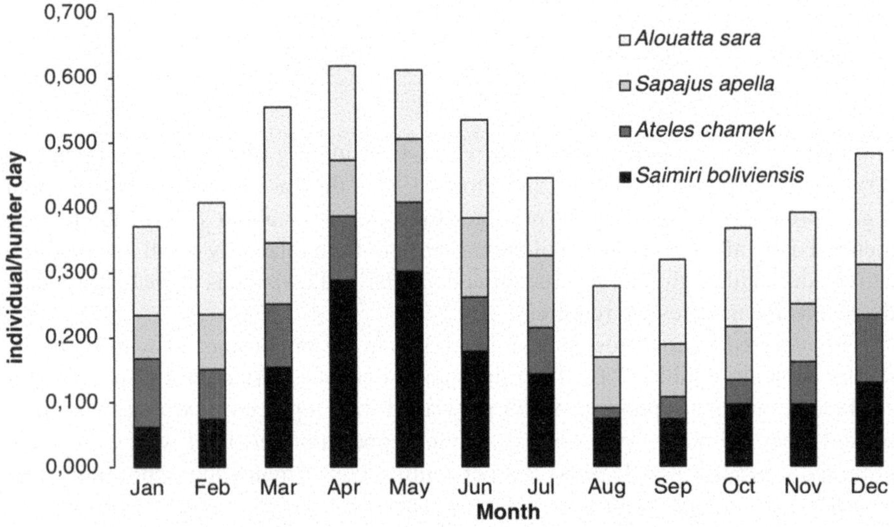

Fig. 17.3 Comparison of monthly harvest rate (individuals hunted/hunt day) for the four most frequently hunted primate species in the self-registered Tacana game harvest

fruit resources. Also travel upstream to habitats where *Ateles* remain is more difficult. The hunting rate of primates was similar to overall wildlife hunting rates (Fig. 17.4).

Tacana hunters also reported on the location they hunted each individual. That location was registered on a 1 km^2 grid placed over a satellite image, and distance to the community was then calculated from the center of each grid block. Average distances of reported kill sites of individuals for each species were calculated (Fig. 17.5). The ranges of distances overlap across species, which is to be expected since they were harvested from the same hunting trips. Despite the overlap in standard deviations (Fig. 17.5), the average distance suggests a tendency for capturing *Ateles* at greater distances than other species. Howler monkeys were taken closer to the community than the average. There is a similarity in hunting distances for *Ateles* and *Tapirus*, the largest and most vulnerable ungulate (Medici et al. 2007), and for *Alouatta* and *Mazama*. Primate species that averaged less than 5 km from the com-

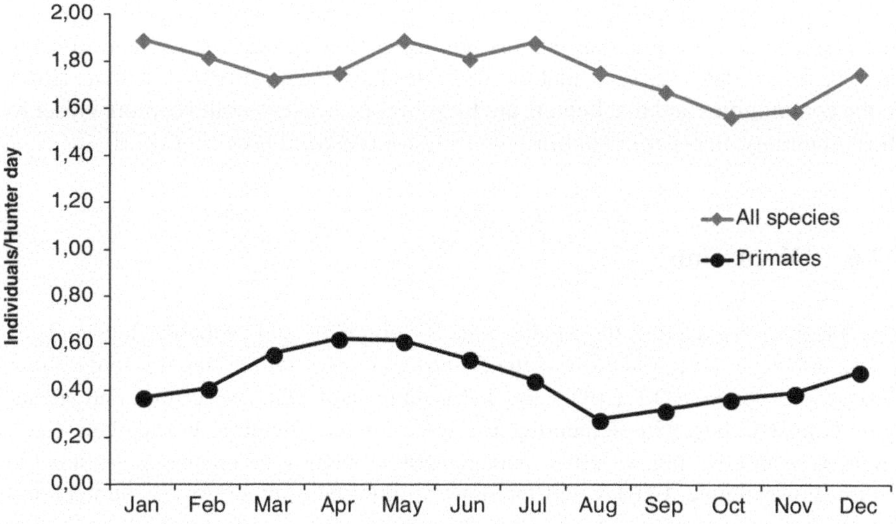

Fig. 17.4 Tacana hunting rate (individual animals/hunter day) for all species and primates hunted by month between 2001 and 2008

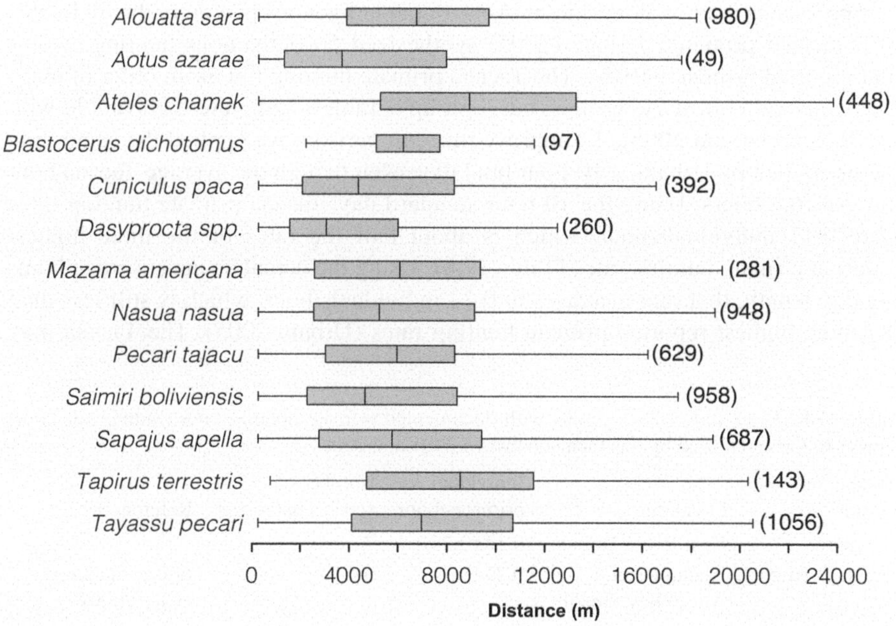

Fig. 17.5 Distance (m) from the community of average hunt location per species, showing mean, standard deviations, and range and comparing primates with other frequently reported species in the Tacana hunting results from 2001 to 2008. The number in parenthesis is the number of individuals with distance estimate evaluated

munity were *Saimiri* and *Aotus*, comparable to two caviomorph rodents *Cuniculus* and *Dasyprocta*. The fact that hunters are going further away on average to bring back larger primates suggests that these species have been extirpated in areas closer to the communities and that hunters are forgoing species they could capture closer to the community, in order to hunt further away for preferred prey (Fig. 17.5).

17.6 Discussion

The Tacana have lived in the Madidi area for hundreds and probably thousands of years, although the population of the 5 participating communities was only about 300 people. Since 2000, CIPTA has led a number of planning efforts and subsequently in 2014 began implementing a second 10-year official territorial plan, which includes strengthening wildlife management strategies to ensure a sustainable future. For example, CIPTA will facilitate an internal dialog between hunters and communities on wildlife status after two devastating floods in 2011 and 2014 (Espinoza et al. 2014). The data presented here represents a historical baseline before those floods which according to Tacana hunters and communities seriously affected the populations of terrestrial wildlife species.

The Tacana are not alone among Amazonian indigenous people in the high rates of harvested primates. Urbani (2005) synthesized 56 indigenous hunting studies that reported primate harvest. The Tacana primate hunting rate is an order of magnitude above 53 of these groups, but 3 groups (Table 17.8) were comparable with the Tacana (Urbani 2005). To improve this comparison, we applied the same time factor applied by Urbani, a 10-hour hunt day, even though the average Tacana hunt day was 6.8 hours. Using the 10-hour standard day, Tacana primate hunting rates were 0.17 individuals/hour, which is about half the rates of the three highest reported primate hunting rates (Table 17.8). Using the actual Tacana average hunting day length, that rate increases to 0.24 individuals/hour, which is still less than the three highest reported primate hunting rates (Urbani 2005). The Tacana may

Table 17.8 Three indigenous groups with documented primate hunting rates (individuals/hour) similar to those reported by the Tacana between 2001 and 2008

Indigenous group	Location	Individual primates/hour	Number of species harvested	References
Tacana	Beni River, Bolivia	0.17 (0.24)[a]	7	
Waimiri-Atroari	Alalau River, Amazonas, Brazil	0.294	6	Souza-Mazurek et al. (2000)[b]
Siona Secoya	Shushufindi, Ecuador	0.399	2	Vickers (1980)
Huaorani	Napo, Ecuador	0.363	8	Mena et al. (2000)

Time estimator from Urbani (2005) of 10 hours per hunt day
[a]Based on the average 6.8 hours Tacana hunt day
[b]Cited in Urbani 2005

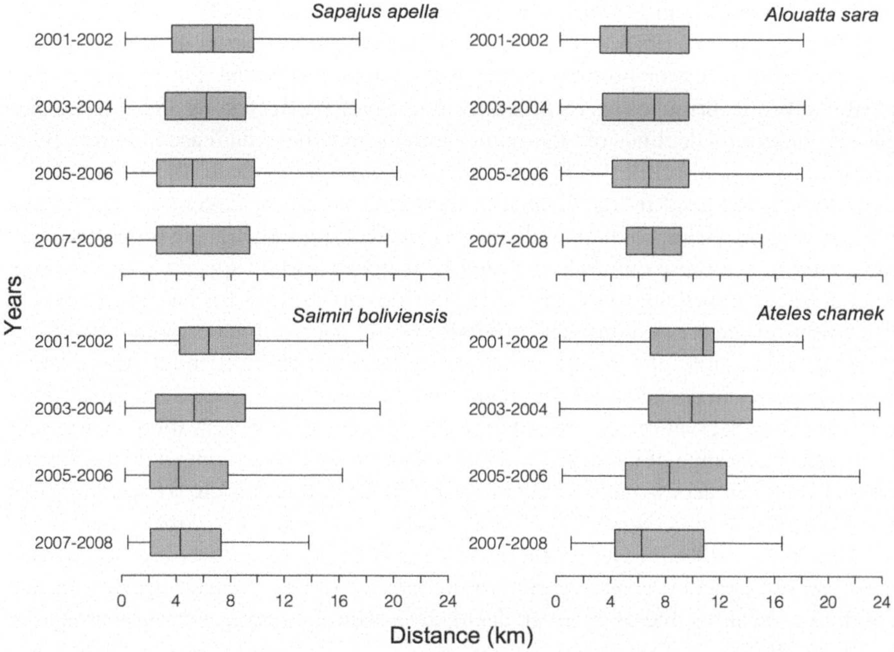

Fig. 17.6 Comparison of average hunting distances of four primate species over time

invest more time specifically looking for primates, and thus hunt more primates, or there may simply be more primates available in the Tacana's forest. Even though overall the Tacana seem to be travelling further for primates than other game species, and they have a higher success rate than 53 of 56 indigenous groups (Urbani 2005), during the 8 years of the study, *Ateles* and *Saimiri* hunted distances decreased as time progressed, whereas *Alouatta* and *Sapajus* hunting distances remained fairly constant (Fig. 17.6).

Analysis of the age and gender structure of the Tacana primate harvest revealed relatively few young were registered, perhaps representing a healthy population, because if only younger animals are taken, it could reflect a limited number of breeding aged animals to replace the population. For most primate species, females were less harvested than males by the Tacana; however, *Ateles* had a higher ratio of females to males harvested. It is quite possible that this reflects the behavioral ecology of this species which typically occurs in female-biased sex ratios (Shimooka et al. 2008). The fact that such a small percentage of harvested female *Ateles* or other primates were carrying young suggests that hunters were not focusing on mothers for their babies. Indeed, harvested primate females displayed a percentage of productivity (baby, milk, pregnant) similar to other studies (Townsend 2000; Bodmer 1994).

Various circumstances may impact the sustainability of the Tacana primate harvest. Primate demand probably is not growing as the human population of the participating communities grew only slightly (three persons) between the 2000 and the 2012 censuses (CIPTA 2012). But two extreme flooding years in 2011 and 2014 may represent a bigger problem to primate populations in the Tacana indigenous territory. While primates were probably not directly affected by the floods, perceived population declines of other game animals may have increased hunting pressure on primates to fulfill the Tacana subsistence needs. As such, the sustainability mechanisms designed by the Tacana in their internal dialogs may require external input to include visions of habitat disturbance and climate change, in order to ensure the continuity of these culturally and nutritionally important species. Conservation scientists can contribute to the dialog to consider predictions for the future and the limitations of the ecosystems. The groundbreaking partnership between CIPTA and WCS is an example of a fruitful relationship between an indigenous organization with autonomy in decision-making regarding a formally recognized and titled territory and a conservation organization with a landscape conservation vision that encompasses a range of land uses and management categories. Indeed, in the Tacana natural resource access and use regulation (CIPTA 2008), article 31 declared that the hunting of three species, wattled curassow (*Crax globulosa*), giant otter (*Pteronura brasiliensis*) and southern Amazon river otter (*Lontra longicauda*) perceived as threatened was no longer permitted. Additionally, the participating Tacana communities along the Beni river decided to reduce hunting pressure on spider monkeys and lowland tapir.

From a primate conservation perspective, it is also important to consider the broader situation in the northern La Paz Department for globally threatened taxa such as the black spider monkey (*Ateles chamek*), which is considered Endangered on the IUCN red data list (Wallace et al. 2008). Spider monkey populations within the Madidi National Protected Area have been stable or increasing (WCS, unpublished data). CIPTA and the Tacana people have been effective at showing significantly lower deforestation rates than other actors in the landscape (Painter et al. 2013), thereby retaining crucial habitat for threatened wildlife, including primates. Thus, even if larger, more vulnerable primate species have been extirpated from the forest in the immediate vicinity of Tacana communities, forest cover is still retained in large portions of the Tacana Indigenous Territory, parts of which are not hunted at all. These are additional sources for primate populations to repopulate forests where primates are hunted. On balance then, the vision and actions of the Tacana people and their Life Plan to date have contributed significantly to the conservation of primate populations, as well as ecosystem integrity and connectivity, and a broader biodiversity conservation viewpoint. Currently, the human population density within the Tacana TCO is less than 1 person per km^2 (3500 people in 3893 km^2), which strongly suggests hunting sustainability in tropical forests (Robinson and Bennett 2000). The challenge for the Tacana will be to maintain sustainable hunting in the face of increasing threats to the region and potential incursions into the Tacana TCO.

Wildlife is an integral part of the Tacana subsistence livelihood and must be respected, but if it is to remain a key resource and a cultural touchstone, hunting will require an internal dialog about its sustainability in the indigenous territory. The Tacana are fortunate to have relatively well-organized grassroots representative organizations (CIPTA and CIMTA) and a strategic plan to ensure the sustainable management of their territory and its natural resources. Previously, in an effort to manage hunting and ensure long-term sustainability, the Tacana hunters who participated in this study agreed to reduce pressure on spider monkeys and lowland tapir (CIPTA-WCS unpublished), as well as declare giant otters as fully protected species in the natural resource management regulation (CIPTA 2008). Facilitating dialog to continue these efforts will be crucial into the future, as well as support to control and vigilance efforts to deter third-party incursions in the Tacana Indigenous Territory.

It is in the best interests of the Tacana culture that the dialog results in mechanisms to ensure that *Ateles*, as well as other primate species, and wildlife, in general, will be available into the future to present in honor to a person one holds in high respect.

Acknowledgments Firstly, we would like to thank the Tacana Indigenous Council (CIPTA) for promoting and facilitating this management activity in the Tacana Indigenous Territory and especially the participating hunters and their families. We thank the Gordon and Betty Moore Foundation, the MacArthur Foundation, the Blue Moon Fund, and the Wildlife Conservation Society (WCS) for support of the Greater Madidi-Tambopata Landscape Conservation Program and specific financial support for indigenous territorial management and community-based natural resource hunting management efforts. During the write-up of this document, RBW was supported by the Yale Institute for Biospheric Studies and the Yale School of Forestry and Environmental Studies. We are especially grateful to the Bolivian Biodiversity and Protected Area Directorate and CIPTA for research permits and logistical support. We would also like to than Maria Copa, Paola Gismondi, Humberto Gomez, Pablo Justiniano, and Alfonso Llobet who worked with Tacana hunters to systematize subsistence hunting data during the management period. Ana Maria Paredes and various students from the Noel Kempff Natural History Museum of the University Gabriel Rene Moreno helped in cleaning and editing the data base.

References

Aguirre L, de Urioste R, Galarza I, Miranda J, Guayo E, Vaca D (1999) El monitoreo de aprovechamiento de fauna en la EBB, un análisis criticó. In: Fang T, Montenegro O, Bodmer R (eds) Manejo y conservación de fauna silvestre de América Latina. Instituto de Ecología, La Paz, pp 97–108

Ayala JM (1997) Utilización de la fauna silvestre del grupo étnico Ayoreode en la comunidad Tobité, Santa Cruz. BSc thesis. UAGRM, Santa Cruz

Baker M (1992) Capuchin monkeys (*Cebus capucinus*) and the ancient Maya. Anc Mesoam 3:219–228

Baleé W (1985) Ka'apor ritual hunting. Hum Ecol 13:485–510

Basso EB (1973) The Kalapalo Indians of Central Brazil. Holt, Rinehart and Winston, New York

Bodmer R (1994) Managing wildlife with local communities: the case of the Reserva Comunal Tamshiyacu-Tahuayo. In: Western D, Wright M, Strum S (eds). Natural connections: Perspectives on community-based management. Island Press, Washington DC, pp 113–134

Braakhuis HEM (1987) Artificers of the days: functions of the howler monkey gods among Mayas. Bijdragen tot de Taal, Land e Volkenkunden 143. Floris Publications, Dortrech

Cárdenas CC (2003) Territorialization y etnicidad: narraciones orales sobre el cuidado de los animales entre los cazadores Tacanas. Bachelor thesis in Literature. Universidad Mayor de San Andres, La Paz

Cárdenas C, Jemio L (2001) Voces de la provincia Iturralde San Buenaventura y Tumupasa. UNIR-UMSA, La Paz

Castro N, Revilla J, Neville M (1975) Carne de monte como una fuente de proteínas en Iquitos con referencia especial a monos. In: Castro-Rodriguez N (ed) La primatologia en el Perú. Proyecto Peruana de Primatologia, Lima

Chapman CA, Peres CA (2001) Primate conservation in the New Millenium: the role of scientists. Evol Anthropol 10:16–33

Chiovoloni M (1996) Caracterización y evaluación de las estrategias de manejo de recursos naturales del pueblo Tacana. PNUD/SAE, La Paz

CIPTA (2001) Estrategia de desarrollo sostenible de la TCO Tacana con base en la conservación y el manejo sostenible de los recursos naturales. Wildlife Conservation Society, La Paz

CIPTA (2008) Reglamento de acceso, uso y aprovechamiento de los recursos naturales renovables del Territorio Indígena Tacana. Wildlife Conservation Society, La Paz

CIPTA, CIMTA (2014) Plan de gestión territorial indígena del pueblo Tacana. Kema Ejudhes'a jakuastas'iati s'aidha enime 2015-2025. Consejo Indígena del Pueblo Tacana (CIPTA), Consejo Indígena de Mujeres Tacanas (CIMTA), Wildlife Conservation Society (WCS), La Paz

CIPTA, WCS (2012) Datos censo de la TCO Tacana para su PGTI. Unpublished document

Constantino P (2016) Deforestation and hunting effects on wildlife across Amazonian indigenous lands. Ecol Soc 21(2):3. https://doi.org/10.5751/ES-08323-210203

Cormier LA (2003) Kinship with monkeys: the Guajá foragers of eastern Amazonia. Columbia University Press, New York

Cowlishaw G, Dunbar R (2000) Primate conservation biology. University of Chicago Press, Chicago

Diaz Astete A, Murillo D (1998) Pueblos indígenas de tierras bajas - características principales. MUAF, La Paz

Espinoza JC, Marengo JA, Ronchail J, Molina Carpio J, Noriega Flores L, Guyot JL (2014) The extreme 2014 flood in south-western Amazon basin: the role of tropical-subtropical South Atlantic SST gradient. Environ Res Lett 9(2014):124007. (9pp). https://doi.org/10.1088/1748-9326/9/12/124007

Estrada, A., Garber, P. A., Fernandez-Duque, E, Heymann, EW.,Setchell, JM., de Guinea, M, Boyle, S A., Meyer, ALS Kone, I, Rylands, A B., Di Fiore, A, Lambert, J E., Gillespie, T R., Gouveia, S, Fuentes, A,Wich, S, Roos, C, Nekaris, KA, Rovero, F, Mittermeier, R A., Dobrovolski, R, . MacKinnon, K C, Sussman, R W., Nijman, V, Barelli, C, Verde Arregoitia L, Shanee, S, Shanee, N, Amato, K R., Pan, R, Li, B. (2017). Impending extinction crisis of the world's primates: why primates matter. Sci Adv, 3(1), e1600946. https://doi.org/10.1126/sciadv.1600946

García del Cuero H (1989) Acerca de la connotación simbólico-ritual del mono en la sociedad prehispánica (Antiplano Central). In: Estrada A, Lopez-Wilchis R, Coates-Estrada R (eds) Primatologia en México: comportamiento, ecología, aprovechamiento y conservación de primates. Universidad Autónoma Metropolitana, Iztapalapa, Mexico City, pp 144–159

Gismondi P, Llobet A, Wallace RB (n.d.) Evaluación de la confiabilidad del automonitoreo de cacería como una herramienta de conservación. Unedited report, Wildlife Conservation Society, La Paz

Gobierno Departamental Autónomo de Santa Cruz (2009) Propuesta para el manejo de fauna silvestre y lineamientos para promover el biocomercio en el Departamento de Santa Cruz. Secretaria Departamental de Desarrollo Sostenible-Dirección de Ordenamiento Territorial, Cuencas PLUS Proyecto Plan Departamental de Ordenamiento Territorial, Santa Cruz

Hissink K, Hann A (2000) Los Tacana. Datos sobre la historia de su civilización [1952-54]. Edición APCOB, La Paz

Iwamura T, Lambin EF, Silvius KM, Luzar JB, Fragoso JMV (2014) Agent-based modeling of hunting and subsistence agriculture on indigenous lands: understanding interactions between social and ecological systems. Environ Model Softw 2014:1–19

Karadimas D (1999) La constellation des quatre dignes. Interpretation etno archeoastronomique des motifs de "El Carchi-Capuli" (Colombie, Équator). J Soc Américanistes 85:115–145

Lara-Delgado K (2003) Vamos de cacería. El acceso al espacio y la distribución de la carne de monte en las comunidades Tacana de Tumupasha y San Silvestre. MA thesis. Universidad de la Cordillera, La Paz

Lehm Z, Lara K, Solares T (2017) Ingresos económicos en hogares de la TCO Tacana I. Wildlife Conservation Society (WCS) and Consejo Indígena del Pueblo Tacana (CIPTA), La Paz, 99 pp

Medici EP, Desbiez ALJ, Gonçalves da Silva A et al. (2007) Lowland tapir (*Tapirus terrestris*) population and habitat viability assessment (PHVA). Final report, IUCN/SSC Tapir Specialist Group (TSG) and IUCN/SSC Conservation Breeding Specialist Group (CBSG), Sorocaba

Mena VP, Stallings JR, Regalado BJ, Cueva LR (2000) The sustainability of current hunting practices by the Huaorani. In: Robinson JG, Bennett EL (eds) Hunting for sustainability in tropical forests. Columbia University Press, New York, pp 57–78

Morales TM (1999) Bases para un manejo sostenible de la fauna en Lomerío, mediante el monitoreo de la cacería. BSc thesis. UMSA, La Paz

Navarro AJ, Redford KH, Bodmer RE (2000) Effect of hunting in source sink systems in the Neotropics. Conserv Biol 14:713–721

Noss AJ, Cuellar E, Cuellar RL (2003) Hunter self-monitoring as a basis for biological research: data from the Bolivian Chaco. Masto Neotrop 10(1):49–67

Noss AJ, Cuellar E, Cuellar RL (2004) An evaluation of hunter-self monitoring in the Bolivian Chaco. Hum Ecol 32:685–702

Ojasti J (1993) Utilización de la fauna silvestre en América Latina: situación y perspectivas para un manejo sostenible. Guia FAO Conservación # 25, Rome

Painter L, Siles TM, Reinaga A, Wallace R (2013) Escenarios de deforestación en el Gran Paisaje Madidi-Tambopata. Wildlife Conservation Society, La Paz, 41 pp

Paredes X (1997) El efecto socioeconómico de la explotación forestal en el Pueblo Tacana (Provincia Iturralde, Departamento de La Paz). BSc thesis, UMSA, La Paz

Peres CA (1990) Effects of hunting on western Amazonian primate communities. Biol Conserv 54:47–59

Peres CA (1991) Humboldt's woolly monkeys decimated by hunting in Amazonia. Oryx 25:89–95

Reichel-Dolmatoff G (1971) Amazonian cosmos: the sexual and religious symbolism of the Tukano indians. University of Chicago Press, Chicago

Reichel-Dolmatoff G (1976) Cosmology as ecological analysis: a view from the rain forest. Man 11:307–318

Reichel-Dolmatoff G (1989) Biological and social aspects of the Yuruparí complex of the Colombian Vaupés territory. J Lat Am Lore 15:95–135

Ribera MO (1992) Regiones ecológicas. In: Marconi M (ed) Conservación de la diversidad biológica de Bolivia CDC y USAID/ Bolivia, La Paz, p 9–72

Robinson J, Bennett E (2000) Hunting for sustainability in tropical forests. Columbia University Press, New York

Roosevelt AC, Housloy RA, Imazio da Silveira M, Marenca S, Johnson R (1991) Eight millennium pottery from a prehistoric shell midden in the Brazilian Amazon. Science 254:1621–1624

Salinas E (2010) Valor cultural de los mamíferos en Bolivia. In: Wallace RB, Gomez H, Porcel ZR, Rumiz DI (eds) Distribución, ecologia y conservación de los mamíferos medianos y grandes de Bolivia. Fundación Simón Patiño, Santa Cruz, pp 5–51

Shimooka Y, Campbell C, DiFiore A, Felton A, Izawa K, Link A, Nishimura A, Ramos-Fernandez G, Wallace R (2008) Demography and group composition of *Ateles*. In: Campbell C (ed) Spider monkeys: behavior, ecology and evolution of the genus *Ateles*. Cambridge University Press, Cambridge, pp 329–348

Souza-Mazurek R, Pedrinho T, Feliciano X, Hilario W, Geroncio S, Marcelo E (2000) Subsistence hunting among the Waimiri Atroari Indians in central Amazonia, Brazil. Biodivers Conserv 9:579–596

Stearman AM (1990) The effects of settler incursion on fish and game resources of the Yuqui, a Native American society of eastern Bolivia. Hum Organ 49(4):373–385

Strecker M (1987) Contribución al estudio de arte rupestre sudamericano: introducción, bibliografía comentada e índice de sitios. Sociedad de Investigación del Area Rupestre de Bolivia (SIARB), Boletín 1, La Paz, 72 pp

Townsend WR (1995) Living on the edge: Sirionó hunting and fishing in lowland Bolivia. PhD Dissertation, University of Florida, Gainesville

Townsend WR (2000) Holding on to the land: the long journey of the Sirionó indians of eastern lowland Bolivia. In: Weber R, Butler J, Larson P (eds). Indigenous peoples and conservation organizations: Experiences in collaboration, The Biodiversity Support Program (BSP), pp 73–89

Townsend WR (2009) Usos no comerciales de la fauna silvestre. In: Gobierno Departamental Autónomo de Santa Cruz (eds). Propuesta para el manejo de fauna silvestre y lineamientos para promover el biocomercio en el Departamento de Santa Cruz. Secretaria Departamental de Desarrollo Sostenible-Dirección de Ordenamiento territorial, Cuencas PLUS Proyecto Plan Departamental de Ordenamiento Territorial, pp 65–82

Townsend WR, Macuritofe-Ramírez V (1995) Cultural teachings as an ecological data base: Murui (Witoto) knowledge about primates. University of Florida, Gainesville. http://www.latam.ufl.edu/latinoamericanist95/townsend.html

Townsend WR, Borman R, Yiyoguaje E, Mendua L (2005) Cofán Indians' monitoring of freshwater turtles in Zábalo, Ecuador. Biodivers Conserv 14:2743–2755

Urbani B (2005) The targeted monkey: a re-evaluation of predation on New World primates. J Anthropol Sci 83:89–109

Urbani B, Gil E (2001) Consideraciones sobre restos de primates de un yacimiento arqueológico del oriente de Venezuela (América del Sur): cueva del Guácharo, estado Monagas. Munibe (Antropol-Arkeol) 53:135–142

Van Akkeren R (1998) The monkey and the black heart. Polar north in ancient mesoamerica. In: Izquierdo AL (ed) Memorias del Tercer Congreso Internacional de Mayistas. Universidad Nacional Autonomo de Mexico, Mexico City, pp 165–185

van Vliet N, Quiceno-Mesa PM, Cruz-Antia D, Tellez L, Martins C, Ribeiro de Oliveira M, Adams C, Morsello C, Valencia L, Bonilla T, Haiden E, Yagüe B, Nasi R (2015) From fish and bushmeat to chicken nuggets: the nutrition transition in a continuum from rural to urban settings in the Colombian Amazon region. Ethnobiol Conserv 5:4–6. https://doi.org/10.15451/ec2015-7-4.6-1-12

Vickers W (1980) An analysis of Amazonian yields as a function of settlement age. Work Pap Am Indians 2:7–30

Wallace RB, Mittermeier RA, Cornejo F, Boubli J-P (2008) *Ateles chamek*. The IUCN red list of threatened species 2008: e.T41547A10497375. https://doi.org/10.2305/IUCN.UK.2008.RLTS.T41547A10497375.en. Downloaded on 4 April 2017

Wentzel S (1991) Contribución para la segunda edición de "En busca de la loma Los Tacana" Alemania

Wu Leung WT, Flores M (1961) Tabla de composición de alimentos para uso en América Latina. ICNND Instituto Nacional para Artritis y Enfermedades Metabólicas, Instituto Nacional de Salud de Estados Unidos, Institución de nutrición de Centro América y Panama (INCAP), Guatamala

Chapter 18
When Monkeys Were Humans: Narratives of the Relationship Between Primates and the Qom (Toba) People of the Gran Chaco of Argentina

Celeste Medrano and Valentín Suárez

18.1 Introduction

The Qom (previously known as Toba, an exonym used from the conquest and in gradual disuse due to indigenous reinvindicatory struggles) constitute an indigenous group whose subsistence practices originally consisted of hunting, fishing, and gathering crops. They currently reside in rural communities in northeastern Argentina or in settlements surrounding large urban areas such as Buenos Aires, Santa Fe, Rosario, Resistencia, and Formosa, among others. Those who still dwell in the regions of their ancient territory do not live entirely off the forest and its resources since the plundering of the land, the sedentarization, and the colonization restricted the access to former territory to complement their subsistence. The Qom, who together with other Amerindians like the Pilagá and the Mocoví people integrates the Guaycuru linguistic family, constitutes the largest indigenous society in the Gran Chaco ecological region, accounting for around 65,000 people throughout Argentina (INDEC 2004–2005). In the communities located in the countryside, the Qom people live surrounded by relatives forming extended families (that is to say, the family composed of several generations and by the husbands of the children) usually live in the same land, in one or more houses (the new couples usually coexist in the field of the parents of the young wife) (López and Tola 2016). In these communities there are bilingual schools where young indigenous people attend and usually small health posts. People agree on the evangelic, the religion now more expanded, with some shamanic practices in reemergence.

C. Medrano (✉)
Instituto de Ciencias Antropológicas, CONICET, Buenos Aires, Argentina
e-mail: celestazo@hotmail.com

V. Suárez
Escuela Rural N° 295, Comunidad Riacho de Oro, Formosa, Argentina

B. Urbani, M. Lizarralde (eds.), *Neotropical Ethnoprimatology*, Ethnobiology,
https://doi.org/10.1007/978-3-030-27504-4_18

The Gran Chaco is the third greatest biogeographic and morphostructural region in Latin America after the Amazon and the South American Savannah System and the second in terms of area covered by forests after the Amazon and Pacific tropical rainforests of Colombia and Ecuador. Its more than 1,000,000 km^2 stretch along four countries (Argentina, Paraguay, Bolivia, and Brazil), being the area in Argentina the largest (Morello et al. 2009).

This study contemplates relationships between the Qom people and their fauna. We can find the first information about these relationships in the legacy written by the Jesuits, who, like Florian Paucke or Martín Dobrizhoffer, were missionaries in the Chaco during the eighteenth century. In the works of these priests, the data related to subsistence prevailed. Although, there were also an introduction of the cosmological aspects, specifically the transformation of men into animals, in the texts of Paucke and Dobrizhoffer.

These references are followed by those of the first travelers who visited the territories of the Chaco at the end of the eighteenth century and the beginning of the nineteenth century, particularly the Spanish military and naturalist, Azara (1998 [1809]), who mentions that the Guaycurues (name of the linguistic family which includes the Toba) were engaged in hunting activities, highlighting that none developed crops and some took care of small herds of cows, sheep, and rams or robbed the Spaniards. The French naturalist, d'Orbigny (1998 [1835–1847]), who also gives news about the life and customs of the Qom people, mentions the possession of some domestic animals such as cows that, according to the author, had been given by the governor of the province of Corrientes. An interesting story is the one that refers to the trade of leather of mammals that the natives maintained with the Spaniards.

Continuing with this historical development, we find at the beginning of the twentieth century the works of the first ethnographers. In those ones there is abundant data linked to hunting and fishing activities, especially the techniques and weapons used which are described (Karsten 1932; Palavecino 1936; Métraux 1996 [1946]; Chaparro 1947). With regard to indigenous cosmology, Karsten (1932), Palavecino (1936), and Métraux (1996 [1946]) refer that, for the Toba/Qom, animals possess a spirit that is both responsible for diseases and the auxiliary shaman. Also, Métraux (1996 [1946]) provides background on the role of animal owners. In addition, we find in his work some evidence about the property that shamans would have for transforming themselves into animals.

The investigations that followed those of these ethnographers focused on deepening specific sociocultural aspects. Specifically, when referring to the "worldview and religious beliefs" of the Guaycurús groups, Cordeu and Siffredi (1971: 14) argued that "the religious organization rested, respectively, in an animalistic complex… [Said] animalistic or hunter complex consisted of a highly developed hierarchical scheme of animal owners, closely linked with the cultural principles of space classification, with hunting regulations and with initiation and shamanic practice."

Later, the debate on the link between the Qom people, these owners of the animals and other non-humans was retaken to explain in depth aspects of indigenous sociocosmology. In this regard, Wright (2008: 142) described the power relationship between owners and other non-humans with shamans and gave some details on "general criteria of animal classification," while describing the intervention of animals in the dream world.

It's from the decade of the 1980s that the first specific monographs on the link between animals and Qom people emerged. Specifically, Vuoto (1981a) and Balducci (1982) observed that these indigenous people gave the animals certain human characteristics that allowed them and others to communicate. Vuoto (1981a: 19) concluded that an "accurate frontier separating human nature from animal" cannot be established. These authors explored the ability of certain species to transmit messages to humans, the link that exists between shamans and their auxiliaries (non-humans in general and animals in particular) and the forms of "contagion" of animal properties to indigenous people.

Although these works were significant to understand the relationship between the Qom people and the animals, the theme was scarcely retaken in the following decades. In this sense, it is worth mentioning the works of Vuoto (1981b) and Martínez Crovetto (1995) on the *zoonomia* of the Toba, Cuneo, and Porta (2009) on the vocabulary of fish and birds and those of Arenas and Porini (2009) and Medrano et al. (2011) on Toba knowledge related to birds and mammals, respectively.

We have specifically focused on studying the zoology of different Qom groups settled in the eastern part of the Province of Formosa (Medrano 2013, 2014, 2016a) (see Fig. 18.1). However, this work will illuminate the relationship between

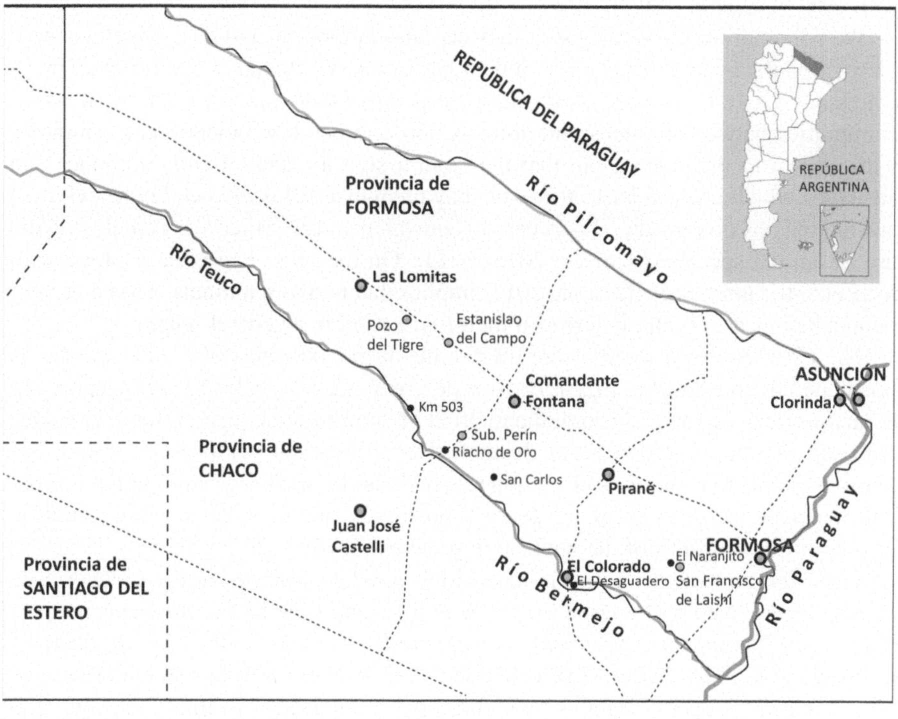

Fig. 18.1 Qom indigenous communities where the fieldwork was carried out (black dots)

indigenous people and monkeys. Two species of non-human primates are distributed in this area: the *carayá* or black-and-golden howler monkey (*Alouatta caraya*: Atelidae), and the *mirikina* (night monkey, *Aotus azarae*: Cebidae) (Zunino and Kowalewski 2008; Ojeda et al. 2012). The Qom relate mainly with the black-and-golden howler monkey (*A. caraya)*, who are refered to as *huoỹem* in Qom/Toba language (as a specific and generic taxa). The frequency of relationship with the night monkey (*A. azarae),* in turn, is very low, and which is called *huoỹem capio'olec*, meaning literally "small monkey" (Medrano et al. 2011). Our hypothesis is that this lack of animal names is related to the behavior of these primates. While the *carayá* (black-and-golden howler monkey) is a diurnal monkey, the habits of the *mirikina* (night monkey) are typically nocturnal (Canevari and Vaccaro 2007). We will concentrate on understanding the relationships between the Qom and the black-and-golden howler monkey.

The black-and-golden howler monkey, one of the largest American monkeys (adult male of *A. caraya* mean weight is 6.42 kg, adult female mean weight is 4.33 kg, cf. Rumiz 1990). This primate is a tree-dweller and lives in family groups made up of as many as ten individuals (Chebez et al. 2005; Canevari and Vaccaro 2007). In Argentina, it is distributed in the provinces of Misiones, Corrientes, eastern Chaco and Formosa, and the banks and islands of the Paraná River in northeastern Santa Fe. A great part of this population lives outside protected areas, where forests face changes in composition, fragmentation, and reduction due to human activity (Ojeda et al. 2012).

We will analyze the relationship between these monkeys and the indigenous peoples within the general framework of the Qom zoology, which has been studied by us (Medrano 2014). One of the most relevant aspects of these studies is the existence of continuities between humans and animals. On one hand, we identified a similarity with respect to the anatomy in that the Qom assign the animal body aptitudes and attributes which are similar to the ones they recognize for their own bodies. Human and animal physiology also shows certain equivalence we verified when analyzing the use of animal species (Medrano 2013, 2014). On the other hand, the analogy with respect to the *interiority* (Descola 2012) implies that humans, animals, and other non-human beings are similar in terms of their *lqui'i* (soul in a general sense).

The Qom zoology, at the same time, hints at the existence of equivocations, as postulated by the anthropologist Viveiros de Castro (2004). In his view, the approach to an *other* can be made through the method of "controlled equivocation." This idea consists of a controlled reading through two ontological perspectives which use homonymic terms: "equivocation appears here as the mode of communication par excellence between different perspectival positions –and therefore as both condition of possibility and limit of the anthropological enterprise" (2004: 3).

Thus, the aim of this essay is to ask: are howler monkeys in eastern Formosa Province perceived in the same way to both indigenous and nonindigenous people? To answer this question, we will first examine the Qom mythology which links monkeys to humans. At the end will be to track the origin of humanity, animality, and their convergences. Then, we will analyze current scenarios that reflect the relationships between human and non-human primates in order to understand those patterns in the context of the Qom indigenous sociocosmology.

18.2 Methodology

This essay was written in co-authorship with one of my Qom indigenous colleague, Valentín Suárez. Specifically, we collected mythological narratives published in well-known compilations (Wilbert and Simoneau 1983, 1989; Terán 2005) and read them together. Being not only a bilingual teacher, a Qom leader, and also a reflective thinker of his own society, Valentín became an authoritative voice to discuss ideas about politics, history, and other aspects from an indigenous perspective. Thus, the methodology allowed us to exchange interpretations about the narrative in these myths and to contrast the way in which indigenous and nonindigenous people relate with monkeys.

It must be made clear that to the Qom, as to many Amerindian societies, myths are not understood as a closed symbolic system (Hill 1988), as a corpus of fictitious time. On the contrary, just like historical discourses are conceived as the past, mythic discourses are conceived as the past of the past (Lima 1999). Particularly, Qom mythology comprises a body of tales "par excellence 'true' for knowledge itself" (Cordeu 1969–1970: 68). With this in mind, we will read the stories that follow.

18.3 When Monkeys Were Humans

Within the corpus of Qom mythology, there is a tale that refers to the origin of the black-and-golden howler monkey which Mr. Santo González told ethnographer Buenaventura Terán: "A long time ago when the world began, there was a great fire and everything was burned. The whole world was on fire, and everything that lived on it was consumed. People had hidden in a hole deep in the earth, and, when the fire had burned out, they were instructed how they were to emerge. They were not to look toward the front until they had walked so many meters. However, one man paid no attention. He looked straight ahead, and, as he did so, he was turned into the howler monkey. Howler monkeys used to be people, and they descend from the man" (*apud* Wilbert and Simoneau 1989: 93–94).

The previous narration refers not only to the origin of the monkey but of all animal species. According to what the elderly Qom say, people who emerged from the earth had to look ahead. Whoever looked back or sideways was turned into a bear, a monkey, or an iguana depending on the size of their bodies. "So the origin of all the animals is the human person," concludes Valentín Suárez (pers. comm., 2017).

In order to analyze the origin of the monkeys, we discussed the Western theory of evolution. A simplified image portrays human primates evolving from nonhuman primates. Valentín claimed that, for the Qom, the origin of the monkey is different. He draws conclusions from the systematization of observations: "I always compare with the monkeys, they live for ages in circuses but they never learn [to speak] a word, and how is it possible that they [humans] come from them [monkeys]? If [you tell me that] we come from the parrot, maybe. Because the parrot at least speaks, imitates people, makes gestures, hisses, shouts. It speaks, sings.

And the origin of the monkey is different for the Toba. To me it depends on each people, each tribe, the origin that each person, the human being, has" (Valentín Suárez, pers. inf., 2017).

Just like the Qom, many Amerindian societies believe monkeys – and all animals – share a human origin. Viveiros de Castro (2013) states that these explanations, rooted in the origin mythology, rest on a fundamental assumption according to which the common background of humanity and animality is humanity. This cultural ground on which the indigenous peoples place both humans and non-humans is the condition for the maintenance of social relationships between both classes of beings – understood, ultimately, as subjects. Inversely, learned science builds the whole chain of beings that inhabit the planet on a common animal origin. Thus, modern society is built by segregating the animals in order to construct the rational, educated group of speaking humans. It is in this context that we venture to suggest that the *carayá* (black-and-golden howler monkey) and the *huoỹem* are not the same thing. This is the first clue for establishing an equivocation.

For Valentín, a hypothetical ancestor of the humans could be the parrot. The bird would have been human first, then animal, then human, following a chain of completely feasible transformations for the Qom. These transformations are possible in an epistemological context where becoming in transformation is the norm (Medrano 2013). Far from intending to mark the "eccentricity of the indigenous evolutionary theory," we bring this up to show the internal coherence of the logic that defies the paradigms of modern science. By suggesting that the origin of the monkey is linked to "each people," Valentín is projecting the existence of other epistemologies. Hegemonic science mockingly refuses to accept other explanations and supports uncontrolled equivocations; by not going beyond its own conventions, "it remains more an ideology than a science" (Wagner 2010: 29).

18.4 Thou Shalt Not Kill

The reading of mythological narratives led us to examine stories that the Qom consider more contemporary. Among them there is one called "the monkeys protect a village from an epidemic" (Terán 2005: 57–58), which can be circumscribed to the context of the massive contagion of European illnesses suffered by the indigenous peoples as of the seventeenth century (Rosso 2011). The narrative accounts for the existence of a shaman who has as a non-human companion a *huoỹem lta'a* (literally meaning "the father of the monkeys"), an entity that has the tutelage of all the howler monkeys. These animals also have the capacity to communicate with clouds, lightning, rain, and wind. To the Qom, the plague "is like a cloud coming from the mountain" (Terán 2005: 58) so the father of the monkeys of the tale could prevent the arrival of such cloud and divert it with the help of the shaman.

Valentín reflected upon the connection between the clouds and the monkeys and added that to the Qom the *huoỹem* are in permanent communication with *Qasoxonaxa*, a non-human being who inhabits heaven and has the capacity to control weather events, mainly rain, thunder, and lightning (see Fig. 18.2). This non-human entity –

Fig. 18.2 Communication between the *huoỹem* and *Qasoxonaxa* (drawing Valentín Suárez). There are no closed hypotheses about why this non-human is represented as an elephant. However also *Qasoxonaxa* means "mountain" so that when Qom people met, the elephant could have put the same name in reference to the size of the animal. Then I had the analogy with regard to the image

one among the many that populate the Qom universe – has intentionality capable of performing actions which have profound implications for the indigenous social life. It is part of a series of entities that "far from being considered spirits, characters, or gods who live in the nontemporal dimension of myth, suspended in time, or pertaining to remote scopes of the universe, ...coexist with past and present human beings" (Tola 2014: 71).

According to the academic information regarding the biology of the *A. caraya*, "the male produces its loud howls at dawn and at dusk but also when it is about to rain or when there is imminent danger" (Canevari and Vaccaro 2007: 79). To the Qom, the *huoỹem* (black-and-golden howler monkey) also announce weather events. As Valentín says, "[nowadays] anyone can hear the [black-and-golden howler] monkey sing early in the morning; a strong north wind will probably blow that day. Because it is not normal, if the monkey sings it is because a heavy north wind will blow and, after that, rain." So far, the ecology and the indigenous' data seem to coincide, but the Qom explanation goes beyond. The Qom consider that the monkey carries out its meteorological activities due to a bond – between subjects – with *Qasoxonaxa*, the non-human master of meteorological phenomena.

As we will explain later, this social network made up of humans, monkeys, and other non-humans accounts for the reason why hunting monkeys is permanently forbidden for the Qom. However, during the late 1980s and early 1990s, Argentina went through a deep economic crisis which forced many indigenous peoples to look for alternative subsistence practices. By then, the black-and-golden howler monkey (*A. caraya*) was one of the species most traded as pets (Canevari and Vaccaro 2007) and was significantly represented in the illegal traffic routes of fauna (Bertonatti 1995; Chebez 2009). Chebez (2009) describes that in order to capture the offspring, the adults – especially the dominant male and the female bearer – were killed with firearms or by bringing down the trees were the troop sheltered. The *carayás* (black-and-golden howler monkey) were reduced of their distribution area and could also integrate a circuit of precarious transactions which offered the animals "on national routes with improvised signs" (Chebez 2009: 331). The Qom longed for joining this circuit, which coincided with the distribution area of the *A. caraya*.

Valentín explains that back then people went to the indigenous communities and bought monkeys. Given this economic possibility, the Qom started going to the forests to look for *huoỹem*: "The man wanted to have the monkey's offspring but it happened that out of tiredness he thought of killing the little monkey's mother and when he killed it the weather, the storm, immediately came into being, right there where he was. Lightning, all those things, because of that monkey which was killed. That's why it has a lot to do with that. With the relationship with nature, water, thunder, all those things. That is why it is strictly [prohibited for] the Qom to kill monkeys" (Valentín Suárez, 2017, pers. inf.).

The man of the preceding tale was a victim of *Qasoxonaxa*'s anger for having wrongfully attacked a monkey. The prohibition to kill determines then the modes of relationship among subjects – the Qom, the *huoỹem*, and other non-humans. The "nature" Valentín mentions in the narration is therefore reconstructed as the structure of social precepts and rules of etiquette which regulate the link between more-than-human beings. Thunder, rain, and wind may be simple weather events beyond human control or, also, transcendent messages which the subjects of the Qom universe use to communicate extralinguistically. This reinforces our question: Is the black-and-golden howler monkey (*A. caraya*) the same to both indigenous and non-indigenous people? So far, if we answered affirmatively, we would not be acknowledging an equivocation.

18.5 Thou Shalt Do No Harm

The Qom also keeps the offsprings of the black-and-golden howler monkey as pets (for the concept of "pet", see Medrano 2016a, b). When reflecting upon the precautions that should be observed if a *huoỹem* was raised, Valentín mentioned that "when there's no mistreatment there's no danger, when animals are mistreated, yes [there is danger], [therefore] a *nauaga* ['contagion' or 'influence'] can occur. That's why animal mistreatment is very much prevented. No mistreating [should happen to the nigh howler monkeys]." Within the domestic context, the monkey is integrated into the human social network without losing the attributes linked to his interiority. So we recalled together the piece of advice received by an elderly Qom: "Pregnant women must refrain from looking at the *huoỹem* in the face, and from feeding or striking it because it can infects their baby. Otherwise, the baby could come out with the howler monkey's face or with its same gestures and movements; and when adults speak [to the human baby], she/he will not understand and just laughs [like the monkey]. According to Qom tradition, mothers and grandmothers must offer their young daughter's advice for them to respect this prescription" (Mauricio Maidana in Medrano et al. 2011: 70).

This tales account for a phenomenon called *nauoga* in Toba, which has been broadly studied from diverse perspectives and translated as "contagion" or "influence." For classic readings on the process of *nauoga*, see Karsten (1932) and Métraux (1946). For more contemporary readings about the topic, see Vuoto (1981a), Balducci (1982), Wright (2008), Tola (2012), and Medrano (2013). Through this process, certain characteristics present in animals are likely to be assimilated by humans. Thus, formal or behavioral properties circulate between humans and non-humans given the porosity of their bodies (Tola 2012) and the possibility of permanent transformation that humans, animals, and other beings experience.

In this context, one of the aspects that transforms primates into the realm of pets link more complex is related to this continuity between humans and animals which is characteristic of a zoology that attributes analogies with respect to the interiority. In 1979, ethnographer Luis Vuoto was told by a Qom that when a monkey "enters a child's heart,... the monkey' vices enter the heart... [then] every day they [start] shouting [or howling] and so on, they can't keep calm, [and behaving] all just like the monkey" (A. A¢ilaj in Vuoto 1981a: 106).

This human-monkey combination that can only be undone by the shaman power is possible due to the resemblance that humans and animals have regarding their *lqui'i*. The *lqui'i*, commonly translated as soul, provides the capacity to feel, think, move, and walk around, therefore being linked to the idea of regimes of corporality (Tola 2012). When a contagion like the one described by A¢ilaj in Vuoto (1981a) takes place, the monkey enters the heart because this organ is the "instrument" of the *lqui'i* and since it is the site of emotions-thoughts, which it allows for the connection between human persons (Tola 2012) and between the latter, animals, and other non-human beings. Thus, the analogous *lqui'i* of the monkey and the child combines giving as a result a hybrid that embodies the illness.

If in the indigenous universe the possession of a *lqui'i* – and the subsequent condition of person – extends to non-humans, the relationships between the Qom and the monkeys become links between social subjects. This acquires its condition of possibility in the context of a cosmology that – like the Qom's – is inscribed in an animist ontology. We make use of Descola's definition of ontology, that is, "the different ways of expressing continuities and discontinuities between humans and nonhumans" (Descola 2014: 440). A cosmology, as the author suggests, "is simply the form of distribution in space of the components of an ontology and the kind of relations that conjoin them" (2014, 437). The animism is "the assumption that, under certain circumstances, non-humans of various kinds behave as if they had an intentionality analogous to the one humans believe they are endowed with" (Descola 2010: 338).

At the opposite end, we find the Western cosmologies that place humans and non-humans in two watertight ontological domains, thus laying the basis for naturalistic ontology. P. Descola (2010: 338) mentions that naturalism, opposed to animism, characterizes the modern world and "insists on the differences between humans and non-humans on the interiority axis: humans alone are supposed to have a meaningful selfhood whether individual (mind, capacity for symbolism) or collective (Volksgeist, cultures). In contrast, …humans and non-humans are linked by their shared physicality: they belong to a continuum where the same laws of physics, biology and chemistry apply." Based on the dualism that this ontology founds, modern thought segregates culture and nature fetishizing this last domain as a transcendental object, thus inaugurating the "modern Constitution" according to which the "human subjects" objectify their illusion of nature (Latour 2007).

At this point of our exposition, the equivocation becomes self-evidently ratified. The howler monkey – subject to the Qom, object to the modern imaginations – embodies at least two ontologies. The indigenous peoples consider it a member of a sociocosmological network which includes humans, the *huoỹem*, *huoỹem lta'a*, *Qasoxonaxa*, other non-humans and animals, the rain, thunder, lightning, shamans, etc. The imaginations of modern citizens keep to themselves an image of the monkey which raises scientific questions and awakens ecological sensitivities (Descola 1998). The latter are linked to the anthropocentric responsibility of the human primate, the arbiter of life and its continuity in the entire planet.

18.6 *Huoỹem/Carayá* Is Multiple

In 2009, anthropologist Mario Blaser began a substantial study about the indigenous Innu in collaboration with them. By 2013, the government of Newfoundland and Labrador (Canada), the two provinces where the Innu communities live, announced a 5-year hunting ban of the caribou (*Rangifer tarandus*). Blaser wrote that, "while for the wildlife managers in the provincial government hunting could mean the disappearance of the caribou, for the Innu hunters and elders being prevented from hunting according to protocol almost assuredly would mean the disappearance of *atîku* [Innu word for the caribou]" (Blaser 2016: 546). For hunters and elders, the hunting ban would make it impossible to repair the relationship with *atîku* and its spirit master. Rethinking the concept of cosmopolitics – initially put forward by Isabelle Stengers and Bruno Latour – to analyze this case, Blaser concluded that the caribou and the *atîku* do not refer to different cultural perspectives of the same "thing" but to "different things": "*Atîku* emerges from an assemblage that involves *atanukan*, hunters, sharing of meat, generosity, a 'spirit master,' and so on; caribou emerges from an assemblage that involves the discipline of biology, wildlife managers, predictive modeling, calculations to balance environmental and economic concerns, and so on" (Blaser 2016: 558).

In this case, *atîku*/caribou is multiple, argues Blaser. In our case, *huoỹem*/*carayá* is also multiple. *Huoỹem* (black-and-golden howler monkey) emerges from an assemblage that involves *huoỹem lta'a*, *Qasoxonaxa*, the Qom who sell monkeys and those who make them into mascots, the shamans, and so on; the *carayá* (black-and-golden howler monkey) emerges from an assemblage made up of biologists and conservationists, wildlife managers, statistical models, genetic explorations, and governmental bureaucracies.

But these examples not only bring multiple ontologies into play but also multiple "worldings," "a form of enacting a reality" as Blaser (2013: 23) defined. And, as both cases show, forcing the reality toward a common world by hiding or suppressing other ways of enacting reality puts species, territories, humans, and non-humans inhabiting them in danger (cf. Povinelli 2001; Nadasdy 2007; Blaser 2009; Cayón 2012; Martínez Dueñas 2012; Di Giminiani 2013; among others). This is at the same time inscribed in a context where the dominant tendency in leading conservation research/action circles is to qualify indigenous environmental practices and [local ecological] knowledge as a supply "that can be integrated into the toolkit of conservation practitioners, often as mere informational inputs" (Blaser 2009: 15).

Huoỹem is not the same thing as *carayá*, and *atîku* is not the same thing as caribou, but how can the worlds in which both things enact be outlined? As we showed throughout the text, one of the ways is through the founding of equivocations. As suggested by Viveiros de Castro (2004), controlled equivocations are the conceptual places where the ontological differences are expressed, and they pose misunderstandings which make the anthropological question possible. But, more importantly, they allow striping modern science of its monopoly of nature representation and, by bringing the multiple worldings to the fore, to move toward a scenario of legitimate dialogues.

18.7 Final Words

Bacigallupo (2013: 77) claims that mythohistory "is a mixed genre that mediates among different memoralizations of the past to obliterate dominant... history and to create alternative indigenous histories". Similarly, Qom mythology constitutes the touchstone that allows us to reveal the foundations of an indigenous zoology and to object to the supremacy of academic epistemological discourses. However, our intention is not to convince our contemporaries of the existence of the *huoỹem* nor persuade the Qom about the objectivity of the *carayá* (black-and-golden howler monkey). Our desire is to create zoologies – in plural – where the misunderstandings inaugurate different knowledge practices. We aspire to find a "pluriverse" where the condition of possibility of multiple worlds be guaranteed.

We finally wish to highlight an important aspect connected not only with the Qom zoology in the singular but with the indigenous zoologies in the plural and with the relationships with non-human primates in particular. In this respect, multiple contributions revealing the social relationship between human and non-human primates have been published (cf. Bruner and Cucina 2005; Urbani 2005; Cormier 2006; Fuentes 2006, 2010; Cormier and Urbani 2008; Sá 2013; Urbani and Cormier 2015). Moreover, in animist indigenous societies, monkeys are included in the human set and integrate parental networks (cf. Cormier 2003a, b). Rereading such investigations, we are concerned with highlighting the outcome that the dispossession of territory and the species decline has for the indigenous peoples. We conclude then by asserting that when the Qom and many other indigenous societies face the dilapidation of what we call "nature," they suffer a double dispossession: of the "resources" and of the social relationships established with them. These pillages are as intertwined in life sustenance as our conservation plans and fauna protection projects. Different but assembled: if the *huoỹem* becomes extinct, so does the *carayá* (black-and-golden howler monkey). Let's sail then, guided by the compass of the equivocations, toward the worlds where all worlds be possible.

Acknowledgments To all Qom people for teaching us their zoology. To Florencia Tola for guiding us with generosity. To Lilián Amestoy for helping us with the translation.

References

Arenas P, Porini G (2009) Las aves en la vida de los tobas del oeste de la provincial de Formosa (Argentina). Editorial Tiempo de Historia, Asunción

Bacigallupo AM (2013) Mapuche struggles to obliterate dominant history: mythohistory, spiritual agency and shamanic historical consciousness in southern Chile. Identities 20(1):77–95

Balducci MI (1982) Códigos de comunicación con el mundo animal entre los Toba-Taksik. Dissertation. Universidad de Buenos Aires, Buenos Aires

Bertonatti C (1995) El comercio de primates en la República Argentina. Neotro Primates 3(2):35–37

Blaser M (2009) The threat of the Yrmo: the political ontology of a sustainable hunting program. Am Anthropol 111(1):10–20

Blaser M (2013) Notes towards a political ontology of 'environmental' conflicts. In: Green L (ed) Contested ecologies: nature and knowledge. Human Science Research Council of South Africa Press, Cape Town, pp 13–27

Blaser M (2016) Is another cosmopolitics possible? Cult Anthropol 31(4):545–570

Bruner E, Cucina A (2005) Alouatta, Ateles, and the ancient Mesoamerican cultures. J Anthropol Sci 83:111–117

Canevari M, Vaccaro O (2007) Guía de mamíferos del sur de América del Sur. LOLA, Buenos Aires

Cayón L (2012) Plans de vie et gestion du monde: cosmopolitique autochtone du développement en Amazonie colombienne. Rech amérind Québec 42(2–3):63–77

Chaparro F (1947) Los guaycurúes. Establecimiento Gráfico Molachino y Scarabino, Rosario
Chebez JC (2009) Otros que se van. Fauna argentina amenazada, Albatros, Buenos Aires
Chebez JC, Pereira J, Massoia E, Di Giacomo AG, Heinonen Fortabat S (2005) Mamíferos de la Reserva El Bagual. In: Di Giacomo AG, Krapovickas SF (eds) Historia natural y paisaje de la Reserva El Bagual, Formosa, Argentina. Aves Argentinas/Asociación Ornitológica del Plata, Buenos Aires, pp 467–499
Cordeu E (1969–1970) Aproximación al horizonte mítico de los tobas. Runa 12(1–2):67–176
Cordeu E, Siffredi A (1971) De la Algarroba al Algodón, movimientos Milenaristas del Chaco Argentino. Juárez Editor, Buenos Aires
Cormier LA (2003a) Kinship with monkeys: the Guajá foragers of eastern Amazonia. Columbia University Press, New York
Cormier LA (2003b) Animism, cannibalism, and pet-keeping among the Guajá of Eastern Amazonia. Tipití 1(1):81–98
Cormier LA (2006) A preliminary review of neotropical primates in subsistence and symbolism of indigenous lowland South American peoples. Ecol Environ Anthropol 2:14–32
Cormier LA, Urbani B (2008) The ethnoprimatology of the spider monkey (*Ateles* spp.): from past to present. In: Campbell C (ed) Spider monkeys: the biology, behavior and ecology of the genus *Ateles*. Cambridge University Press, Cambridge, pp 377–403
Cuneo P, Porta A (2009) Vocabulario toba sobre peces y aves. In: Braunstein J, Messineo C (eds) Hacia una nueva carta etnica del Gran Chaco VIII. Centro del Hombre Antiguo Chaqueño, Buenos Aires/Las Lomita, pp 237–252
D'Orbigny A (1998 [1835–1847]) Viaje por la América Meridional. Emecé, Buenos Aires
de Azara F (1998 [1809) Viajes por la Amética Meridional. El Elefante Blanco, Buenos Aires
Descola P (1998) Estructura ou sentimento: a relação com o animal na Amazônia. Mana 4(1):23–45
Descola P (2010) Cognition, perception and worlding. Interdiscip Sci Rev 35(3–4):334–340
Descola P (2012) Más allá de naturaleza y cultura. Amorrortu, Buenos Aires
Descola P (2014) The difficult art of composing worlds (and of replying to objections). Hau J Ethnogr Theory 4(3):431–443
Di Giminiani P (2013) The contested rewe: sacred sites, misunderstandings, and ontological pluralism in Mapuche land negotiations. J R Anthropol Inst 19:527–544
Fuentes A (2006) Human-nonhuman primate interconnections and their relevance to anthropology. Ecol Environ Anthropol 2(2):1–11
Fuentes A (2010) Naturalcultural encounters in Bali: monkeys, temples, tourists, and ethnoprimatology. Cult Anthropol 25(4):600–624
Hill J (ed) (1988) Rethinking history and myth: indigenous south American perspectives on the past. University of Illinois Press, Urbana
INDEC (2004–2005) Encuesta Complementaria de Pueblos Indígenas (ECPI) 2001
Karsten R (1932) Indian tribes of the Argentine and Bolivian Chaco. Soc Scientiar Fennica 4:10–126
Latour B (2007) Nunca fuimos modernos. Ensayo de antropología simética. Siglo XXI, Buenos Aires
Lima T (1999) O pássaro de fogo. Rev Antropol 42(1–2):1–12
López A, Tola F (2016) Tobas (qom) y moqoit (mocovíes). Antiguas y nuevas andanzas por el Gran Chaco. Ministerio de Educación y Deportes de la Nación, Ciudad Autónoma de Buenos Aires
Martínez Crovetto R (1995) Zoonimia y etnozoología de los pilaga, toba, mocovi, mataco y vilela. In: Universidad de Buenos Aires (ed) Facultad de Filosofía y Letras, Archivo de Lenguas Indoamericanas, Colección Nuestra América, Buenos Aires
Martínez Dueñas WA (2012) Quand H^2O et esprit de l'eau se rencontrent: coexistence de plusieurs mondes à Puracé, Colombie. Rech amérind Québec 42(2–3):39–47
Medrano C (2013) Devenir-en-transformación: debates etnozoológicos en torno a la metamorfosis animal entre los *qom*. In: Tola F, Medrano C, Cardin L (eds) Gran Chaco. Ontología, poder, afectividad. IWGIA/Rumbo Sur, Buenos Aires, pp 77–101
Medrano C (2014) Zoo-sociocosmología *qom*: seres humanos, animales y sus relaciones en el Gran Chaco. J Soc Am 100:225–257
Medrano C (2016a) Los no-animales y la categoría 'animal'. Definiendo la zoo-sociocosmología entre los toba (*qom*) del Chaco argentino. Mana 22:369–402

Medrano C (2016b) Hacer a un Perro. Relaciones entre los *qom* del Gran Chaco Argentino y sus compañeros animales de caza. Anthropos 111:113–125
Medrano C, Maidana M, Gómez C (2011) Zoología Qom. Conocimientos tobas sobre el mundo animal. Ediciones Biológica, Santa Fe
Métraux A (1946) Indians of the Gran Chaco. Ethnography of the Chaco. In: Steward JH (ed) Handbook of South American Indians, 1, The marginal tribes. Smithsonian Institute, Bureau of American Ethnology, Washington, pp 197–370
Morello JH, Rodríguez AF, Silva M (2009) Clasificación de ambientes en áreas protegidas de las ecorregiones Del Chaco húmedo y seco. In: Morello J, Rodríguez A (eds) El Chaco sin bosques: la Pampa o el desierto del futuro. Orientación Gráfica, Buenos Aires, pp 53–91
Nadasdy P (2007) The gift in the animal: the ontology of hunting and human-animal sociality. Am Ethnol 34(1):25–43
Ojeda RA, Chillo V, Diaz Isentath G (eds) (2012) Libro Rojo de Mamíferos Amenazados de la Argentina. Sociedad Argentina para el Estudio de los Mamíferos (SAREM), Argentina
Palavecino E (1936) Las culturas aborígenes del Chaco. En: De la Historia de la Nación Argentina, vol I. Junta de Historia y Numismática Americana, Imprenta de la Universidad, Buenos Aires
Povinelli E (2001) Radical worlds: the anthropology of incommesurability and inconceivability. Annu Rev Anthropol 30:319–335
Rosso C (2011) Epidemias de viruela en las reducciones chaqueñas de abipones y mocovíes durante el siglo XVIII. Eä 2:3. http://www.ea-journal.com/art2.3/Epidemias-de-viruela-en-reducciones-chaquenas-de-abipones-y-mocovies-siglo-XVIII.pdf. Accessed 18 Oct 2016
Rumiz DI (1990) Alouatta caraya: population density and demography in northern Argentina. Am J Primatol 21:279–294
Sá G (2013) No mesmo galho: antropologia de coletivos humanos e animais. 7 Letras, Rio de Janeiro
Terán B (ed) (2005) Lo que cuentan los tobas. Ediciones del Sol, Buenos Aires
Tola F (2012) Yo no estoy solo en mi cuerpo. Cuerpos-personas múltiples entre los tobas del Chaco argentino. Biblos/Culturalia, Buenos Aires
Tola F (2014) The materiality of "spiritual presences" and the notion of person in an Amerindian Society. In: Blanes R, Espírito Santo D (eds) The social life of spirits. The University of Chicago Press, Chicago, pp 69–92
Urbani B (2005) The targeted monkey: a re-evaluation of predation on New World primates. J Anthropol Sci 83:89–109
Urbani B, Cormier LA (2015) The ethnoprimatology of howler monkeys (*Alouatta* spp.): from past to present. In: Kowalewski MM, Garder PA, Cortés-Ortiz L, Urbani B, Youlatos D (eds) Howler monkeys. Springer, New York, pp 259–280
Viveiros de Castro E (2004) Perspectival anthropology and the method of controlled equivocation. Tipití 2(1):3–22
Viveiros de Castro E (2013) La mirada del jaguar. Tinta Limón, Buenos Aires
Vuoto DL (1981a) Aspectos de la interrelación entre la fauna y la cultura toba-taksek. Dissertation. Universidad de Buenos Aires, Buenos Aires
Vuoto DL (1981b) La fauna de los Toba-Taksek. In: Notas etnográficas de los Toba-Tasek. Entregas del Instituto de Tilcara 10:77–138
Wagner R (2010 [1975]) A invenção da cultura. Cosac Naify, São Paulo
Wilbert J, Simeneau K (eds) (1983) Folk literature of the Toba Indians (Volumen one). UCLA, Los Angeles
Wilbert J, Simoneau K (eds) (1989) Folk literature of the Toba Indians (Volumen two). UCLA, Los Angeles
Wright P (2008) Ser-en-el-sueño. Crónicas de historia y vida toba. Biblos/Culturalia, Buenos Aires
Zunino GE, Kowalewski MM (2008) Primate research and conservation in northern Argentina: the field station Corrientes (Estación Biológica de Usos Múltiples – EBCo). Trop Conserv Sci 1(2):140–150

Subject Index

B. Urbani, M. Lizarralde (eds.), *Neotropical Ethnoprimatology*, Ethnobiology,
https://doi.org/10.1007/978-3-030-27504-4

Q

R

Taxonomy Index

B. Urbani, M. Lizarralde (eds.), *Neotropical Ethnoprimatology*, Ethnobiology,
https://doi.org/10.1007/978-3-030-27504-4

N

O

P

The manufacturer's authorised representative in the EU is Springer Nature Customer Service Centre GmbH, Europaplatz 3, 69115 Heidelberg, Germany. If you have any concerns regarding our products, please contact ProductSafety@springernature.com

Printed and bound by CPI Group (UK) Ltd, Croydon, CR0 4YY

30/09/2026

02225462-0016